MOLECULES AND LIFE

An Introduction to Molecular Biology

MOLECULES AND LIFE

An Introduction to Molecular Biology

Mikhail V. Vol'kenshtein

Institute of Molecular Biology
Academy of Sciences of the USSR, Moscow

Translated from the Russian by
Serge N. Timasheff
Graduate Department of Biochemistry
Brandeis University
Waltham, Massachusetts

A PLENUM/ROSETTA EDITION

Library of Congress Cataloging in Publication Data

Vol'kenshteĭn, Mikhail Vladimirovich, 1912-
  Molecules and life.

  Translation of Molekuly i zhizn'.
  "A Plenum/Rosetta edition."
  Bibliography: p.
  1. Molecular biology. I. Title. [DNLM: 1. Molecular biology.
QT34 V917m 1970a]
[QH506.V613 1974]   574.8'8   74-773
  ISBN-13: 978-1-4615-8596-1   e-ISBN-13: 978-1-4615-8594-7
  DOI: 10.1007/978-1-4615-8594-7

The Russian text, originally published by Nauka Press in Moscow in 1965, has been extensively revised by the author for this edition. The present translation is published under an agreement with Mezhdunarodnaya Kniga, the Soviet book export agency.

Михаил Владимирович Волькенштейн

Молекулы и жизнь
MOLEKULY I ZHIZN'
MOLECULES AND LIFE

A Plenum/Rosetta Edition
Published by Plenum Publishing Corporation
227 West 17th Street, New York, N.Y. 10011

First paperback printing 1974

© 1970 Plenum Press, New York
A Division of Plenum Publishing Corporation

United Kingdom edition published by Plenum Press, London
A Division of Plenum Publishing Corporation, Ltd.
4a Lower John Street, London W1R 3PD, England

This book is dedicated to
S. I. Alenikova-Vol'kenshtein

PREFACE TO THE TRANSLATION

The Russian edition of this book was published in the Fall of 1965; the manuscript had been finished a year earlier. Since then, many changes have taken place in molecular biology. No other field of science is currently undergoing as rapid a development. As a result, this book has been extensively revised for the translation, and much new material has been introduced. And yet, even this version will to some extent be obsolete at the time the translation appears in print. This, of course, cannot be helped. I believe, however, that the basic concepts of molecular biophysics which are the subject of this book will remain significant for a considerable length of time.

M. V. Vol'kenshtein

Nature is not what you believe——
It is neither a mold nor a spiritless face.
It has a soul, it has freedom,
It has love, it has a tongue.

F. I. TYUTCHEV

Nature is not a temple, but a workshop, in which man is the laborer.

I. S. TURGENEV

PREFACE TO THE RUSSIAN EDITION

Nature has been described as a "work of beauty and love" by the Russian poet Tyutchev, and as a "workshop" by Turgenev. Are these two descriptions mutually contradictory? Must we wonder at the beauty and charm of living nature, at the soft aroma of a rose, at the flight of a butterfly, at the thoughts of man, or are these amazing phenomena the rightful subjects of a scholarly examination?

Nature is a magnificent workshop in which the development of matter under strict laws has led to the creation of living organisms, of highly complicated systems, characterized by an amazing structural and functional coordination. The culmination of this development on earth is the human mind, which at present can not only understand the laws of life, but can actively interfere with the surrounding nature and modify it to man's benefit. This is the aim of material science. True science, however, cannot exist without an esthetic appreciation of the objects examined. That very soul and freedom of nature, of which the Russian poet sings, from a scientist's point of view are expressions of the laws of biology, physics, and chemistry. This does not, however, render them any less attractive and interesting.

This book has for its purpose the introduction to molecular biophysics. Molecular biophysics is, on one hand, a part of molecular physics and, on the other hand, a part of molecular biology, a new realm of science, developed during the last decades. The joint efforts of biologists, physicists, and chemists have made it possible to unravel the essence of a number of the basic phenomena of life, starting from the structure and properties of molecules which form organisms, and principally of proteins and nucleic acids. The achievements of molecular biology testify to the success of material science in a realm which, until recently, appeared totally enigmatic and mysterious. Further scientific developments should bring to mankind vast developments both in theoretical knowledge and in practical applications, namely, in agriculture, medicine, and technology.

The purpose of this book is to explain molecular biophysics to all who might wish to learn about it, to biologists, to physicists, to chemists. This book contains descriptive sections, as well as sections devoted to rigorous mathematical treatment of a number of problems, some of which have been studied by the author and his collaborators. These sections may be omitted during a first reading. Each chapter has a selected bibliography.

This book is far from an exhaustive treatise on molecular biophysics. It deals principally with questions related to the structures and functions of proteins and nucleic acids.

M. V. Vol'kenshtein
Leningrad, September, 1964

CONTENTS

Chapter 1

PHYSICS AND BIOLOGY

PHYSICS AND LIFE

Living organisms which exist on earth are open self-regulating and self-replicating systems. Their most important functional components are biopolymers, namely, proteins and nucleic acids. Quite general scientific considerations indicate that life ought to exist beyond the limits of the earth, somewhere in the universe, if not on other planets of the solar system. At present, we know only organisms which exist on earth; reports on the finding of microorganisms in meteorites have not been confirmed up to now.

The term *open system* when applied to an organism has the same meaning as in thermodynamics. An open system is a system which exchanges matter and energy with the environment. Such an exchange is an essential characteristic of life; its irreversible cessation means death. One of the basic problems of biochemistry is the study of the chemical properties of metabolism. The term *self-regulation* points to the high degree in which the regulation of all the physical, chemical, and biological processes inside the organism are coordinated. This regulation is the result of the specific structure of the organism and of its components right down to molecules.

The term *self-replicating* is related to the properties of growth and reproduction which are characteristic of life. These processes, which can be followed down to the self-replication of molecules, follow strictly the path of covariant reduplication.[1] The point is the following: When organisms and their component parts multiply, they reduplicate themselves, i.e., they produce their own copies. These copies are not absolutely exact, since they reflect changes and variations in the object which occur before

1

the copying. It is clear that, being an open system, a reduplicating object is subject to the influence of the environment, which may bring about various changes.

It should be emphasized that living organisms must contain interacting substances of different natures and that chemical heterogeneity is essential for life. No isolated chemical compound, when taken by itself, can be living; this is true whether it is a protein or a nucleic acid. It makes no sense to talk of living molecules. Life arises as a result of a special interaction between a number of molecules of different types.

It is quite evident that organisms are much more complicated than any known nonliving matter. To what extent can physics investigate the structure and properties of living organisms? For a long time, there had been strong interaction between physics and biology. Then at one point it was somehow lost. At present, developments in biology have led us again to a close contact with physics.

Let us point out that the basic law of physics, namely, the law of conservation and transformation of energy, was discovered in the course of investigations on living organisms. Meyer noted that humans who live in tropical climates have venous blood with a color quite similar to arterial. He concluded from this that when the temperature of the environment is raised, a lower expenditure of energy is necessary to maintain constant body temperature. Further reasoning led Meyer to the formulation of a general law and to the calculation of the mechanical equivalent of heat. It is less well known that another codiscoverer of the law of conservation of energy, Helmholtz, also started from biological information.

Biophysics developed almost independently of the general problem of the relation between physics and biology. Until recently, its realm was, on one hand, the development of the general laws of physics in living nature, and, on the other hand, physiological phenomena more or less directly related to evident physical interactions within organisms. Particularly detailed investigations have been carried out on the sensory organs; the purpose of these was an examination of the reaction of organisms to physical and chemical actions. Helmholtz had already subjected a number of related problems to a rigorous physical and mathematical analysis. Sechenov has made the statement that physiology is the physical chemistry of the living organism. The structure of the traditional biophysics course was, as a rule, similar to that of a physics course: mechanics of organisms, heat phenomena, bioacoustics, action of light on organisms, electrical phenomena, etc. The emphasis was either on the properties of organisms, regarded as physical bodies, or on the interaction of physical fields with organisms. Thus, traditional biophysics did not study the fundamental phenomena of life; it accepted the existence of organisms as a fact which does not require any analysis. The subject of this book is a new realm of biophysics, namely, molecular biophysics.

In the middle of the twentieth century, along with atomic physics, the leading fields of science have become cybernetics and molecular biology. The rapid development of biology has resulted in a much stronger interaction with physics and chemistry than before. Biology has passed from macroscopic observations and experiments to a detailed investigation of cellular structures in relation to their functions; it has, thus, uncovered the molecular nature of a number of basic phenomena of life. Molecular biology examines the structure and behavior of those molecules which are responsible for the most important functions of organisms, for heredity, for mutations, for metabolism, and for the motions both of organisms and of their component parts. Such molecules, first of all, encompass proteins and nucleic acids. Molecular biology is a new field of science; it is on the border between molecular physics, organic chemistry, and biology.

MOLECULAR PHYSICS

First, let us examine those basic principles of molecular physics which will be necessary for the discussion that follows.

Molecular physics is concerned with those physical phenomena which are closely related to the atomic and molecular nature of matter, phenomena in which this nature finds a direct or indirect expression. Thus, molecular optical phenomena encompass both the refraction and scattering of light, and those processes of interaction of light with matter which are determined by the structure of the latter. On the other hand, diffraction and interference of light can be studied *per se* as optical phenomena.

In this manner, molecular physics gives us information first on the structure of matter, i.e., on the structure of molecules, liquids, crystals, derived from studies of its physical properties. Second, from studies on the structure of matter, it gives us information on the nature of the physical (and physical-chemical, since it is impossible to draw a line between molecular physics and physical chemistry) processes which take place within it. It is possible to list three groups of problems in molecular physics.

The first group covers the determination of structure in the broad meaning of the word. Physical (and chemical) methods enable us to establish the geometric distribution of atoms within molecules, crystals and liquids. The principal role is played here by X-ray diffraction and electron microscopy. The interference of X-ray and electron waves, scattered by atoms, depends on interatomic distances, which enables us to determine the latter. In both cases, such investigations use information on the structure of molecules obtained by chemical methods; in many (and quite complex) cases, however, exhaustive information on the geometric structure of molecules has been obtained strictly by the method of X-ray diffraction analysis. This is true, for example, of the structure of penicillin, and of structures of important proteins as well, such as myoglobin and hemoglobin. The geometry of simple molecules can also be studied by the method of radiospectroscopy. The distances between the levels of the rotational energies of molecules depend on their moments of inertia, i.e., on interatomic distances. Therefore, using rotational spectra, it becomes possible to determine these distances with great accuracy.

Hydrogen atoms are weak scatterers of X-rays and electrons. The distribution of hydrogen atoms can be determined better by the scattering of neutrons.

The mechanical characteristics of molecules, i.e., forces of interatomic interactions, determine the frequencies and patterns of vibration relative to each other of atoms within a molecule. They can be determined, therefore, from studies of the vibrational spectra of molecules, i.e., of their infrared and Raman spectra. The space distribution of electrons in molecules, the set of the energy levels of electrons, and the probability of transition between them determine the optical, electric, and magnetic properties of molecules. The electric and optical molecular properties are determined primarily by two fundamental molecular constants, namely, the polarizability and the dipole moment. Polarizability describes the ability of the electrons to become displaced under the influence of a constant or alternating electric field; it is reflected in the dielectric permeability, in refraction, in the scattering of light, etc. The dipole moment of a molecule or of its component parts (of an individual chemical bond or group of atoms) expresses the polarity of the system, the equilibrium distribution of charges. Properties of the electron shell of a molecule are reflected directly in the electronic, i.e., the ultraviolet and visible spectra. Quantum mechanics is the theory of the structure of the electron clouds of molecules and of the phenomena determined by it. The very existence of a chemical bond and, thus, the existence of a molecule is determined by quantum-mechanical laws. In many cases, however, problems related to the electromagnetic properties of a molecule can be solved quite successfully by using a semiempirical classical theory, which allows one to circumvent the extreme difficulties of quantum mechanical calculations encountered in the case of more or less complicated multielectron systems. Questions relating to the structure of molecules have been discussed elsewhere.[2]

The second group of problems in molecular physics is related to the equilibrium properties of molecular systems, namely, of gases, crystals, and liquids. Thermodynamics give a general phenomenological description of these properties.

Let us examine, for example, the crystallization of a liquid, a phase transition of the first order, i.e., a thermodynamic transition of such nature that it is accompanied by discontinuities in a number of basic thermodynamic quantities, such as the volume V, the enthalpy H, and the entropy S. The thermodynamic condition for the transition is that the thermodynamic potentials of the liquid and the crystal be equal,

$$F'_l = F'_{cr} \tag{1.1}$$

$$F' = H - TS = E + pV - TS \tag{1.2}$$

From (1.1), it follows that

$$\Delta H - T_m \Delta S = 0 \tag{1.3}$$

where ΔH and ΔS are the differences in enthalpy and entropy between the liquid and the crystal. From this, the temperature of crystallization and melting, T_m, is equal to

$$T_m = \frac{\Delta H}{\Delta S} \tag{1.4}$$

An actual calculation of ΔH and ΔS requires knowledge of the dependence of these quantities on the structure of the molecules in the liquid, as well as knowledge of the molecular nature of the process of crystallization.

The relation between the thermodynamic quantities and the structure of the molecule is given by statistical mechanics (e.g., see Ref. 2). The thermodynamic functions of a substance, which characterize its equilibrium properties, are expressed by the partition function

$$Q = \sum_i g_i e^{-E_i/kT} \tag{15}$$

where E_i is the energy of the ith energy level of the system; g_i is the corresponding statistical weight, i.e., the number of states of a system which possess energy E_i; and $k = 1.37 \times 10^{-16}$ erg/deg is the Boltzmann constant. Summation is carried out over all levels of the system. For quite dilute gases, energies E_i are the energies of isolated molecules. These can be calculated with a knowledge of the atomic masses and the geometric structure of the molecule, as well as the levels of the rotational, vibrational, and electronic energies. These levels can be determined by spectroscopic methods.

With a knowledge of Q, it is easy to find the fundamental thermodynamic functions. The free energy per molecule is equal to

$$F = E - TS = -kT \ln Q \tag{1.6}$$

The internal energy is

$$E = kT^2 \frac{\partial \ln Q}{\partial T} \tag{1.7}$$

and the entropy is

$$S = \frac{E - F}{T} = kT \frac{\partial \ln Q}{\partial T} + k \ln Q \tag{1.8}$$

In this way a direct relationship is established between thermodynamics and molecular theory.

The properties of condensed systems, however, cannot be expressed through the properties of isolated molecules. It is essential to take into account the interactions between the molecules which, in turn, are determined by the structures of their electron shells and their relative positions. In this process, the calculation of the partition function becomes considerably more complicated and, in practice, it is not always possible. Some extremely important processes which take place in a condensed state are cooperative, i.e., they are a strong function of the interactions between molecules. Crystallization is exactly such a process. If molecules did not interact, they would not be able to form crystals.

It is not at all obvious how expressions (1.5) to (1.8) can explain a discontinuity in the thermodynamic quantities during a phase transition. If the totality (ensemble) contains a finite number of molecules, Q can be differentiated with respect to T any number of times without singularities and breaks at any value of T, since the function $e^{-E_i/kt}$ changes monotonically with temperature. On the other hand, for a very large and practically infinite ensemble, a series in Q or some of its derivatives with respect to T can be nonconvergent at some value of T. A phase transition is possible only in a very large ensemble. It is defined by the relation between E (or H) and S [see (1.2)].

The Van der Waals equation

$$\left(p + \frac{a}{V^2}\right)(V - b) = RT \tag{1.9}$$

is the result of the application to a real gas of a simplified molecular theory with interactions taken into account. It gives the conditions for the gas–liquid transition described by the molecular parameters a and b. The transition is related to the term a/V^2, which, in turn, is a function of V. As the volume decreases, the magnitude of this term increases and, at some temperature, a sharp decrease in volume becomes favorable, since the corresponding loss of entropy is compensated for by a gain in the energy of attraction between the molecules. Such an interaction is cooperative; it helps itself. (The term *cooperative*, as applied to molecular phenomena, was introduced by Fowler.)

The fraction of molecules present in a state with energy E_i at equilibrium is

$$x_i = \frac{g_i e^{-E_i/kT}}{\sum\limits_i g_i e^{-E_i/kT}} = \frac{g_i e^{-E_i/kT}}{Q} \tag{1.10}$$

It follows from (1.6) that

$$Q = e^{-F/kT} \tag{1.11}$$

The term $g_i e^{E_i/kT}$ can also be written in the form $e^{-F_i/kt}$, since the statistical weight g_i is an entropic factor,

$$g_i = e^{S_i/k} \tag{1.12}$$

Consequently,

$$g_i e^{-E_i/kT} = e^{(S_i T/kT)-(E_i/kT)} = e^{-F_i/kT} \tag{1.13}$$

The equilibrium constant for the transition of a molecule from one state into another $1 \rightleftharpoons 2$, is equal to

$$K = \frac{x_2}{x_1} = \frac{g_2 e^{-E_2/kT}}{g_1 e^{-E_1/kT}} = e^{-(F_2-F_1)/kT} = e^{-\Delta F/kT} \tag{1.14}$$

Hence

$$\Delta F = -kT \ln K \tag{1.15}$$

This relationship is valid at constant volume. At constant pressure, we have

$$K_p = e^{-\Delta F'/kT} \tag{1.16}$$

and

$$\Delta F' = -kT \ln K_p \tag{1.17}$$

In a study of equilibrium systems, the aims of molecular physics are to find the thermodynamic conditions of equilibrium and to interpret them statistically on the basis of molecular constants. To solve these problems, it is essentially necessary and sufficient to find such a molecular model as would make it possible to calculate the partition function Q.

The third group of problems in molecular physics is related to the examination of the kinetic properties of the systems under investigation. Classical thermodynamics is essentially thermostatistics; it ignores the time dependence of the processes, regarding them as essentially infinitely slow. Thus, neither thermodynamics nor statistical calculations give answers to questions related to the rates of the changes which occur in matter. Such problems are solved by the methods of physical kinetics and of the thermodynamics of irreversible processes.

Any kinetic phenomena, both physical and chemical, consist in transitions of molecules from one equilibrium state into another. During the viscous flow of a liquid, the transition means a change in the position in space of a molecule; during the course of a chemical reaction, for example, during the tautomeric transformation of a ketamino group into an imino group

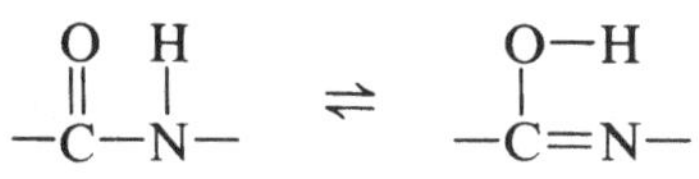

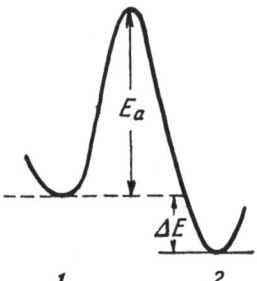

Fig. 1. Energy schematic diagram of a chemical reaction.

the molecule passes from one equilibrium state into another. The equilibrium states correspond to minima in the free energy; these states are separated by an energetic barrier, which must be surmounted for the transition to take place (Fig. 1). It is evident that transitions occur less frequently (the rate of the process is slow) if this barrier is high.

The probability that the molecule has acquired an energy equal to the height of the barrier is expressed by the Boltzmann factor. For a molecule in state 1 (Fig. 1), it is

$$w_{1 \to 2} = A_1 e^{-E_a/kT} \tag{1.18}$$

where E_a is the activation energy of the process, A_1 is a proportionality factor which has a dimension of sec^{-1} and contains an entropic multiplier $e^{S_a/k}$ (S_a is the entropy of activation). The rate of the transition, i.e., the number of transitions from the left to the right during 1 second is

$$v_{1 \to 2} = n_1 w_{1 \to 2} = n_1 A_1 e^{-E_a/kT} \tag{1.19}$$

where n_1 is the number of molecules in state 1. This is the law of Arrhenius, which expresses the dependence of the rate of the process on temperature. The higher the temperature, the higher the rate of the process.

For a transition from the right to the left,

$$w_{2 \to 1} = A_2 e^{-(E_a + \Delta E)/kT} \tag{1.20}$$

$$v_{2 \to 1} = n_2 A_2 e^{-(E_a + \Delta E)/kT} \tag{1.21}$$

At thermal equilibrium,

$$v_{1 \to 2} = v_{2 \to 1} \tag{1.22}$$

The equilibrium constant is equal to

$$K = \frac{n_1}{n_2} = \frac{A_2}{A_1} e^{-\Delta E/kT} = e^{-\Delta F/kT}$$

We have again obtained Eq. (1.14). Thus, the height of the barrier E_a controls the rate of the transition but not the equilibrium amounts of the molecules. The magnitude of E_a is responsible for the kinetics of the process; to the contrary, the difference of the energy of the final and initial states, ΔE, determines the conditions of thermodynamic equilibrium.

In cooperative processes the picture is considerably more complicated. For example, the viscous flow of a liquid is cooperative, since the transported particles of the liquid interact with their neighbors. Therefore, in order to become displaced, particles must push their neighbors and, thus, must overcome a certain barrier, i.e., they must possess an activation energy. The situation is similar to that of a man who tries to push his way through a filled bus. In order for him to go through, other passengers must move. The cooperativity is expressed as a dependence of the energy of activation on the numbers of particles which have already passed over the barrier. In fact, the more people that have come out of the bus, the easier it is to reach the exit. In the absence of cooperativity, the dependence of the logarithm of the rate of the process on the reciprocal of the temperature is linear; it follows from the law of Arrhenius (1.19) that

$$\ln v = \ln n_1 A_1 - \frac{E_a}{kT} \tag{1.23}$$

In the case of cooperativity, the law of Arrhenius and condition (1.23) are no longer valid, and E_a is no longer a constant.

MOLECULAR BIOPHYSICS

The discovery of some molecular processes which lie at the very foundation of life phenomena have brought biology in direct contact with molecular physics. Molecular biophysics studies the physical nature of various phenomena which are examined by molecular biology in their biological aspects. We find here the same three groups of problems, namely, structures, thermodynamic equilibrium, and kinetics.

Modern molecular biophysics is centered mostly on those problems related to the structure and physical properties of proteins and nucleic acids, which determine their biological function. Molecular biophysics investigates the conditions of equilibrium and the kinetics of molecular biological processes involving these biological macromolecules and other molecules and ions. We are interested in those structural and chemical transformations of proteins and nucleic acids which are responsible for a number of extremely important life phenomena, such as the synthesis of proteins and enzymatic catalysis. As we shall see, molecular biophysics is based on the theoretical concepts and experimental methods of modern molecular physics. Thus, molecular biophysics forms a part of molecular biology which encompasses a wider spectrum of knowledge. Molecular biophysics is its physical part.

It is, of course, impossible to separate molecular biophysics from biophysical chemistry, just as it is impossible to separate molecular physics from physical chemistry. It is clear that, on the molecular level, processes which take place both in living and nonliving nature can be simultaneously both physical and chemical.

Thus, then, we may fairly ask the question: Is the term molecular biophysics useful? If one accepts the existence of biophysics as a science (and nobody has any doubts about that), then it is quite reasonable to talk of its molecular part, i.e., of molecular biophysics. Today it is the most important and most interesting part of biophysics.

Molecular biology and molecular biophysics do not exhaust the realm of biology. If genetic problems can be solved at present on the molecular level, this certainly does not apply to a number of other biological phenomena, such as problems of the behavior of organisms in their entirety. The simplest functional unit of an organism, namely, the cell, is immeasurably more complicated than the molecules from which it is constructed. It is a system with high order, consisting of supramolecular structures, the properties of which cannot be obtained by a simple summation of the properties of the component molecules; it is important to take into account their interactions. Biology penetrates ever deeper into the details of the structure and physiology of cells, while physics and chemistry study the molecules of biologically functional substances. Today, these two paths coincide to a great extent; the development of the physical chemistry of the complicated supramolecular structures found within cells is a problem for the future.

As we shall see, the existence of proteins is essential for life, since the most important functions of organisms are carried out by proteins. Biochemical reactions and the processes of metabolism take place always with the participation of catalysts, i.e., enzymes; these are not altered as a result of the reaction, but they greatly accelerate the reaction. Enzymes are proteins. Proteins are responsible for the transport of matter into cells; they also perform the special functions of the transport of oxygen in the organism (the hemoglobin of blood) and of the protection of the organism from foreign pathogenic materials in the processes of immunity. Life is impossible without mechanical motion, without the displacements in space of organisms and their parts. Such mechanical motions, from molecular to muscle movement, are performed by special proteins.

There is only one thing that proteins cannot do; they cannot synthesize themselves. It turns out that, for this, it is necessary to have molecules of another type, namely, nucleic acids. Their function consists strictly in synthesis of proteins of a rigidly determined structure. The disruption of such a synthesis results in most serious consequences for the organism. The hereditary replication of organisms means, first of all, the replication of its entire set of essential proteins. Thus, substances which are necessary for the synthesis of proteins, namely, nucleic acids, are responsible both for heredity and for variability in organisms.

The foundations of atomic-molecular biology and of biophysics were laid down in the studies of Timofeev-Resovskii, Delbrück, and others, who have established some of the basic laws of radiobiology and who have obtained important information on the molecular nature of the gene.[1] Timofeev-Resovskii has pointed out that the first attempt to introduce molecular concepts into genetics was made in 1893 by Kolli, in Moscow. Later, Koltsov formulated quite clearly the basic positions of a molecular hypothesis of genetic phenomena.[1] Now, we no longer have a hypothesis but a complete physical-chemical theory of heredity. This theory is not only based on vast experimental verifications, but it also permits conscious interference in genetic phenomena.

Proteins and nucleic acids are large chain molecules which contain hundreds and thousands of atoms. They are polymers. Consequently, molecular biophysics is the physics of biopolymers, a new realm in the physics of polymers, which, up until recent times, was interested essentially in synthetic polymers (rubbers, plastics, fibers). The theoretical foundations of the physics of polymers and of molecular biophysics are thermodynamics and statistical physics. Since biological processes are, first of all, processes of ordering, their physical interpretation must be based to a great extent on the theory of phase transitions and the theory of cooperative processes. As a result, molecular biophysics is based on the theory of cooperative processes. What is the relation between molecular biophysics and the physics of microsystems, namely, quantum mechanics?

It is only through quantum mechanics that it is possible to explain the very fact of the existence of molecules, the phenomenon of chemical bonds between atoms and molecules. In this sense, quantum mechanics lie at the foundation of molecular physics. Any changes in the structures of molecules and their energetic states have a quantum, discrete nature.

By starting from the fact of the existence of molecules, it is possible, however, to examine molecular phenomena on a classical basis, since organisms consist of a very large number of atoms; the same applies to molecular biopolymers. The application of quantum-mechanical concepts requires the interpretation of a number of special processes, in particular, those which are related to the interaction of shortwave radiation with organisms. A number of scientists hold another point of view; they consider that the quantum-mechanical processes of the migration of energy and of resonance have a decisive role in molecular biology. The question of the relation between molecular biophysics and quantum mechanics is examined in detail in the last chapter of this book.

Molecular biology and biophysics give the promise of future discoveries of enormous practical importance. Physical-chemical genetics must result in the conscious direction of heredity, in the production of new types of agricultural plants and animals. The understanding of the molecular nature of heredity and of variability, of metabolic processes, and of enzymatic catalysis must lead to the solution of most important problems of medicine, such as the biological problem of cancer. On the other hand, molecular biology opens unfathomable perspectives for chemistry and chemical technology. As a result of studies of the physical chemistry of molecular biological processes, chemistry has acquired amazing possibilities for highly specific syntheses of new substances; energetics will give to man mechanochemical appliances with a high coefficient of useful work, similar to muscles.

Today molecular biology is still principally concerned with theoretical studies. It is hard to state the significance of such investigations in better words than those of K. A. Timiryazev:[3] "The question is not whether scientists and science must serve society and humanity; such a question cannot even be raised. The real question is: which path is the shortest and surest way to this goal? Should a scientist follow the lead of social wizards and myopic leaders or should he, rejecting these, follow the only possible road, determined by the internal logics of facts which guide the development of science.... The criterion of true science is not the outward appearance of a narrow immediate benefit, which is exploited most adroitly by the practitioners of pseudoscience, who, without effort, obtain for their parodies recognitions of their practical importance and even of national usefulness".

Molecular biology was founded as a result of the irrefutable logic of science. Molecular biology opens a direct path to such future discoveries in the fields of agriculture and medicine as we can hardly imagine at present. This statement will look less like an assertion when the reader reaches the end of this book.

THERMODYNAMICS AND BIOLOGY

It is natural that the physical investigation of life, namely, the study of the organism, must start with an examination of its most general properties as a molecular system. Such an examination, which must precede a complete molecular physical analysis, is based on the laws of thermodynamics.

At first glance, it would appear that a living organism exists in spite of thermodynamics. According to the second law of thermodynamics, ordering in an isolated system must decrease with a concomitant increase in the measure of this order, i.e., in entropy. Expressions of this fundamental law are found in the equilibration of temperature as a result of thermal conductivity, the equilibration of the concentrations of gases and liquids as a result of diffusion, and in a number of other processes during which the entropy of the system tends to a maximum.

To the contrary, an organism is continually creating order out of order in the sense of self-replication and order out of disorder as a result of metabolism. Using a disordered system of small molecules, which it obtains in the processes of nutrition and respiration, the organism creates the complicated highly ordered macromolecules of biopolymers, and builds perfect supramolecular structures; the organism either grows and develops or it remains in a stationary (but not static) state of high order. This is what was meant by Schrödinger when he said that an organism is an aperiodic crystal.

A building left to itself gradually crumbles; sooner or later it becomes a heap of rubble. The intelligent activity of a living organism, namely, of man, is directed at overcoming the second law of thermodynamics. Man constructs buildings and keeps them in good repair; he does not increase but he decreases entropy.

Does this mean that living organisms do not obey the laws of physics?

The laws of classical thermodynamics are valid for systems which are isolated from the outside world. An organism is an open system; it exchanges both matter and energy with the surrounding medium. In such a system, the laws of conservation of mass and energy are no longer obeyed. A thermodynamic description of an organism requires the use of the thermodynamics of open systems, which have been developed during the twentieth century, in particular, by the Belgian physicists, De Donder,[4] and Prigogine.[5] Independently, the same ideas have been applied brilliantly to some questions of molecular physics by Mandel'shtam and Leontovich.[6]

As could be expected, once it is realized that organisms are open systems, all contradictions with thermodynamics turn out to be imaginary.

Let us examine a few principles of the thermodynamics of open systems.

The entropy of an open system changes as the result of irreversible processes which take place inside of it and of the inflow, or outflow, of entropy from the outside medium

$$dS = d_iS + d_eS \qquad (1.24)$$

where dS is the total change of entropy, d_iS is its production inside the system, and d_eS is the inflow, or outflow, of entropy. According to the second law of thermodynamics

$$d_iS \geqslant 0 \qquad (1.25)$$

the sign of d_eS can be both positive and negative. In organisms, entropy d_iS is produced principally by chemical reaction. The masses of chemical components change during their reactions. The change of the mass m_γ of components γ during time dt can be written in the form

$$dm_\gamma = v_\gamma M_\gamma \, d\xi \qquad (1.26)$$

where v_γ is a stoichiometric coefficient, M_γ is the mass of component γ in moles, and ξ is the extent of the reactions. For example, in the oxidation of glucose,

$$C_6H_{12}O_6 + 6O_2 \longrightarrow 6CO_2 + 6H_2O$$

$$v_{gl} = -1 \qquad v_{O_2} = -6 \qquad v_{CO_2} = 6 \qquad v_{H_2O} = 6$$

and

$$\frac{dm_{gl}}{-M_{gl}} = \frac{dm_{O_2}}{-6M_{O_2}} = \frac{dm_{CO_2}}{6M_{CO_2}} = \frac{dm_{H_2O}}{6M_{H_2O}} = d\xi$$

In a closed system, the law of conservation of mass is valid,

$$m = \sum_\gamma m_\gamma \qquad dm = 0 \tag{1.27}$$

and consequently,

$$dm = d\xi \sum_\gamma v_\gamma M_\gamma = 0 \tag{1.28}$$

The equation

$$\sum_\gamma v_\gamma M_\gamma = 0 \tag{1.29}$$

which follows from (1.28), is the equation of the chemical reaction. By replacing the masses m_γ by numbers of moles n_γ instead of (1.26), we can write

$$dn_\gamma = v_\gamma \, d\xi \tag{1.30}$$

The rate of the reaction can be expressed in the form

$$\frac{dn_\gamma}{dt} = v_\gamma \frac{d\xi}{dt} = v_\gamma v \tag{1.31}$$

The entropy of a system may change both as a result of a change of the amount of heat

$$dS' = \frac{dQ}{T} = \frac{dE + p \, dV}{T}$$

where dE is the change of the internal energy and $p \, dV$ is mechanical work, and as a result of a chemical reaction, the entropy of the products of a reaction is different from that of the reactants. This change is expressed by

$$d_iS = -\sum_\gamma \frac{\mu_\gamma}{T} dn_\gamma \tag{1.32}$$

The quantities μ_γ, which are called *chemical potentials*, are, in turn, expressed by

$$\mu_\gamma = -T \left(\frac{\partial S}{\partial n_\gamma} \right)_{E,V,n'_\gamma}$$

The subscript n'_γ indicates that the derivative with respect to the number of moles n_γ of a given component is taken with the amounts of all the other chemical components kept constant.

Following De Donder, we can write

$$d_iS = \frac{A \, d\xi}{T} > 0 \tag{1.33}$$

Quantity A is the affinity of the chemical reaction. Combining (1.33), (1.32), and (1.30), we find

$$A = -\sum_\gamma v_\gamma \mu_\gamma \tag{1.34}$$

At equilibrium $A = 0$.

The production of entropy per unit of time is

$$\frac{d_iS}{dt} = \frac{A}{T} \frac{d\xi}{dt} = \frac{Av}{T} > 0 \tag{1.35}$$

If several chemical reactions take place simultaneously within a system (and there are many of them in an organism), then

$$d_iS = \frac{1}{T} \sum_\rho A_\rho \, d\xi_\rho \tag{1.36}$$

where A_ρ is the affinity of the ρth reaction. Consequently,

$$\frac{d_iS}{dt} = \frac{1}{T} \sum_\rho A_\rho v_\rho > 0 \tag{1.37}$$

This result is very important. Let us assume that an isolated reaction cannot take place, since

$$A_1 v_1 < 0 \tag{1.38}$$

The first reaction becomes possible when a second reaction 2 occurs simultaneously, if for 2, $A_2 v_2 > 0$ and if the condition

$$A_1 v_1 + A_2 v_2 > 0 \tag{1.39}$$

is fulfilled, since Eq. (1.37) refers to the sum of the two processes. It is just such a pattern which is found for the chemical reactions within an organism. The isolated reaction of the synthesis of a protein from the amino acids which compose it is impossible, since it results in an increase in free energy. But the linkage of this process in the organism with other thermo-dynamically advantageous reactions, described by Eq. (1.37), makes it possible. The same applies to other phenomena. Active transport of a number of substances into living cells takes place in a direction opposite to a decrease in concentration. In other words, even where there is a low concentration of the necessary substance in the external medium, its concentra-tion in the cell can increase; the cell sucks in the dissolved substance from the solution, contrary to the law of normal diffusion, which is a particular case of the second law of thermodynamics. This transport is possible as a result of its conjugation with other processes, since their totality fulfills the requirements of the second law of thermodynamics.

In a state which is not at thermodynamic equilibrium, but close to it, the rate of the chemical reaction is proportional to the affinity

$$v = \frac{LA}{T} \tag{1.40}$$

If several chemical reactions occur in the system, then the rate of any given reactions can actually be a function of the affinity of each reaction,

$$v_\rho = \frac{1}{T} \sum_\tau L_{\rho\tau} A_\tau \tag{1.41}$$

and consequently, the relation (1.37) can be rewritten in the form

$$\frac{d_iS}{dt} = \frac{1}{T^2} \sum_\rho \sum_\tau L_{\rho\tau} A_\rho A_\tau > 0 \tag{1.42}$$

The proportionality coefficients $L_{\rho\tau}$, which are called *phenomenological coefficients*, are sym-metric, i.e., they satisfy the Onsager relation

$$L_{\rho\tau} = L_{\tau\rho} \tag{1.43}$$

As has been pointed out, all of this is valid close to the state of equilibrium. The quantitative condition for such a closeness has the form

$$\frac{|A|}{RT} \ll 1 \qquad (1.44)$$

It is, of course, not always fulfilled. It is possible, however, to use the linear relations (1.41) and (1.42), if the condition (1.44) is fulfilled for individual reactions, i.e.,

$$\frac{|A_\rho|}{RT} \ll 1 \qquad (1.45)$$

even if in the system as a whole

$$\frac{|A|}{RT} = \frac{\left|\sum_\rho A_\rho\right|}{RT} > 1 \qquad (1.46)$$

Biological processes which take place in organisms frequently fulfill these conditions and admit the application of linear relationships. The development of thermodynamics for nonlinear regions is, at present, one of the active problems of science.

The production of entropy per unit of time d_iS/dt is the principal quantity of the thermodynamics of irreversible processes. It describes the deviation from thermodynamic equilibrium, in which $A = 0$, $v = 0$, and $d_iS/dt = 0$. In an isolated system, the entropy continuously increases, tending toward a maximum, which corresponds to equilibrium; if the system is open, however, it can remain for a long time in a state of low entropy as a result of its flow from the system into the surrounding medium. The quantity dS in (1.24) can also be negative if $d_eS < 0$ and $|d_eS| > |d_iS|$.

Let us examine an isolated system which consists of an organism and some environment. The organism receives from this environment its nutrition and the oxygen which it breathes. At the same time, the environment receives all the substances which are secreted by the organism. For example, an astronaut exists under such conditions. He is an open system with relation to his spaceship, but the ship as a whole is well isolated. The total change of entropy of the system is

$$dS = dS_1 + dS_2 \qquad (1.47)$$

where dS_1 is the change of entropy of the astronaut, and dS_2 is the change of entropy of the environment in the ship. Furthermore,

$$dS = d_iS + d_eS \qquad (1.47a)$$

where d_iS is the production of entropy as a result of internal chemical processes, and d_eS is the flow of entropy from the astronaut to the environment (or in the reverse direction). By applying the second law of thermodynamics, in an irreversible process $dS \geqslant 0$ and $d_iS > 0$. The change of entropy of the astronaut may be set equal to the sum of dS_i and part of the flow d_eS

$$dS_1 = d_iS + d_eS_1$$

The entropy of the astronaut may remain constant or decrease under the condition that $dS_1 \leqslant 0$, that is, $d_eS_1 < 0$ and $|d_eS_i| > |d_iS|$. Furthermore, $dS_2 > 0$ and

$$dS_2 = d_eS_2 > d_iS + d_eS_1 = dS_1$$

In other words, the increase in order in the organism of the astronaut is due to a negative flow of entropy, i.e., due to an increase in entropy in the environment. This increase must be compensated for by a decrease of the entropy of the astronaut. The materials excreted by the astronaut have a higher entropy than those that enter the organism.

The same applies to the entire biosphere of our planet. If we regard the entire solar system as an isolated system (which, of course, is quite far from true), its entropy must be constantly increasing at the expense of the radiation of the sun. The decrease of the entropy in all living organisms is negligibly small against the background of this grandiose process.

An open system attains a stationary state if the production of entropy within it compensates exactly for the flow of entropy into the environment. In a stationary state, all functions of the state are independent of time; this is true also of entropy. Thus,

$$\frac{dS}{dt} = \frac{d_iS}{dt} + \frac{d_eS}{dt} = 0 \qquad (1.48)$$

that is

$$\frac{d_iS}{dt} > 0 \qquad \frac{d_eS}{dt} < 0$$

so that

$$\frac{d_eS}{dt} = -\frac{d_iS}{dt}$$

It is evident that the stationary state is not a state of equilibrium in which

$$\frac{d_iS}{dt} = 0$$

i.e., the entropy has reached a maximum. To maintain the nonequilibrium state of an open system, it is necessary to have a flow of entropy through it. The entropy of a substance which enters into the system is less than the entropy given out by the system to the surrounding medium. In this sense, an open system degrades the substance: the astronaut degrades the food, transforming it into substances which are excreted from his organism. Such phenomena can be described quite well by a model in which a liquid flows from one vessel into another. The flow of the liquid represents the flow of the chemical reaction. If the system is closed, i.e., no liquid enters the vessel from the outside or flows out, then all the liquid will flow from the upper vessel into the lower one at a rate determined by the size of the faucet and a state of equilibrium will be established. (Fig. 2.) The level of liquid in the lower vessel will represent the extent of the reaction of such a state.

If the system is open, then a definite level of liquid, which does not correspond to equilibrium, will be established both in the upper and lower vessels (Fig. 3). This level will depend on the degree of opening the faucet. The faucet can serve as a model for a catalyst, namely, a substance which does not participate in the reaction but which affects its velocity. In a closed system, the final extent of the reaction does not depend on its velocity, i.e., on the degree the faucet is opened; eventually, a constant level of liquid is established in the lower level. In an open system, both the velocity and the extent of the reaction are found to be functions of the catalyst. When the concentration of the catalyst is changed (with a turn of the faucet), new levels of the liquid, new stationary states, are established.

The ability of organisms to maintain a constant concentration of materials (homeostasis) is due to the fact that organisms are open systems which have reached a stationary state. Thermodynamic analysis shows that an open system which has been displaced from the

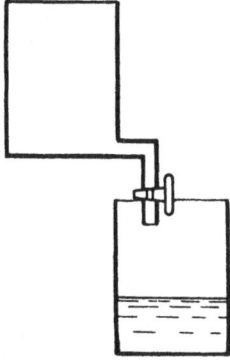

Fig. 2. Equilibrium state of a liquid in a closed system of vessels.

stationary state tends to return to that state just as a closed system which has been displaced from a state of equilibrium returns to that state. An important property of the stationary state is that this state corresponds to the minimal production of entropy. Let us assume that chemical processes are taking place in the system and there is a flow of heat. Then

$$\frac{d_i S}{dt} = -\frac{W_x}{T^2}\frac{\partial T}{\partial x} + \frac{Av}{T} > 0 \tag{1.49}$$

The first term expresses the change of the internal entropy as a result of heat conductance; W_x is the flow of heat. In the linear region

$$W_x = -\frac{L_{th}}{T^2}\frac{\partial T}{\partial x} + \bar{L}_{12}\frac{A}{T}$$

$$v = -\frac{L_{21}}{T^2}\frac{\partial T}{\partial x} + L_{ch}\frac{A}{T} \tag{1.50}$$

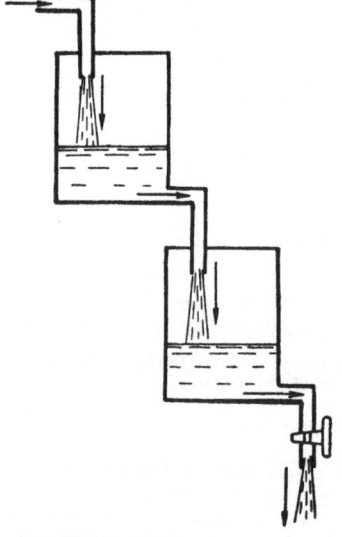

Fig. 3. Stationary state of a liquid in an open system of vessels.

Furthermore, $L_{12} = L_{21}$ [see Eq. (1.43)]. Consequently,

$$\frac{d_iS}{dt} = \frac{L_{th}}{T^4}\left(\frac{\partial T}{\partial x}\right)^2 - 2\frac{L_{12}}{T^2}\frac{\partial T}{\partial x}\frac{A}{T} + L_{ch}\left(\frac{A}{T}\right)^2 \qquad (1.51)$$

and

$$\frac{\partial}{\partial\left(\frac{A}{T}\right)}\left(\frac{d_iS}{dt}\right) = -2\frac{L_{12}}{T^2}\frac{\partial T}{\partial x} + 2L_{ch}\frac{A}{T} = 2v \qquad (1.52)$$

In a stationary state, however, the total flow of material which is described by v, is equal to zero. There is no accumulation nor loss of material. Consequently, the derivative d_iS/dt in Eq. (1.52) is equal to zero and d_iS/dt is minimal in this case (the second derivative, as can be easily seen, is positive and is equal to $2L_{ch}$).

The thermodynamic analysis of an individual organism or of the whole process of evolution[7] turns out to be highly instructive.

INFORMATION THEORY

Information theory is one of the bases of cybernetics, namely, of the science which studies processes of the regulation and control in regulating systems. Living organisms and their functional parts are most perfect examples of directing self-regulating systems. As a result, they are subject to cybernetic investigation. Processes of regulation and control are possible only as a result of the transfer and conversion of the corresponding information. The mathematical investigation of these processes consists in the algorithmic description of the transformation of information, of the study of the structure of the transforming algorithms, and of the principles of its realization.[8] Here we are interested, not in the mathematical aspects of the question, but in the physical bases of the biological processes which we are studying.

The everyday meaning of the term *information* does not require explanation. In science, however, this concept has taken on a strict quantitative character. It turned out to be closely related to the concept of entropy, that is, to the measure of disorder.

The information contained in any message whatever is a quantitative measurement of the knowledge which such a message contains. Let us assume that we are dealing with dice. As long as a die is not thrown, there is no event and there is no message about an event. All six results have equal probability and information is absent, i.e., it is equal to zero. Once the die has been thrown, and, let us assume, a three has turned up, the experiment has been carried out, a definite event has taken place, and information is different from zero. How can we measure it quantitatively? It is evident that in an experiment in which we throw two dice, we obtain twice as much information. In the message, the first die has a 3 and the second a 5; there is twice as much information than in the message that one die turned up with a 3. The probability that two independent events will occur is equal to the product of their probabilities; the information about two such events is equal to the sum of information about these events individually. Consequently, it is rational to relate the information to the logarithm of the probability. In this particular example, the probability that a definite number will turn up on the die is $P_{01} = \frac{1}{6}$. The probability that a definite number will fall on the first and second die is

$$P_0 = P_{01}P_{02} = \frac{1}{36}$$

The information is

$$I = -K \log P_0 = -K \log P_{01} - K \log P_{02}$$

where K is a proportionality factor. It is evident that I is proportional to the logarithm of the ratio of the number of possible events in the initial situation (before the casting of the die, $1/P_0 = 36$ possible results) to the number of actually realized events in the final situation (in this particular example, one event is 3 on the first die, 5 on the second). It is reasonable to use a logarithm to the base 2, and K can be set equal to unity. Then

$$I = \log_2 P \tag{1.53}$$

Such a definition is valid for problems in which only two events are possible, i.e., only two answers are possible about the result of any individual experiment, namely, yes or no. Let us assume that in a maternity ward, n infants are born during a given period of time. The probability that a boy or a girl is born is equal to $\frac{1}{2}$. The initial situation contains no information; it is possible to have 2^n individual events (the first infant is a boy and the others $n - 1$ are girls; the first infant is a girl and the others are boys; etc.). The report on the result of all the births contains information equal to

$$I = -\log_2 \frac{1}{2^n} = \log_2 2^n = n \tag{1.54}$$

i.e., with such a definition, the information is equal simply to the number of choices with two alternatives. A unit of information when $n = 1$ is called a bit.

Let us give one more example. Let us assume that we have taken a three-digit number. What information will we obtain from a knowledge of this number? In all, it is possible to have 900 three-digit numbers. Each one of the three digits can have 10 values from 0 to 9. Consequently,

$$I = 3 \log_2 10 = 3 \times 3.32 = 9.96 \text{ bits}$$

This definition of information corresponds to a binary system of counting in which any number can be written out in the form of powers of 2 by means of the numbers 0 and 1 (for example, $7 = 111$, since $4 = 100$, that is, 2^2; $2 = 10$, that is, 2^1; and $1 = 1$, that is, 2^0). Each decimal unit gives 3.32 bits of information, that is, the binary system of numbers requires 3.32 times as many numbers as a decimal system (see also Refs. 9 and 10).

We should emphasize that a quantitative definition of information is not related at all to its value for man. Thus, life on Mars can exist or not exist. There are only two possibilities. It is evident that a positive answer on the question about the existence of life on Mars contains one bit of information, i.e., just as much as the knowledge that when a coin is tossed, heads turn up. On the other hand, the values of the obtained information in the two cases are not comparable. If, however, the tossing of a coin decides some important question, then the value of the obtained information is much greater than when the tossing is done just as a simple game. The qualitative determination of the value of information may be linked to the reserve of information already stored in the receiver. Here, we do not encounter the problem of the value of the information.

We see that the scientific concept of information is closely related to probability. As is known, the statistical, i.e., molecular, interpretation of entropy, given by Boltzmann, is also related to the probability of the state of the system. This makes it possible to establish a quantitative relationship between information and entropy, which was first done by Szillard,[11] an outstanding physicist, who recently has been studying biological questions.

Entropy is expressed by Boltzmann's equation

$$S = k \ln W \tag{1.55}$$

where W is the thermodynamic probability of a given state of the system, i.e., the number of various ways of obtaining the given state, and k is the Boltzmann constant equal to 1.37×10^{-16} erg/deg. It is evident that W is inversely proportional to the probability P which we have just used.

Let us examine the crystallization of a liquid. A liquid possesses a greater entropy than the crystal since its state can be realized by a larger number of ways in which molecules may be distributed than the state of a crystal. The crystal is ordered; its molecules can be located only at the knots of the crystal lattice. During crystallization, the entropy decreases by the quantity

$$\Delta S = S_l - S_{cr} = k(\ln W_l - \ln W_{cr}) \tag{1.56}$$

In passing from the liquid to the crystal, we have selected from all the possible states of the molecule only those which correspond to an ordered lattice. In this manner, contrary to the liquid, a crystal contains a certain amount of information on the distribution of the molecules. If we introduce into the definition of the information a proportionality coefficient equal to

$$K = k \log_2 e \tag{1.57}$$

we can equate the information contained in the crystal with the decrease of entropy S or with an increase of "negentropy" (negative entropy) N:

$$I = -k \log_2 e \log P = k \ln \frac{W_l}{W_{cr}} = \Delta S = -\Delta N \qquad (1.58)$$

Thus, an increase of information contained in the system means a decrease in its entropy. This is not a formal analogy but a description of concrete physical properties.

In analyzing the second law of thermodynamics, Maxwell has proposed the following fantastic experiment. A vessel containing gas is divided into two sections by a barrier with a shutter which is controlled by a demon, namely, by a conscious being of molecular dimensions. When a rapidly flying molecule approaches the shutter, the demon opens it. He does not admit slow molecules into his half of the vessel. As a result, fast molecules become concentrated in one-half of the vessel and slow ones in the other. Now there is a difference in temperatures and the second law of thermodynamics becomes violated.

This paradox has long ago been solved in science. Brillouin has examined it from the point of view of the theory of information.[12] To see molecules, the demon must first illuminate them, i.e., he must possess a source of radiation which is not in equilibrium with the surrounding medium. Such a source has a low entropy, i.e., it possesses negentropy from which the demon obtains information. Once information has been obtained, it can be used for lowering the entropy of the gas, i.e., for increasing the negentropy. The complete calculation of the change of entropy, taking into account the balance, negentropy → information → negentropy, shows that the entropy of the system as a whole (the source of the light plus the gas plus the demon) does not decrease, but it increases and the second law of thermodynamics is fully observed.

Any experiment which gives information on the physical system leads to an increase of the entropy of the system or of its surroundings. This increase in entropy is greater than or equal to the obtained information.

An increase in information (or a decrease of entropy) during a transition from an embryonic cell to a multicellular organism is possible only in an open system and it is similar to the increase of information during the crystallization of a liquid. Crystallization is also impossible in isolated systems since, to transform a liquid into a crystal, it must be cooled, i.e., the entropy of the refrigerator must be increased. The decrease of the entropy both during the growth of an organism and during the crystallization of a liquid is paid for by an increase in the entropy of the surrounding medium, in total agreement with the second law of thermodynamics.

The information on a future crystal is already coded into the structure of each one of its molecules, into the structure of the electron cloud. A knowledge of the structure of molecules makes it possible to say in advance what the crystal will be like, whether it will have a cubic lattice like NaCl or a hexagonal one like H_2O. All the data of modern biology indicate that the molecules of the embryonic cells contain in coded form the information on the properties of the future organism, i.e., on the aperiodic crystal, so that the analogy remains complete in this sense as well.

During the development of the organism the total amount of information increases but not information per individual cell. Exactly the same thing happens during the crystallization of a liquid. A liquid contains less information, it possesses a higher entropy, than the molecules which compose it; this is due to thermal motion. In crystallization, the excess entropy of the liquid is given to the refrigerator and the entropy of the crystal is decreased, roughly speaking, down to the entropy of the immobile molecules. In this manner, information increases. Since both entropy and information are additive quantities, the larger the amount of liquid that crystallizes, the greater will be the gain of information. However, the increase of information per molecule will not change during this process.

During the growth of a multicellular organism from a single cell, the information per cell does not increase. Furthermore, part of this information becomes blocked. It is evident that a larger volume of information is used in the embryonic cell than in any specialized cell found in any organ of the whole organism; the purpose of the cells of an organ is to perform only some definite function, such as the function of muscle cells or nerve cells. It is the entire organism which develops from the embryonic cell.

The informational aspects of biology are quite important. Information theory and cybernetics as a whole yield descriptions of organisms, and a clear discussion not only of their detailed properties but also of their behavior as a whole. The translation of biological laws into the language of information theory enables us to pose quite clearly problems of scientific investigation and to reject a number of false concepts. A number of such examples have been cited by Lyapunov.[8] Let us take one. The assertion that acquired traits may be transmitted by hereditary means assumes that information flows from somatic to sex cells. Not a single one of the proponents of the theory of the heredity of acquired traits has ever cited data to support such a flow; they have not even attempted to obtain such information.

Of what information should we speak when dealing with a living organism? First of all there is the genotypic, hereditary information, which defines the species and the individual with his characteristic traits. The ability to store genotypic information guarantees the stability of hereditary traits. The phenotypic information, the phenotypic memory, signifies the storage of information on the conditions for the existence of a given individual, i.e., acquired traits, including conditioned reflexes. Finally, there is memory in the usual sense of the word, i.e., the storage of information in the brain cells.

Information is always stored and coded in some way either in an automatic system or in the organism. Thus, any number in a binary system can be coded with two symbols, 0 and 1. The Morse code is a four-letter code; its symbols are dot, dash, space between symbols, and double space between words. Information may be transmitted from one part of the system to another part also by a code.[13] Frequently, this code is further translated into another code (for example, the written deciphering of a telegraphic message transmitted in the Morse code); a number of codings is possible. Any transfer of information is accompanied by an increase in entropy; i.e., by a partial destruction of the information. In the theory of communications, this is called noise. Thus, in the well-known children's game of the broken telephone, its whole charm consists in the distortion of information due to noise. A well-designed machine or an organism reduces these noises to a minimum, by using, for example, excess information (such as multiple repetition of the same message). However, it is impossible to eliminate noise completely. Various external interactions increase the noise level in a system. Both the stored and transmitted information become distorted. This is the essence of the variability of organisms, i.e., the impossibility of the existence of hereditary materials and genotypic information in an absolutely immutable state.

Feedback plays a very important role in regulating systems in living organisms and their communities. The system carries out certain actions and obtains information on their result. Depending on this information, the system changes its behavior. It is obvious that any regulation is based on the feedback. Feedback is carried out directly in reactions of organisms, in physiological processes, in reflex, in instinctive or conscious behavior. The homeostasis which exists in organisms, for example, the maintenance of constant temperature, of osmotic pressure, and of concentration of ions in the protoplasm, on one hand, signifies a stationary thermodynamic state and, on the other hand, the presence of feedback. In other words, any change of the corresponding physical-chemical quantities changes the functioning of the regulating mechanism in such a way that it restores such values of these quantities as are necessary for normal living. Feedback takes place not only in individual elements of the organism or in the organism as a whole but also in population and in the evolutionary process.

A vivid example of a regulating system with a feedback based on instinctive behavior is given by the regulation of temperature in a beehive.[14] If the temperature increases above normal, the worker bees cease to bring honey to the hive, but instead bring water. Other bees which are in the beehive, take this water from the returning bees and disperse it, increasing its evaporation by a constant beating of their wings. The feedback is manifested in that, as soon as the temperature in the beehive decreases, the bees cease to take the water from the returning workers and these once again return to collecting honey.

Cybernetics and the theory of information have a fundamental meaning for theoretical biology, since they make possible a direct physical mathematical analysis of living processes. The amazing advances in this realm and the possibilities of its future development have been described in a number of papers.[15,16] Such an analysis, however, is phenomenological and, to a great extent, it can be reduced to a description by models of biological processes. This explains much, but it is far from exhausting the problem. In order to study the essence of living processes, along with cybernetics, we need molecular biology and molecular biophysics.

The problem of molecular biophysics consists in the discovery of concrete molecular mechanisms for storing, coding, transferring, and transposing biological information, in the elucidation of the nature of noises which destroy this information, in regulating mechanisms, and in feedback. Modern physics of molecules has deciphered the code which relates the structure of the molecule to the structure of crystals. The development of molecular biology gives serious reasons for believing that, in due time, by knowing the molecular structure of the embryonic cells, we shall be able to predict all the hereditary properties of the organism, i.e., we shall learn to decipher completely the information which has been placed in the cell. Furthermore, even today, we already speak with confidence about the molecular basis of evolution as a whole.[17] In order to carry out the entire program, it will be necessary to exert much effort for a long time; but this program can be carried out.

Molecular biology has had its greatest success up to now in the deciphering of the genetic code, since the genetic information is coded on the molecular level within the structure of the molecules of nucleic acids. Still much remains to be done here, but solid bases of molecular genetics have already been laid. These questions will be discussed in detail in Chapters 6 and 7.

To the contrary, the nature of the regulating processes is only now beginning to be studied from the points of view of molecular physics and physical chemistry. The role of cybernetics remains leading here as well. This is most valid for phenomena of higher nervous activity, for thought, the physical chemical nature of which is still completely unknown.

As has been pointed out already, cooperative processes play an important role in molecular biophysics. It is important that they may be controlling the feedback. This is shown by the Van der Waals equation. A decrease of the volume of the real gas V results in the increase of pressure which, in its turn, decreases the volume. Further continuation of the process depends on the results of the preceding stage. Such a situation is possible only with an effective interaction between the particles which participate in the process, i.e., with cooperativity. As we shall see, processes of regulation in cells, the molecular nature of which is becoming known now, are based on feedback that is carried out by a cooperative process.

Molecular biophysics does not contradict biocybernetics and thermo-dynamics but joins them in the solution of great problems of modern science. The very posing of these problems becomes possible as a result of the development of biology. Further advances take place as a result of the simultaneous efforts of biologists, physicists, chemists, mathematicians, and cyberneticists, efforts based on the optimistic conviction in the un-limited possibilities of science.

Chapter 2

CELLS, VIRUSES, AND HEREDITY

THE LIVING CELL

Before turning directly to molecular biology and biophysics, it is necessary to present some basic concepts of cytology, i.e., the science of the cell, and of genetics, the science of heredity and changes in organisms. This chapter will contain a brief survey of the most important facts which are related to these topics of biology, and it is directed at readers who are physicists or chemists.

The basic and relatively simplest unit of any living organism is the cell. A cell lives, it carries out all the living functions. There are a number of single-cell organisms, animal, vegetable, and bacterial, which play an important role in nature. In these cases, cells live completely independently and their behavior can be very complex. On the other hand, any multicellular organism not only develops from a single embryonic cell, the zygote, formed as a result of the union of the male and female sex cells, the gamete, but it is also composed of cells. Some types of cells of multicellular organisms exist as independent entities without being joined in any general structure. Such cells are usually found in a liquid medium. These encompass, for example, the red blood cells or erythrocytes, the sperm or the male sex cell, etc. Most cells of multicellular organisms are part of complex systems in which the cells act cooperatively in tissues which, in their turn, form organs and systems of organs.

Noncellular forms of life do not exist on earth at present. Assertions which could be found so recently on the existence of such forms have turned out to be based on crude experimental errors. If a primitive cellular form of life were to arise at the present time, it would be immediately absorbed and destroyed by the more perfected cellular organisms. What has been said, of course, does not concern viruses.

An elementary picture of the cell (Fig. 4) can be given as follows. A cell consists of a nucleus and cytoplasm and it is surrounded by a membrane. One can already see in an optical microscope that cytoplasm has a complex structure, it contains a number of organoids—centrosomes, mitochondria, and also bubbles or vacuoles. The nucleus contains chromatin and nucleoli. The liquid substance of the cell, which is called the *protoplasm*, consists basically of water, but also contains complicated organic compounds and salts.

Notwithstanding the enormous accomplishments of optical microscopy, the inherent limits of the resolving ability of the microscope do not permit the determination of the details of the structure of cellular organoids. The discovery of the wave nature of matter has led to the development of the electron microscope; it permits us to observe not only the supermolecular structure of cellular organoids, but also individual large molecules.

At present we know that the structure of a cell is very complicated. Not only does the organism consist of cells but, in essence, each cell is an individual organism, the constituent parts of which act cooperatively, carrying out different life functions. In this sense, it is possible to speak of

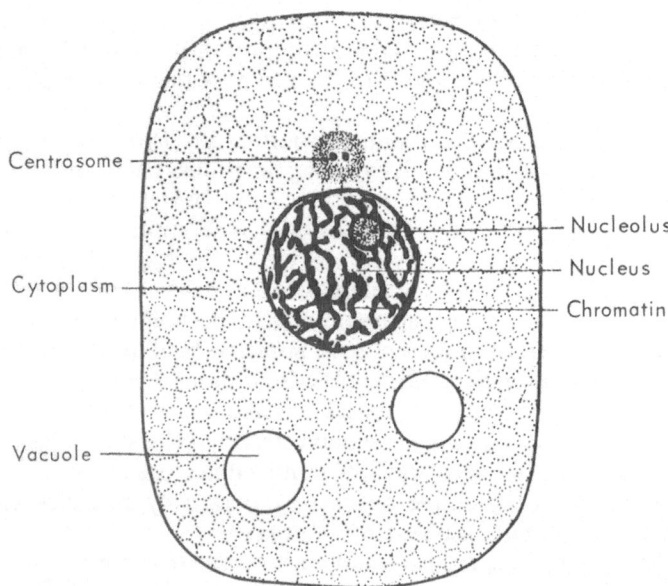

Fig. 4. Schematic representation of a cell, based on observations with an optical microscope. (From "The Living Cell," by Jean Brachet, ©1961 by Scientific American, Inc. All rights reserved.)

the physiology of the cell. A multicellular organism is built up of cells, while the cell is built up of organoids which are supermolecular structures. It is essential to understand the functions of a cell and the role which is played in the execution of these functions by its structural elements.

Living cells are living because they can (a) store and transform energy; (b) synthesize molecules of biologically functional substances and, first of all, of proteins; (c) divide (to reproduce); (d) differentiate (during the division of the embryonic and later cells of multicellular organisms); (e) carry out various mechanical motions, such as the displacement of the component parts of the cell and of the cell as a whole in space; and (f) perform adaptive reactions, which guarantee the conservation of the basic intracellular processes during changes in the state of the external environment.

The first function, i.e., the energetic one, is necessary for carrying out all the others. The chemical and mechanical processes require energy; otherwise, the cell would be a *perpetuum mobile* of the first kind. Where does this energy come from?

A cell is a thermodynamically open system which is capable of exchanging energy and matter with the surrounding medium. The energy enters into the cell from the outside. On earth, there are organisms of two types, autotropic and heterotropic. The first encompass purple bacteria and all the green plants. They receive their energy from solar light. In the cells of autotropic organisms one can find chloroplasts, i.e., organoids which contain the green substance chlorophyll. The energy of the light quanta absorbed by chlorophyll is stored in the chemical bond of definite organic compounds. This process is called *photosynthesis*. During this process, complex organic compounds are formed from simple inorganic ones. Thus, glucose is synthesized from water and oxygen

$$H_2O + CO_2 + h\nu \longrightarrow C_6H_{12}O_6 + O_2$$

($h\nu$ is the energy of the light quantum).

The second type of autotropic organisms are chemically synthesizing bacteria, which obtain their energy during the oxidation of some mineral compounds.

Heterotropes are organisms which are unable to synthesize nutritional substances from inorganic compounds. They obtain energy from the chemical substances of other organisms, receiving as nutrition either these organisms or products of their decomposition. The class of heterotropes encompasses all animals, mushrooms, and the majority of bacteria.

To carry out useful work, it is necessary to liberate the energy contained in chemical bonds. This process occurs as a result of respiration, i.e., of oxidation. This is performed in specialized organoids, namely, in the mitochondria, which are present in all the plant and animal cells.

The process of oxidation, i.e., the release of energy, is the opposite of the process of energy storage, for example, of photosynthesis,

$$C_6H_{12}O_6 + 6O_2 \longrightarrow 6CO_2 + 6H_2O + 686\,\text{kcal/mole}$$

Oxidation can occur also without the participation of oxygen, under anaerobic conditions. Thus, the fermentation of sugar into alcohol by yeast is an incomplete oxidation,

$$C_6H_{12}O_6 \longrightarrow 2C_2H_5OH + 2CO_2$$

The amount of free energy liberated during this process is equal to 55 kcal/mole.

An amazing property of the living cell is its ability to store and transform energy under conditions of a constant and relatively low temperature; furthermore, the chemical compounds which participate in these processes are present in part in the form of dilute aqueous solutions within a narrow interval of pH.

The principal substance storing energy in all cells is called *adenosine triphosphate*, abbreviated by the letters ATP. The structural formula of this compound is shown in Fig. 5. During the splitting of the bond which connects the end phosphate group to the rest of the molecule, energy is liberated to the extent of about 12 kcal/mole. In this process ATP is transformed into adenosine diphosphate, namely, ADP. The breakdown of the next phosphate bond releases energy of the same order of magnitude; the breakdown of the third bond liberates only 2.5 kcal/mole. The first two phosphate bonds are energy-rich. Storage of energy in any other compound, for example, in glucose, takes place with a participation of ATP, which becomes dephosphorylated to ADP. Above, we wrote out only the total reaction for the formation of glucose; in actuality, it passes through a whole series of intermediate steps.

The free energy which is formed in the cell is expended for the synthesis of proteins and other biologically functional substances, i.e., for the

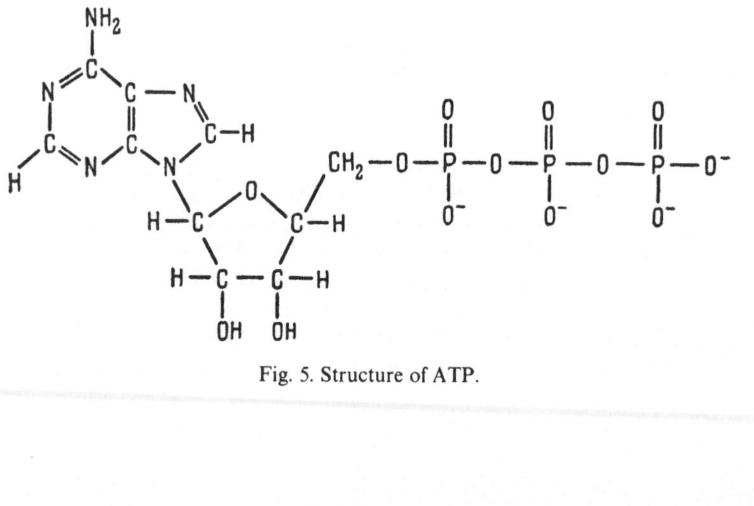

Fig. 5. Structure of ATP.

chemical work. Mechanical work is carried out during cell division, during the intracellular motion of protoplasm, during the motion of cilia and flagella which are found in many cells, and during muscle action. Transport of nutritional and other substances from outside the cell and from the cell into the environment takes place through membranes. Furthermore, quite frequently, such transport occurs from a region in which the concentration of the transported material is low into a region in which it is high, i.e., into a direction contrary to diffusion. Therefore, the cells perform osmotic work. If such a transport is accompanied by the separation of oppositely charged ions, then the cell also performs electric work. In cells of specialized tissues, for example, in the electric eel, the electric work can be quite large. Finally, luminescent organisms transform the energy stored in ATP into the energy of light. We see, thus, that the cell is a complicated physical-chemical system which performs a variety of functions.

Notwithstanding the essential similarity of all cells, there is no such thing as a typical cell. Bacterial cells and the specialized cells of plants and animals are quite different from each other.

A typical bacterial cell has a length of about $3\,\mu$ and a diameter of about $1\,\mu$. Its volume is approximately $2.5\,\mu^3$. Such a cell consists of water to the extent of 70%; the dry weight of the cell is about 6.7×10^{-13} g. The greater part of this weight, 70%, can be accounted for by protein; about 15% by nucleic acids; about 10% by fatty materials, namely, lipids and phospholipids; and about 5% by polysaccharides (carbohydrates). The total number of molecules of these compounds (not counting the small molecules of inorganic salts, etc.) adds up to approximately 40 million.

There exist also very small single-cell organisms, called mycoplasmas (organisms similar to those which bring about pleuropneumonia). The smallest cell of this type is *Mycoplasma laidlawii*; it has a diameter only 1,000 times greater than that of the atom, namely, about $0.1\,\mu$. Such a cell contains about 20,000 large molecules of biopolymers; a number of enzymes are synthesized in it. These organisms, similar to bacteria, are capable of multiplying on a nonliving nutritional medium. It is evident that here we are faced with the lower limits of the possibilities for cell dimensions; if we take into account that the cellular membrane must have a definite thickness, we find that the absolutely minimal dimension of a cell is about $0.05\,\mu$.

Some cells can be seen even with the naked eye; the cells of multicellular organisms are also considerably larger than bacterial cells.

Let us look at the structure of the cell. This is shown schematically in Fig. 6.[2]

The cell is surrounded by a membrane which has a thickness of about 80 Å, that is, 8×10^{-7} cm. The membrane consists of proteins and lipids the molecules of which are distributed in the form of ordered layers; between two layers of lipids there is a layer of protein. It would seem that

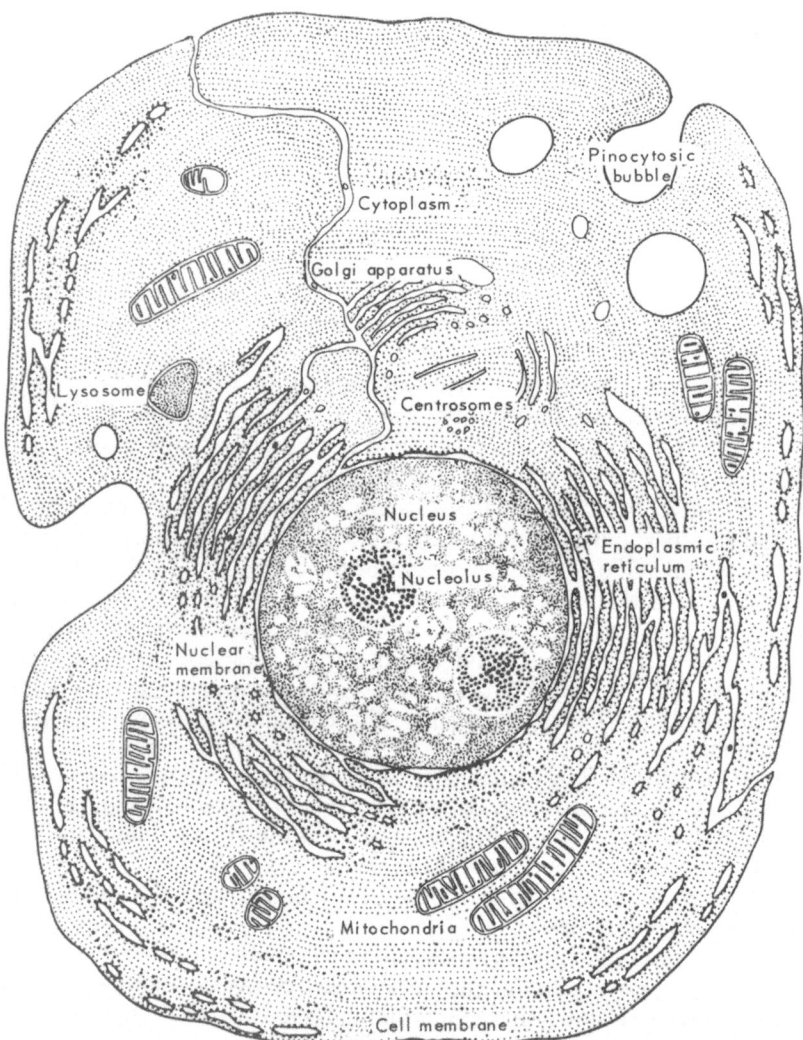

Fig. 6. Modern schematic drawing of the structure of the cell based on electron-microscope information. (From "The Living Cell," by Jean Brachet, ©1961 by Scientific American, Inc. All rights reserved.)

the membrane has pores which permit various substances to enter the cell (and also the layers of the lipids and proteins). The cellular membrane behaves in a very special manner in relation to these substances. There are special arrangements in the cells which permit an unequal permeability of the membrane to ions of Na^+ and K^+, notwithstanding the fact that these ions are very similar to each other in charge and dimensions. The excitation of nerves is the result of the penetration of these ions into the nerve cells and their exit from them.

The cytoplasm of the cells has a complicated structure; it contains internal membranes and various organoids.

We have already mentioned the mitochondria, which are present in all cells, as well as the chloroplasts found in plant cells. The mitochondria are the power stations of that factory of proteins and other substances which is the cell. The number of mitochondria per cell varies within broad limits, from 50 to 5,000. These are relatively large organoids; they have a length of 3 to 4 μ (in the liver cells of mammals) and consequently can be seen under a normal microscope. The electron microscope has revealed the internal structures of mitochondria; these have turned out to be quite complicated (see Fig. 7). Mitochondria are vessels with internal barriers. The external and internal membranes have a thickness of about 185 Å; they consist of two layers of protein separated by a double layer of lipids. Molecules which perform the oxidizing processes and which are related to electron transfer, and the biological catalysts, enzymes, without which these processes cannot occur, seem to be arranged in a regular order in the protein layer.

Chloroplasts in plants are solar-energy stations. The chlorophyll molecules which absorb the light are disposed in the regular structure of the chloroplasts in strict order and form something similar to electric batteries (Fig. 8).

The cytoplasm contains lysosomes, i.e. organoids which contain those enzymes that have the function of breaking down into small fragments the large molecules of fats, proteins, and nucleic acids; these fragments are then oxidized in the mitochondria. The action of these degrading enzymes is so strong that when the membrane of the lysosome is broken, they destroy the cells, splitting the proteins and lipids of the external membrane.

Cytoplasm contains the endoplasmic reticulum which is a complicated membrane system and forms a number of channels along which

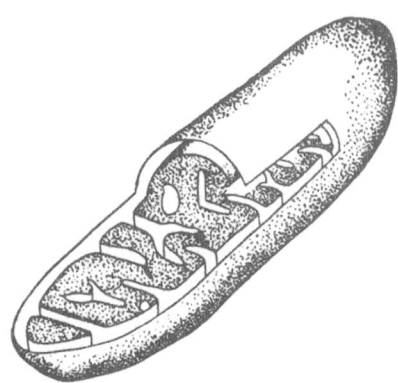

Fig. 7. Structure of mitochondria.

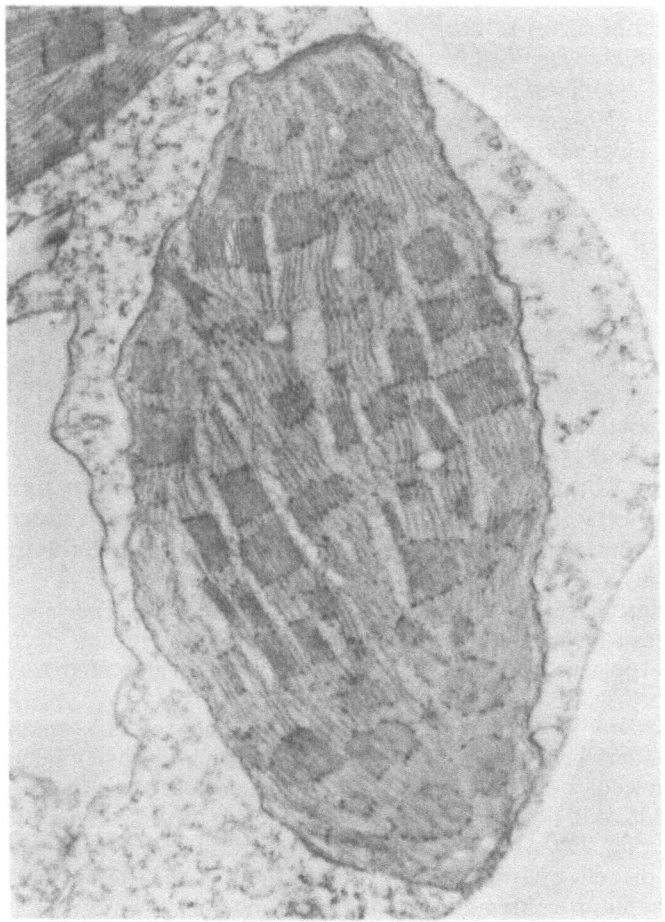

Fig. 8. Chloroplast of a corn cell (40,000 × magnification, electron microscope).

matter is transported from the external membrane to the nucleus. The Golgi apparatus is closely related to this system; its role appears to be the synthesis of new membranes. Another possible function of the Golgi apparatus is protective, i.e., the removal of foreign harmful particles.

A large number of particles of molecular dimensions (with diameters of about 100 to 150 Å), called *ribosomes*, are located along the walls of the endoplasmic reticulum and also directly in the cytoplasm. These granules, which can be seen only through the electron microscope, contain one of the nucleic acids, namely, ribonucleic acid, abbreviated to RNA. The role of the ribosome is extremely important, since protein is synthesized on it.

If we continue the analogy between a cell and a factory, the ribosomes are the assembly line which carry out the main production.

Finally, the cells contain the centrosomes, or centrioles. In many organisms these appear as cylindrical particles, which are perpendicular to each other. The height of the cylinder is 3,000 to 5,000 Å, its diameter is about 1,500 Å. It consists of nine groups of parallel tubules; two or three tubules are found in each group (Fig. 9). The centrioles play an important role in cell division. It is interesting that the cells which are capable of independent motion, i.e., which have flagella and cilia, have components, namely, kinetosomes, located at the base of flagella and cilia, which have structures similar to that of the centrioles. The centrioles and kinetosomes are devices that take part in mechanical motion. Further details on cell structure can be found in specialized monographs.[2,4]

We have described the principal components of the factory cell. How is this factory directed?

The directing center of the cell is its nucleus. It is separated from the cytoplasm by a membrane which contains pores through which the nucleus communicates with the cytoplasm. The nucleus contains chromatin fibers, long structures which can be observed with great difficulty in a non-dividing cell, i.e., in a cell between two divisions. The chromatin is spread in the nucleus of such a cell in a rather diffuse manner. However, when the cell is ready to divide, the chromatin forms well-visible (even under the usual microscope) structures, namely, the chromosomes. The number and

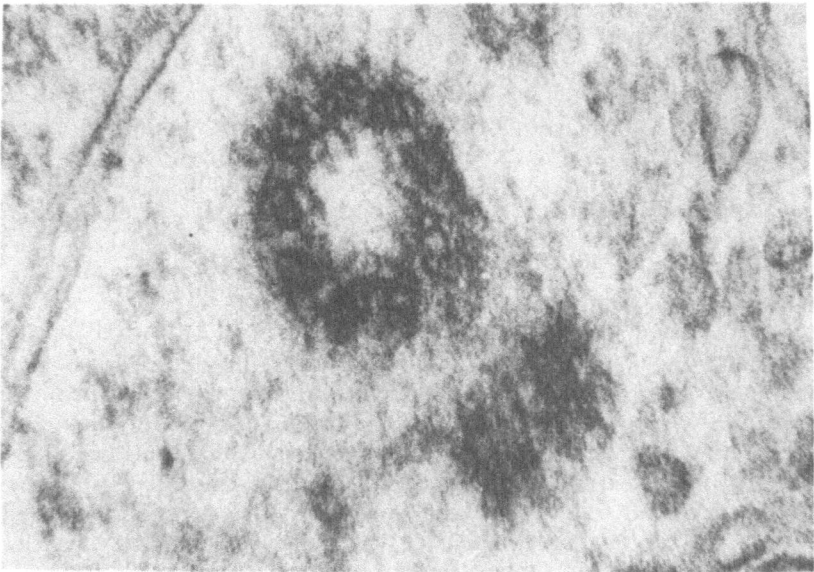

Fig. 9. Structure of centrioles (cross section of an electron micrograph, magnified 175,000 ×).

shapes of the chromosomes are absolutely constant for any given form of organism. Usually the chromosomes are found in pairs, i.e., there are two chromosomes of each type. Such cells are called diploid. In each somatic cell of the human body (but not in the sex cells) there are 23 pairs of chromosomes, i.e., a total of 46; in the somatic cells of the drosophila fly, there are four pairs of chromosomes; the somatic cells of soft wheat contain 21 pairs of chromosomes. In the sex cells, the gametes, the number of chromosomes is reduced by a factor of 2, that is, there is only a single chromosome from each pair. Such cells are called haploids. During the normal division of diploid cells, mitosis, the number of chromosomes is doubled and each new cell contains a complete diploidal set of chromosomes analogous to the parent cell. During cell division, which results in the formation of gametes, the number of chromosomes is decreased by half.

Chromosomes contain another nucleic acid, namely, deoxyribonucleic acid, DNA, and special proteins. (A chromosome is drawn schematically in Fig. 24.) Getting ahead of ourselves, let us point out that it is the chromosomes which are primarily responsible for the life of the cell, the development of the organism, and the transmittal of hereditary properties. It is in there that the hereditary genetic information is stored. They direct the synthesis of proteins; they direct all the work of the factory. The cell without a nucleus can live for a certain amount of time. But it can no longer survive. The chromosomes are the so-called planning offices of the factory. In some manner, they transmit the necessary information to the working assembly line.

Bacterial cells do not have a visible nucleus, but they contain chromosomes, i.e., nuclear structure. The bacteria of the intestine, namely, *Escherichia coli*, have a single chromosome; they have been studied in particularly great detail from a genetic point of view. Bacteria are haploidal cells.

In addition to chromatin, the nucleus also contains the nucleolus, i.e., a spherical body rich in RNA. The nucleolus contains granules which are similar to ribosomes and, correspondingly, protein synthesis and the synthesis of RNA take place in it as well. The chromatin and nucleoli are immersed in a liquid medium, the nuclear juice.

It is evident that the knowledge of the structure of the cell and its functions raises a whole number of physical and chemical questions. What are the general principles of activity of the complicated cooperative organization? What induces the cells to divide and how does this take place? In which manner is the genetic information included in their chromosomes, and how is it transferred to the ribosomes that synthesize proteins? Why do cells differentiate? Answers to these questions are to be found in biochemistry and molecular biophysics. By analyzing the cells, we have approached the molecular level of organization directly and now we are dealing with molecules.

CELL DIVISION

Cell division is now well understood. It has been photographed with the help of a motion picture camera. The normal mitotic splitting of cells takes place in the following manner (Fig. 10).[5]

As has been said already, between two divisions, the chromatin fibers, which compose the chromosomes, are distributed in a rather diffuse manner in the nucleus. Each of the two pairs of centrioles consists of one larger parent centriole and its smaller offspring centriole placed at right angles to the parent (Fig. 10a). At a certain moment before the beginning of division, the chromosomes double, that is, they replicate. In other words, next to each chromosome there appears a copy of the original one. Simultaneously, a mitotic spindle is formed between pairs of centrioles, i.e., fibers stretched between the centrioles (Fig. 10b). In the processes of mitosis, (Fig. 10c and d) the chromosomes wind into helices and become denser, the nuclear membrane and nucleolus disappear, and the centrioles migrate to opposite ends of the cell and form poles to which the chromosomes will move. In this phase, a bond is formed between the specific central portions of the chromosome, the centromeres, and the poles, by means of the fibers of the mitotic apparatus. In the metaphase (Fig. 10e), the chromosomes are displaced toward the equator of the cell. During the anaphase (Fig. 10f), the double chromosomes divide and move to the opposite poles. Finally, in the final stage of mitosis, in the telophase (Fig. 10g), the chromosomes again lose their spiral structures and again nuclear membranes and nucleoli are formed. Each centriole has "given birth" to a progeny. The division and formation of a new intercellular membrane and the two newly formed cells are now present in the state of the interphase. In such manner, as a result of mitosis, two new cells are formed from the original one; these new cells are completely analogous to the original one. The most important characteristics of this similarity is the exact reproduction of the original set of chromosomes.

The duration of the states of mitosis is different for different tissues and forms of organisms. In living human cells, the prophase lasts 30 to 60 minutes; the metaphase, 2 to 6 minutes; the anaphase, 3 to 15 minutes; and the telophase, 30 to 60 minutes. The entire process occurs over 1 to 2 hours.

These are the facts. The observer of mitosis sees a completely regular process, the biological expediency of which is beyond any doubt. Should we limit ourselves to a recognition of this regularity and regard it just as a support for the concept of vitalism?[6] It seems desirable to investigate the details of this process further.

It is evident that at the basis of mitosis there must be profound physical-chemical laws. As has been pointed out by Aleksandrov,[7] there are three basic problems in the study of cells: evolution, self-regulation, and cell replication. In the study of mitosis, it is the third problem which

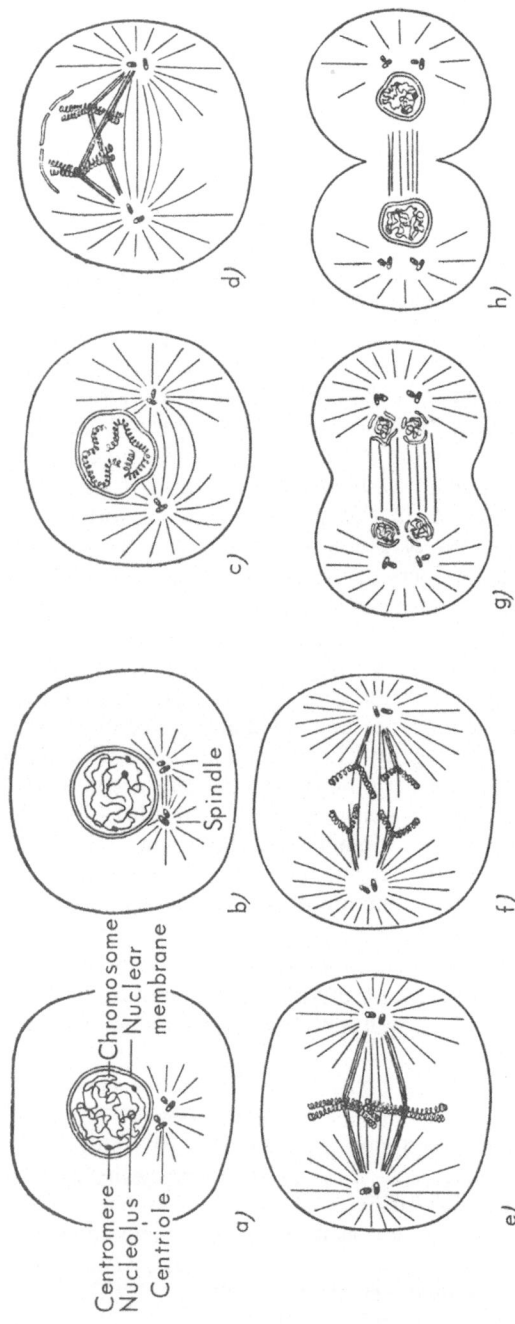

Fig. 10. Scheme of mitosis: (a) the period between two divisions, interphase; (b) the spindle begins to form; (c, d) the prophase; (e) metaphase; (f) anaphase; (g, h) telophase.

comes up immediately. What are the physical-chemical mechanisms of duplication of the chromosomes and centrioles and of their separation in the progeny cells? A chromosome is a supramolecular structure and it is absolutely evident that the replication of such a structure is directly related to the replication of the molecules which form it and, first of all, of DNA. Consequently, this problem can be reduced to a great extent to molecular physics. Recently, much has become clear in this field.

The problem of the displacement of chromosomes and centrioles is related to the more general question on the nature of the mechanical processes in cells and organisms. These are mechanochemical processes. They occur at conditions of constant temperature and pressure; the source of the mechanical work is the chemical energy stored within the cell. Mazia, who has studied mitosis in particularly great detail in recent years, has discovered a number of factors which explain the structure and function of the mitotic apparatus on a molecular level.[8] These questions are discussed in Chapter 9 of this book.

The problem of the self-regulation of mitosis, i.e., the explanation of the coordination of behavior of the nucleus and cytoplasm of a divided cell, has been studied to a much lesser extent. This problem has been stated very clearly by Aleksandrov.[7] At present, we do not understand what physical-chemical factors induce the cell to divide and which molecular interaction determines the coordinated behavior of the entire cell during mitosis.

The problem of evolution is related to mitosis in the following manner. Biological science starts from the basic position of the formation of life from nonliving matter. The cell and all its properties, and among them the ability for mitotic division have resulted from long evolution during which noncellular forms of life became extinct on the earth. The biological advantage of mitosis indicates the definite advantages in natural selection. It is very difficult to attain a knowledge of the process of evolution which resulted in mitotic structures, as there are no traces of precellular life on earth. However, we should be contradicting the principles of science if we were to consider these difficulties as insurmountable. If life was generated on earth from inorganic matter, then there are no reasons in principle why scientists could not repeat this process in their laboratories.

Let us look now at the reduction cell division which results in the formation of gametes, namely, meiosis.

Meiosis consists essentially of two cell divisions. As a result, the names of the phases are the same as in mitosis. In the early prophase of the first meiotic splitting (Fig. 11a), a spindle forms and chromosomes appear in the nucleus. Then a specific process follows, the synapsis, or conjugation, of homologous chromosomes. Two chromosomes of each pair combine with each other (Fig. 11b). The chromosome is a nonhomogeneous structure and its individual parts differ from each other both morphologically and functionally. During synapsis, the parent chromosomes

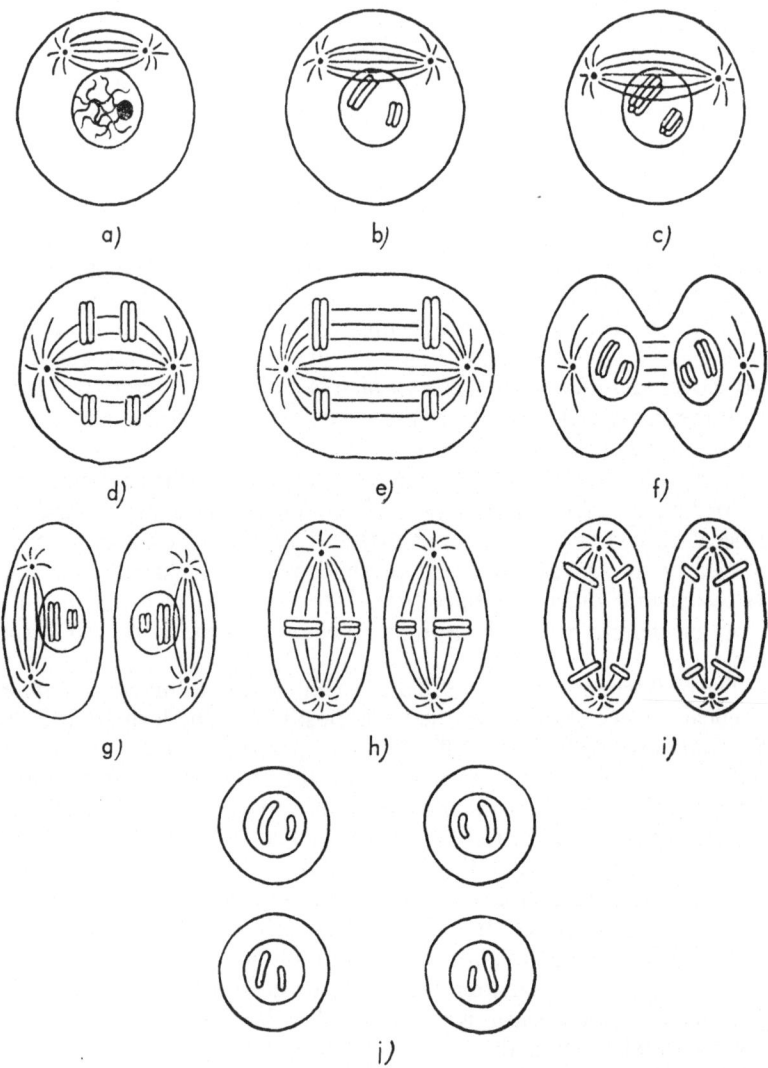

Fig. 11. Scheme of meiosis. The number of chromosomes has been taken as four and the initial cells are diploidal. (a) Early prophase of the first meiotic division; (b) synapsis of chromosomes; (c) formation of the tetrads; (d) metaphase of the first division; (e) anaphase of the first division; (f) telophase of the first division; (g) prophase of the second meiotic division; (h) metaphase of the second division; (i) anaphase of the second division; (j) mature gamete.

unite by their congruent parts. Then a doubling of the synaptic chromosome takes place, as the result of which there are formed groups of four identical chromosomes, the tetrads (Fig. 11c). In the metaphase of the first meiotic division, the tetrads are located along the equator of the cell (Fig. 11d). In the anaphase, the chromosomes migrate to the poles in pairs, contrary to what is found in mitosis (Fig. 11e). In the telophase of the first meiotic division, two cells are formed with paired synaptic chromosomes (Fig. 11f). Then the second meiotic division starts. The prophase is shown in Fig. 11g; the metaphase and anaphase are shown in Fig. 11h and i. As a result, there are mature gametes (Fig. 11j) which contain a number of chromosomes equal to half of those found in the initial cell. From a single diploidal cell four haploidal ones have been formed.

The scheme of meiosis just described is valid for spermatogenesis, i.e., the formation of the male gametes, namely, the sperm. In ovagenesis, namely, the formation of ova, the four gametes are not identical; three of them are what are called the *polar bodies*; they degenerate and only one becomes a mature ovum.[4,9]

Meiosis also brings up the same general scientific problems as mitosis, plus a number of special problems. For physics, the most interesting one of these is the process of synapsis. What forces cause the identical chromosomes to be attracted at homologous loci? This question has intrigued physicists for a long time and quantum mechanics have been applied for its solution. It is evident that meiotic division, which is the most important stage of sexual reproduction, guarantees the maintenance of the basic properties of the species in their progeny. When the ova combine with the sperm, the diploidal type of chromosome specific for the given organism is restored. Half of the chromosomes come from the father, half from the mother. As a result, there is a definite possibility (and certainly not unique) of variation of traits as a result of the combination of heredity from the father and the mother.

It is important that the chromosomes of any one pair during meiosis separate independently from the chromosomes of other pairs. This gives rise to the possibility of a large number of combinations of paternal and maternal chromosomes; the number of combinations in a man is 2^{23}. The sexual reproduction determines the most important advantages in natural selection, in the evolution of species.

Fertilization consists in the penetration of the sperm nucleus into the ovum. The very approach of the sperm, of a freely moving cell, to the ovum and their interaction is certainly brought about by physical-chemical factors. The same factors, but in a more enigmatic manner, permit only a single sperm to penetrate into the ovum; after penetration of the sperm into the ovum, a thick membrane is formed on the surface of the ovum which prevents penetration by other sperm. The haploidal set of chromosomes of the sperm and ovum combine, a zygote is formed; this, then, divides and differentiates.

One should point out that the spermatozoid brings into the ovum practically only chromosomes. Its role is secondary in everything else. The entire cytoplasmic organisation of the zygote is obtained from the mother, but it does not follow from this that the further development of the organism follows along maternal lines. The fertilization renews the planning office of the factory cells and the production which will follow depends equally on both parents.

However, a nonfertilized ovum can also develop right up to a mature organism. Such a multiplication is called parthenogenic. For example, in the case of Hymenoptera insects (bees, ants, wasps, and gall wasps), the male develops from nonfertilized eggs. Parthenogenesis can be brought about artificially by adding some chemical substances to the water in which the egg is immersed or by cutting with a thin needle. It has been possible to obtain parthenogenetic rabbits and frogs.

Even before fertilization the egg grows a great deal; in the case of the frog, its volume increases 1,600,000-fold. A number of structural changes take place in the egg, as well as an increased synthesis of lipids, glycogen, proteins, and nucleic acids. The radial symmetry of the egg is replaced by bilateral symmetry (Figs. 12 and 13); this corresponds to the general symmetry of the future tadpole. In such manner, differentiation of the egg material starts even before fertilization. After fertilization, the division of the zygote results in differentiated specialized cells and, finally, a complete multicellular organism is formed.

Problems of cell differentiation are studied in the realm of biology called *dynamics of development*. The description of these most interesting and important phenomena is outside the realm of this book. We are still very far from understanding the molecular processes which determine differentiation, from understanding the reasons which make a given cell

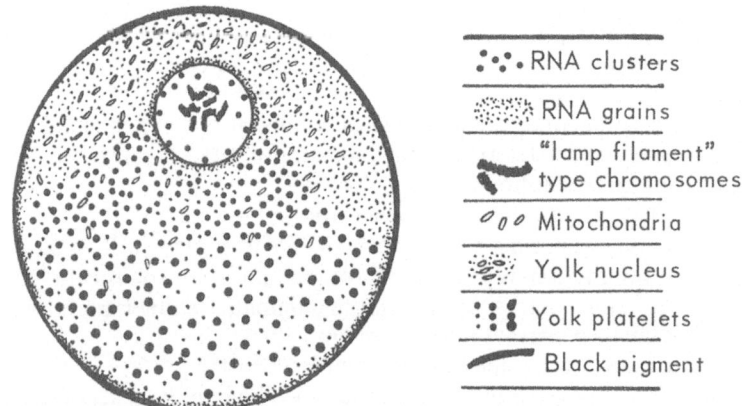

Fig. 12. Ovum at the end of growth.

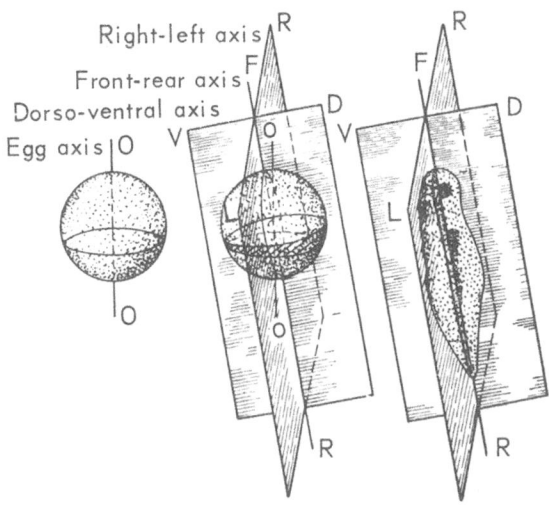

Right-left axis
Front-rear axis
Dorso-ventral axis
Egg axis

Fig. 13. Bilateral symmetry of the egg and symmetry of the
tadpole.

the embryo of a future eye or of a future appendage. Let us point out only
that direct experiments show that the differentiation of the cells is already
predetermined by the structure of the nucleus and cytoplasm of the initial
zygote.[10,11,12]

It is quite evident that, in the final account, the entire mechanism of
the development of a multicellular organism is determined by the physical
and chemical interactions of compounds synthesized in the cell. This
synthesis is controlled genetically; the totality of the fundamental characteristics of an organism is transmitted by heredity. Questions on the
mechanics of development turn out to be much more complicated than
questions of genetics as such; we must consider here not only the synthesis
of individual compounds but also their extremely complicated interactions, which manifest themselves in the structures of definite supramolecular entities and by the interactions which take place within such
structures. These interactions must certainly be cooperative and they must
obey a complicated system of direct and reverse relations. It is not only the
structure which is predetermined, but also a strict regulation and, in
particular, the synchronization of the growth and division of cells, since
its violation would absolutely lead to the preclusion of the formation of a
normal organism. The genotypic information sees light in the mechanics
of development. At present, molecular biology and molecular biophysics are already crossing the frontier beyond which lie the problems of
differentiation. Up to now, such investigations were of a purely biological
character.

The principal achievements of molecular biology have been in the realm of the investigation of single-cellular organisms, namely, bacteria. It is evident that problems of differentiation do not come up in this realm.

VIRUSES AND BACTERIOPHAGES

In 1892, the Russian botanist Ivanovskii observed that the juice of a tobacco plant that was diseased with the mosaic disease, which manifests itself by the appearance of spots on the leaves, could infect healthy plants even after this juice had been passed through the finest filters. This marked the discovery of the first filterable virus, namely, the tobacco mosaic virus TMV. Later, it was possible to study with the electron microscope particles of TMV and of other viruses. The TMV particles have the shape of a long cylinder with a diameter of 17 mμ and a length of 300 mμ. The hollow cylinder contains RNA, while the membrane of the cylinder consists of a number of layers of proteins. At present, we know a large number of plant and animal viruses, which cause a vast number of diseases. The low potato yields, observed in the southern regions of the USSR, are due to a virus disease of the plant, to a continuous infection, which is maintained by the vegetative reproduction, by the tubers. A study of the viruses and quarantine measures are resulting now in a sharp increase of the potato yield.

Virus diseases which afflict man include influenza, measles, small pox, poliomyelitis, etc. A number of tumor diseases are caused by viruses. Virus particles can be of different shapes and sizes. The particles of the virus of psittacosis, a disease transmitted by parrots and very dangerous to man, has a diameter of approximately 275 mμ; particles of the virus of foot and mouth disease of large horned cattle have a diameter of 10 mμ.

Viruses are not cells. Their structure is immeasurably simpler; they are devoid of a nucleus and a cytoplasm and can essentially be regarded as molecular complexes of protein with RNA (and also in a number of cases with DNA). Thus, contrary to bacteria, viruses cannot multiply on a non-living nutritional medium. They multiply only in living cells. Viruses are parasites of the cell. Thus, viruses are not capable of reproduction; they are reproduced only by living cells. Stanley showed in 1935 that a TMV virus can be obtained in crystalline form. Other viruses have also been crystallized.

Are viruses living or nonliving? This question was the subject of the lively discussion at the Moscow symposium on the origin of life, held in 1957.[13] In fact, this is the question of the definition of life. If such a definition includes the ability of reproduction in a nonliving medium, then viruses are nonliving. If such a rigid requirement is not included in the definition, then viruses can be regarded as living. In the case of viruses, their importance to science is certainly independent of the definition and we shall not pursue this question.

The existence of viruses is absolutely linked to the existence of cells and organisms and, disregarding the definition, the evolutionary problem relative to viruses is quite obvious. In this problem, a number of various exclusive hypotheses have been proposed; we cannot examine these in detail. Oparin[14] regarded viruses as direct descendants of some primitive precellular form of life. On the other hand, the degeneration hypothesis considers that viruses have developed from a cellular form of life by shedding their entire cytoplasm, which they no longer need for their existence. Just like chromosomes, a virus particle is a nucleoprotein, i.e., a complex between a protein and a nucleic acid. Further details can be found in the literature.[15] The studies of Gierer and Schramm[16] and of Fraenkel-Conrat[17] have shown that the RNA extracted from TMV has in itself the ability to infect, i.e., when a plant is infected with it, entire virus particles appear in the plant. In other words, the nucleic acid is sufficient to cause the infected cells to synthesize both the nucleic acid and the protein membrane of the virus. Still earlier, Stanley had obtained hybrid viruses, by combining the RNA of one type of virus with the protein of another. These hybrids multiplied and possessed new properties.

Let us examine in greater detail the properties of virus particles with the example of bacterial viruses, namely, of bacteriophages, which were discovered by Twart in 1915 and D'Hérelle in 1917. The history of the discovery of a bacteriophage has been convincingly described in the delightful novel of Sinclair Lewis, "Arrowsmith." The hero of the novel, a young bacteriologist, makes his discovery almost simultaneously with D'Hérelle, but the paper of the latter appears in press earlier and Martin Arrowsmith can do nothing more than publish a paper containing some additional details. Bacteriophages, or bacterial viruses, are bacterial parasites. Just like viruses, they cannot multiply in a nonliving nutritional medium, but they multiply in bacterial cells. At present, the use of the electron microscope and a number of biochemical and bacteriological techniques has made possible a thorough investigation of a large number of bacteriophages, their morphology, their functionality, and their genetics. A particle of one of the best known phages, phage T2, which multiplies in *E. coli*, has the shape shown in Fig. 14; the dimensions in angstroms are also shown in this figure. Figure 15 shows an electron-microscopic photograph of such phage particles. The T2 phage consists of DNA present inside the particle and of a protein membrane. Furthermore, it contains a small number of cellular proteins, which are probably present as a complex with DNA, as well as of carbohydrates which are also linked with DNA. Thus, a phage consists primarily of DNA and a protein membrane. In this, it differs from those viruses which contain RNA rather than DNA. The process of infection of a bacterium by a phage takes place in the following manner (Fig. 16). The phage particles moving with Brownian motion meet a bacterial cell. The phage particle becomes adsorbed on the cell, being attached to its wall by the end of its tail. This process is absolutely

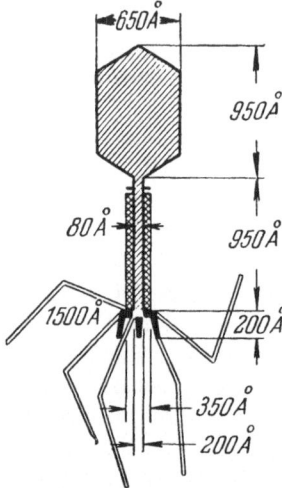

Fig. 14. Schematic representation of the structure of phage T2.

specific; a given strain of bacteria interacts only with a definite phage. The proper concentration of ions in the surrounding medium is essential for adsorption. Certain phages require definite compounds without which adsorption does not take place. Thus, for example, for one of the phages of *E. coli*, it is necessary to have the amino acid tryptophane; the trypto-

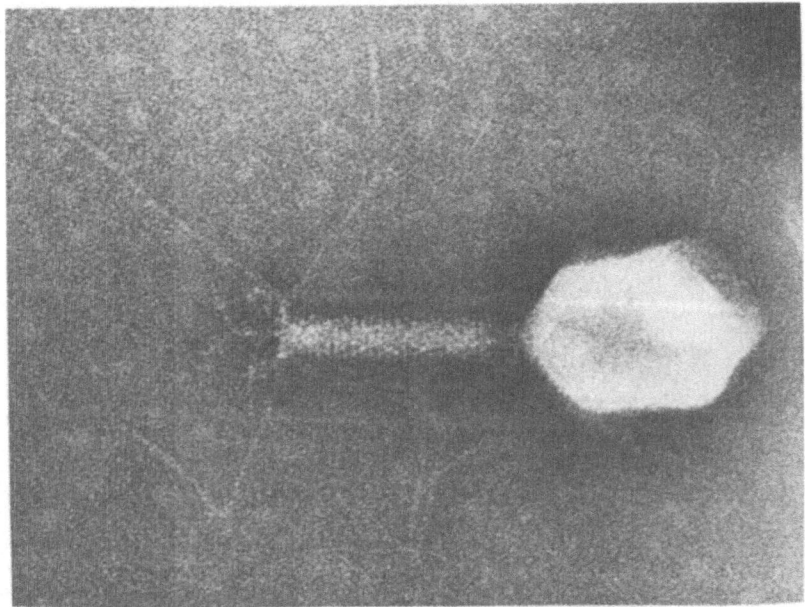

Fig. 15. The electron micrograph of T-even phage particles.

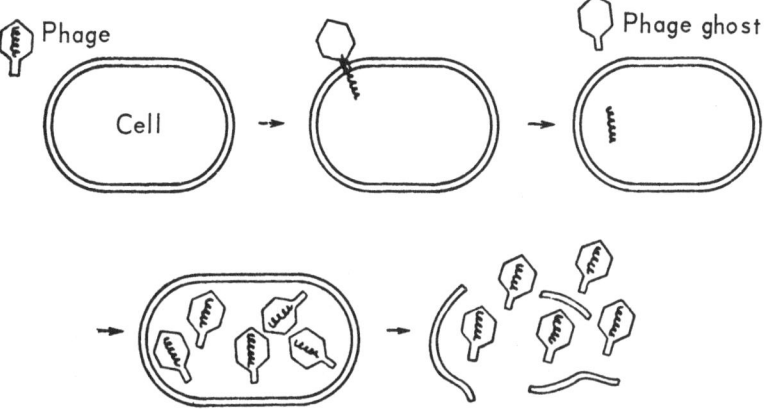

Fig. 16. Schematic representation of the development of a phage, from the infection of the cells up to cell lysis.

phane enters into a reversible reaction with the phage and transforms it from a nonadsorbing to an absorbing entity.

The adsorption process is a physical-chemical one, and with an excess of bacteria, the rate of adsorption is expressed by the simple equation (Krueger)

$$-\frac{dP}{dt} = kBP \tag{2.1}$$

i.e., the rate of decrease of the phage concentration P is proportional to its concentration and to the concentration of the bacteria, B; k is the rate constant. The concentration of free phage particles decreases with time according to the equation

$$P = P_0 e^{-kBt} \tag{2.2}$$

The rate constant k can be expressed through the diffusion constant of the phage particle D and the radius of the bacteria (Schlesinger)

$$k = 4\pi DR \tag{2.3}$$

Taking the radius of a sphere R, with a surface identical with the cell of $E.$ $coli$, equal to 8×10^{-5} cm, and taking the experimental value of the diffusion constant of the T4 phage, $T = 2.4 \times 10^{-6}$ cm^2/min, we find from Eq. (2.3), $k = 0.24 \times 10^{-8}$ cm^3/min. A direct measurement of the rate of adsorption gives, for T4 phage, $k = 0.25 \times 10^{-8}$ cm^3/min. The agreement between theory and experiment is excellent.

Adsorption seems to consist of two stages, a reversible and an irreversible one. The second stage is a strong function of temperature. The first stage consists in the formation of ionic bonds between charged groups on the surface of the phage and of the bacterium, principally of NH_3^+ and COO^-, and also of intermolecular interactions of other types. This is followed by the development of the more specific chemical interaction, which enables the phage, or more precisely part of it, to penetrate into the cell.

The second stage consists in the penetration by the phage. Its nature was unravelled by Hershey and Chase by a study of this process with labeled atoms. The protein contains sulfur but does not contain any phosphorus. The DNA contains phosphorus but does not contain sulfur. The phage was labeled with ^{32}P and ^{35}S. After penetration, practically all of the radioactive phosphorus was found in the cell, while the radioactive sulfur remained in the external medium. Therefore, it was only the DNA of the phage which penetrated into the cell, while its protein membrane remained outside. This was confirmed by direct observations made with the help of an electron microscope. It is possible to see ghosts of the phage, namely, empty protein shells, which were left after the penetration by the phage DNA into the bacterium (Fig. 17).

We see that the phage particle acts like a hypodermic syringe. It injects its contents into the bacterium. This process can be regarded as a

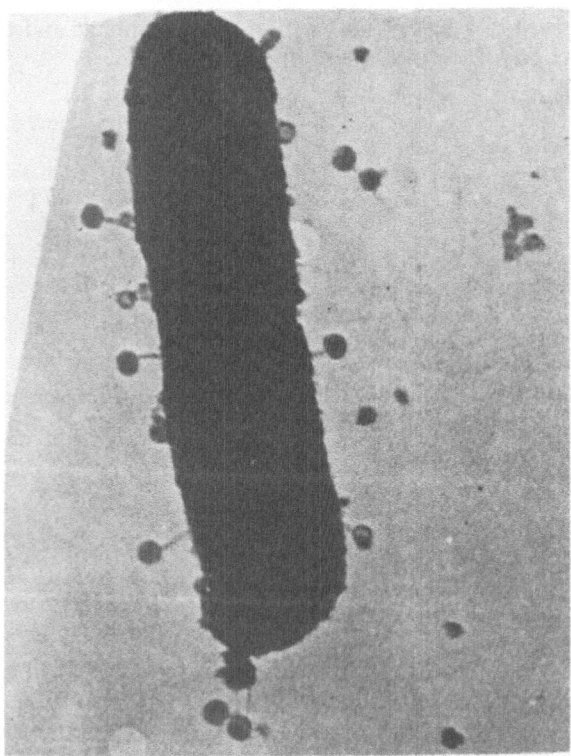

Fig. 17. Electron micrograph of an *E. coli* bacterium with particles of T5 phage adsorbed to its walls.

chemical reaction if the phage particle is regarded as a molecule of nucleo-protein, i.e., as a complex of protein with DNA: Protein–DNA + Bacterium → Protein + Bacterium–DNA. After injection of the phage DNA into the cell and the passing of some time (the latent period), entire phage particles arise in the cell (Fig. 16); furthermore, seeding with the DNA of a single starting particle results in an almost hundredfold harvest. During the multiplication of the phage inside the cell, serious changes take place in its cytoplasm; it stops dividing and finally it perishes as a result of lysis, i.e., of the destruction of the cell wall by new phage particles (Fig. 16). These particles go out into the medium and can now infect new cells.

The ability of the phage DNA to organize within the cytoplasm of the host cell the synthesis of the phage protein membrane is of extreme interest. This process is a model of the most important molecular-biological phenomena which are described in Chapter 7 of this book.

The lysis of bacteria by a phage can be observed directly. If a bacterial culture is grown on a dense nutritional medium, on agar, the bacteria form an opaque film. When the bacteria are infected with a phage and it multiplies, round transparent sterile plaques appear on the film; the shape and dimension of these plaques are characteristic for a given type of phage. A study of these plaques gives very valuable information on phage genetics.

The process described of the interaction between the phage and the bacteria is not unique. There is also the amazing phenomenon of lysogeny in which the phage does not kill the bacteria but it influences both their behavior and properties and those of distant progeny of the bacteria. Some definite so-called "moderate" phages can, in some cases, kill the cell and, in others, form stable complexes with surviving bacteria, remaining inside of them in the form of harmless prophages. During this process, the genetic material of the phage combines with the genetic material of the host, the physiological properties of the cell change, and these changes are maintained in further generations. If a lysogenic cell is acted upon by some strong agent, for example, by ultraviolet radiation, it begins to produce phage and perishes. Lysogeny is a hereditary property of the bacteria, acquired during the infection with the phage.

The general scheme of interaction of bacterial cells with a lysogenic phage is shown in Fig. 18.

Another phenomenon which is very important for molecular biology is transduction. In a number of cases, moderate phages can transport a part of the genetic information from one bacterial cell to another. Zinder and Lederberg have shown that if bacterial cultures which cannot ferment certain hydrocarbons such as arabinose and rhamnose are treated with filtrates of phage lysates (phage P22) of a culture which was capable of such fermentation, bacteria appear which now can split these carbohydrates. Three components take part in the phenomenon of transduction: the donor bacteria, the transducing phage, and the receptor bacteria. It is

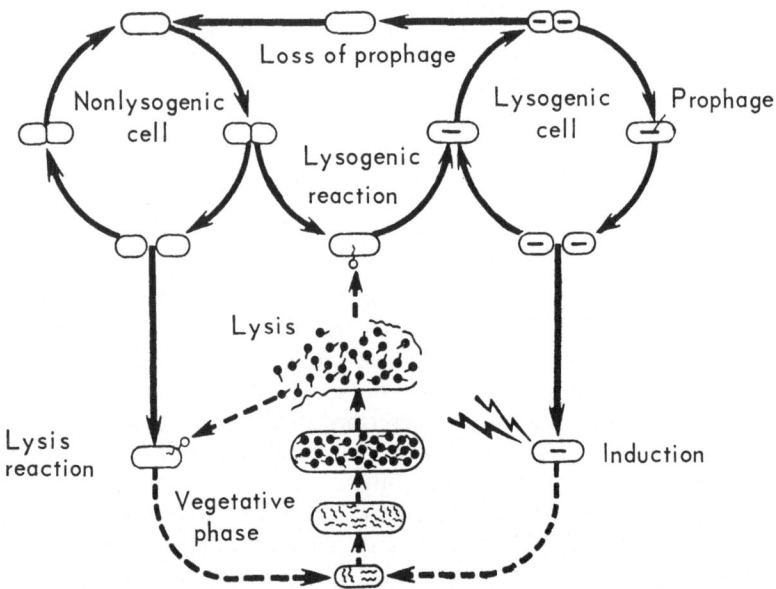

Fig. 18. Schematic representation of the relation between the cell and the phage during lysogeny.

possible to distinguish between nonspecific and specific transduction. In the first case, the phage transduces different portions of the genetic system of the bacteria, different characteristics; in the second case, only definite genes are transported from the donor. In order to understand this phenomenon, it is necessary to turn to bacterial genetics.[18-20]

The genetic properties of the phages are particularly important for molecular biology. In what follows, we shall frequently come across these. At present, let us point out that, in phage genetics, we can distinctly see all the laws which are general for all organisms; in this sense, nature acts according to a single principle both in the parasite *E. coli* and in the cells of man.

BASIC LAWS OF GENETICS

The laws of the transfer of hereditary information from parent to offspring which are part of the very basis of the life of the species and of the individual, as well as of the evolution of species, were discovered first on the level of multicellular organisms by observations of results obtained in the crossing of plants. This great discovery, which was of fundamental importance for all of biology, was made 100 years ago by Gregory Mendel.

In crossing two types of sweet pea (with white and purple flowers), Mendel established that, in the first generation of offspring, all the flowers were purple. Therefore, the property of having purple color introduced into the zygote by one of the parents dominated over that of the white flower. Purple color was a dominant characteristic, while the white color was a recessive one. The existence of such a dominance was already extremely important and very interesting.

The question was asked then: What will happen in the second generation during the crossing of purple plants of the first generation? The cells of such plants, in some way, contain the characteristics of both starting parents, with the purple color dominant. What is the further fate of these characteristics? Having studied a large number of plants of the second generation (it is essential that this number be large; we shall show later the reason why), Mendel established his first law, namely, the law of distribution. Of the plants, 75% had purple flowers, while 25% had white flowers, i.e., the ratio of flowers with dominant and recessive characteristics was 3:1. These experiments have shown that the appearance of a hereditary characteristic is related to some material structures, which can exist in two forms giving purple and white, respectively. Such forms are called alleles. In the first generation, they exist without mixing, while in the second generation, they separate. Indicating the dominant alleles by R, and the recessive ones by r, we obtain the scheme shown in Fig. 19.

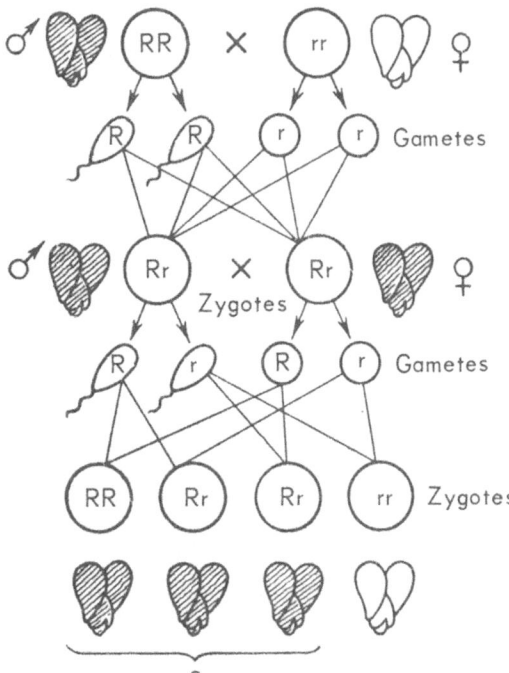

Fig. 19. Scheme of inheriting dominant and recessive characteristics; ♂ male, ♀ female.

This scheme is in complete agreement with what has already been said about chromosomes. In haploid cells, in gametes, there is only one chromosome which carries the characteristic R or r. During fertilization pairs are formed, RR, Rr, rR, and rr. It is evident that the probability of these four zygotes in the second generation are equal. But since the R allele dominates, the plants which grow from the zygotes RR, Rr, and rR, have purple flowers while those which grow from the rr zygote are white. Therefore, the ratio is $3:1$.

The difference between plants RR and Rr and rR is that only purple flowers will be formed after any crossing involving the first pair; when Rr or rR are crossed, however, there will again be a distribution according to the $3:1$ law. The RR, Rr, and rR plants are identical macroscopically, but their gametes differ. They belong to a single phenotype but the RR and the Rr and rR genotypes are different. The RR and rr plants are called homozygotes, since their zygotes contain only a single type of allele. The genotype does not change during their crossing. The Rr and rR are heterozygotes, since their alleles are not identical; when these are crossed, distribution follows.

A similar examination of crosses between RR and Rr shows that, in the first generation, all the plants will have purple flowers (RR, RR, Rr, rR), and the ratio of homozygous to heterozygous species will remain the same as in the parents, that is, $1:1$.

Thus, the hereditary factors, the genes, exist in living organisms in pairs; during the formation of gametes, each gene separates from the other one in a given pair and becomes transferred into the gamete as a single entity. As a result, each gamete contains only one gene of each type.

Now it becomes clear why it is necessary to have a large number of individuals in order to establish these laws. During fertilization many gametes can meet, and if we have a small number of plants, it is possible, while the probability is quite low, that all the plants of the second generation will be white (rr). Statistical laws are followed the more precisely, the greater the number of experiments. Thus, when we toss a coin, the $1:1$ distribution of heads and tails will occur with greater certainty, the greater the number of tosses. On the other hand, the theory of probability makes it possible to evaluate departures from the $3:1$ law with a rather small number of samples.[21]

The second law of Mendel was established during the crossing of genotypes which differed by several independent characteristics. This is the law of independent distribution. Let us illustrate it by an example taken not from the plant, but from the animal kingdom.[9] Let us cross black short-haired guinea pigs with brown long-haired ones. Both characteristics of the first one are dominant. Let us indicate them by B and S and the corresponding recessive alleles by b and s. The parent gametes are BS and bs. In the first generation, we obtain zygotes Bb and Ss, heterozygotic individuals with dominant genotypes in both characteristics.

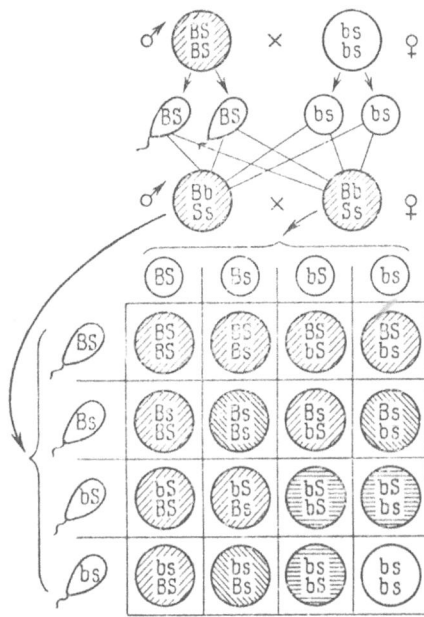

Fig. 20. Heredity scheme for two independent characteristics.

The gametes of the second generation are *BS*, *Bs*, *bS*, *bs*. To determine the distribution of characteristics in the second generation, we can use the scheme of Fig. 20. It is found that for every nine black short-haired ones, there are three black long-haired, three brown short-haired, and one brown long-haired one. The ratio is 9:3:3:1. These results show that different chromosomes are responsible for independently distributing characteristics. The adherence to the laws of Mendel in such simple form is certainly not true of all characteristics and certainly not for all species. In most cases, the picture turns out to be much more complicated, since the characteristics do not follow dominance. For example, in certain types of plants, the heterozygotic individuals have flowers with a color intermediate between the homozygotic ones (or let us say, that the flowers of *Rr* are pink), i.e., the characteristics can be, not independent, but linked with each other in a definite manner, etc. However, the laws of Mendel and their interpretation on the basis of knowledge about chromosomes always remain as the key to an understanding of heredity.

It is said at times that Mendel was fortunate, that he stumbled upon plants in which the genetific laws are obeyed in this particularly simple form. This should not be taken seriously. It was necessary to carry out a long study in order to find such plants and to start from pure species, from homozygotic genotypes. Had the starting plants with purple flowers been not homozygotes but heterozygotes, it would not have been possible to establish these laws. It is also very important that Mendel gave a correct interpretation of his results, concluding, as a result of a vigorous logical analysis, that they indicate the existence of material factors of heredity which are present in the gametes. This discovery was made long before the discovery of the genetic role of chromosomes!

In 1911, Johannsen gave the name *genes* to the factors contained in the chromosomes and responsible for the hereditary characteristics. Mendel in fact had already spoken of them in his studies. Later on, we shall discuss the modern concept of genes, the *cistron*. Genetic experiments established with great accuracy a number of the properties of the genes long before their nature had been unraveled by molecular biology.

Further studies on the laws of genetics have been carried out on a most unappealing being, one that has no practical significance, namely, the fruit fly, *Drosophila melanogaster*. The Drosophila is extremely convenient for experiments as a result of several factors. It multiplies fast and gives a very large progeny. It is characterized by a large variety of hereditary characteristics which are easy to observe (the color of the body, of the eyes, the shape of the wings, etc.). Its somatic cells contain only four pairs of chromosomes. The cells of the Drosophila salivary glands contain chromosomes of very large dimensions; their structure can be easily studied under the microscope. In short, the Drosophila represents a most convenient object for genetic experiments. Its study, first carried out by Morgan,[22] has made it possible to establish the basic laws of genetics, which are valid for the most diverse types of organisms, for man[23] and for plants with a great practical importance, for example, for corn. The situation is very typical for the methodology of a scientific investigation; extremely important practical deductions are made as a result of a study of what would seem at first completely useless objects, which, however, are excellent models for the basic laws.

In a number of cases, different characteristics turn out to be linked with each other; they are determined by factors which are localized in a single chromosome. Let us examine, as an example, the heredity of characteristics linked with sex. It is known, for example, that the serious hereditary disease, hemophilia, is found predominantly in males.[23] Sex is a genetically determined characteristic. Diploidal organisms have one exception from the rule of pairing chromosomes. There is one pair of chromosomes which is responsible for sex. In the female, both chromosomes in the pair are identical and they are called the X chromosomes (pair XX). In the male, one chromosome of the pair is X, while the other one, of smaller size, is called Y (pair XY). As a result, the female gametes will always be X, while the male ones will be X or Y in equal proportion. Therefore, the sex of the progeny is determined not by the ovum but by the sperm. In the Drosophila, the gene characteristic for the color of eyes is located within the X chromosome. The red color (R) dominates over white (r). During crossing,

$$♀ RXRX × rXY ♂$$

in the first generation, we find red males and females, $\male$ RXY and $\female$ $RXrX$. In the second generation, the distribution is $\female$ $RXRX$, $\female$ $rXRX$, $\male$ RXY, and $\male$ rXY. The ratio of sexes is $1:1$, the ratio of characteristics is $3:1$. During the reverse crossing,

$$\female\, rXrX \times RXY\male$$

in the first generation we have $\female$ $RXrX$ and $\male$ rXY; the females have red eyes, the males have white eyes. In the second generation, the distribution is

$$\female\, RXrX \qquad \female\, rXrX \qquad \male\, RXY, \qquad \male\, rXY$$

The ratio of sexes and characteristics is $1:1$.[24]

It is evident that in studying the splitting of characteristics, it is possible to establish which of these are independent and which are linked to each other, i.e., which belong to the same chromosome. Without any knowledge about chromosomes, it is possible to establish from experiments on crossing, the number of linked groups, i.e., the number of chromosomes. Furthermore, genetics make it possible to determine the sequence of the distribution of genes in a chromosome.

Drosophila has another sex-linked characteristic: the body color is grey (W) or yellow (w). The yellow color of the body is determined by the X chromosome. In the crossing of a white-eyed yellow female with a red-eyed grey male

$$\female\, rwXrwX \times RWXY\male$$

one can expect, as in the case just described, that in the first generation we shall have $\female$ $RWXrwX$ and $\male$ $rwXY$ namely, red-eyed grey females and white-eyed yellow males. In the second generation there must be 25% $\female$ $RWXrwX$, 25% $\female$ $rwXrwX$, 25% $\male$ $RWXY$, and 25% $\male$ $rwXY$. The experimental results for both male and female (Fig. 21) are:

grey red-eyed	24.725%
yellow white-eyed	24.725%
grey white-eyed	0.275%
yellow red-eyed	0.275%

Thus, notwithstanding the linkage of the two genes in the same chromosome the characteristics split to some extent. This is explained by the phenomenon of crossing over. The combining homologous chromosomes exchange regions which contain the corresponding genes (see Fig. 22). This conclusion, reached on the basis of experiments in crossing, has been confirmed by microscopic observations. The crossing-over takes place at the time of the synapsis of the chromosome during meiosis.

These amazing facts indicate that the genes are distributed in some linear sequence on a chromosome. It is natural to assume that the

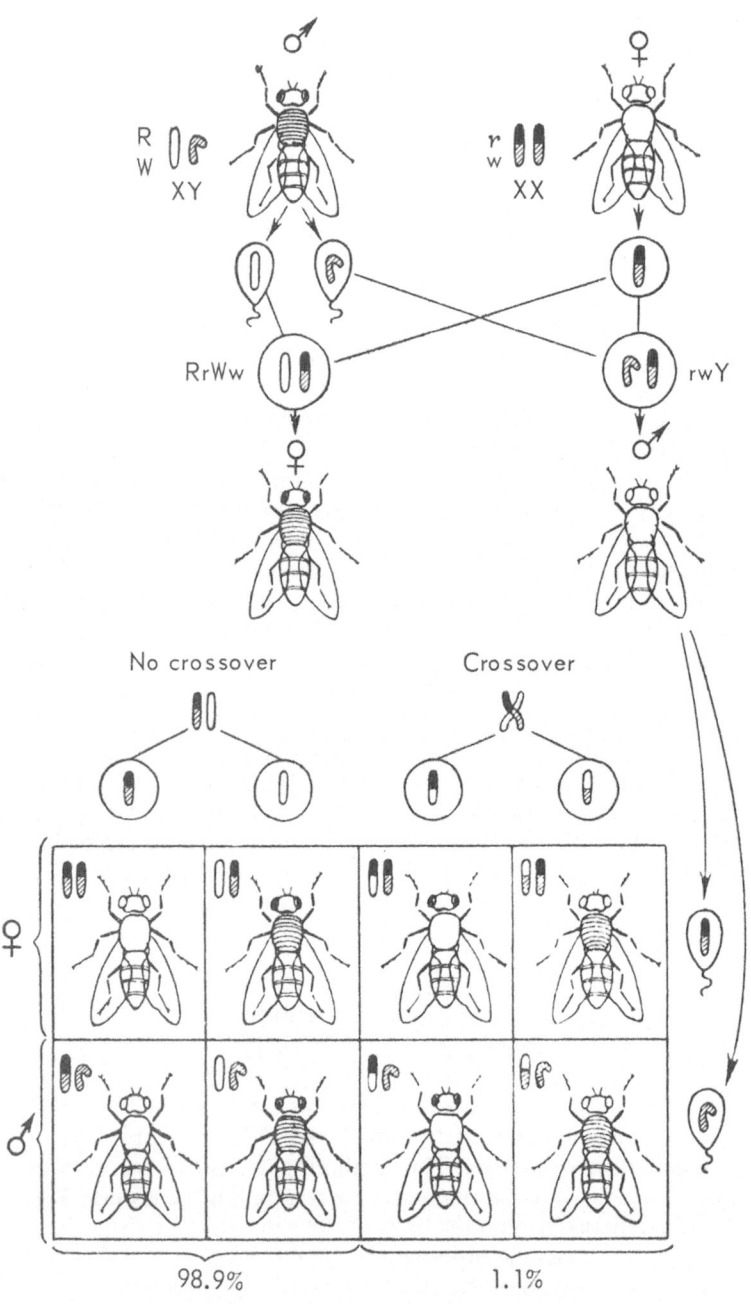

Fig. 21. Hereditary scheme during crossing.

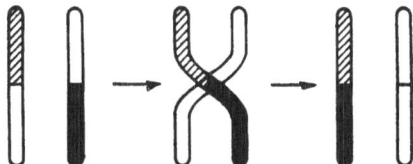

Fig. 22. Scheme of crossing.

frequency of recombination of two linked genes depends on the distance between them; the greater this distance, the greater the chances of crossing over. In this manner, it has been possible to obtain *genetic maps*, as they are called, namely, a representation of the sequence of genes in a chromosome with distances expressed as percentage of recombination (Fig. 23). Let us assume that three genes *A*, *B*, and *C* are present in a single chromosome. The probability of crossing over between *A* and *B* is 5%, and between *B* and *C* is 3%. Then the probability of crossing over between *A* and *C* must be equal either to 5 − 3 = 2% or 5 + 3 = 8%; the latter value is observed. In the first case, the gene sequence is *ACB*; the second is *ABC*. These conclusions have been fully confirmed by direct observation on the giant chromosome of the salivary gland of two-winged insects (and in particular, of Drosophila). The structural nonuniformity of these chromosomes is in agreement with the genetic map (Fig. 24).

We have examined only the simplest cases of heredity. The situation becomes much more complicated in those quite widely distributed cases in which the appearance of a characteristic depends on several different genes. These cases have been discussed in the literature.[24,25]

The genes and organisms in their entirety are under a direct strong influence of the environment. This interaction results, on one hand, in the change of the genes themselves (of hereditary factors), in mutations. Along with such changes of the genotype, the influence of the medium also determines the phenotype, i.e., one or another manifestation of the action of genes. By changing the external conditions, it is possible to direct the individual development of organisms. Very important results in this were obtained by Michurin who wrote, "I do not deny the values of the laws of Mendel; to the contrary, I only insist on the necessity of introducing some corrections and additions."[27] Michurin has shown that it is possible to direct the dominating characteristics during the development of hybrids of fruit and berry-bearing plants. The experiments of Michurin have resulted in important practical applications. It should be emphasized that, unlike Mendel and Morgan, Michurin worked with very complicated objects; cultured plants have a complicated hybrid origin and the genetic analysis is very complicated in this case. The studies of Michurin are based apparently on the specifics of individual phenotypic development and on long term modifications. Long term modifications are determined by changes in the properties of the enzyme system of the organism which occur under the influence of the environment. During this, the genotype does not

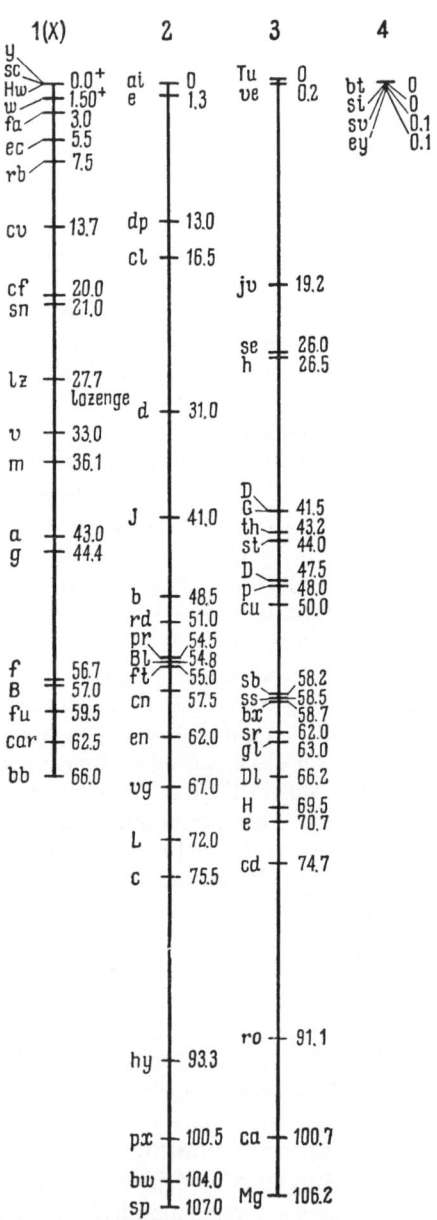

Fig. 23. Genetic map of Drosophila
chromosomes.

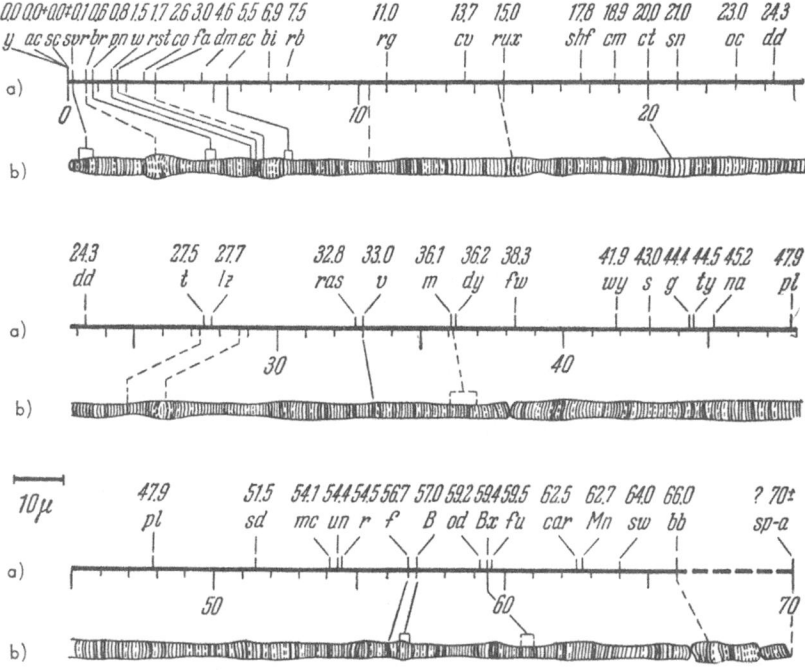

Fig. 24. (a) The genetic map and (b) structure of the Drosophila X chromosome.

change, and therefore, such modifications are not inherited, unlike the mutational changes of genes, which will be discussed in the next section.

There is no such thing as inheritance of acquired characteristics. Acquired characteristics are those characteristics which appear as a result of the action of the environment on adult organisms; these characteristics are somatic. From what has been said, it is evident that they cannot be adequately transferred to the progeny through the sex cells; changes which have taken place in various somatic cells cannot touch the genes contained in the gamete.

Statements to the contrary, which were first really formulated by Lamark, are still made even at present by individuals who are far removed from science. Attempts to prove experimentally the inheritance of acquired characteristics have invariably led to negative results; positive results which are contained in the literature turn out to be either consequences of poor methodology or even of simple falsification. This is a situation which is quite reminescent of attempts at the construction of perpetual motion, the examination of which was rejected by the French Academy of Sciences in the eighteenth century.

MUTATIONS AND MUTABILITY

As we have said, genes are subjected to the influence of the environment. It is absolutely wrong to regard the genes contained in the chromosomes as immutable, to consider that all of the diversity of individuals of a given species and all the variations of species are related only to different combinations of a once-created genetic pool. Such opinions, which have been expressed in the early stages of the development of genetics, have been refuted long ago. The mutability of genes can be explained on the basis of physical-chemical concepts.

Already purely genetic investigations show that heredity is determined by real material structures which are part of the chromosomes. These are certainly molecular structures. However, there are no immutable and immobile molecules. All molecules, and particularly molecules of organic compounds built from the light C, H, O, N, P, S atoms (nucleic acid and proteins), are alive in the sense that they are subjected to thermal motion, which does not stop even at absolute zero. As a result of thermal motion (and first of all of the vibrational motion of atoms within molecules), chemical bonds may statistically acquire enough energy to become ruptured. Consequently, even thermal motion can lead to changes in the chemical structure of the molecule of the hereditary structure. Even greater changes must take place as a result of various interactions with this substance, for example, of chemical and radiational action. Thus, it is impossible to reconcile the molecular concept with the absolute immutability of genes.

On the other hand, the molecular structure of the genes must be relatively stable since, if it could easily be violated as a result of thermal motion, or, let us say, of the cosmic radiation which falls on the surface of the earth, it would be absolutely impossible to speak of regular heredity, and life on earth would look completely different.

De Vries has shown that new forms occur in plants as a result of sudden changes and these are inherited by the progeny. Such changes were given the name of *mutations*. At present, it is clear that hereditary variations of a given characteristic are the results of mutation. Thus, the genes which determine the body or eye color in Drosophila give different colors in mutants. It is customary to call the more frequently occurring, i.e., the dominant, type the *wild type*, and the changed type the mutant. A grey red-eyed drosophila belongs to the wild type. Red hair in man is the result of a mutation. The number of such examples is enormous; just in drosophila, more than 400 different mutants are known.

The term wild type is not totally arbitrary, although it would have certainly been possible to call the rarer form wild, and the form which is particularly widespread a mutant. In the present sense, the wild form was formed as a result of long evolution, of selection of such mutant characteristics as guaranteed to the species the most favorable conditions of existence in a given environment. Consequently, mutant forms are usually less adaptive. In many cases, dominance means the presence of some favorable characteristic, recessivity means the absence of this characteristic. It is natural then that the phenotype of the Rr heterozygote is similar to the phenotype of the RR homozygote and not to that of rr homozygotes. R characterizes the presence of some characteristics, r characterizes its absence.

It follows from what we have said that the appearance of new mutants has a much higher probability of resulting in the appearance of characteristics which are harmful to its existence than for those that are beneficial to it. Consequently, the mutants will perish in the subsequent fight for their existence, and it is the wild type and not the mutated ones which will continue to exist. In this manner, a wild genotype becomes stabilized by natural selection until the time when a significant change takes place in the environment.

Quite frequently mutation is simply lethal, i.e., it results in an organism which is not capable of living and which perishes during some stage of its development. It is possible, for example, that mutant gametes will arise which will not be capable of fertilization. This is evidently accompanied by a change in the Mendel ratios. Thus, if r is a lethal allele, homozygote rr will not appear at all. In the generation formed as a result of crossing Rr with rR, we shall find only species RR, Rr, and rR instead of RR, Rr, rR, and rr. They are all similar phenotypically. If R is lethal, we shall find, to the contrary, Rr, rR, and rr in the ratios of $1:2$. This is the result obtained during the crossing of the yellow (R) and black (r) mice.

Two types of mutations are known, *gene*, or *point*, *mutations* and *chromosomal mutations*. In the first case, we speak about changes of the genetic substance on the molecular level of organization. Such changes cannot be observed under the microscope. On the other hand, chromosomal mutations consist of breaks and reconstructions of chromosomes which are visible under the microscope; in most cases, these are lethal for the organism.

Let us discuss first the nature of point mutations. They must result in the formation of discrete new characteristics relative to the nonmutated organism. The result of such a mutation must be not a gradual but a discontinuous change in the gene. The reason for this is the discrete nature of the state of molecules and of chemical bonds.

The principal results of quantum mechanics, as applied to atoms and molecules, are that systems, which consist of atomic nuclei and electrons, may exist only in discrete energetic states. The chemical bond between atoms either exists or does not exist; the intermediate state is impossible. In order for mutations to take place, it is important to have tautomeric transformations of molecules, related to the migration of a hydrogen atom from one position to another. As a result, one set of chemical bonds is replaced by another.

In this manner, a molecular point mutation must occur according to the all-or-none principle. In other words, a mutation either occurs or it does not occur; intermediate changes cannot take place. The discontinuous rather than gradual change of hereditary characteristics has quite a strong quantum-mechanical foundation. Here again, we see the universality of the laws of nature.

Certainly, chromosomal mutations, as well, are discrete, simply because a reconstruction of the supramolecular structure (chromosomes) can occur only as a result of changes in chemical bonds.

Mutations can occur spontaneously or under the influence of external action. Spontaneous mutations are caused by thermal motion. It is evident that an increase in temperature (i.e., a change in the conditions of the environment) must result in an increase in the probability of thermal mutation. What is the law?

A mutation must be treated as the result of some chemical reaction, i.e., of the change of molecular structure. In the simplest case, the rate of such a process obeys the Arrhenius law

$$v = Ae^{-E_a/kT} \qquad (2.4)$$

where v characterizes the number of transitions per second over the activation barrier with height E_a. Obviously the considerable stability of genetic material must be related to high energy barriers which protect the organisms from mutations. An assumption of the immutability of genes means an infinitely high barrier, that is, $E_a \to \infty$ and consequently $v \to 0$. On the other hand, it is evident that with $E_a \to 0$, transitions would occur constantly and, instead of a discrete distribution of states of a molecule and, thus, also of mutant forms, there would be some smeared continuity of these forms. For organic molecules, however, E_a values are high but not infinite. Physicochemically determined values of E_a are just of such an order of magnitude that they account for the stability of the genetic material and for its ability to mutate occasionally as well.[7] As a first approximation, point mutations do obey the law of Arrhenius (Eq. (2.1)]; the mutability of the Drosophila increases with temperature in accordance with this law.

An exact testing of this law is rendered difficult by the fact that any given organism can exist only over a narrow temperature interval. Furthermore, a change of temperature influences not only the gene molecule but also its environment; this changes the state of the environment which, in turn, must be reflected in the gene. The scheme shown in Fig. 1 and Eq. (2.4) are simplified representations; they do not take into account the cooperative nature of the processes under investigation. These questions are extremely interesting and important and are the subjects of further investigations.

The frequency of mutation per gene per generation in Drosophila varies under normal conditions between 10^{-5} and 10^{-6}. This value can be greatly increased by external interaction. In 1925, Nadson and Philippov observed for the first time the effect of radioactive radiation on the hereditary changes of the lowest fungi.[28] Two years later, Muller discovered radiational mutagenesis by showing that X-rays increase the frequency of mutations in Drosophila several hundred times relative to the frequency of spontaneous mutations.[29] These processes were studied quantitatively for the first time by Timofeev-Resovskii and coworkers.[30,31,32] This study showed that the number of mutations is proportional to the total dose of radiation measured by the degree of ionization caused by it, independent of the frequency of the quantum. Radiational mutagenesis appears to be caused by the ionization either of the genetic molecules themselves or of neighboring molecules, which leads to a reconstruction of the genes. With the accumulation of experimental material the picture has become very complicated; at present, radiational genetics is an important, wide field of biology, which is based entirely on physical-chemical investigations. It is impossible to present here the vast amount of data and their generalizations, obtained in radiobiology during its short existence (the reader is referred to monograph literature).[33,34,35]

It is necessary to emphasize that the action of ionizing radiation is related to changes in chromosomes and not in the cytoplasm. This was particularly well demonstrated in the experiments of Astaurov,[36] who studied the effect of radiation on the hereditary properties of the butterfly, the larvae of which are the silkworms. They make silk. Astaurov has shown that in this insect, it is possible to obtain androgenesis, namely, the development of an embryo from the ovum deprived of its own nucleus and containing only the father's chromosomes introduced by one or several sperms. The individuals developed from such a zygote are always male. The eggs were irradiated with X-rays and then fertilized with nonirradiated sperm. Shortly after penetration by the sperm (the silkworm has polysperms, i.e., the egg can be penetrated by several sperms), the eggs were heated up to 40°C for 135 minutes. During this, the female nuclei lost their ability to participate in fertilization. A male fertilization took place, i.e., the blending of two male nuclei with the formation of quite viable androgenetic embryos. Thus, irradiation of the cytoplasm had no effect on their fate. On the other hand, normal individuals, both of the male and female sexes, which had received chromosomes both from an irradiated egg and a nonirradiated sperm, were less viable the higher the irradiation dose of the ovum. Astaurov obtained for the first time mature organisms after the interaction of the nucleus and cytoplasm of different species. It was possible to carry out interspecies androgenesis, namely, to fertilize, with several sperm of the silkworm of one type, the egg of a silkworm of another type, the nucleus of which had been subjected to thermal shock. As was to be expected, the resulting organisms were similar to the father and not to the mother.

The experiments of Astaurov have proved absolutely that the genetic action of radiation is on the nucleus, i.e., on the chromosomes, and not on the cytoplasm. Astaurov makes a thorough comparison between the quantum of radiation and a bullet. A bullet destroys equally the muscles both of the heart and of the legs. However, if it hits the heart, it kills the organism, while a hit in the leg is relatively not dangerous. The nucleus with its chromosome plays in the cell the same role as the heart, namely, it controls all the biosynthetic and species-forming properties, while cytoplasm performs the metabolism, carries out the adaptive reactions, etc.

Here we meet one of the principal questions of biology. It is evident that "it is only the cell as a complete system, closely linked by intimate nucleus–plasma interactions, and furthermore, only within a definite range of environmental conditions necessary for life, that can guarantee the succession of the organization, which we call heredity" (Astaurov[37]). It is the nucleus of the cell with its chromosomes which is the leading informational and directing system.

The plan of the differentiated cytoplasm is genotypic; consequently, it is located in the nucleus.

In addition to nuclear there is also cytoplasmic heredity.[37,25] In some rare cases, not only chromosomes but also some cytoplasmic structures are duplicated. These phenomena evidently do not obey the laws of Mendel and must be regarded as exceptions to the general rule. Certainly, they are of great interest. Furthermore, it has been established that the number of factors which have been regarded as cytoplasmic heredity in actuality have a completely different character.

There are strains of *Paramecium aurelia* which are called killers. They secrete into the surrounding medium a substance called *paramecin* which kills the cells of other strains; the killers themselves, however, are insensitive to this substance. Such strains are homozygous in the dominant factor K, and their plasma contain what are called *kappa particles*. During the conjugation of a paramecium of type K with a paramecium of type K^+, which is sensitive to paramecin, KK^+ species are formed; these contain kappa particles. Had these particles been absent, the KK^+ species would have remained sensitive to paramecin. The kappa particles are transferred from one cell to another over a plasma bridge during conjugation, and it would seem that they are hereditary plasma structures, which are capable of multiplying in the plasma only in the presence of the K allele. It has been found, however, that these particles, which contain DNA and are rod-like bodies with a length of 0.6 to 1.0 μ, are foreign bodies; possibly, they are modified viruses or bacterial particles. The question here is not of plasma heredity but of genotypically controlled symbiosis.

Good evidence for the existence of nonchromosomal genes, i.e., of carriers of cytoplasmic heredity, have been found in genetics. During the process of meiosis, these genes are transferred to the progeny from the mother and not from the father; this establishes their presence in the cytoplasm. In fact, the zygote contains maternal cytoplasm, while the spermatozoid introduces into the ovum practically only a nucleus. Mutations in nonchromosomal genes have been found in the green algae, *Chlamydomonas*. These genes influence the development of chloroplasts and mitochondria.[38] The cytoplasmic nonchromosomal heredity has, as yet, not been sufficiently investigated.

The strong mutagenic action of short-wavelength radiation is one of the greatest dangers of experiments with nuclear weapons. Investigations of the radiobiological phenomena, on one hand, give a scientific basis for the necessity of stopping all experiments with nuclear weapons and, on the other hand, they indicate pathways for counteracting radiational mutagenesis.

The same questions come up in relation to space flight. The level of cosmic radiation increases sharply in certain zones beyond the limits of the earth's atmosphere. To be able to fly without danger to other planets and return to earth, it is necessary to have effective protection from the action of radiation.

Radiationally induced mutations are harmful as a rule. However, they can also be put to the benefit of mankind. As a result of the fact that radiation results in a vast number of different mutations, it is possible to choose from these such mutations of agricultural plants and animals as will be of great interest for selection. In this way, it has been possible to obtain new high-yield cultures of wheat, beans, and rye which are resistant to a number of fungal diseases.[48] By using radiation mutagenesis, it has been possible to obtain mutants of molds which produce great yields of penicillin and other antibiotics.

Chemical mutagenesis was first investigated by Sakharov,[39] Lobashev,[40] and especially Rapoport.[41] A whole number of substances result in chemical reactions which lead both to gene and chromosomal mutations. Among these substances one may cite cholchicine, mustard gas, etc. These phenomena will be discussed later.[25,42]

One of the genetic changes brought about, for example, by cholchicine is the appearance of polyploidism. Polyploidal cells are cells (and organisms) which contain more than two homologous chromosomes, i.e., not pairs, but triplets, quadruplets, etc., of homologous chromosomes. Such a multiplication of chromosomes is very interesting for genetics, since, in this case, each gene is present in the form of several copies. Polyploidal organisms usually have greater dimensions than diploidal ones. Polyploidism is a property of cultured plants and microorganisms (yeast) and is quite important for agriculture.

Up to now, we have talked about mutations in gametes, namely, in sex cells. It is evident that only such mutations have a genetic significance and can appear in the progeny.

It is evident that the chromosomes and genes of any cells are susceptible to mutation, both sex and somatic cells. Somatic mutations can be quite important for the organism. Those called *mosaics* are of great interest. When a dividing cell mutates at the stage of formation of the first two cells and if the mutation occurs in only one of these, an organism is formed whose right and left halves can have some different characteristics. Such are humans who have one brown and one blue eye, parrots which have blue feathers on one half of the body and green ones on the other.

A much more important fact is that malignant tumors, which arise as a result of the pathological growth and multiplication of cells, are also related to somatic mutations. This is understandable since it is chromosomes which direct biosynthesis, i.e., the development of the cells. Somatic mutations of cells which multiply rapidly by their nature (the cells of blood-forming organs, of intestinal epithelium, etc.) must have particularly serious consequences for the organism. This apparently is one of the principal reasons of radiation sickness.

To control mutations in an intelligent manner and direct heredity and changes, it is necessary to study the phenomena of heredity and mutations on the molecular level by the methods of molecular biology and molecular biophysics. In the case of bacteria and viruses, these possibilities have already been brought to life. It is with such possibilities that the immense perspectives of molecular biology are connected.

GENETICS OF BACTERIA AND PHAGES

At present it is no longer Drosophila but much simpler organisms, fungi, bacteria, and even viruses, which are the principal materials of genetic investigation. It is not very important for human life whether an individual's eyes are blue or brown (these are hereditary characteristics). It is important that the entire system of biochemical reactions in his cells be correctly regulated and that the necessary enzymes be synthesized in them. It is exactly the synthesis of biologically functional substances and, first of all, of proteins which is determined genetically.

How can we study the direct chemical actions of genes? In the case of multicellular organisms, this is quite difficult. The red color of the eyes in a Drosophila is the result of a vast number of chemical processes and it is extremely difficult to follow them.

Bacteria are extremely convenient materials for studies of this nature. They multiply rapidly and form on a dense medium (argar) colonies of 10^7 to 10^9 cells which are descended from a single cell. What is most important is that their genetic characteristics are directly linked with metabolism. Beadle and Tatum found new possibilities in biology when they started the genetic studies of bacteria. Bacteria grow on a nonliving nutritional medium which contains inorganic salts and some sugar which acts as the source of energy and of carbon.

The composition of a nutritional medium for *Escherichia coli* is 1,000 ml of water, 1.0 g NH_4Cl, 0.13 g $MgSO_4$, 3.0 g KH_2PO_4, 6.0 g

Na_2HPO_4, and 4.0 g glucose. The phosphate salts are sources of phosphorous which is necessary for the synthesis of nucleic acid; at the same time, they maintain the necessary concentration of hydrogen ions. Ammonium chloride is a nitrogen source. Magnesium, potassium, and sodium are necessary for life. For rapid multiplication of the cells of *E. coli*, it is desirable to aerate the system; without oxygen, the multiplication is very slow. Genetic variations, that is, mutations, are established directly by the biochemical properties of the bacteria. The mutants can differ in their ability to assimilate various sugars, their ability to synthesize from inorganic nitrogen compounds and sugar amino acids and other substances necessary for life, in their ability to resist such poisons as antibiotics and sulfamides, and in their degree of stability relative to bacteriophages. All these are chemical characteristics.

For any type of bacterium, it is possible to establish the *minimal medium*, which contains the smallest number of substances necessary for the multiplication of the given type, called the *prototrophic strain*. The same applies to a number of fungal organisms. On the other hand, there also exist mutant strains of bacteria and fungi which are deprived of their ability to synthesize one or several amino acids or other biologically important compounds from simpler precursors. These are *auxotrophic strains*. When a growth factor which the auxotroph is not able to synthesize independently is added to the medium, it begins to multiply, as well as the prototroph.

Before examining bacterial genetics, i.e., their biochemical genetics, it is necessary to establish the character of their hereditary changes.

Quite unfortunately, bacteria mutate very easily. Thus, we observe the adaptation of bacteria to antibiotics. In the treatment of pneumonia by penicillin, in drawn-out cases it stops helping and it becomes necessary to turn to other antibiotics, namely, to streptomycin. A new strain of pneumococci is formed; it is resistant to penicillin and is, thus, more dangerous to man.

The appearance of stable strains of bacteria is explained sometimes by their adaptation, i.e., by the fact that they get used to the changed conditions of the medium. Furthermore, it is natural to assume that adaptation is transferred by hereditary means. At first glance, such an assumption does not contradict the law of non-inheritance of acquired characteristics; the bacterium is a single-cell organism, it is both a somatic and a sexual cell.

Another possibility is determined by mutations. If a given population of bacteria contains mutants which are stable relative to some poisons, then, if the medium contains such poisons, only the stable mutants will survive and multiply. In this case, the medium acts on the bacteria as a factor of selection, i.e., just as in the natural selection of multicellular organisms. Selection takes place, the bacteria are chosen artificially, and a new stable resisting strain develops; this strain is very dangerous to the sick.

Thus, there are two ways of changing bacteria, by adaptation and by mutation. Which one of these takes place?

This question can be examined experimentally. Such experiments have been carried out by Delbrück and Luria.[43]

Hereditary adaptation occurs gradually and it is directed; it becomes stronger from generation to generation, while mutation has a random character. A statistical population analysis must give different results in the two cases. The experiments have been carried out in the following manner.

A large number N of batches of the nutritional medium are infected with small amounts of cells of a single strain of the microorganisms. After a certain time of growth, the average number $\bar{n}$ of stable resistant cells is determined in each of these by seeding a medium that contains the poison (let us say, penicillin, if we are talking of pneumococci). If the phenomenon is adaptation, then the probability of development of resistance in cells is equal in all of the cultures. Furthermore, in this case, it is also possible to determine the distribution of deviations from the mean number $\bar{n}$ of stable cells. This is the usual Poisson distribution. The mean square of deviations from the mean is equal to the mean itself,

$$\overline{(n - \bar{n})^2} = \bar{n} \tag{2.5}$$

It has turned out that, within the limits of a single culture, the Poisson distribution is obeyed. The mean values $\bar{n}$, in different cultures differ at times from each other by one or two orders of magnitude. This result rejects the assumption of adaptation. The remaining assumption is that getting used to the poisons is the result of mutation.

Since mutations occur randomly and with low probability during the growth of each culture, statistical fluctuations occur in their numbers; these are particularly important at the beginning of growth. The fluctuations increase sharply during the multiplication of cells, which results naturally in great differences between the values of $\bar{n}$ in different cultures. Delbrück and Luria carried out a rigorous mathematical analysis of these experiments; they showed that these results are in complete accord with the hypothesis of mutation rather than an adaptational change of bacteria.[43,44]

Thus, resistant cells can arise in any portion of the bacterial cultures; the medium containing the poison acts only as a developer for these cells.

Similar results have been obtained in all the cases in which the changes in the microorganisms have been studied by vigorous quantitative methods. Single-cellular organisms also do not inherit acquired characteristics. Their difference from multicellular organisms in this sense can be reduced to a greater variability and their ability to produce new mutant strains very quickly.

This does not mean, certainly, that in nature there is no adaptation, i.e., nonhereditary adaptation, to the conditions of the environment. To the contrary, any organism possesses a definite genotype and has, thus, a definite ability to adapt and it must adapt. Variations in the phenotypic characteristics are determined both by hereditary mutations and by modifications which arise as a result of the action of the environment. It is only the first, however, which are inherited. Modifications include characteristics which pass from generation to generation during uninterrupted existence under constant conditions, but which vanish immediately once these conditions are changed. There are also long-lived modifications which are kept over several generations living under changed conditions; these also vanish finally. Long-lasting modifications are normally observed with nonsexual reproduction: conjugation, i.e., renewal of the nuclear material, removes them immediately. This is a situation, for example, with paramecia which are able to adapt themselves to existing in a 5% solution of arsenic acid. This adaptation remains; it decreases gradually over several hundred generations obtained by division and disappears immediately during conjugation.[45]

Let us return to bacteria. How can we study the genetics, how can we obtain a genetic map? To do this, it is necessary to carry out crossing experiments; bacteria, however, normally multiply by simple cell division.

Along with cell division, however, one also observes conjugation of bacteria, namely, their temporary combination, followed by division. It turns out that, in certain cases, such a conjugation is similar to fertilization, i.e., one of the cells behaves like a male and the other like a female sex cell. In other words, bacteria can have sex. This was found quite recently in genetic investigations.[44,46,47]

The favorite object in the genetics of bacteria and in molecular biology is the intestinal bacterium, *E. coli*. These bacteria are normally not pathogenic; it is easy to grow them and to study them. *Escherichia coli* have a number of mutant forms. Thus, for example, the wild strain can live in a medium containing lactose. This strain Lac^+ ferments lactose. The Lac^- mutant does not have such an ability; it does not grow on a minimal medium which contains only lactose as a carbon source. The P^+ strain can synthesize the enzyme phosphatase, the P^- strain cannot do this. The *E. coli* strains differ in their stability toward phages, antibiotics, etc.

Lederberg[48] was the first to establish that when certain different *E. coli* strains are mixed together, genetic recombinations occur with a probability of the order of 10^{-7}; these are similar to the ones that take place during the crossing-over in multicellular organisms. It was further shown that wild strains of the type HFR (high-frequency recombinants) induce genetic recombinations in cells of the type F^-. The first cells have male donor-cell characteristics, the second are female receptor cells. During conjugation, there is transfer of hereditary materials from the first to the second, but not *vice versa*. Genetic recombinations are rarer than conjugations. In any conjugations, however, it is possible to have asymmetric changes of the genetic properties of the conjugating cells.

Fig. 25. Scheme of formation of a zygote
in bacteria.

Consequently, bacteria are also subject to crossing over, or similar phenomena. Bacteria are haploidal cells possessing a single chromosome.
During conjugation, the chromosome passes from the male cell to the
female and the diploidal zygote is formed with the chromosomal structure
shown in Fig. 25. The male chromosome is HFR; F^- is the female chromosome. The notations S and L represent some genetic markers. Recombinations can take place by crossing-over (Fig. 26a) or by other mechanisms.[48]

Lederberg studied the so-called "copy choice" (Fig. 26b). The newly
synthesized chromosome in the progeny cell copies partly the father and
partly the mother chromosome. This process is random and, therefore, the
probability that crossing-over by the copy will occur in the S–4 segment,
as shown in Fig. 26b, is proportional to the length of the segment. Thus,
copy choice is similar to crossover in its results.

Finally, recombinations can take place as a result of the fragmentation
of chromosomes, namely, the fragmentation of chromosomes and the
introduction into a specific position of a fragment of another chromosome
with a subsequent duplication of the genetic material. Thus, it would
seem that recombination in the phages T2 and T4 takes place in this
manner. At present it is not yet clear how this occurs in bacteria; however,
with any mechanism, recombinations make it possible to draw a genetic
map of the chromosomes in which the distances between the genes are
expressed in frequencies of recombination. These distances can also be
expressed in units of time, for example, in minutes.

The chromosome in the cell of *E. coli* has the shape of a closed ring.
This fact has been established genetically and confirmed by electron
microscopy. Let us indicate by the letter F some material factor present

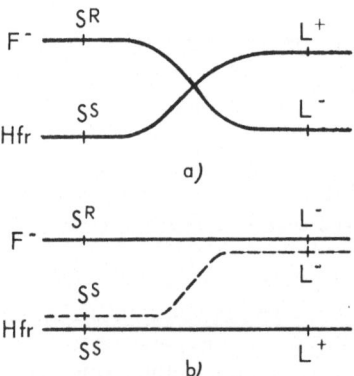

Fig. 26. Recombination: (a) crossover, (b) copy
choice.

in the cytoplasm, which under certain conditions can confer on the cells the ability to be a donor, i.e., male. This factor exists in the type F^+. The F factor can be inactivated by X rays. This results in the formation of F^- mutants which can play the role only of acceptors, i.e., of female cells. The presence of the F factor in type F^+ does not mean that all the F^+ cells will be male and able to transmit to chromosomal acceptors, i.e., to female cells. Wollman and Jacob have established that the closed-ring chromosome can break open in F^+ cells, with the result that the process $F^+ \rightarrow$ HFR takes place. The ring, broken in some position, combines with factor F and becomes a linear chromosome capable of passing from a male cell to a female F^- during their conjugation (Fig. 27). The F factor is called the sex factor.

The breaking of the chromosome can be regarded as a spontaneous mutation $F^+ \rightarrow$ HFR; its probability is 10^{-4} to 10^{-5}. Wollman and Jacob have carried out experiments with an interrupted conjugation of cells. A chromosome requires some time for passing from cell HFR into cell F^-. If conjugation is interrupted before the passage of this time, only part of the HFR chromosome will pass into the F^- cell; it is possible to cause such interruption simply by vigorous mixing of the culture in a Waring blender. As a result, what is called a *merozygote* is formed, namely, a cell containing an entire maternal chromosome and only part of the paternal one. A merozygote has a ploidity greater than 1 and less than 2. By studying the properties of the progeny of the merozygote, it is possible to determine how much time is necessary for transfer of any given marker (gene) from HFR to F^-. It has been established that the entire chromosome is transferred over 2 hours, a very long time. The F factor is transferred during the same time; consequently, it is attached to the far end of the chromosome and enters last into the cell. Other markers are obviously transferred over a shorter time. As a result, it is possible to construct a genetic map of the chromosome with distances expressed in transfer times.

The situation is complicated by the fact that the breaking of the ring can occur, in principle, in any place and, therefore, the markers distributed between the initial point 0 and the end point F of the resulting linear chromosome can follow in different order. This has been studied and taken into account during the construction of the genetic map. The genetic map of the *E. coli* chromosome is shown in Fig. 28.

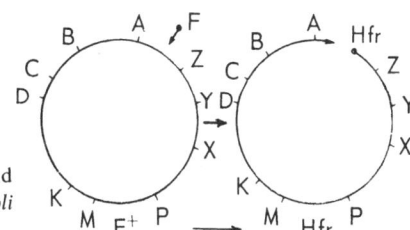

Fig. 27. Combination of the F factor and breaking of the chromosome in the *E. coli* bacterium.

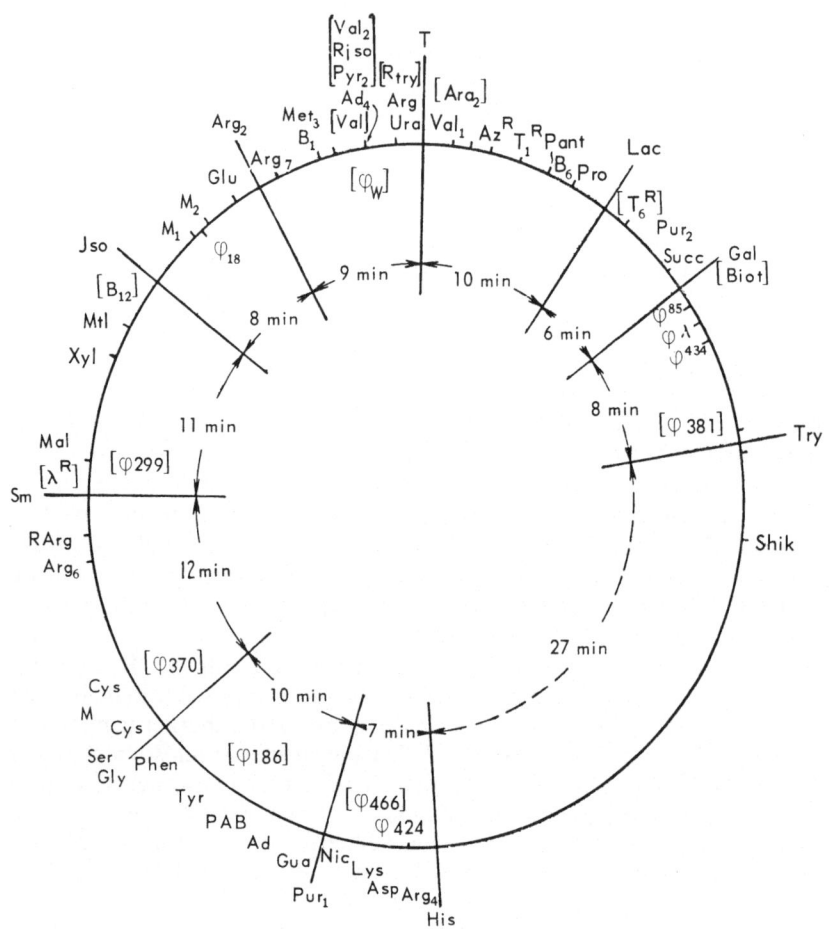

Fig. 28. Genetic map of the *E. coli* chromosome (strain *K*12). The genetic markers indicate: (*a*) *The ability to synthesize* T, threonine; L, lysine; Pant, pantothenic acid; Pro, proline; Pur, purines; Biot, biotine; Pyr, pyrimidines; Try, tryptophane; Shik, shikinic acid; His, histidine; Ara, arginine; Lys, lysine; Nic, nicotinamide; Gua, guanine; Ad, adenine; PAB, paraaminobenzoic acid; Tyr, tyrosine; Phen, phenylalanine; Gly, glycine; Ser, serine; Cys, cysteine; M, methionine; B_{12}, vitamin B_{12}; Iso, isoleucine; B_1, thiamine; Val, valine. (*b*) *The ability to ferment* Ara, arabinose; Lac, lactose; Gal, galactose; Mal, maltose; Xyl, xylose; Mtl, mannitose. (*c*) *The requirement of* Succ for succinic acid; Asp for aspartic acid; Glu for glutamic acid. (*d*) *The resistance of* Az^R to Na azide; Sm^R to streptomycin; T_1^R to phage T_1; T_6^R to phage T_6. (*e*) *The repression by* R_{Iso} of isoleucine; R_{Try} of tryptophane. (*f*) *The loci of prophages induced by ultraviolet light*, 85, λ, 434, 381, 424, and 466; *the loci of noninduced prophages*, 186, 370, 299, 18, and w.

Of very great significance for the genetics of bacteria and of phages are the phenomena of lysogeny and transduction. As has been stated already, lysogeny is a hereditary ability of the cell to produce a phage under certain inductional interactions. This ability is acquired by the cell during its infection by a moderate phage. Genetic investigations based on the study of recombination have shown that the phage genome (the totality of its genes) is directly linked in this case to the chromosome of the bacteria. In particular, the division of prophages is exactly balanced with the division of the bacterial cell. Each type of prophage is located in a definite way in the bacterial chromosome. Consequently, there is an interaction between the phage genome and the bacterial genome. The existence of such an interaction has been shown in direct experiments. On the other hand, it is not possible to regard the prophage as part of the chromosome. It is attached to the chromosome but most of it remains outside the chromosome. It is characteristic that the prophage is acquired or lost as a single entity.[19]

Above, we have discussed cancer, malignant tumors, as the result of some somatic mutation. Another theory on cancer was proposed long ago; a number of scientists believe that it is caused by a viral infection. The phenomenon of lysogeny shows that the difference between these points of view is not very significant. The existence of prophages bridges the difference between heredity and infection.[46] It is not only in the case of lysogenic bacteria that viruses lead to changes in the genetic properties of the cells. Malignant neoplastic growths caused by viruses are known; such are the Rous sarcoma in chickens and polyoma in a number of rodents. These are moderate viruses which are able to multiply inside cells over a number of generations like moderate bacteriophages. It is clear that the integration of a phage or viral genome in the chromosome of a bacterium is similar to mutation in its consequences.

A prophage is similar to the sex factor of bacteria from the point of view that it can exist in two different states, namely, in the autonomic state within the cytoplasm of the cell and in an integrated state, attached to the chromosome. Genetic elements which possess such properties are called *episomes*. Obviously, episomes are not necessarily present within cells. If episomes are absent, they can become introduced into the cell only from without. In an integrated state, episomes are localized on the bacterial chromosome, but they are not part of its linear structure. On the other hand, genetic recombinations are possible between the episomes and neighboring elements of the chromosome; as a result, typical bacterial genes can acquire the properties of episomes.

For a detailed presentation of the results obtained on the genetics of bacteria and viruses, the reader is referred to the monographs of Jacob and Wollman[46] and of Hayes[47] (see also Lederberg[48]). Let us examine now in greater detail the concept of the gene. Up to now, we have introduced this concept as some material part of the chromosome, responsible for the appearance of definite hereditary characteristics. On the other hand, in the phenomena of crossover and copy choice, the chromosome undergoes recombination and the gene functions as some unit of recombination, i.e., as a unit in the structure of the chromosome which does not split into constituent parts during recombination. Furthermore, in the phenomena of mutation, the gene appears as a region of the chromosome which changes its structure and properties as the result of a spontaneous or induced mutation; it acts as a unit of mutation. Benzer[49] introduced the concepts of recon and muton which represent, respectively, the smallest element in the linear arrangement which can be exchanged but can divide during genetic recombination, and the smallest element in which a change results in the appearance of a mutant form of the organism. Further analysis has shown, however, that there is no significant difference between a recon and a muton.

It is quite evident that the biosynthetic ability and the ability for recombination and mutation are the results of real molecular structures. Molecular biology deals with the question of the nature of these structures.

As has already been pointed out, the principal function of the genes is biosynthetic. Beadle has said "one gene, one enzyme." In this remark, he was considering the functional unit of the hereditary material which is responsible for the synthesis of a given protein. The statement of Beadle can be made more precise in the following manner: "one gene, one protein chain." The point is that many enzymes and other proteins consist of several polypeptide chains, each of which is controlled by a given gene. Modern science starts from the functional definition of a gene. Furthermore, the old indeterminate term *gene* is practically no longer used. At present, the functional unit is called, not a gene, but a *cistron*.[49] The first problem that comes up is the number of cistrons in a chromosome. The resolution of this problem would make it possible to establish whether a cistron coincides with a muton and a recon, or if these are different types of units.

The number of cistrons and their dimensions expressed in units of recombination or in time (in the case of bacteria) can be determined by genetic methods independent of biochemistry. Another phenomenon that should be discussed is pseudoallelism. Definite sections of the chromosome control closely related processes. This can be found, for example, in the *lozenge* gene of Drosophila. A mutation in such a segment results in a change of the color of the eyes and of other characteristics of the insect. The muton is recessive and, consequently, the heterozygote has a normal dominant genotype. It has been established in crossover experiments that in the *lozenge* segment there are three mutational loci, i.e., a mutation in any of these three loci brings about the appearance of a recessive allele. The distance between the three loci is very small; it is less than 0.1 of a recombination unit. It is possible, then, to obtain zygotes of different types, as shown in Fig. 29. In this figure, segments of two homologous

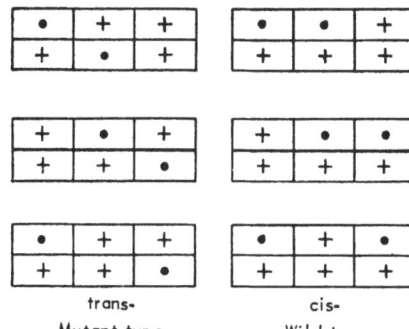

Fig. 29. Cis-trans test.

chromosomes of the zygote are shown; these have been divided into three segments which correspond to the mutating loci, i.e., to the mutons. The dot indicates *lozenge* mutations; the + indicates nonmutated mutons. In all the six cases shown, the zygote contains two mutations. If these two mutations take place in different homologous chromosomes (which is called the *trans configuration*), then the insects belong to a mutant phenotype; if they occur in only one chromosome (cis *configuration*), then the normal wild genotype is preserved. How can this be explained? It is evident that mutation in any of the three mutons means a disruption of the normal function of the gene. If the mutation takes place in both chromosomes (trans), this function is absent. If one of the chromosomes, however, has remained intact, then the function is preserved. In this manner, it is possible to isolate the functional segment of the chromosome; this genetic experiment is called the *cis-trans test*. This is the origin of the term cistron.

The concept of a cistron contains a certain difficulty. Once one is familiar with the recombination phenomena, it might appear that recombined chromosomes participate in a trans configuration. This is not so; here we are discussing only the two parental chromosomes without any recombination.

We see that the cistron, i.e., the gene, possesses fine structure; it contains a number of mutons. An extremely detailed investigation of the fine structure of the cistron has been carried out by Benzer, who worked with bacteriophage T4. Ignoring the question of whether phages are living or nonliving, it is of interest that they follow all the genetic laws.

Phage T4 is a parasite of *E. coli* cells. The wild type of the phage forms small scattered plaques on a colony on the B strain of *E. coli*. Nine different mutants, indicated by the letter *r*, give us different types, i.e., larger plaques. The same mutants behave differently on a colony of another strain, namely, strain *K*, two give plaques of type *r*, one gives plaques of the wild type, and six do not give any plaques since they do not multiply in these cells. In the case of the *E. coli* strain *S*, the situation is still different.

The genetic map of a phage can be constructed as a result of experiments on the infectivity of a single cell by two phage particles of different types. In this process, they cross over and recombination takes place. It turns out that the three groups of mutations fall in different segments of the map. A morphological characterization of the three types of mutants is shown in Table 1; a genetic map (depending on the choice of hosts) is shown in Fig. 30.

Table 1[27]   **Morphology of Plaques Formed by Different Strains of T4 Phage, during the Infection of Various Hosts**

| Phage strain | Morphology of the plaques after infection of the strain | | |
	B	S	K
Wild type, w	Wild type	Wild type	Wild type
rI	r	r	r
rII	r	Wild type	Does not multiply
rIII	r	Wild type	Wild type

Among the mutants of group rII, there are such which cannot be points on a genetic map. They correspond to changes of entire sections of the map and never give reverse mutations of the type $r \to w$ (wild type), while normal mutations can revert. These mutations correspond to deletions, i.e., the loss of genetic material and its disappearance from the map. As a result of deletions, Benzer could greatly simplify the problem, since these exclude the necessity of crossing the mutants in question with others in which the mutations occurred in a section of the map that is inside the region of deletion. Mutations are first classified into groups which are located in different regions of the map, on the basis of crossing-over, with the use of deleted mutants. Further crossovers take place within each group.

Since the *E. coli* cell into which the genetic material was introduced from two phage particles was similar to a diploidal zygote, it was possible to carry out the cis-trans test. It turned out that the mutants belonged to two different cistrons. Benzer examined the fine structure of the cistron

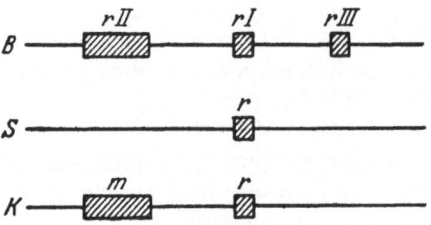

Fig. 30. Genetic map of T4 phage as a function of the *E. coli*.

by investigating 923 mutants of r! As a result of this outstanding work, it has been established that recons and mutons have an order of magnitude of hundredths of a percent in frequency, while cistrons contain hundreds of mutons.

Let us point out that, as a result of this type of investigation, it has been possible to determine the molecular dimensions of the gene cistron and its structural details, its mutons and recons. The point is that the genetic material of the phage is contained in one long DNA molecule, the dimensions of which can be measured by physical methods. This makes it possible to translate the units of recombination into numbers of definite atomic groups. Studies on phage genetics which were started by Benzer had quite far-reaching consequences. They have led to the discovery of the basic laws of the genetic code.

As a result of the investigation of the genetic properties of a number of phages, it was possible to discover that the chromosomes are cyclic.[19]

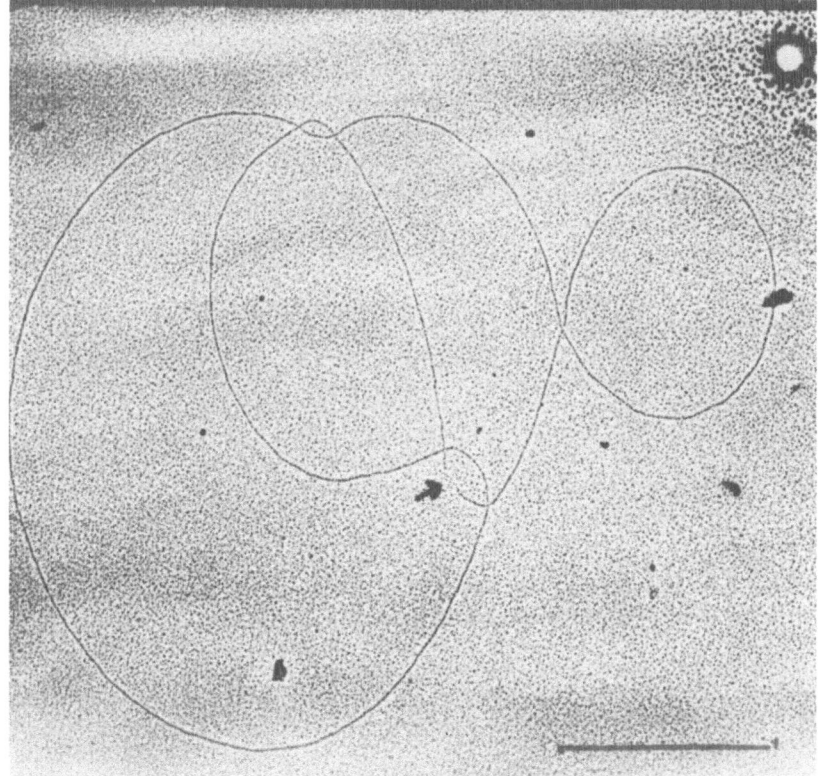

Fig. 31. Electron micrograph of the cyclic chromosome of λ phage. Magnification 44,000 ×. The break is marked by the arrow.

This has been directly confirmed. In Fig. 31 is shown the electron micrograph of a chromosome, i.e., of the DNA chain, of λ phage, which was made by Ris and Chandler.[50] The phage chromosomes are cyclic just like bacterial chromosomes.

Thus, at present, a gene (cistron) is no longer a hypothetical but an absolutely concrete molecular entity, the structure and properties of which have been elucidated by science.

The history of the gene is similar to the history of the atom. The concept of the atom arose as a necessity in chemistry long before its investigation by physicists. The concept of the gene came as a necessity in genetics long before its investigation by specialists in the field of molecular biology. A number of physicists and physical chemists, and among them Ostwald, denied the existence of atoms right up to the time of their direct proof by physical methods. Even at present, there are still people who deny the existence of genes and who defend a heredity deprived of concrete material carriers.

The experimental and theoretical investigation of the structure of atoms was started by Bohr in 1913; at present, the structure of the atom is well known. The investigation of the structure of genes started with the discovery of the genetic role of DNA and the studies of its structure; at present, we know the basic molecular characteristics of the gene. One must say that, once direct concrete evidence was available for the existence of atoms, Ostwald withdrew his earlier point of view.

Chapter 3

BIOLOGICAL MOLECULES

AMINO ACIDS AND PROTEINS

Let us examine now the chemical structures and properties of the principal substances which carry out the living functions of organisms. We have already said much about these molecules; it is now time to examine them in greater detail.

While the physical-chemical structure of an organism is extremely complicated, it is possible to approach an understanding of this structure on the molecular level by studying the properties of a relatively small class of organic compounds. As we shall see, these investigations are still in a rather early state of development; molecular biology is a young science. It is important, however, that organisms do not contain an infinite variety of compounds. The inexhaustible variety of forms and individual properties of living matter turns out to be related to the variety in which a limited number of atomic groups in biological molecules can be combined. In this sense, nature acts conservatively. The conservatism, of course, is a result of the systematic process of natural selection and is not the result of any special vital forces which are inaccessible to the usual scientific approach. All the protein molecules which exist in nature are chains of different lengths; they are made up of links of a limited number of types. These links are amino acid residues. In the overwhelming majority of cases, one finds that there are 20 types of amino acids; in addition, in nature there are a few other rarely occurring amino acids.

Amino acids are compounds which contain a carboxylic group COOH and an amino group NH_2. The structure of a molecule of an

α-amino acid is

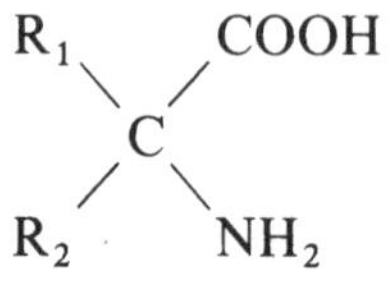

$$
\begin{array}{c}
R_1 \diagdown \quad \diagup COOH \\
C \\
R_2 \diagup \quad \diagdown NH_2
\end{array}
$$

where R_1 and R_2 are some radicals. Thus, amino acids always contain C, O, N, H; the radicals may contain other atoms, such as S.

The amino acids in a protein are linked together by polycondensation. Water is eliminated and a peptide bond $-CO-NH-$ is formed. The reaction between two amino acids leading to the formation of a peptide bond proceeds according to the scheme

$$
\begin{array}{c}
R_1 \diagdown \quad \diagup COOH \qquad R_3 \diagdown \quad \diagup COOH \\
C \qquad\quad + \qquad\quad C \\
R_2 \diagup \quad \diagdown NH_2 \qquad R_4 \diagup \quad \diagdown NH_4
\end{array}
$$

$$
\longrightarrow \quad
\begin{array}{c}
R_1 \diagdown \qquad\qquad\qquad\qquad \diagup R_3 \\
C-CO-NH-C-R_4 + H_2O \\
R_2 \diagup \; | \qquad\qquad\qquad\; | \\
\quad NH_2 \qquad\qquad\qquad COOH
\end{array}
$$

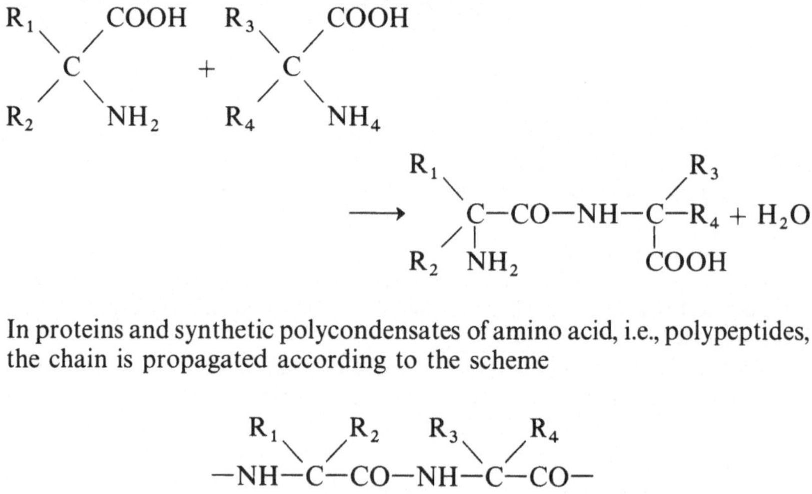

In proteins and synthetic polycondensates of amino acid, i.e., polypeptides, the chain is propagated according to the scheme

$$
\begin{array}{c}
R_1 \diagdown \quad \diagup R_2 \qquad R_3 \diagdown \quad \diagup R_4 \\
-NH-C-CO-NH-C-CO-
\end{array}
$$

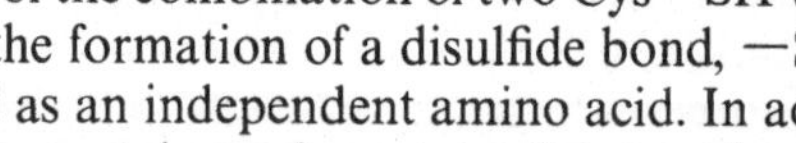

Table 2 contains the formulas of all 20 amino acids which enter into the composition of proteins.[1]

The Cys$-$S$-$S$-$Cys residue, shown after residue 13, Glu$-NH_2$, is formed as a result of the combination of two Cys$-$SH with the elimination of hydrogen and the formation of a disulfide bond, $-$S$-$S$-$. As a result, it is not examined as an independent amino acid. In addition to the listed residues, proteins contain end groups, the terminal carboxyl $-CO_2^-$ (indicated by $-$C), and amino, $-NH_3^+$ (indicated by $-$N), groups.

In Table 2 the amino acid residues have been divided into three classes. The first class contains those residues which do not carry electric charges; the second and third classes contain those residues which acquire such charges in the proper medium. At this point, it should be indicated that in a neutral medium, the amino acids do not exist in the form shown

Table 2 Amino Acid Residues Normally Encountered in Proteins

No.	Amino acid	Abbreviation	Formula
		I Neutral residues	
1	Glycine	Gly	$-CO-CH_2$ with NH below
2	Alanine	Ala	$-CO-CH-CH_3$ with NH below
3	Valine	Val	$-CO-CH-CH\big<^{CH_3}_{CH_3}$ with NH below
4	Leucine	Leu	$-CO-CH-CH_2-CH\big<^{CH_3}_{CH_3}$ with NH below
5	Isoleucine	Ileu	$-CO-CH-CH\big<^{CH_2-CH_3}_{CH_3}$ with NH below
6	Phenylalanine	Phe	$-CO-CH-CH_2-$ (benzene ring) with NH below
7	Proline	Pro	$-CO-CH-CH_2$; $N-CH_2$ ring $>CH_2$
8	Tryptophan	Trp	$-CO-CH-CH_2-$ (indole ring) with NH below
9	Serine	Ser	$-CO-CH-CH_2-OH$ with NH below

Table 2   (continued)

No.	Amino acid	Abbreviation	Formula
10	Threonine	Thr	$-CO-CH-CH\begin{smallmatrix} OH \\ CH_3 \end{smallmatrix}$, with NH on the CH
11	Methionine	Met	$-CO-CH-CH_2-CH_2-S-CH_3$, NH
12	Asparagine	Asp–NH$_2$	$-CO-CH-CH_2-CO-NH_2$, NH
13	Glutamine	Glu–NH$_2$	$-CO-CH-CH_2-CH_2-CO-NH_2$, NH
	Cystine	Cys	$-CO-CH-CH_2-S-S-CH_2-CH-CO-$, NH ... NH

II   Acid residues

No.	Amino acid	Abbreviation	Formula
14	Aspartic acid	Asp	$-CO-CH-CH_2-COO^-$, NH
15	Glutamic acid	Glu	$-CO-CH-CH_2-CH_2-COO^-$, NH
16	Cysteine	Cys	$-CO-CH-CH_2-S^-$, NH
17	Tyrosine	Tyr	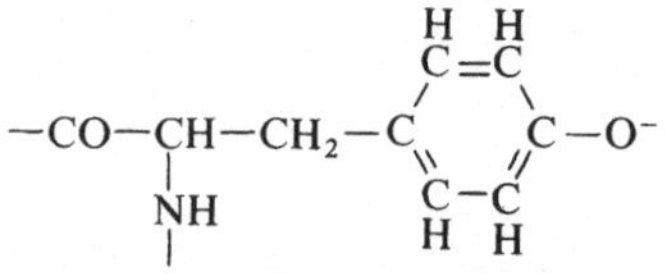 $-CO-CH-CH_2-\langle\text{ring}\rangle-O^-$, NH

III   Basic residues

No.	Amino acid	Abbreviation	Formula
18	Histidine	His	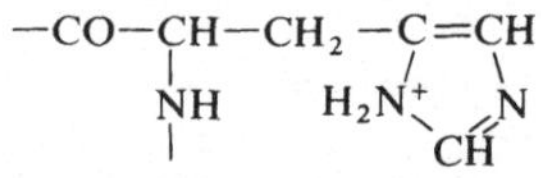 $-CO-CH-CH_2-C=CH$, NH , imidazole with H_2N^+, N, CH

Table 2 (continued)

No.	Amino acid	Abbreviation	Formula
19	Lysine	Lys	$-CO-\underset{\underset{NH}{|}}{CH}-CH_2-CH_2-CH_2-CH_2-\overset{+}{N}H_3$
20	Arginine	Arg	$-CO-\underset{\underset{NH}{|}}{CH}-CH_2-CH_2-CH_2-NH-\underset{\underset{NH_2}{|}}{C}=\overset{+}{N}H_2$

in the first structure given in this chapter, but in the form

$$\underset{R_2}{\overset{R_1}{\diagdown}}C\underset{\overset{+}{N}H_3}{\overset{COO^-}{\diagup}}$$

Such a molecule, which, as a whole, is neutral but which has localized positive and negative charges, is called a *dipolar ion* or a *zwitterion*.

Of the 20 amino acids, 2 (methionine and cystine) contain sulfur and 4 (phenylalanine, tryptophan, tyrosine, and histidine) contain an aromatic or heterocyclic group. Proline is actually not an amino but an imino acid; its residue contains, not an $-NH-$, but an $-N$ group. Serine and threonine contain hydroxyls. We can see that these 20 amino acids are at times called magic, because of their universality; they are very diversified both in their chemical structures and in their ability to acquire electric charges.

The rare amino acids (in addition to the 20 which are normally found) are derivatives of some of the fundamental ones. These encompass the neutral residue hydroxyproline

$$-CO-\underset{\underset{\diagup}{N}-CH_2}{\overset{|}{CH}}-CH_2\diagdown \atop CH-OH$$

The acid residues diiodotyrosine and dibromotyrosine are encountered only in marine organisms

$$-CO-\underset{\underset{\diagup}{NH}}{CH}-CH_2-C\diagdown\diagup\overset{CH-CI}{\underset{CH=CI}{\diagdown}}C-O^-$$

and

$$-CO-\underset{\underset{/}{NH}}{CH}-CH_2-\underset{\diagdown}{C}\overset{\diagup CH-CBr}{\underset{\diagdown CH=CBr}{\diagdown}}C-O^-$$

The thyroxyl residue has the structure

$$-CO-\underset{\underset{/}{NH}}{CH}-CH_2-\underset{\diagdown}{C}\overset{\diagup CH-Cl}{\underset{\diagdown CH=Cl}{\diagdown}}C-O-\underset{\diagdown}{C}\overset{\diagup CH-Cl}{\underset{\diagdown CH=Cl}{\diagdown}}C-O^-$$

Collagen contains the basic residue hydroxylysine

$$-CO-\underset{\underset{|}{NH}}{CH}-CH_2-CH_2-\underset{\underset{|}{OH}}{CH}-CH_2-\overset{+}{N}H_3$$

There are reasons to believe that these residues are secondary, i.e., they are formed as a result of the corresponding chemical reactions in already formed protein chains.

The number of amino acid residues in protein molecules varies within very broad limits. In a protein essential for human and other mammalian life, namely the hormone insulin, we find 51 amino acid residues; in the muscle protein, myosin, this number reaches 4,000. Consequently, the molecular weights of proteins lie within a broad range, namely, from several thousand to hundreds of thousands and even to tens of millions. (The average weight of an amino acid residue in a protein is 100 to 110.)

In studying the structure of a protein, it is necessary first of all to establish its amino acid composition. The protein is split down to amino acids by hydrolysis, namely, by heating it in a $6M$ solution of hydrochloric acid for 10 to 15 hours. A similar process occurs in the stomach of man or any other animal. The stomach contains hydrochloric acid as well as enzymes, catalytic substances, and the proteins derived from food are hydrolyzed in the stomach down to amino acids. Then, these amino acids are again absorbed by cells and new proteins are synthesized from them; these new proteins are now of such a structure and composition as are essential for the existence of the organism.

The hydrolysate, i.e., the solution which contains a mixture of all the amino acids of the protein, is analyzed chromatographically. This method was discovered by the Russian botanist Tsvet in 1906. This magnificent discovery by Tsvet has played a major role in biochemistry and molecular biology; without chromatographic analysis, the development of many spheres of this science would have been considerably slowed down.

The chromatographic method consists in the following. The mixture under investigation (the solution) is filtered through a column of adsorbing material. Since different components of the mixture are adsorbed in different ways, they distribute themselves in different zones on the column. Then, they can be eluted from these zones by various solvents and separated from each other in this manner. Modern chromatography represents a broad field of physical chemistry; it has very important scientific and technological applications.[2,3] The term *chromatography* (from the Greek word *chromos*, which means color) stems from the fact that Tsvet was trying to separate chlorophyll into its components and the zones in his experiments had different colors.

Chromatography can be carried out not only on columns but also on filter paper. If a corner of a blotting paper is immersed in ink, separation takes place; there is adsorption, and capillary forces lift the water higher than the dyeing material of the ink. A drop of the protein hydrolysate is introduced on one corner of a rectangular piece of dense filter paper. This sheet of filter paper is then hung from a trough which contains special

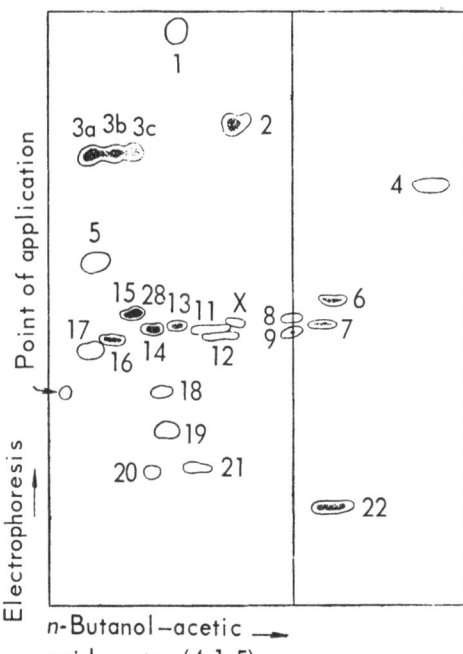

Fig. 32. Chromatogram (finger-print) of a protein hydrolysate, oxidized ribonuclease.

solvents and is washed with these solvents. After some time the paper is dried, it is turned by 90°, and it is now washed with another solvent. The sheet of paper is again dried, it is sprayed with a solution of a substance which gives a color reaction with amino acids, and then it is heated. Colored spots appear in various places on the paper; these correspond to various amino acids (Fig. 32). Knowing the conditions of the experiment, it is possible to determine the acid composition of the protein from the position of these "fingerprints." If segments of the paper are cut out and eluted by extracting from them the amino acid, it is possible to establish the quantitative amino acid composition.

At present, amino acid compositions are determined on automatic instruments which give the information on the protein composition in the shape of a curve registered by a recorder. Thus, the determination of an amino acid composition presents no difficulties.

With knowledge of the structure of the simplest proteinlike compounds, namely, of the hormones adrenocorticotropin and insulin, it has been possible in recent years to synthesize them from substances of non-living origin.

Furthermore, there are a number of ways of chemically synthesized polypeptides, i.e., polycondensates of amino acids. Thus, polyglutamic acid, which has the structure

$$\text{HOOC}-\underset{\begin{array}{c}|\\ CH_2\\ |\\ CH_2\\ |\\ COOH\end{array}}{CH}-NH-CO-\underset{\begin{array}{c}|\\ CH_2\\ |\\ CH_2\\ |\\ COOH\end{array}}{CH}-NH-CO-\cdots-NH-CO-\underset{\begin{array}{c}|\\ CH_2\\ |\\ CH_2\\ |\\ COOH\end{array}}{CH}-NH_2$$

is obtained as a result of a chemical reaction *in vitro* under normal laboratory conditions. The synthetic polypeptides are similar to proteins; they are, however, devoid of any biological function which depends not only on the composition but also on the sequence of the amino acid residues. Nevertheless, the investigation of polyamino acids is very important for molecular biophysics, since they can be taken as good models for many of the physical-chemical properties of proteins. Both proteins and polyamino acids contain the peptide bond. The space distribution of the atoms which are neighboring this bond is shown in Fig. 33, in which the interatomic distances are also indicated in angstroms. It is important that all the four atoms of the peptide $-NH-CO-$ lie in a single plane; this has been shown by X-ray analysis. This fact, together with the reduced interatomic distance between N and C relative to the normal, indicates that the N$-$C and C$=$O groups are conjugated, i.e., their electron shells overlap.

The manner in which proteins are assembled from amino acids with the formation of peptide bonds has been established exactly. Some other

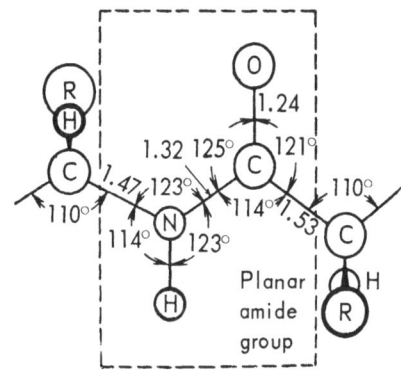

Fig. 33. Geometry of the peptide bond.

hypotheses on the chemical structure of proteins (for example, the "diketopiperazine" theory[4]), which was proposed in the past, have turned out to be wrong.

ASYMMETRY OF BIOLOGICAL MOLECULES

Functionally important biological molecules possess one very interesting property, namely, they are asymmetric.

In 1848, Pasteur was studying tartaric acid, a substance extracted from products of a biological nature, namely, from wine; he established that the crystals of tartaric acid exist in two forms, a right form and a left form (Fig. 34). These forms are mirror images of each other, i.e., they are related to each other just as the right and left hand. The discovery by Pasteur was the foundation of stereochemistry, namely, that branch of chemistry which studies the space structure of molecules and crystals. It was established that there are asymmetric substances of two types. The first class contains crystalline substances which lose their asymmetry during melting or dissolution and which are consequently characterized by an asymmetric structure of the crystal as a whole, but not of the molecule from which these crystals are built. Such, for example, are crystals of quartz. Melted quartz is devoid of asymmetry. The second class contains substances made up of asymmetric molecules. A melt or solution obtained from crystals of tartaric acid of one type, the right or left type, keeps the

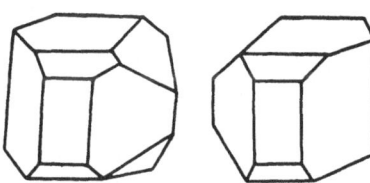

Fig. 34. Crystals of right and left tartaric acid.

property of asymmetry. The same is found to be true with solutions of sugar and with a vast number of materials of biological origin.

How is asymmetry manifested? The right and left forms of tartaric acid have identical chemical properties, their crystals have identical densities, melting point, etc. The principal specific property of asymmetric materials consists in their ability to rotate the plane of polarization of light, i.e., they possess optical activity. A solution of the right type of crystals of tartaric acid rotates the plane of polarization of light in one direction, that of the left type of crystal rotates the light in the other direction. The theory of optical activity will be discussed later.

To rotate the plane of polarization, a medium must be asymmetric, i.e., it must be devoid of a plane or center of symmetry. Figure 35 shows these symmetry elements in a molecule of paradichlorobenzene. A plane of symmetry is such a plane that upon reflection in it all the atoms of the molecule fall on identical atoms. The dichlorobenzene molecule has three such planes, the plane of the molecule itself and two planes normal to it. A center of symmetry is a point in the center of the molecule so located that reflection about it of all the atoms results in their superposition on identical atoms; for example, the first chlorine falls on the second, etc.

The majority of the asymmetric molecules of organic compounds contain asymmetric carbon atoms, i.e., carbon atoms which are covalently bonded to four different groups. When such an asymmetric carbon atom is present, there is neither a center nor a plane of symmetry, since the valence bonds of carbon are distributed at tetrahedral angles to each other, i.e., at angles close to 190° 28′. As a result, such molecules can exist both in a right and a left form which cannot be transformed into each other by any possible rotation. Two such forms (optical antipodes) of the amino acid molecule alanine are shown in Fig. 36. The asymmetric C atom is marked by a star. The antipodes are mutual mirror images just as a right and a left hand.

It is evident that asymmetric carbon atoms are present in many very diversified organic compounds, if these have a somewhat complicated

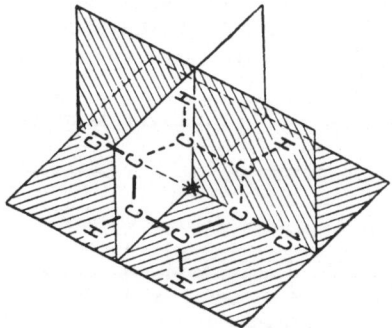

Fig. 35. Planes and center of symmetry in a molecule of paradichlorobenzene.

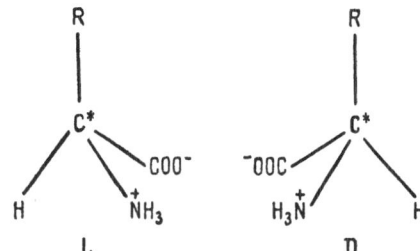

Fig. 36. Optical antipodes of the amino
acid alanine.

structure. In a molecule of tartaric acid, there are two such atoms

$$
\begin{array}{c}
\text{COOH} \\
\text{H} \diagdown \mid \diagup \text{OH} \\
\text{C*} \\
\mid \\
\text{C*} \\
\diagup \mid \diagdown \\
\text{HO} \quad \mid \quad \text{COOH} \\
\text{H}
\end{array}
$$

The C atom bonded with the NH_2 and COOH groups is asymmetric in all amino acids, except for glycine (see Table 2).

Does this mean that a substance which is built up of molecules that contain asymmetric carbon atoms will be asymmetric as a whole, that it will rotate the plane of polarization of light? Let us suppose that we synthesize such a substance by usual chemical means in a test tube. In such a case we should obtain a mixture consisting of 50% right-handed molecules and 50% left-handed ones; such a mixture is called *racemic*. In fact, such a composition corresponds to maximal disorder, to maximal entropy, i.e., it is consistent with the second law of thermodynamics.

When molecules are mixed, entropy increases. Let us assume that we have N_d right and N_l left types of molecules in a gas or in solution. As long as these gases are not mixed, the contribution to entropy which is dependent on the volume is

$$S_0 = N_d k \ln V_d + N_l k \ln V_l \tag{3.1}$$

where V_d and V_l are the volumes occupied by the right and left types of molecules. Let us remove the barrier which separates the gases. In such a case, both types of molecules will occupy the total volume $V = V_d + V_l$. The entropy changes and becomes equal to

$$S = Nk \ln V \tag{3.2}$$

where $N = N_d + N_l$. The entropy difference is

$$\Delta S = S - S_0 = k \left(N_d \ln \frac{V}{V_d} + N_l \ln \frac{V}{V_l} \right) \tag{3.3}$$

If both gases are present under the same pressure, then

$$\frac{V}{V_d} = \frac{N}{N_d} \qquad \frac{V}{V_l} = \frac{N}{N_l} \tag{3.4}$$

and the entropy of mixing is equal to

$$\Delta S = k\left(N_d \ln \frac{N}{N_d} + N_l \ln \frac{N}{N_l}\right) > 0 \qquad (3.5)$$

We may ask for what values of N_d and N_l, if N is constant, will ΔS, and thus S, have a maximal value? To find the maximum, let us write out $N_l = N - N_d$; let us differentiate ΔS with respect to N_d and equate the results with zero,

$$\frac{\partial \Delta S}{\partial N_d} = \ln \frac{N - N_d}{N_d} = 0 \qquad (3.6)$$

It can be seen from this that the entropy is maximal when $N - N_d = N_l = N_d$, i.e., the mixture contains 50% of right and 50% of left types of molecules.

In fact, under normal conditions, chemical synthesis always results in a racemic mixture which does not rotate the plane of polarization of light, since it contains equal amounts of the right and left types of molecules. On the other hand, the most important substances of biological origin contain only a single optical antipode. This again is a manifestation of the general rule of ordering in living matter, of the nonequality of the entropy of an organism to maximum, which, as we have seen, is possible only in an open system.

All the amino acids which can be extracted from proteins belong to one stereochemical type; these are called the L amino acids (L stands for left). This does not mean that they rotate the plane of polarization to the left. The left series of organic compounds originates from levorotating glyceraldehyde, $O=CH-C^*H(OH)-CH_2OH$. All the L compounds can, in principle, be obtained from this compound by substituting the necessary atoms or groups attached to C^*, without changing the total configuration of the molecule. This can result both in dextro- and levo-rotating compounds, and the symbol L indicates not the sign of rotation, but the participation in a definite stereochemical family.

The absolute configuration of natural amino acids is shown in Fig. 37.

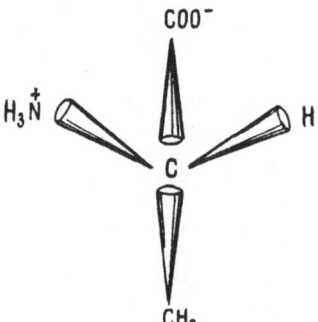

Fig. 37. The absolute configuration of an L-amino acid (L-alanine).

All the most important biological molecules, namely, amino acids, proteins, nucleic acids, carbohydrates, possess optical activity, i.e., they have the structure of a pure antipode. This interesting phenomenon is highly specific for life. Thus, Vernadskii spoke of the asymmetric organization of space by living organisms.

The production of pure antipodes in chemical laboratories is carried out with the help of asymmetric substances of a biological origin (usually of alkaloids). It is possible to isolate the individual antipode by acting on a racemic mixture by asymmetric substances. Let us denote a racemic mixture by (D, L). Let us react this substance with substance L'. We obtain

$$(D, L) + L' \rightarrow DL' + LL'$$

The substances DL' and LL' are no longer mirror image antipodes (such antipodes would have been substances DL' and LD'); consequently, they must differ also in other properties in addition to optical activity. They can be separated, for example, by crystallization.

To separate the right antipode from the left, it is necessary to have a substance (or a being), which senses the difference between right and left, i.e., which interacts differently with the right and left antipodes. Such a substance, or being, must itself have an asymmetric structure. This role was played by Pasteur when he separated the right and left crystals of tartaric acid under a magnifying glass. Man, just as any organism, has an asymmetric structure and detects the difference between right and left. If we look at quartz, however, we find that, no matter what the source of crystalline quartz SiO_2, which is a strictly inorganic substance, in accordance with the second law of thermodynamics there will be generally equal amounts of right and left crystals. Man separates the right crystals from the left ones; this is essential for the construction of optical instruments.

Asymmetry is a property not only of biological molecules but also of organisms as a whole. In Fig. 38, we have shown the external appearance of colonies of *Bacillus mycoides* which form structures, twisted to the left in a counterclockwise direction. Right strains are very rare. In the same figure we have shown the left and right spiral shells of the mollusk *Fructicola lantzi*, from Central Asia.[5] Once again one form is much more widespread than the other.

We are convinced of the abiogenic origin of life, namely, of its appearance from nonliving matter. The theory of Oparin, which, for the first time, presented detailed arguments in favor of this concept,[6] has received widespread recognition.

A number of questions come up in relation to the asymmetry of everything living on all levels of its structure, namely, from an entire organism down to the molecular state. First, let us ask how asymmetry could have arisen, if the second law of thermodynamics predicts a racemic mixture and not pure antipodes. Furthermore, why and in which way does

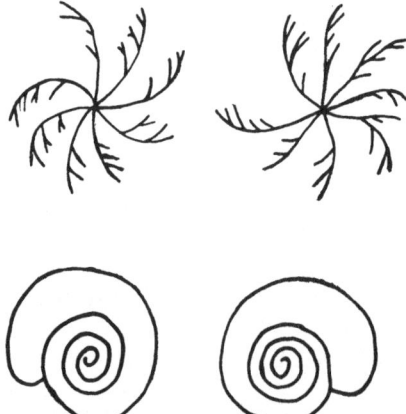

Fig. 38. Colonies of *Bacillus mycoides* and shells of *Fructicicola lantzi*.

asymmetry persist, once it has arisen; what role does it play in the development of life and in evolution?

The first question is, in general, related to the problem of the origin of life as such. As we have seen, the high degree of ordering and organization in living matter contradicts the second law only at first glance. In this sense, asymmetry is only a particular manifestation of the general law; the increase of information (negentropy) related to asymmetry is relatively small. Furthermore, it is possible to conceive of many different concrete ways in which pure antipodes became isolated at the dawn of life.

Optical activity is intimately bound with the Cotton effect, i.e., with circular dichroism. This phenomenon consists in the fact that an asymmetric body, a pure antipode, absorbs to different extents right and left polarized light.[7] Let us recall that in circularly polarized light the electric-field vector of the light waves, which is always perpendicular to the direction of its propagation, is constantly rotating about this direction. If the right antipode absorbed more strongly left polarized light, then the left antipode will absorb more strongly the right polarized light, and *vice versa*. Braun, Kuhn, and Knopf[8,9] used the Cotton effect to separate antipodes partially. They studied substances which are subject to photochemical decomposition under the action of light, i.e., the ethyl ester of α-bromopropionic acid and the dimethyl amide of α-azide propionic acid. The yield of the photochemical reaction is proportional to the intensity of the absorbed light. The racemic mixture was subjected to the action of circularly polarized light. One antipode was preferentially decomposed, namely, the one in which this light was absorbed more strongly because of the Cotton effect. As a result, the mixture was partially enriched with one of the antipodes, it became asymmetric and optically active.

Along with asymmetric photochemical decompositions, it has been possible to accomplish also asymmetric photosynthesis. Thus, when the diethyl ester of fumaric acid is hydroxylated with hydrogen peroxide under right circularly polarized light, we obtain an excess of the dextrorotatory diethyl ester of tartaric acid[10]

$$
\underset{\substack{H \\}}{\overset{\substack{H_5C_2OOC \qquad H}}{\underset{\displaystyle COOC_2H_5}{\underset{\displaystyle \|}{\overset{\displaystyle C}{\overset{\displaystyle \|}{C}}}}}} \quad + 2OH \longrightarrow \quad \begin{array}{l} H \\ | \\ HO-C^*-COOC_2H_5 \\ | \\ H-C^*-COOC_2H_5 \\ | \\ OH \end{array}
$$

Solar light reflected from the surface of the earth has a slight excess of the right circularly polarized component of light. Thus, it becomes possible to make a hypothesis about the abiogenic origin of the primary antipodes in biological molecules.

The second possibility is related to the existence of asymmetric crystals and, in particular, of quartz. In relation with the general spatial scheme, the interaction of an asymmetric crystalline surface with a racemic mixture can result in the resolution of the mixture. The crystal surface may catalyze a reaction during which one of the antipodes will be transformed preferentially. It can carry out asymmetric adsorption, etc. Many experiments of this nature can be carried out quite successfully.[11]

As has been pointed out already, however, the number of right and left crystals of quartz is the same no matter what the location of the source. Thus, on the average, a racemic mixture should be formed.

Finally, while the probability is very low, the appearance of a pure antipode is possible as a result of a local fluctuation, of a local violation of the second law of thermodynamics, which is valid statistically. Experiments on the spontaneous crystallization of racemic mixtures are known. The primary randomly arising nucleus of a crystal growing from a racemic mixture can contain right and left molecules with equal probability. But if the nucleus is right, then, with further growth, mostly right molecules will become attached to it. It has been possible to achieve the complete extraction of an antipode by repeated crystallization of a substance with the use of such asymmetric nuclei.[12]

The analogy between the ordering of matter in living organisms and during the process of crystallization permits the assumption that an accidental asymmetry becomes fixed in organisms and is replicated without limit. It is evident that the asymmetry of biological molecules leads to more specific reactions, namely, an asymmetric molecule interacts differently with right-handed and left-handed objects. A large number of examples of such types are cited in the interesting book by Gauze, "Asymmetry of Protoplasm."[13] Some substances are known to be poisonous in one antipode and harmless in the form of the other antipode. The R aspartic acid is sweet; L aspartic acid is bitter. As has been shown already by Pasteur, bacteria are nourished primarily by one antipode of a given substance, etc.

It is obvious that the entire cycle of substances in living matter must be coordinated from the point of view of the selection of antipodes. Heterotrophic organisms which are nourished by other heterotrophs or by auxotrophs can assimilate only L-amino acids; the D-amino acids do not become incorporated in proteins. Were an organism living on earth to migrate to an antiworld constructed of D-amino acids and the corresponding antipodes of other substances, it would die of starvation, notwithstanding the vast amount of nourishment. In Lewis Carroll's book "Through The Looking Glass," Alice says to the kitten, "Perhaps looking glass milk isn't good to drink." In fact, it follows from what has been said that to drink a mirror-image milk is, at best, useless.

The molecular and supramolecular structures of biologically functional substances are always asymmetric. We shall be greatly concerned with asymmetry in what follows. Racemization, symmetrization would mean the destruction of the highly ordered biological structures. It seems, then, that asymmetry became fixed during the evolutionary process, during which an organism was ridding itself of antipodes of the opposite side. It is quite characteristic that proteins, nucleic acids, lipids, and carbohydrates in organisms are always asymmetric; molecules which are less important for life, namely, products of metabolism, excrements, and reserve substances (organic acids, alkaloids, glucosides, terpenes, etc.) can exist in the form of both antipodes and in their partially racemized form.

It is difficult to imagine that life originated only once and in only one place. It is much more probable that life originated in many locations on the surface of the earth, or more exactly, in the ocean which covered the earth. This springing up of life was not all simultaneous. As a result, competition occurred between more or less developed forms of life. Systems which accidentally broke out in front and which contained an excess of one antipode, eliminated less-developed systems constructed of opposite antipodes; this process is similar to spontaneous asymmetric crystallization.

Thus, the problem of the asymmetry of life is intimately related to the general problem of the origin and development of life. It is not so important how a given antipode came to be in excess in any given location. It could have been the result of a photochemical process, of a catalytic reaction on the surface of quartz or just of fluctuation. What is important is the fact that the asymmetry became fixed in biogenesis and in evolution.[6,14]

PRIMARY STRUCTURE OF PROTEINS

Let us return to proteins. Knowledge of the amino acid composition of a protein is still insufficient to characterize its structure. The situation is the same as in the case of usual compounds. For example, we may have an elementary analysis: C, 52.2%; O, 34.8%; and H, 13.0%. The formula which corresponds to such a composition is C_2H_6O. This formula, however, tells us nothing about the structure of the molecule; even in this very simple case, there are two possibilities: C_2H_5OH (ethyl alcohol) or $(CH_3)_2O$ (dimethyl ether). The problem is that of isomerism. Similarly for any given amino acid composition, the residues may be distributed in different sequences along the protein chain. As a rule, proteins differ from each other both in their amino acid composition and sequence. It is customary to refer to the amino acid sequence as the primary structure of a protein. As we shall see, the primary structure determines all the properties, the entire biological functionality of protein molecules.

In describing the primary structure of a protein and its function, it is customary to use a very simple analogy which has a real scientific meaning. It is difficult to conceive of a better explanation; and one is really not necessary. It is possible to compare a protein molecule with a text written with a 20-letter alphabet. Each amino acid residue plays the role of a letter. Twenty letters are more than sufficient to record any information; for example, the Morse code has only four symbols. Just as the properties and functionalities of a protein are a function of the sequence of amino acids, the meaning and contents of a text depend on the sequence of letters. Short protein chains may be compared with a text of several lines, with some aphorisms; long chains may be compared with a pamphlet, if not with a thick book. This raises a few questions about the text and the typography. Is the text absolutely definite and identical in all the copies of a given book, i.e., are all the molecules of a given protein identical, or is the protein a heterogeneous mixture of different materials? How can we read the text, i.e., how can we find the sequence of amino acids? What is the sense of what has been read, i.e., in which way does the given primary structure determine the biological function of the protein? What is the role of a misprint? How is this text written, i.e., what causes the amino acids to combine into a strict sequence during the biological synthesis of a protein?

It has been definitely established that proteins are individual chemical compounds, and not mixtures of different molecules. The given text is identical in all the copies of the book. This is not accidental. If the text had been different from molecule to molecule, then we should not be able to speak about the biological role of the primary structure of a protein. The ability of the cell to print a large number of identical copies, to produce the corresponding text in large editions has a fundamental meaning.

How do we read the text? When we determine the amino acid composition of the protein, the entire type falls apart, and there remain only individual letters. What can we say about the text after counting the number of letters of each 20 types? We can only indicate the language in which the text had been written, but we cannot say who was the author; it remains absolutely unknown whether the book was written by Shakespeare or Dickens. In fact, the amino acid composition of a whole number of homologous proteins which perform identical functions in different organisms, is very similar. These proteins can differ only in the amino acid sequence.

The first major success in the determination of the primary structure of the protein was achieved by Sanger, who determined the structure of insulin. (Insulin is a protein hormone produced by the Langerhans' islets of the pancreatic gland; in the absence of insulin, there is a serious disturbance of metabolism, namely, diabetes.) The basic idea of Sanger's method is that chromatographic analysis makes possible the identification of short peptides, containing three to four amino acids, namely, syllables or words of the text. First of all, it is necessary to establish the number of lines, namely of polypeptide chains which make up the protein. In insulin, there are two such chains (Fig. 39). This can be established by determining the number of N— terminals of the chains. Then the lines are separated from each other and studied separately. This is done by oxidizing the —S—S— bonds which link together the peptide chains. Each line is broken down into syllables (or short words) by enzymatic hydrolysis. One of the proteolytic enzymes, trypsin, splits peptide bonds only next to arginine and lysine. The fragments are separated chromatographically. There are a number of chemical reactions which permit the end amino acids of the fragments to be determined. The protein is then split into other syllables with other enzymes, etc. The assembly of all the fragments formed in the several splittings makes it possible to establish the text, namely, the primary structure of the protein. Let us assume that, as a result of the first splitting, we have obtained a mixture of fragments which have been identified as

PRO, NSTRU, CTU, TEI, RE

and in the second splitting

PR, STRU, CT, OTEIN, URE

Simple combination of fragments, similar to the process used in the solution of crosswords, makes it possible to read the line as

PROTEINSTRUCTURE

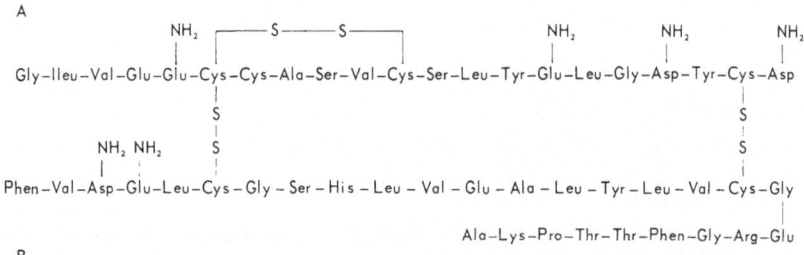

Fig. 39. The structure of insulin.

Obviously these chemical investigations are actually extremely difficult and require much hard work (a detailed description of the method is given in the monograph literature[15]). In recent years, a direct physical method has come to the aid of these chemical methods; this is the method of X-ray analysis (see Chapter 5). As a result of the chemical and physical investigations, at present we know the complete primary structure of a number of proteins, such as insulin, ribonuclease, the TMV protein, ACTH (a hormone secreted by the hypophysis gland). The X-ray method has been used for the investigation of myoglobin, hemoglobin, lysozyme, ribonuclease, etc. The primary structures of chymotrypsin and of 14 different cytochromes C, which belong to different species, have been studied. At present, the primary structures of five so-called "Bence-Jones" proteins, obtained from a single organism, have been determined. These last proteins are part of the composition of gamma globulin, and their investigation is quite important for the problem of the structure of individual antibodies. The structures of single fragments of a large number of proteins have been determined. Knowledge of protein structure is progressing very rapidly.

Let us pass to the next question, namely, to the meaning of the protein text. The amino acid sequence of a protein determines all its properties, its biological function. A classical, particularly well-studied example, is that of the relation between the primary structure and the biological function of hemoglobin.

In some parts of Africa and the Mediterranean countries, as well as in southeast Asian countries, there are serious hereditary diseases of the blood, namely anemias. In the case of the so-called sickle-cell anemia, the red blood cells have the shape of a sickle. In this case, the hemoglobin (S hemoglobin, while the normal hemoglobin is A hemoglobin) forms crystallike structures quite easily; its erythrocytes easily clot and are subject to hemolysis, i.e., to decomposition. The serious disturbances of blood circulation, which result from this disease, lead to death at an early age. Mediterranean anemia (C-hemoglobin) is also fatal. Its symptoms are a rapid decomposition of the erythrocytes, i.e., anemia, a compensating increased growth of the blood-forming tissue of bone marrow which results in skeletal deformations and the enlargement of the liver and pancreas. These diseases are inherited in a recessive manner in accordance with the first law of Mendel. In other words, the S and C hemoglobins are generated by allelic genes, which normally produce A hemoglobin. The normal gene is dominant. Therefore, the disease becomes evident only in homozygous individuals ss and tt, but not in heterozygous ones, As or At. During the crossing $As \times As$, the progeny will have a probability of being 0.25 healthy individuals AA, a probability of 0.5 healthy As, and a probability of 0.25 of sick ss individuals, who are doomed to an early death. An As blood contains both sickle and normal cells. Pauling, Itano, Singer, and Wells[16] have found that the electrophoretic mobilities of S and A

hemoglobins are different; this is due to a difference in the amino acid sequence, which results in a difference in the electric charges of these proteins. Pauling defined the hemoglobin diseases as molecular diseases. The molecular diseases lead to organic diseases. In fact, Ingram has shown that the difference between the anomalous and normal hemoglobins is due to the substitution of only one amino acid out of the entire peptide chain of the protein.[17] The meaning of the text is drastically changed when only one letter in one line is substituted.

Table 3 lists the singular replacements of amino acid residues in human hemoglobin, which result in pathological changes.

The question of the reason why these hereditary diseases have remained for hundreds of years in each region must be asked. Certainly homozygotes of *ss* and others must die before maturity. Yet in Africa, there are regions where the carriers of the *s* genes in a heterozygous form constitute 40% of the population. The probability of marriages between them is equal to $0.4 \times 0.4 = 16\%$, and the probability of the *ss* homozygote is $0.25 \times 16\% = 4\%$. It can be shown that, in order to reach such a high content of the *s* genes, it is necessary that the *As* heterozygote be 20 to 25% more viable than healthy *AA* individuals.[18] What is the reason?

These hereditary anemias are found in those parts of the earth where one of the principal causes of death is malaria. It turns out that the diseased hemoglobins are immune against malaria; the *Plasmodium* in the malaria

Table 3   Amino Acid Substitutions in Human Hemoglobin[18]

Residue No.	Substitution		Name of the hemoglobin	
	α-chain			
16	Lys	Asp	Hb *I*	
30	Glu	GluN	Hb *G*	(Honolulu)
57	Glu	Asp	Hb	(Norfolk)
58	His	Tyr	Hb *M*	(Boston)
68	AspN	Lys	Hb *G*	(Philadelphia)
116	Glu	Lys	Hb *O*	(Indonesia)
	β-chain			
6	Glu	Val	Hb *S*	
6	Glu	Lys	Hb *C*	
7	Glu	Gly	Hb *G*	(San José)
26	Glu	Lys	Hb *E*	
63	His	Tyr	Hb *M*	(Saskatoon)
63	His	Arg	Hb *M*	(Zürich)
67	Val	Glu	Hb *M*	(Milwaukee)
121	Glu	GluN	Hb D_β	(Punjab)
121	Glu	Lys	Hb *O*	(Arabia)

mosquito does not multiply in *As* erythrocytes. This seems to be related to a change in the conditions of interaction between the *Plasmodium* enzymes and hemoglobin.[18] As soon as serious antimalarial precautions will be taken in North Africa, Italy, Greece, and Southeastern Asia, the percentage of *As* heterozygotes will strongly decrease. We see that misprints in the protein text are inadmissible; they are extremely dangerous for the organism. A number of other hereditary molecular diseases are also known.[19]

The primary structure of proteins, which have the same biological function, remains quite similar in different species. Amino acid substitution in a chain of five mammalian animals is cited in Table 4.

Table 4 Amino Acids Numbers 8, 9, 10 in the A Chain of Insulin

Bovine	Ala	Ser	Val
Pig	Thr	Ser	Ileu
Horse	Thr	Gly	Ileu
Sheep	Ala	Gly	Val
Whale	Thr	Ser	Ileu

The sequences are found to be identical in such different animals as the pig and the whale. It is possible to follow, both in this case and in others, how the primary structure of the protein changes with evolution. Sorm[20] has shown that it is possible to establish definite laws of amino acid substitution. The general conclusion is that the primary structures of proteins are far from accidental. In fact, they are determined by a single pathway of development of life on earth, by a single biochemical evolution. Investigations of the changes of the primary structures of similar proteins during evolution are actually quite similar to textological studies. In studies of the histories of cultures and of societies, it is very interesting to follow the manner in which the text of some ancient literary work, for example, of the biographies of Plutarch, changed over centuries.

The next question is related to the manner in which the text is established. The cited examples indicate that this method has a genetic nature; it is dictated by genes (cistrons), which are contained within chromosomes. This is quite consistent with the facts described in Chapter 2.

Physical-chemical considerations show that the primary structure of the protein cannot be formed just by itself. Let us imagine that we have all the amino acids present in a solution, in a homogeneous medium. Even if all the most favorable conditions are present for their combination, the text formed will be meaningless; the chain which will be formed will consist of a disordered sequence of residues. A given amino acid does not understand exactly why it must combine in a given spot in the chain. After all, real chemical and intermolecular forces which decrease rapidly with distance are acting between these molecules. A given amino acid may feel one or possibly two neighboring amino acids, but it has no information whatever on the subsequent chain. A possible solution would have been the introduction into the system of large numbers of specific enzymes. It is well known that these are necessary for biosynthetic processes. However, if the combination of any two amino acids is modulated by a particular enzyme, then the synthesis of a single protein with a definite primary structure would necessitate a vast number of enzymes. But enzymes are proteins. Thus, in order to synthesize any given protein, there is a requirement for many others. How did these come about? Thus, we fall into a vicious circle.

Thus, direct synthesis of a protein in a homogeneous medium is impossible. It is necessary to have a device for printing the text which is similar to a typographical matrix. This device must be present in the chromosome. The logical analysis of biological factors on the basis of physical and chemical concepts has brought us to a quite important conclusion. The process of protein synthesis is described in detail in Chapter 7. At this point, let it suffice to point out that the role of the starting matrix is played by DNA molecules which constitute the gene.

On the other hand, once the primary structure of a protein has been determined, in principle, there are no obstacles for its synthesis in the laboratory by chemical methods. A number of methods of preparing synthetic polypeptides by polycondensation of amino acids are known. Recently, methods have been found for combining amino acids in a given sequence as a result of a gradual increase in the chain length. Thus, for example, synthetic corticotropin has been prepared; this material has the normal biological hormonal activity.[21] This tetracosapeptidamide contains 23 amino acids (including those which differ from natural ones). The formula of the compound is shown in Fig. 40. A number of synthetic polypeptides, which have been synthesized in a special way, are known to perform quite well certain biological functions of proteins, such as catalytic[22] and immunological.[23]

At the end of 1963, Katsoyannis, Tometsko, and Fukuda reported on the successful completion of the synthesis of an insulin similar to sheep insulin.[24] It is important that in this case it was possible to obtain a protein hormone which contained two polypeptide chains linked together

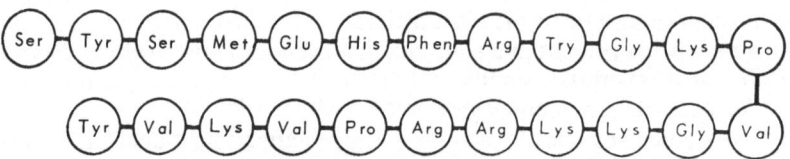

Fig. 40. Structure of synthetic adrenocorticotropin.

in a very definite manner (Fig. 39). Synthetic insulin has the same structure and the same properties as natural insulin.‡

NUCLEIC ACIDS

Nucleic acids were discovered much later than proteins. They were discovered by Miescher in 1872, who studied the chemical composition of salmon sperm. Just as proteins, nucleic acids are high polymeric, macromolecular compounds. Their chains, however, are composed of links of a completely different type. Compared with proteins, the main chain of a nucleic acid is quite uniform in structure; it consists of alternating links of phosphoric acid and a sugar, ribose in ribonucleic acid (RNA) and deoxyribose in deoxyribonucleic acid (DNA). Nitrogen bases are attached to the sugar; there are four kinds, both in DNA and in RNA. The general structural scheme of the chain has the form

$$\begin{array}{ccc} \text{Nitrogen base} & & \text{Nitrogen base} \\ | & & | \\ -\text{Sugar}-\text{Phosphate}-\text{Sugar}-\text{Phosphate} \end{array}$$

The structural formulas of D-ribose and D-2-deoxyribose are

$$\begin{array}{ccccc} \text{HOCH}_2 & & \text{O} & & \text{H} \\ & \text{C*H} & & \text{HC*} & \\ \text{HO} & \text{C*} & - & \text{C*} & \text{OH} \\ & | & & | & \\ & \text{OH} & & \text{OH} & \end{array}$$

$$\begin{array}{ccccc} \text{HOCH}_2 & & \text{O} & & \text{H} \\ & \text{C*H} & & \text{HC*} & \\ \text{HO} & \text{C*} & - & \text{C} & \text{OH} \\ & | & & | & \\ & \text{OH} & & \text{H} & \end{array}$$

Purines and pyrimidines are nitrogen-containing heterocyclic planar molecules; they are basic, since the nitrogen atoms can combine with protons and acquire a positive charge:

$$\text{>}N + HCl \longrightarrow \text{>}\overset{+}{N}-H + Cl^-$$

---

‡In 1966, Chinese workers reported on the complete synthesis of insulin with full biological activity; these workers have criticized the work of Katsoyannis *et al.*

Pyrimidine bases are derivatives of pyrimidine:

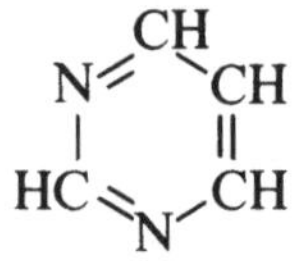

$$\begin{array}{c} \text{CH} \\ \text{N} \diagup \quad \diagdown \text{CH} \\ | \qquad \| \\ \text{HC} \diagdown \quad \diagup \text{CH} \\ \text{N} \end{array}$$

Pyrimidine

DNA contains

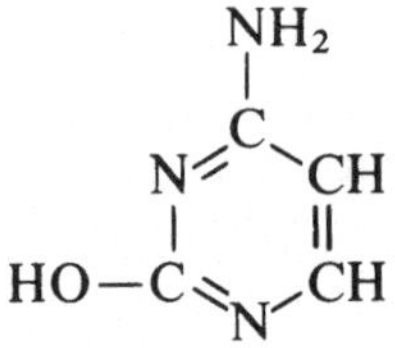

$$\begin{array}{c} \text{NH}_2 \\ | \\ \text{C} \\ \text{N} \diagup \quad \diagdown \text{CH} \\ | \qquad \| \\ \text{HO} - \text{C} \diagdown \quad \diagup \text{CH} \\ \text{N} \end{array} \qquad \begin{array}{c} \text{OH} \\ | \\ \text{C} \\ \text{N} \diagup \quad \diagdown \text{C} - \text{CH}_3 \\ | \qquad \| \\ \text{HO} - \text{C} \diagdown \quad \diagup \text{CH} \\ \text{N} \end{array}$$

Cytosine C 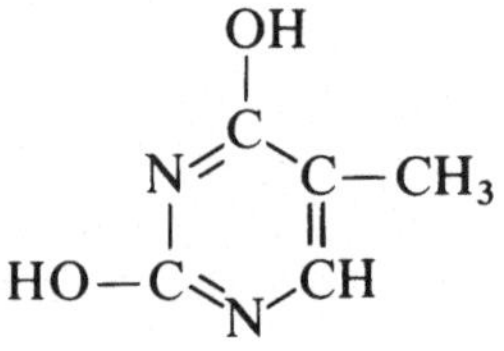   Thymine T

and in a few cases, small amounts of

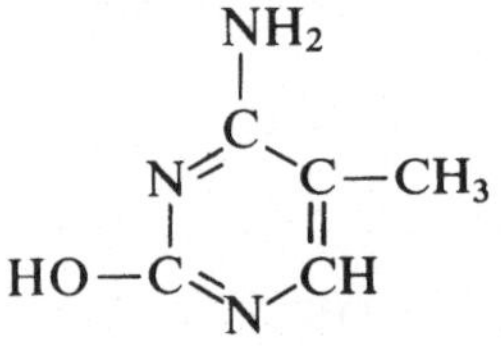

$$\begin{array}{c} \text{NH}_2 \\ | \\ \text{C} \\ \text{N} \diagup \quad \diagdown \text{C} - \text{CH}_3 \\ \| \qquad \| \\ \text{HO} - \text{C} \diagdown \quad \diagup \text{CH} \\ \text{N} \end{array} \qquad \begin{array}{c} \text{NH}_2 \\ | \\ \text{C} \\ \text{N} \diagup \quad \diagdown \text{C} - \text{CH}_2\text{OH} \\ \| \qquad \| \\ \text{HO} - \text{C} \diagdown \quad \diagup \text{CH} \\ \text{N} \end{array}$$

5-Methylcytosine 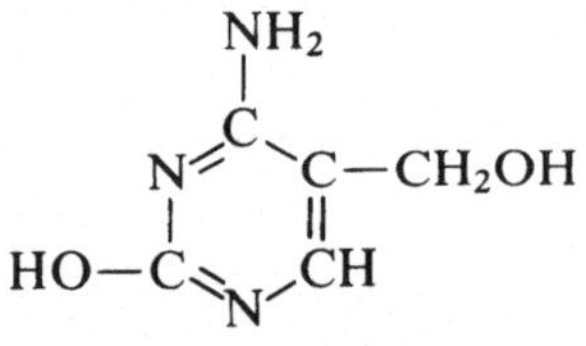   5-Hydroxymethylcytosine
                                (found in phages T2, T4, T6)

RNA contains C and uracil:

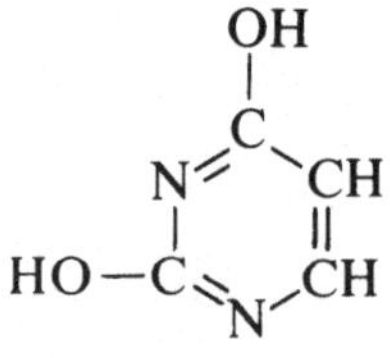

$$\begin{array}{c} \text{OH} \\ | \\ \text{C} \\ \text{N} \diagup \quad \diagdown \text{CH} \\ | \qquad \| \\ \text{HO} - \text{C} \diagdown \quad \diagup \text{CH} \\ \text{N} \end{array}$$

Uracil U

Purine bases contain two rings—a six-membered and a five-membered ring:

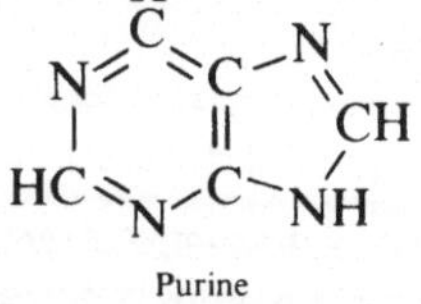

$$\begin{array}{c} \text{H} \\ \text{C} \\ \text{N} \diagup \quad \text{C} - \text{N} \\ | \qquad \| \qquad \diagdown \text{CH} \\ \text{HC} \diagdown \quad \text{C} \diagup \\ \text{N} \quad \text{NH} \end{array}$$

Purine

Both DNA and RNA contain

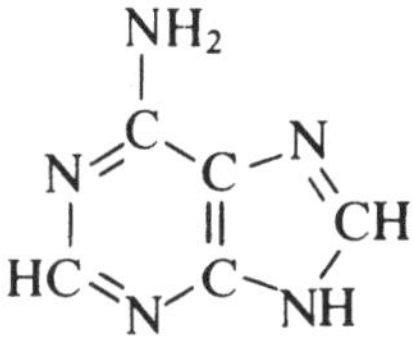

Adenine A 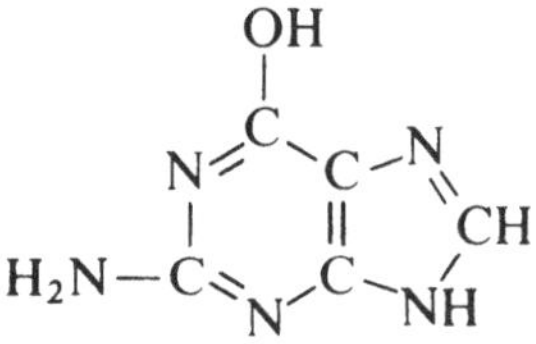   Guanine G

Let us give the formulas of three more purine bases which do not enter into the composition of natural nucleic acids, but are important in molecular biology:

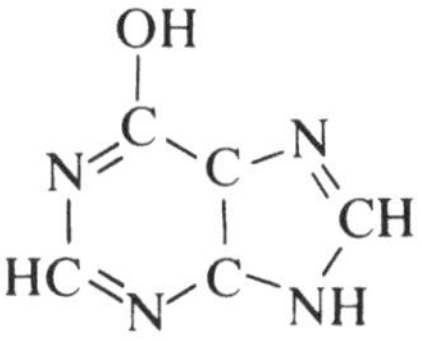

Hypoxanthine 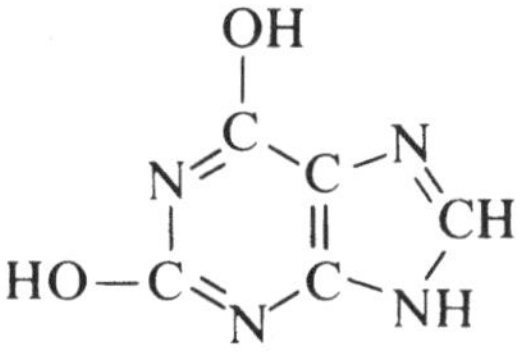   Xanthine 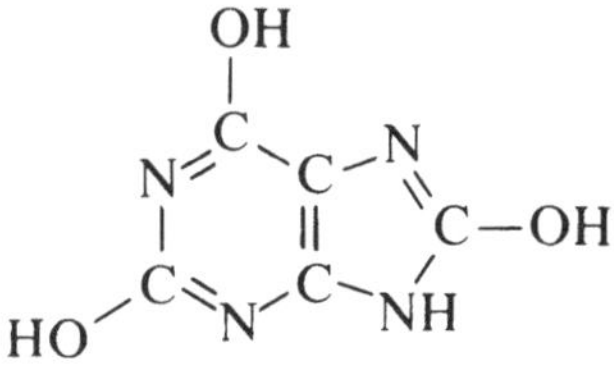   Uric acid

Most purines and pyrimidines may have tautomeric forms. Above, we have shown the enol forms which contain the OH group. Along with these, we may have keto forms

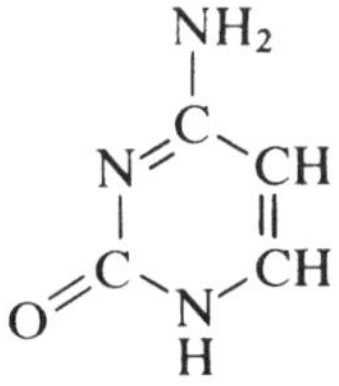

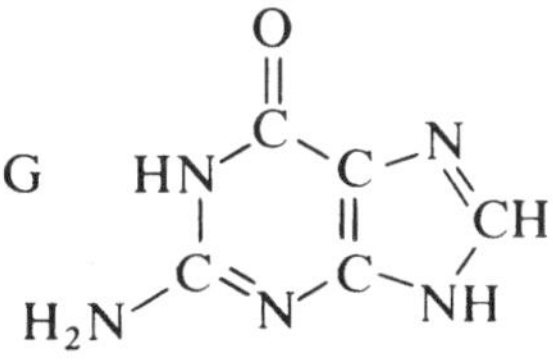

The keto forms of the nitrogen bases predominate in nucleic acid.

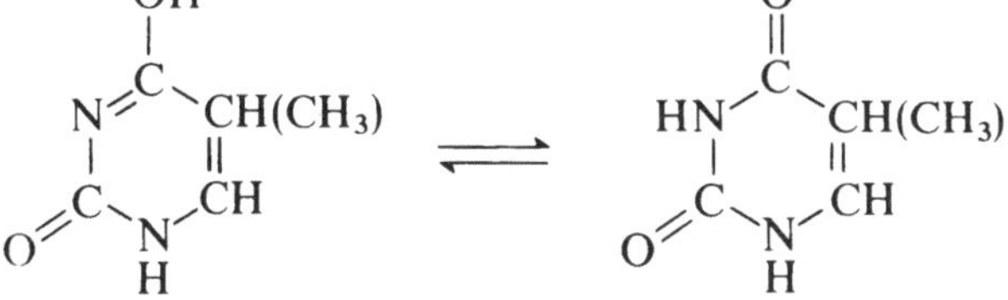

Compounds of the nitrogen bases with ribose and deoxyribose are called *nucleosides, ribonucleosides*, and *deoxyribonucleosides*, respectively. The nitrogen base is always attached to a definite carbon atom of the sugar, for example,

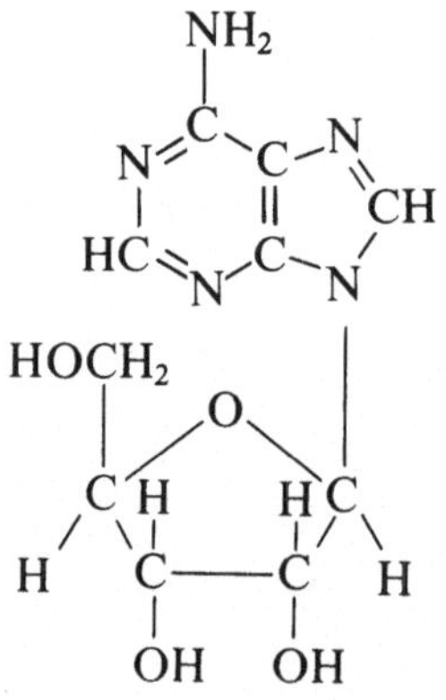

Adenine + Ribose = Adenosine

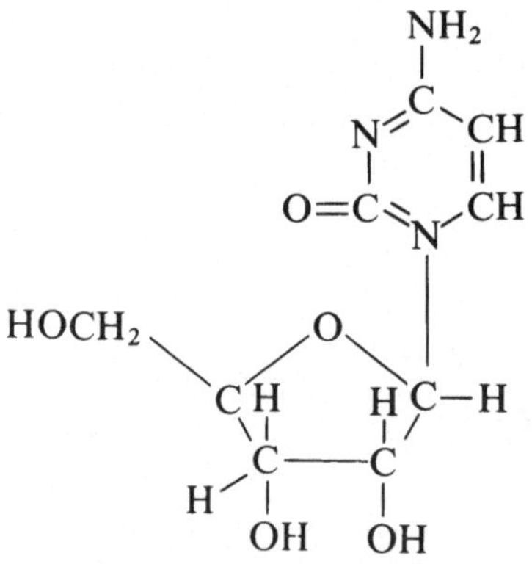

Cytosine + Ribose = Cytidine

Similar compounds of G and T are called *guanosine* and *thymidine*.

The phosphoric esters of nucleosides are called *nucleotides, ribonucleotides*, or *deoxyribonucleotides*. For example,

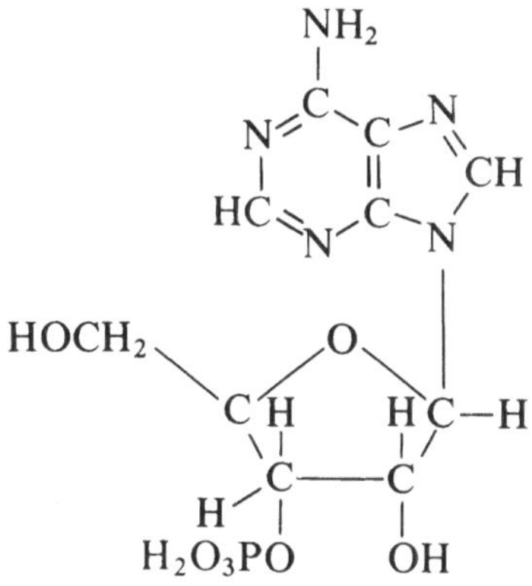

$$NH_2$$

Adenosine-3'-phosphate, adenosine monophosphate AMP

A nucleotide may be phosphorylated further by combining an additional one or two residues of phosphoric acid:

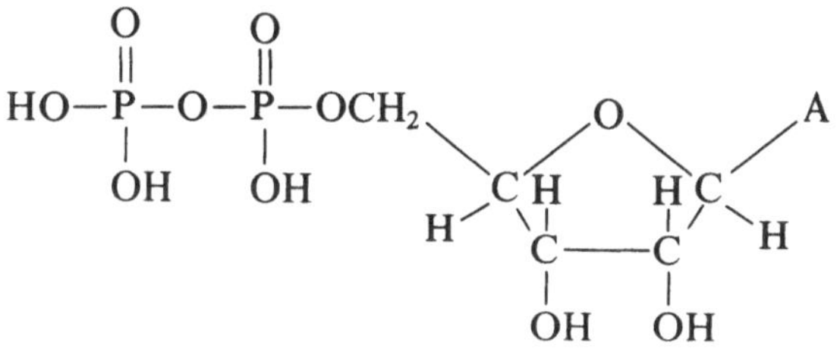

Adenosine diphosphate ADP

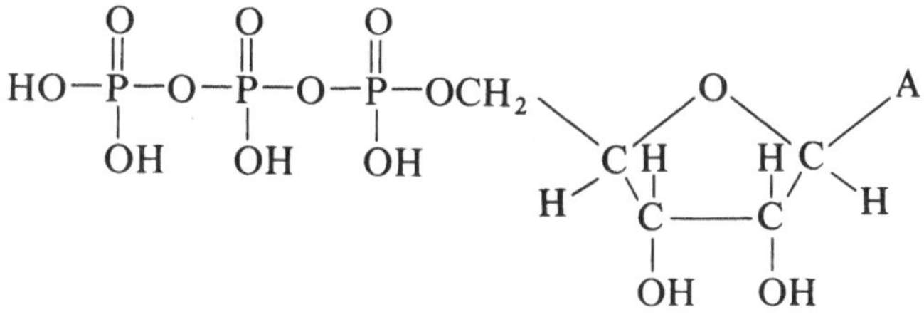

Adenosine triphosphate ATP

These compounds are extremely important in a number of biochemical processes.

Nucleic acids are polymers of nucleotides. The structures of RNA and DNA are shown in Figs. 41 and 42. In absolutely all organisms, from bacteriophages to man, the nucleic acid chains are built from the very same links. If we omit the rarely occurring methylcytosine and hydroxy-methylcytosine, DNA contains A, C, G, and T, while RNA contains A, C, G, and U. Nucleic acids are optically active; this could already be

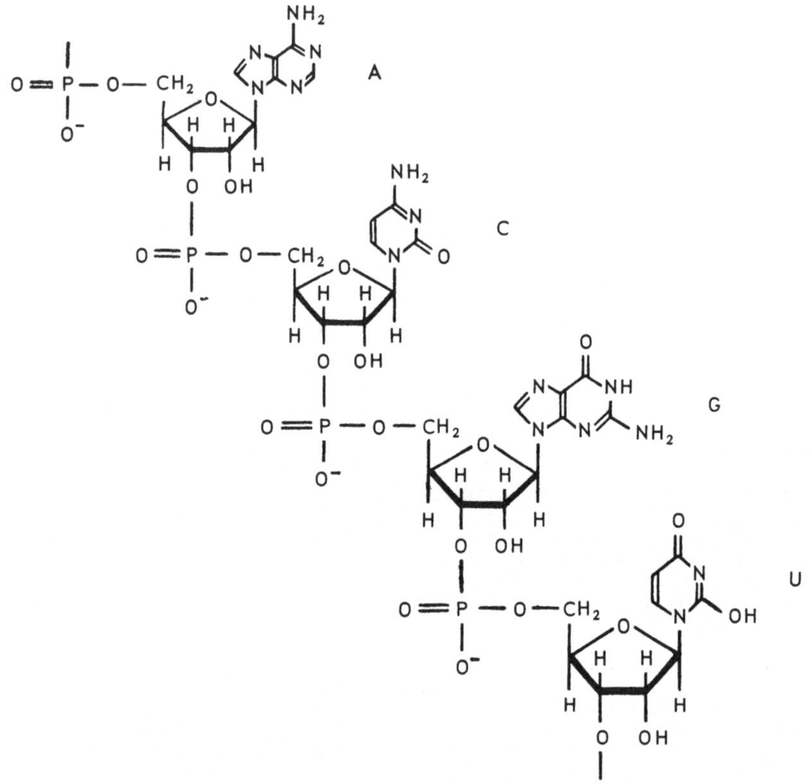

Fig. 41. Structure of RNA.

the result of the fact that both ribose and deoxyribose contain asymmetric carbon atoms and are present as pure antipodes.

The nucleotide composition of nucleic acids is determined by hydrolysis, which breaks the chain into nucleotides, and subsequent chromatography.[15,25] Since the nitrogen bases absorb light in the relatively near ultraviolet region (the absorption maximum is close to 260 mμ), the absorption spectrum may be used to detect nucleic acids and to identify products of their decomposition.

The optical properties of DNA and RNA will be discussed in Chapter 6. The scheme of DNA hydrolysis is DNA → nucleotides → nucleosides + phosphoric acid → A, T, G, C + deoxyribose.

DNA is contained in cellular nuclei and chromosomes, and it also enters into the composition of a number of viruses and phages. The presence of DNA in chromosomes can be detected directly, by microscopy in ultraviolet rays which are absorbed by DNA. It is noteworthy that the DNA contents in the haploidal sex cells is half of that in the diploidal

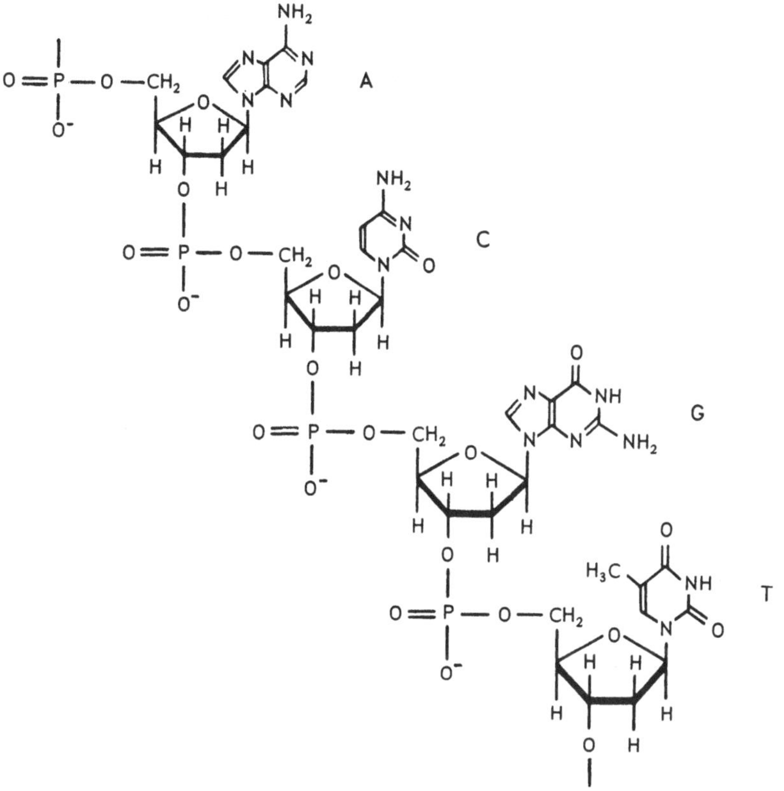

Fig. 42. Structure of DNA.

somatic cells of the same organism. DNA molecules are very large, with molecular weights of the order of 10^7 and greater (see Chapter 6).

Recently some DNA has been detected also in cytoplasmic organoids, namely, in the chloroplasts and mitochondria of the green algae *Chlamydomonas*. This is the DNA of the nonchromosomal genes. The bulk of the DNA, however, is contained in the chromosome, in the cellular nuclei, and we shall not discuss the nonchromosomal DNA in the further presentation.

Of the RNA, 90% is contained in the cytoplasm, and 10% in the cellular nucleus; it is also found in viruses. It has been established that there are four types of RNA: high-molecular-weight RNA which is contained in the ribosomes, with a molecular weight of the order of 2×10^6; transport or T-RNA in the cellular plasma, with a small molecular weight of the order of 25,000 to 30,000; matrix or messenger M-RNA with a large distribution of molecular weights from 30,000 to 50,000 up to 2×10^6, and greater—M-RNA migrates from the nucleus into the cytoplasm, and its relative contents in the cell are the lowest—; and finally, the high-molecular-weight viral RNA, which has a molecular weight of the order of 2×10^6, e.g., TMV-RNA.

The nucleotide composition of DNA follows a definite rule, namely, the rule of Chargaff. The molar concentration of A in DNA must always be equal to that of T, while that of G must always be equal to that of C. Consequently $(A + G)/(T + C) = 1$, that is, the sum of the purine bases in any DNA is equal to the sum of the pyrimidine bases. Furthermore, $(G + T)/(A + C) = 1$, that is, the amount of $6\text{-}NH_2$ group in the DNA bases is equal to that of the $6\text{-}C{=}O$ group. The rules of Chargaff are illustrated in Table 5.

Table 5   Molar ratios (expressed as percent) of Nucleotides in DNAs from Various Sources

Source of DNA	A	G	C	T	Methylcytosine
Bovine, thymus gland	28.2	21.5	21.2	27.8	1.3
Bovine, spleen	27.9	22.7	20.8	27.3	1.3
Bovine, sperm	28.7	22.2	20.7	27.2	1.3
Rat, bone marrow	28.6	21.4	20.4	28.4	1.1
Herring, testicle	27.9	19.5	21.5	28.2	2.8
Paraceutrotus lividus	32.8	17.7	17.3	32.1	1.1
Wheat, germ.................	27.3	22.7	16.8	27.1	6.0
Yeast......................	31.3	18.7	17.1	32.9	
Escherichia coli	26.0	24.9	25.2	23.9	
Mycobacterium tuberculosis	15.1	34.9	35.4	14.6	
Rickettsia prowazekii	35.7	17.1	15.4	31.8	

Belozerskii and Spirin[26] have examined the nucleotide composition of DNA from various forms of plant and animal organisms, and established that the specificity factor $(G + C)/(A + T) = G/A = C/T$ may vary in microorganisms within the broad limits of 0.45 up to 2.8. In bacteriophages, this ratio is close to 0.5. In higher plants and animals, the DNA composition is more or less constant, the specificity factor is equal to 0.43 to 0.45, that is, the contents of A + T are somewhat larger than those of G + C.[27,28] The DNA of phage φX174 and some others are exceptions and do not obey the laws of Chargaff. The molecular interpretation of these rules and the particular properties of these phages will be discussed later.

The total, i.e., essentially the ribosomal, RNA does not obey the rules of Chargaff. The specificity factor for the total RNA, $(G + C)/(A + U)$ in the case of microorganisms changes in a manner parallel to the specificity factor of DNA, within narrower limits, from 1.05 to 1.35.

The development of nucleic acid biochemistry has led to the discovery of a method for the synthesis of their analogs, namely, of polynucleotides. This will be discussed in Chapters 6 and 7.

As has already been pointed out, the biological role of nucleic acid is quite different from that of proteins. The DNA contained in chromosomes is the starting matrix for protein synthesis, it is the substance of the genes. The many and various proofs of this very important fact are summarized in Chapters 6 and 7. A similar role is played by the RNA of viruses. All the other RNAs are related in one form or another to the process of protein synthesis (see Chapter 7).

The primary structures of DNA and RNA are defined as the sequences of the nitrogen bases. These are texts written in a four-letter alphabet. The genetic information is written in the DNA; it is recoded during the synthesis of proteins, namely, the four-letter text is translated into a twenty-letter text. Although the nucleotide text has not been read yet in any case, the code problem has been essentially solved over the last few years. The determination of the primary structure of nucleic acids is much more difficult than the solution of the same problem for proteins, which also is not easy. The point is that the principal chains of DNA and RNA are completely uniform and monotonic; the various letters are attached as side chains. As a result, in this case, it is more difficult to make use of the specific splitting of the chain in definite locations with the formation of definite fragments. However, these difficulties must not be regarded as essentially insurmountable; the nucleotide text will certainly be read.

In 1965, Holley et al. deciphered for the first time the primary structure of one of the T-RNAs, namely, alanine T-RNA obtained from yeast.[29] The molecular weight of this material is 26,600; the molecule contains 77 nucleotides. The structure was determined essentially in the same manner as had been done for the primary structures of proteins. First, small fragments obtained during the complete degradation of RNA by pancreatic ribonuclease and ribonuclease T1 from Tacodistase were determined; then, the structures of progressively larger fragments were established. The success of this important work was due, in part, to the presence in T-RNA of several minor nucleotides, namely, inosine I, pseudouridine ψ, 1-methylguanosine, MeG, dimethylguanosine diMeG, and 5,6-dihydrouridine diHU. The minor bases give additional markers which permit the degradation to be followed. Later on, in Fig. 106, we shall show one of the possible secondary structures of T-RNA, deduced from the primary structure. At present the primary structure of several T-RNAs has been determined. Other chemical methods for the determination of the primary structure of nucleic acids have been examined and used to some extent by Kochetkov et al.[30]

SOME BIOCHEMICAL PROCESSES IN THE CELL

We have discussed the structures of the most important biological molecules, namely, proteins and nucleic acids; we have said nothing, however, about the manner in which they arose in the cell. How are amino acids and nucleotides formed, what is necessary for joining them into chains?

It is impossible to examine in detail here the biochemical reactions which are involved in carrying out these processes. Let us just characterize briefly the fundamental processes involved in the formation of biological molecules.

As we know, the equilibrium constant of a reaction, K, is related to the difference in the free energies, ΔF, of the products and reactants by

$$K = e^{-\Delta F/kT}, \tag{3.7}$$

or

$$\Delta F = -kT \ln K \tag{3.8}$$

with the condition that volume changes in the course of the reaction are negligibly small,

$$\Delta F = \Delta E - T\,\Delta S \tag{3.9}$$

where ΔE is the difference between the internal energies, and ΔS is the entropy difference.

It is obvious that the reaction will proceed in a given direction $1 \to 2$ and that the equilibrium will be displaced in the proper direction, K will be much greater than unity, if ΔF is negative and much larger than kT in absolute value. In other words, a reaction will proceed in the given direction only if there is a decrease in free energy. We may take a simple analogy from mechanics. In purely mechanical processes (neglecting friction), $\Delta S = 0$ and $\Delta F = \Delta E$. Mechanical motion proceeds spontaneously in the direction in which the energy decreases, water flows down and not up.

It should be pointed out that the Eqs. (1.18) to (1.22) are valid for a monomolecular reaction, in the course of which molecules of one type are transformed into molecules of another type in such manner that a single molecule only participates in the primary reaction, $A \leftrightharpoons B$. In the case of a bimolecular reaction (we shall discuss some in what follows) $A + B \leftrightharpoons C + D$, the rates of the forward and reverse reactions are expressed as

$$v_1 = n_A n_B A_1 e^{-E_a/kT} \qquad v_2 = n_C n_D A_2 e^{-(E_a + \Delta E)/kT} \tag{3.10}$$

and the equilibrium constant is defined from the condition $v_1 = v_2$, as

$$K = \frac{n_C n_D}{n_A n_B} = \frac{A_1}{A_2} e^{-\Delta E/kT} \tag{3.11}$$

Relation (3.8) and its consequences remain valid in this case as well.

The majority of reactions which are involved in chemistry actually do proceed with a decrease of free energy. Such reactions are known as exergonic. On the other hand, the principal biochemical processes are accompanied by an increase in free energy; these processes are endergonic. In exergonic reactions, free energy is released; in endergonic ones, it is absorbed. Let us give some simple examples.

The formation of the peptide bond during the combination of amino acids is an endergonic process

Leucine   Glycine   Leucylglycine

The reaction from left to right is accompanied by an increase of free energy of 3 kcal/mole (with molar concentrations). This means that the equilibrium constant is

$$K = e^{-3,000/RT} = e^{-6} \cong 0.01$$

i.e., the reaction will proceed principally not from the left to the right but from the right to the left. If we recall that cells always contain an excess of water, it becomes evident that a spontaneous synthesis of peptides is impossible. In order to carry it out, it is necessary to use work obtained from some other source. The work can be, for example, the removal of the formed water from the area of the reaction.

As another example, let us take the formation of sucrose by the polymerization of two carbohydrate molecules of smaller size,

Glucose   Fructose

Sucrose

In this case as well, the free energy increases (by 5 kcal/mole). The sucrose could be formed spontaneously in the juice of sugar cane or beets if this juice had an excess of glucose and fructose and a shortage of water. In actuality this is, of course, not so.

Sucrose is a disaccharide. In organisms, carbohydrates are represented both by mono- and disaccharides and by polysaccharides. The latter encompass plant starch and glycogen. These polysaccharides have as a role the storage of chemical energy. Part of the structure of the macromolecule glycogen is shown in Fig. 43.[1] In cells, carbohydrates are frequently found in the form of compounds with proteins and lipids. One of the macromolecular carbohydrates is the structural matter of plants, namely, cellulose, a chain fragment of which has the structure

Chitin, the substance which constitutes the external skeleton of insects and crustacea, is also a carbohydrate related to cellulose.

Thus, we see that extremely important biochemical reactions are endergonic. In order to carry them out, it is necessary to force water to flow upward. This is possible by inserting into the system a pump, which, in turn, is activated by the fall of water in another stream. In other words, it is necessary to have a source of free energy, a source of chemical work.

Cells and organisms are open systems. In open systems, it is possible to maintain nonequilibrium concentrations of chemical substances by means of conjugated reactions. Thus, the reaction of the formation of sucrose is carried out by conjugation with the hydrolysis of ATP,[31]

$$ATP + H_2O \rightarrow ADP + H_3PO_4 \qquad \Delta F \cong -12 \, kcal/mole$$

Conjugation takes place by the phosphorylation of glucose and the

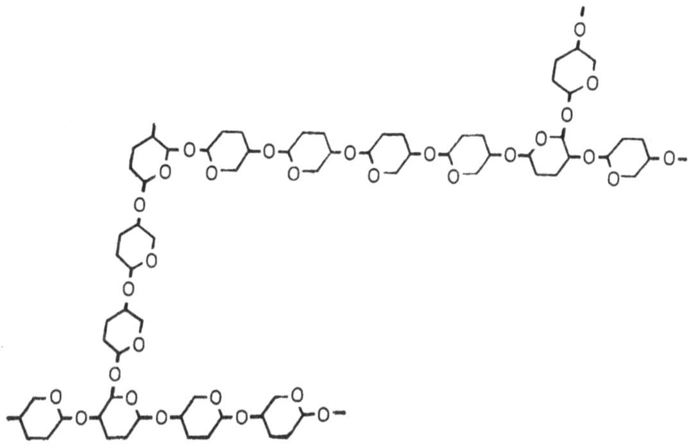

Fig. 43. Structure of glycogen.

subsequent reaction of glucose phosphate with fructose:

$$\text{ATP} + \text{Glucose} \longrightarrow \text{Glucose-1-H}_2\text{PO}_3 + \text{ADP}$$

$$\Delta F = -7 \text{ kcal/mole}$$

$$\text{Glucose-1-H}_2\text{PO}_3 + \text{Fructose} \longrightarrow \text{Saccharose} + \text{H}_3\text{PO}_4$$

$$\Delta F \cong 0 \text{ kcal/mole}$$

The total free-energy balance is -7 kcal/mole, i.e., the process is advantageous as a whole. This process occurs with the participation of catalysts, namely, enzymes. As a result, the formation of one ester link in the disaccharide molecule is accompanied by the loss of one ATP molecule, which degrades the ATP to ADP. In order for the reaction to continue within the cell, it is essential to regenerate ATP continuously; it must constantly be formed from ADP and phosphoric acid; this, in turn, requires 12 kcal/mole of free energy. In an automobile, the battery is charged at the expense of the work of the internal combustion engine, i.e., in the final analysis, at the expense of the chemical energy of the fuel. In living organisms, the charging of ADP takes place by means of the absorption of solar energy (in autotrophs) or of chemical energy obtained from food (in heterotrophs) and from respiration (in plants and animals).

Above we have cited the example of the synthesis of a disaccharide with the participation of ATP. This is a particular case. The role of adenosine triphosphate is universal; it occupies a key position.[32] ATP furnishes the energy for the chemical, mechanical, osmotic, electric, and optical work of the cell. The role of ATP in photosynthesis has been discovered relatively recently.[33] The charging process, i.e., the formation of ATP from ADP takes place in the chloroplasts of the plant cell. In heterotrophic organisms, the charging process takes place in the mitochondria as a result of the oxidative processes which take place during respiration by the organisms. In many cases, the fuel is glucose which is oxidized finally to carbon dioxide and water.

Multicellular heterotrophic organisms require a fuel and materials for growth and development, namely proteins, carbohydrates, and lipids. Another type of material required is vitamins, which have a different significance. Vitamins are involved in the synthesis of what are called the *prosthetic groups* and *coenzymes*, namely, substances which activate enzymes. Many fermentation reactions require as catalysts complexes of enzymes and coenzymes (see Chapter 8). Thus, vitamins are catalytic and not nutritional substances; therefore, they are required in small amounts.

Energy is released from nutritional substances in three phases.[32] In the first phase, macromolecular chains are split to single links; in the second phase, new substances are formed from these links—acetyl coenzyme A and α-ketoglutaric and oxaloacetic acids. In the third phase, oxidation and charging of ATP take place (Table 6). All these processes are carried out in a series of consecutive reactions; their detailed investigation is an extremely important achievement of biochemistry. It is necessary to accumulate a large amount of energy. Using the earlier analogy, this means going against the current to a great height. However, when a ship rises to a considerable height at a large dam, this is done by a series of locks. Similarly, biochemical reactions take place in steps; moreover, the free-energy change in each step is not very large.[34]

Let us now examine in greater detail the second and third phases, starting with glucose (for many bacteria, glucose and other hexoses may serve as the only source of energy and of organic raw material). The reactions of the second phase are

Glucose + ATP $\rightleftharpoons$ Glucose-6-phosphate + ADP

Glycose-6-phosphate $\rightleftharpoons$ Fructose-6-phosphate

Fructose-6-phosphate + ATP $\longrightarrow$ Fructose-1,6-diphosphate + ADP

Fructose diphosphate $\longrightarrow$ Glyceraldehyde-6-phosphate
$\searrow$
Dihydroacetone phosphate

Glyceraldehyde-3-phosphate + $2H_3PO_4$ + 2 DPN $\longrightarrow$
2-glycerol-1,3-diphosphate + 2 $DPNH_2$

2-Glycerol-1,3-diphosphate + 2 ADP $\longrightarrow$ 3-glycerol-3-phosphate + 2 ATP

2-Glycerol-3-phosphate $\longrightarrow$ 2-Enolphosphopyruvate

2-Enolphosphopyruvate + 2 ADP $\longrightarrow$ 2-Pyruvate + 2 ATP

2-Pyruvate + 2 $DPNH_2$ $\longrightarrow$ 2-Lactate + 2 DPN

As a result of this process, 2 moles of ADP are transformed to ATP for each mole of glucose; glucose is transformed into lactic acid $H_3C-CHOH-COOH$:

$$\text{Glucose} + 2\,\text{ADP} + 2H_3PO_4 \longrightarrow 2\text{-lactate} + 2\,\text{ATP}$$

Pyruvate is $H_3C-CO-COOH$, DPN is diphosphopyridine nucleotide, which plays the role of a coenzyme. In the course of the cited reactions it is reduced, while here it is reoxidized (Fig. 44). The process of the transformation of glucose, which contains six carbon atoms, into two molecules of lactic acid, which contain three carbon atoms each, is called glycolysis. The glycolysis process releases 49.7 kcal/mole of free energy. Of these, 24 kcal/mole are used up for charging ATP. It is evident that glycolysis occurs without the participation of oxygen, namely, it is anaerobic. Here the phosphorylation of ADP is not oxidative.

Table 6[32] Three Principal Phases of Energy Release from Nutritional Substances

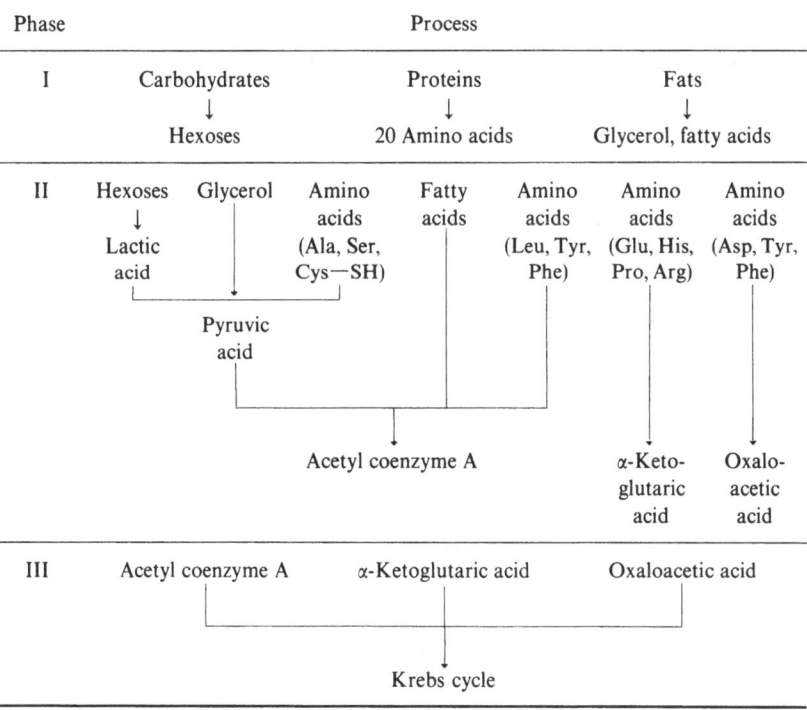

Phase	Process					
I	Carbohydrates		Proteins		Fats	
	↓		↓		↓	
	Hexoses		20 Amino acids		Glycerol, fatty acids	
II	Hexoses → Lactic acid	Glycerol	Amino acids (Ala, Ser, Cys—SH)	Fatty acids	Amino acids (Leu, Tyr, Phe)	Amino acids (Glu, His, Pro, Arg)	Amino acids (Asp, Tyr, Phe)
		Pyruvic acid				α-Keto-glutaric acid	Oxalo-acetic acid
		Acetyl coenzyme A					
III	Acetyl coenzyme A		α-Ketoglutaric acid		Oxaloacetic acid	
			Krebs cycle			

The further oxidation of the products of glycolysis down to CO_2 and H_2O, which is accompanied by the release of the major part of the free energy and by the simultaneous charging up of the major part of ATP, takes place in the three-carbon acid Krebs cycle.[31,32] This cycle is shown schematically in Fig. 45. Thus, the Krebs cycle is accompanied by the release of 2 moles of CO_2 and oxidation, i.e., the release of 8 atoms of hydrogen and, thus, of 8 electrons. The oxidation takes place at the expense of molecular oxygen which penetrates into the cell during respiration. The transfer of each pair of electrons to oxygen

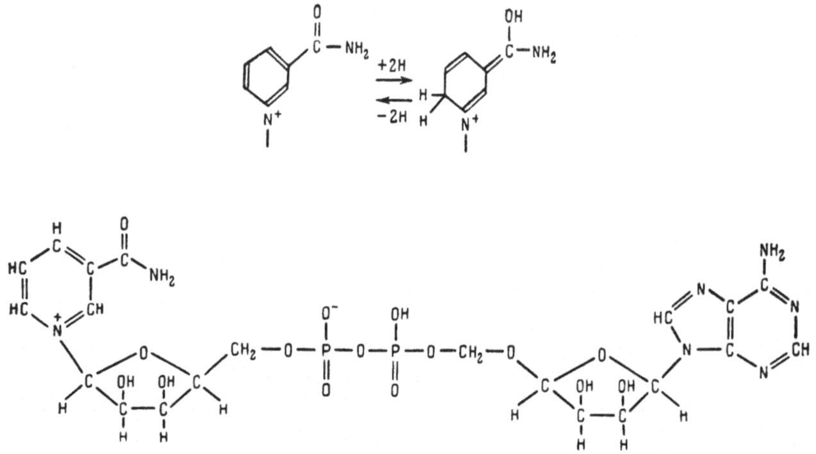

Fig. 44. Diphosphopyridine nucleotide. The scheme of the reduction and oxidation of the pyridine group is shown at the top.

is accompanied by the phosphorylation of 3 moles of ADP. The overall reaction of pyruvate oxidation is

$$CH_3-CO-COOH + 15H_3PO_4 + 15\,ADP + 5O \longrightarrow 3CO_2 + 15\,ATP + 2H_2O$$

or

$$CH_3-CO-COOH + 5O \longrightarrow 3CO_2 + 2H_2O \qquad \Delta F = -273\,\text{kcal/mole}$$

$$15\,ADP + 15H_3PO_4 \longrightarrow 15\,ATP \qquad \Delta F = 12 \times 15 = 180\,\text{kcal/mole}$$

Consequently, oxidation is accompanied by the gain of 273 kcal/mole of free energy, of which 180 kcal/mole are used for charging ATP. The efficiency of the process is 180/273 = 66%.

How does oxidative phosphorylation occur? It has been assumed by Belizer that the role of the carrier of electrons from reduced DPN to O_2 is played by enzymes of the

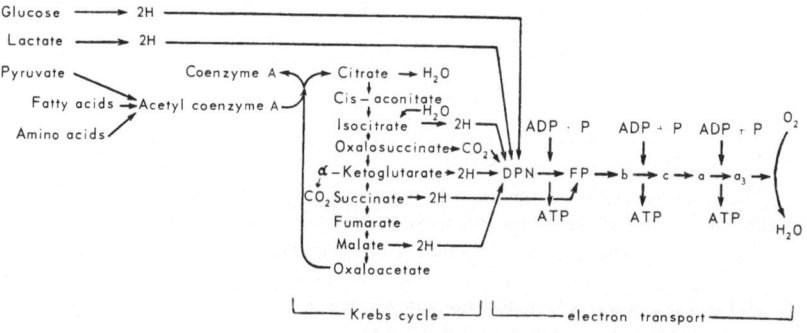

Fig. 45. Krebs cycle.

respiration chain, namely, flavoprotein (FP) and cytochromes b, c, a, and a_3. The overall reaction is

$$DPNH_2 + O + 3H_3PO_4 + 3ADP \longrightarrow DPN + H_2O + 3ATP$$

or

$$DPNH_2 + O \longrightarrow DPN + H_2O \qquad \Delta F = -55 \text{ kcal/mole}$$

$$3H_3PO_4 + 3ADP \longrightarrow 3ATP \qquad \Delta F = 12 \times 3 = 36 \text{ kcal/mole}$$

Flavoproteins are proteins which contain riboflavin, a yellow substance with an intense green fluorescence,

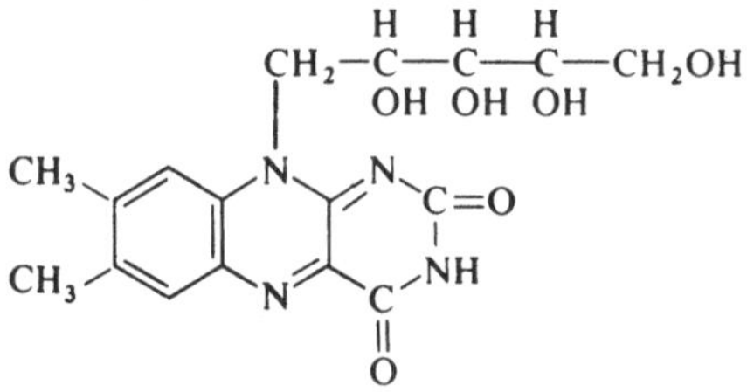

Cytochromes are proteins which contain a heme group, similar to that found in hemoglobin. The structure of bovine cytochrome c is shown in Fig. 46.

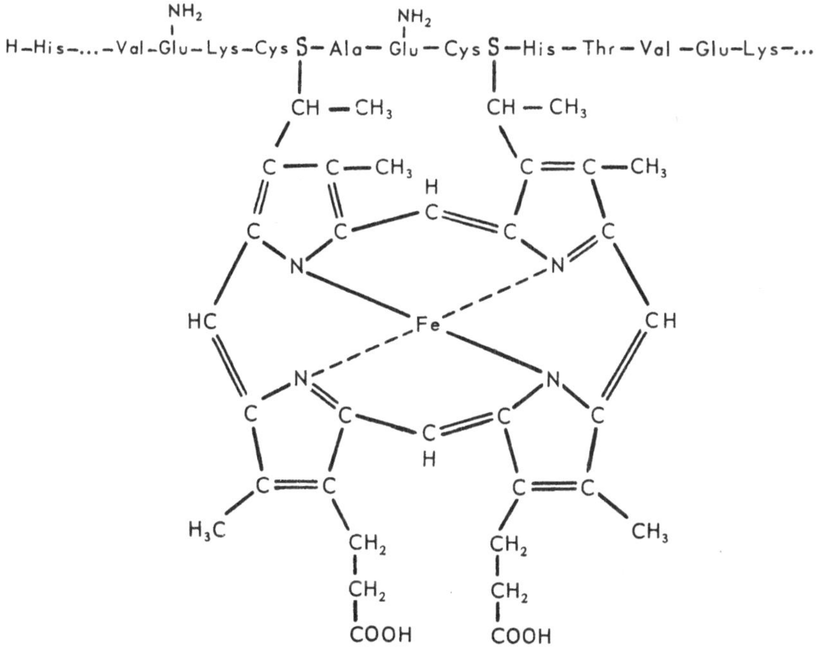

Fig. 46. Bovine cytochrome c.

To carry out the transfer of electrons, it is essential to have an iron atom in the heme group; this atom can become oxidized and reduced, changing from a bivalent to a trivalent state and *vice versa*: $Fe^{2+} \rightleftharpoons Fe^{3+}$ + electrons. Thus, respiratory oxidation is accompanied by phosphorylation.

The role of cytochromes may be illustrated by the following reaction scheme from one of the stages of the Krebs cycle, namely, the oxidation of oxalic acid to fumaric acid:

Succinate + $2Fe^{3+}$ (cytochrome b) $\longrightarrow$ Fumarate + $2Fe^{2+}$ (cytochrome b) + $2H^+$

$2Fe^{2+}$ (cytochrome b) + $2Fe^{3+}$ (cytochrome c) $\longrightarrow$

$$2Fe^{3+} \text{ (cytochrome } b) + 2Fe^{2+} \text{ (cytochrome } c)$$

$2Fe^{2+}$ (cytochrome c) + $2Fe^{3+}$ (cytochrome a) $\longrightarrow$

$$2Fe^{3+} \text{ (cytochrome } c) + 2Fe^{2+} \text{ (cytochrome } a)$$

$2Fe^{2+}$ (cytochrome a) + $2Fe^{3+}$ (cytochrome oxidase a_3) $\longrightarrow$

$$2Fe^{3+} \text{ (cytochrome } a) + 2Fe^{2+} \text{ (cytochrome oxidase } a_3)$$

$2Fe^{2+}$ (cytochrome oxidase a_3) + $\frac{1}{2}O_2$ $\longrightarrow$ $2Fe^{3+}$ (cytochrome oxidase a_3) + O^{--}

$$2H^+ + O^{--} \longrightarrow H_2O$$

The last reaction sums up the process. A series of cytochromes (oxidation–reduction enzymes) are used to transfer two electrons from an atom of oxygen to two atoms of hydrogen with the formation of water. All stages of the process occur with the participation of specific enzymes.

Urry and Eyring[35] have proposed an interesting theoretical model which explains the transport of electrons within the cytochrome chain during oxidative phosphorylation. This is called the model of the *imidazole pump*. The iron atom of the heme group in the cytochrome is linked to the imidazole ring of histidine in the protein. The Fe—His bond is directed perpendicularly to the plane of the porphyrin ring. Forty years ago, Chernyaev had already discovered the effect of the transinfluence in complex compounds of such a type. The negative ligand (in this case, the imidazole ring), attached to the metal atom, renders labile the ligand attached on the opposite side. This ensures the transport of electrons in the cytochrome chain, which are regularly located within the mitochondria. In the presence of an electron donor and an electron acceptor, phosphorylation takes place.

The conjugation of an oxidative process with phosphorylation is specific for biological systems. It is still not clear what its detailed nature is. Blyumenfel'd and Temkin proposed a hypothetical scheme for such a reaction.[36] It is known that A, ADP, and ATP can be reduced by electrochemical means. Such a reduction is accompanied by the capture of an electron and the formation of a free radical of the type of a semiquinone (Fig. 47). Concomitantly, the basisity of the amino group increases sharply; it is higher in the aliphatic amines than in aromatic amines (the dissociation

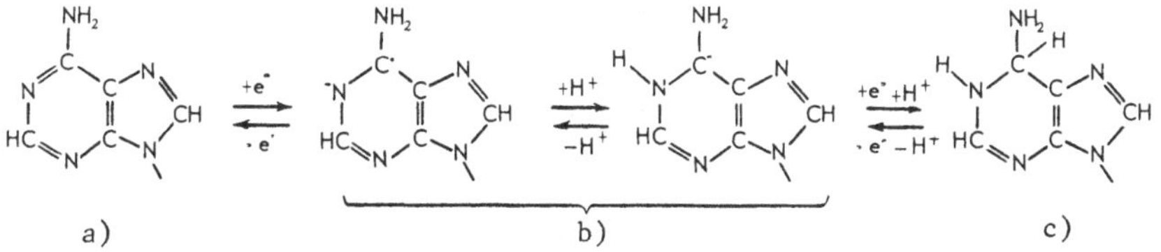

Fig. 47. (a) Oxidized form of adenine, weak base; (b) half-reduced form, strong base; (c) reduced form, strong base.

constant of aniline is 3.5×10^{-10}, that of methylamine is 5×10^{-4}). It is assumed that half-reduced ADP combines with phosphoric acid through an amino group with the formation of a stable ionic bond

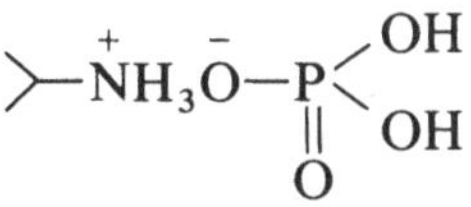

$$\rangle\!-\!\overset{+}{N}H_3\overset{-}{O}\!-\!P\!\underset{\underset{O}{\parallel}}{\overset{OH}{\diagdown}}_{OH}$$

Following the binding of the phosphoric acid to adenine, the third high-energy phosphate bond is formed by closing the ring (Fig. 48). During oxidation, the purine group returns to its initial state and the ionic bond is broken with the formation of ATP. Such a mechanism for charging the battery links the phosphorylation of ADP directly with the oxidation-reduction process in the adenine residue. This attractive and reasonable hypothesis will be confirmed if it is possible to prove the periodicity of the oxidation and reduction.

Further details on oxidative phosphorylation can be found in the specialized literature.[31,32] We have described here the purely biochemical aspect of the process, which at present can be regarded as classical. It had recently been found, however, that enzyme molecules, which participate in the respiratory cycle and are located within the mitochondrial membranes, carry out not only chemical but also osmotic and mechanical work. Mitochondria breathe; they contract and expand under the action of ATP and thus perform the operation for a regular supply and removal

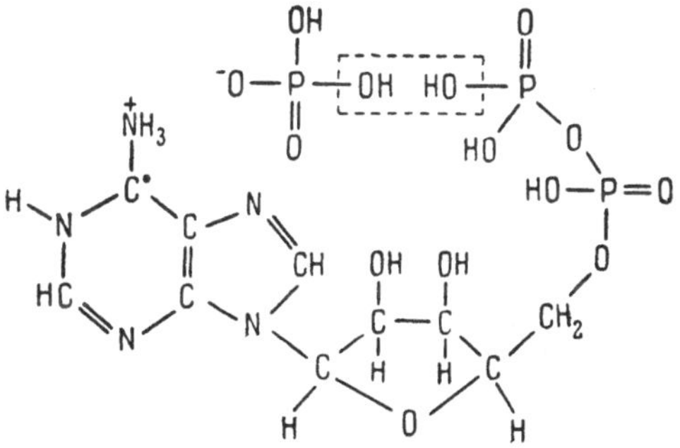

Fig. 48. The formation of ATP.

of water and small ions. There are data which point to the similarity of the mechanochemical processes in mitochondria with processes that determine muscle action,[37] described in Chapter 9.

As we shall see in Chapter 6, polynucleotides and nucleic acids are synthesized from nucleoside triphosphates, compounds similar to ATP and thus containing a reserve of energy. This means that nucleotides must become phosphorylated and activated for the formation of the most important biological molecules, namely, RNA and DNA. In these processes as well, ATP and other triphosphonucleotides play an important role.

We shall not discuss here the process of photosynthesis.[38] The structure of chlorophyll is shown in Fig. 49. Just as in cytochrome, here again there is a specific group of atoms in the center of which is located a single metal atom, in this case, magnesium. This group is called a porphyrin ring. Porphyrin compounds play an extremely important role in biology. Blood hemoglobin contains hemes, an iron-containing porphyrin ring similar to that found in cytochrome. The transport of oxygen in hemoglobin is carried out by attaching an oxygen molecule to the iron atom. In CO poisoning, the gas becomes bound to the iron atom and interferes with its function. The blood of a number of invertebrates contains porphyrin groups with other metal atoms. In centipedes the role of iron is assumed by copper. In the case of *Tunicata*, the tetrapyrrole porphyrin ring is replaced by a dipyrrole structure which binds vanadium. Pyrrole is a simple

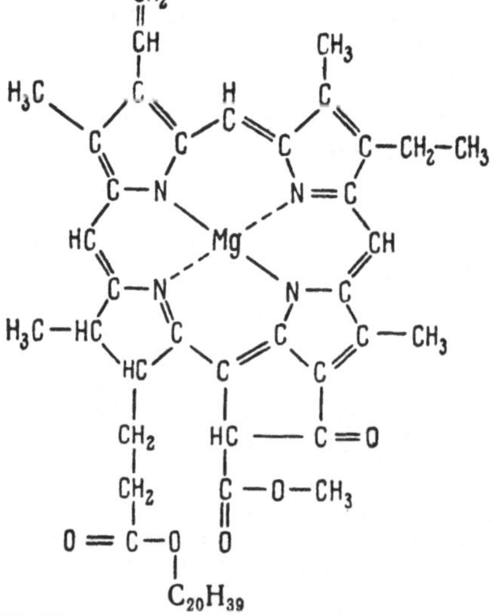

Fig. 49. Structure of chlorophyll.

heterocyclic compound

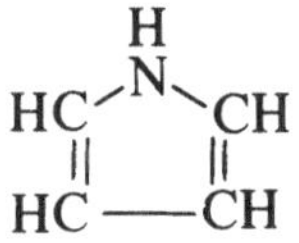

$$
\begin{array}{c}
\text{H} \\
\text{HC} \overset{\text{N}}{\diagup \diagdown} \text{CH} \\
\parallel \quad \parallel \\
\text{HC} \longrightarrow \text{CH}
\end{array}
$$

The general conclusion which must be made from this brief examination of the biochemical processes and biological molecules is the amazing unity of the chemical foundations of life in all organisms, both plant and animal.

Chapter 4

PHYSICS OF MACROMOLECULES

PHYSICAL PROPERTIES OF MACROMOLECULES

Proteins, nucleic acids, and polysaccharides (carbohydrates) are macromolecules; they are compounds with high molecular weights. To understand the nature of their biological function and become acquainted with ways in which they are studied, it is important to establish the general properties of macromolecules.

It is quite natural that the first polymers which were encountered in chemistry and physics were polymers of a biological nature. At this point, in addition to the principal biological macromolecules, we should also mention some compounds which are found in the sap of some plants, namely, rubber and gutta percha. In later years, synthetic chemistry developed methods for preparing a vast number of polymers, which today play an important role in technology and in every day life. Polymers include synthetic rubber, plastics, and fibers. The physics of polymers and macromolecules have been developed over the last 10 or 20 years, mostly with synthetic polymers in mind. The general laws which were later found in this field turn out to be applicable to biological molecules. In other words, synthetic macromolecules, which have a much simpler structure than that of proteins, DNA, and RNA, can serve as models for a number of properties of these biological substances. In this sense, the physics of biological macromolecules is a part of the physics of polymers. Polymer physics has two principal paths of development. The first of them leads to technology, the second to biology. Here we are interested in the second path, and we shall follow it henceforth.

The simplest synthetic polymers are hydrocarbons, such as

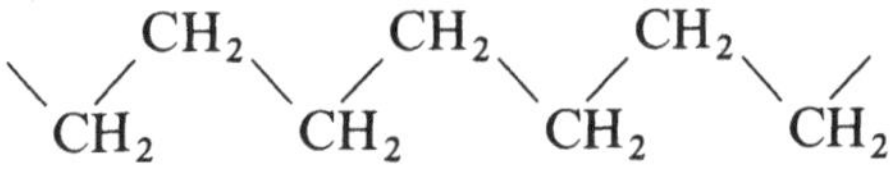

$$\ce{-CH2-CH2-CH2-CH2-CH2-CH2-CH2-CH2-}$$

Polyethylene

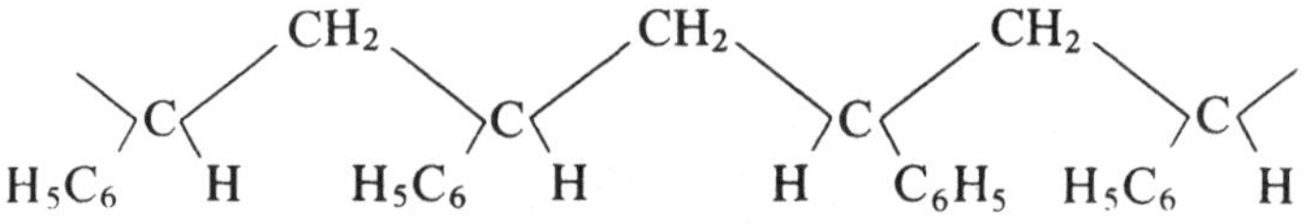

Polystyrene (atactic)

and

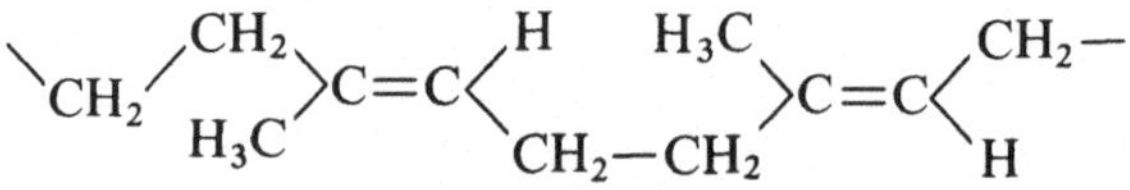

trans-Polyisoprene (gutta percha)

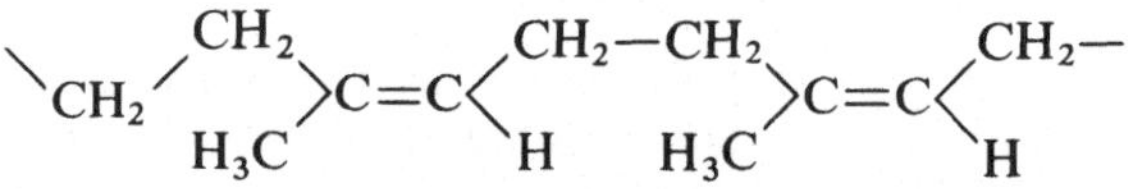

cis-polyisoprene (natural rubber)

These and other polymers are soluble in organic solvents, giving true and not colloidal solutions. The physical properties of isolated macromolecules must be studied in solutions of polymers, since there is no other way of obtaining a single macromolecule; polymers become decomposed before they vaporize.

First, it should be noted that, contrary to proteins, synthetic polymers consist of nonuniform macromolecules. The formulas of the chains which are shown are idealized in the sense that the real chains may also contain small amounts of other groups; some H atoms in polyethylene may be replaced by methyl, CH_3, etc. Linear chains may be branched in some positions. Any sample of a synthetic polymer contains a mixture of macromolecules with different chain lengths. Consequently, the molecular weight of a synthetic polymer is the average value over the entire group of polymer homologs. Polymer chains of the type of $(-CHR-CH_2-)_n$, for example, polystyrene, where $R = C_6H_5$, may have various configurations. Each R group may be oriented to the right or to the left relative to the principal chain, i.e., it may have two possible positions. If the number of monomer links in a chain is equal to 1,000, which is still not large, then the number of chains with different configurations is $2^{1,000}$; this number, consequently, is much greater than the number of macromolecules in the given sample. In other words, all the macromolecules in the sample will be different. In this sense, synthetic macromolecular substances do not obey one of the fundamental laws of chemistry, namely, the law of constancy of composition and structure. In this sense, nature works much more carefully than a chemist in the laboratory. All the molecules of a given protein have an absolutely determined structure and composition. In recent years, chemistry has started approaching nature. The Italian scientist, Natta, has found methods of preparing regular polymers, in which the configurations of the monomer units are absolutely determined. The structures of chains of isotactic and syndiotactic polystyrene are shown in Fig. 50. In the first case, all the phenyl groups C_6H_5 are located on one side of the principal chain; in the second case, they alternate to the right and to the left.

The specificity of the physical properties of macromolecules is determined by the large number of identical links in the chain. This applies to any macromolecules, both synthetic and biological. The amino acids in a protein chain differ from each other, but these differences refer to the side chains, while the principal links, $-C-NH-CO-$, are identical. Since the physical behavior of macromolecules is determined primarily by the chain structure, it is possible to develop a general physical theory of the behavior of macromolecules, which would permit simultaneously taking into account the nature of the atoms and the groups which form the chain. Such a theory is applicable to polyethylene and to proteins, to rubber and to nucleic acids. Thus, a study of macromolecules is not only useful, but actually essential for an understanding of the physical properties of biological polymers.

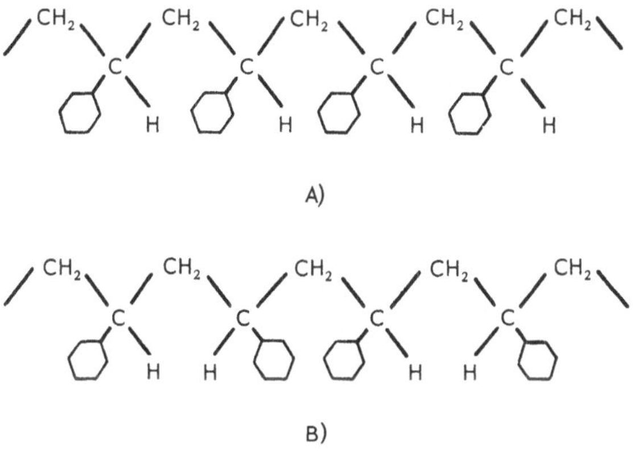

Fig. 50. Structure of isotactic (a) and syndiotactic (b) polystyrene.

A vivid expression of the specificity of macromolecules is the property of high elasticity; this is a property found only in polymers, namely, in natural and synthetic rubbers. At this point, we are talking no longer of individual macromolecules but of polymers in a mass. As we shall see, however, the high elasticity of rubber in the final account is determined by the structure and properties of its macromolecules. High elasticity is the ability to obtain high elastic deformations, attaining hundreds of percent, with a small modulus of elasticity. At reasonable deformations, rubber, just like other elastic bodies, obeys Hooke's law, namely, the deformation is proportional to the applied stress,

$$\sigma = \varepsilon \frac{L - L_0}{L_0} \tag{4.1}$$

Here σ is the stress, i.e., the force per unit of cross section of nondeformed samples, L is the length of the stretched sample, L_0 is the length of the unstretched sample, and ε is the modulus of elasticity. While for steel $\varepsilon = 20,000$ to $22,000 \text{ kg/mm}^2$, for rubber $\varepsilon = 0.02$ to 0.8 kg/mm^2, that is, it is smaller by a factor of 1 million. In this sense, rubber is similar to an ideal gas. When gas is compressed, an elastic force sets in, i.e., the gas starts to expand. Let us calculate the corresponding modulus of elasticity. The ideal gas obeys the equation of state

$$pV = RT \tag{4.2}$$

Let us compress a gas at a constant temperature by increasing the pressure by dp. The volume simultaneously decreases by dV. Differentiation of Eq. (4.2) gives

$$p \, dV + V \, dp = 0 \tag{4.3}$$

or

$$dp = -p \frac{dV}{V} = p \frac{L_0 - L}{L_0} \tag{4.4}$$

where L_0 is the initial position of the piston, L is the position of the piston after compression of the gas. Equation (4.4) is similar to (4.1); here the part of the modulus of elasticity, ε, is played by p. Atmospheric pressure corresponds to $\varepsilon = 1 \text{ kg/cm}^2 = 0.01 \text{ kg/mm}^2$, that is, a value of the same order of magnitude as in rubber. The similarity between the properties of rubber and an ideal gas are not limited just to this. During adiabatic stretching, rubber becomes heated just as an ideal gas becomes heated during adiabatic compression (an adiabatic process is a process which proceeds without the exchange of heat with the surrounding medium). The internal energy of the rubber, practically, does not depend on its stretching, just as the internal energy of an ideal gas is independent of its expansion or compression.

The elastic properties of the usual solid bodies have a somewhat different character. An elastic force arises during the deformation of a steel spring because the deformation results in a change in the interatomic distances and in a corresponding increase in the internal energy. Consequently, the elasticity of steel and of any other nonpolymeric material is energetic in nature. It is evident that the cause of high elasticity is something quite different. In order to understand this phenomenon, let us examine the principal thermodynamic relations which characterize the process of the stretching of rubber.

The work necessary for the isothermal deformation of rubber is equal to the change of its free energy at constant temperature,

$$dA = dF = dE - T\,dS \qquad (4.5)$$

On the other hand, the work of stretching is equal to the product of the stretching force and the elongation of the sample

$$dA = f\,dL \qquad (4.6)$$

Neglecting a very small change in the volume of rubber during deformation, we may write

$$f = \left(\frac{\partial F}{\partial L}\right)_T = \left(\frac{\partial E}{\partial L}\right)_T - T\left(\frac{\partial S}{\partial L}\right) \qquad (4.7)$$

$$\left(\frac{\partial E}{\partial L}\right)_T \cong 0 \qquad (4.8)$$

$$f = -T\left(\frac{\partial S}{\partial L}\right) \qquad (4.9)$$

In fact, the elastic force in rubber is proportional to the absolute temperature T, that is, Eq. (4.9) is valid. Thus, the appearance of the elastic force is determined by a change of the entropy, and not of the internal energy.

This is phenomenological thermodynamics. What is the concrete meaning of what has been said?

The same conclusion is valid for an ideal gas; its elasticity has an entropic nature. This means that, when the volume of an ideal gas is decreased, the number of collisions of molecules with the walls increases, i.e., the elastic force is related to the thermal motion of the molecule. Compression of a gas lowers its entropy, since the gas passes from a more probable dilute state to a less probable compressed state. It is evident that a similar process must also take place in rubber, i.e., rubber consists of a large number of elements which possess some independent thermal motion; the stretching of rubber means transition from a more probable distribution of these particles to a less probable one, i.e., to a decrease in

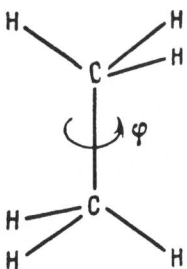

Fig. 51. Structure of the ethane molecule.

entropy. This is the only way in which the analogy between rubber and an ideal gas may be valid.

What are the independent elements of which we might speak in the case of a macromolecular substance, namely of rubber? After all, the atoms are bonded in a chain, while in elastic vulcanized rubber, the chains are linked with each other by separate cross links.

The macromolecules of natural and synthetic rubbers contain single C—C bonds (also C—O, C—N, etc.); free rotation can take place around these bonds. This phenomenon has been thoroughly studied for small molecules.[1] Let us examine an ethane molecule H_3C—CH_3 (Fig. 51). It has been shown from the thermodynamic and other properties of ethane, that the CH_3 groups in its molecules are capable of executing turns about the C—C axis. The internal rotation in ethane is not free; for a rotation of 120° it is necessary to pass over an energetic barrier of approximately 3 kcal/mole in height. The dependence of the energy of the molecule, U, on the angle of rotation in ethane may be expressed by an approximate equation

$$U = \frac{U_0}{2}(1 - \cos 3\varphi), \qquad (4.10)$$

where U_0 is the height of the barrier. The corresponding curve $U(\varphi)$ is shown in Fig. 52. The cause of the hindrance of internal rotation (U_0 is not zero) is due to the interaction of the electron shells of the nonbonded atoms (in this case of H atoms).

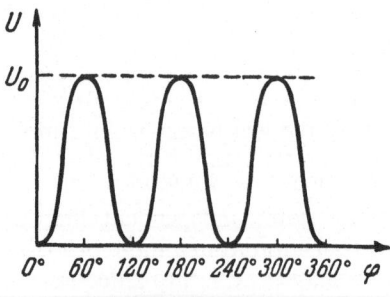

Fig. 52. Potential of the internal rotation for ethane.

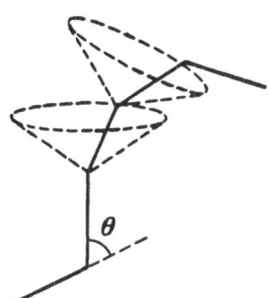

Fig. 53. Internal rotation in a macromolecular chain.

In a macromolecular chain which consists of single C—C bonds, internal rotation occurs in each link. Let us fix the position of the first two links (Fig. 53). The third link can have a number of positions on the surface of a cone with an aperture angle of 2θ, where $\pi - \theta$ is the fixed valence angle between C—C bonds. The position of the fourth bond relative to the first two is still less determined, since it is located on a cone described about each of the positions of the third bond, etc. As a result, a far removed bond is located in a completely random manner relative to the initial or first bond. Therefore, the macromolecule as a whole acquires a random conformation.‡ Since the rotations of the bond occur randomly, the macromolecules are constantly changing; they fluctuate. The macromolecules behave as a flexible string which accidentally has become coiled into a ball. The dimensions of such a ball may be described by the distance h from its beginning to its end. What is this distance?

Let us mentally decompose the macromolecule into segments, i.e., into regions the positions of which are no longer correlated with each other (Fig. 54). Let the number of such free conjugated segments be Z, and the length of each segment be b. It is obvious that the value of the vector

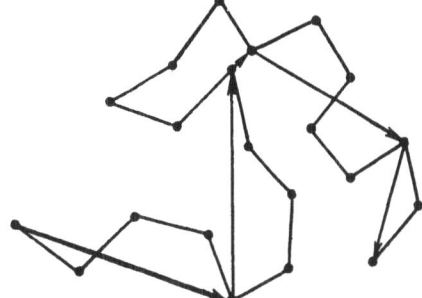

Fig. 54. Decomposition of a macromolecule into segments.

‡The term *conformations* of a molecule means its different forms which can be obtained one from another without breaking chemical bonds, as, for example, by internal rotation. On the other hand, in order to pass from one *configuration* to another, it is necessary to break such bonds (e.g., from a right to a left stereoisomer).

h, averaged over all conformations, is equal to zero, since all directions of the vector have equal probabilities.

This can be clarified by the following analogy. A stone-drunk man is wandering about a public square starting from some point. The length of his step is *b*, and he has made *Z* steps. The question is: Where did he stop, what is the average direction of his path? It is evident that the probable mean length of his path is equal to zero, since he steps with equal probability in any direction both forward and back. The average of the square of the length of his path from the initial to the final point, $\overline{h^2}$, will be different from zero. Solution of the problem of such a random walk (a problem very important in the theory of Brownian motion and in the theory of diffusion) gives, for $Z \gg 1$,

$$\overline{h^2} = Zb^2 \tag{4.11}$$

If the man had been sober and had walked all the time in a single direction, i.e., if the macromolecular chain had been stretched, the length of his path *h* would have been equal to *Zb*, and consequently,

$$h_{max}^2 = Z^2 b^2 \tag{4.12}$$

i.e., the coil is rolled $\sqrt{Z}$-fold. If, for example, the number of segments had been equal to 100, then the ball would have been one-tenth as long as the stretched chain.

The question which comes up is: What is the distribution of the probability of various values of *h*, which may vary from zero to the maximal length of a stretched chain, *Zb*? Once again it is the theory of diffusion which gives an answer to this question. The probability that the end of the chain is displaced from its beginning by a distance between *h* and *h* + *dh* is equal to

$$W(h)\,dh = \left(\frac{3}{2\pi Zb^2}\right)^{3/2} 4\pi h^2 e^{-3h^2/2Zb^2}\,dh \tag{4.13}$$

Relation (4.11) follows directly from (4.13). In fact

$$\overline{h^2} = \int_0^\infty h^2 W(h)\,dh = Zb^2 \tag{4.14}$$

In this manner, free macromolecules are coiled into a ball. Such a state is most probable, since a stretched conformation of the chain can be accomplished only in one way, while a coiled conformation may be accomplished in many ways. Investigations of the dimensions and shapes of macromolecules in solution have confirmed these theoretical results, which were first obtained by Kuhn.

Now, we can understand also the entropic nature of the high elasticity of rubber. When rubber is in a nonstretched state, the chains are coiled into statistical coils. When a coil is stretched, its entropy decreases since the probability decreases; thus, a stretched conformation of the macromolecule is less probable than a coiled conformation. The role of the independent elements, which are subjected to thermal motion, is played by the segments of a freely jointed chain.

The flexibility of macromolecules is their specific physical property. This is the general theory. Physical theory, however, must not only give a general explanation of a phenomenon, but also a quantitative interpretation and a prediction of experimental data. The statistical theory of macromolecular conformations must take into account the idiosyncrasies of their real structures. Therefore, it is not sufficient to describe the macromolecules in terms of freely jointed chains.

If the valent angles are fixed and internal rotation is not free, the mean square of the chain length is given by the following equation,[2] the equation of Oka :‡

$$\overline{h^2} = Nl^2 \frac{1 + \cos \theta}{1 - \cos \theta} \frac{1 + \eta}{1 - \eta} \tag{4.15}$$

where l is the length of a monomer unit and not of an arbitrary segment, N is the number of monomer units in the macromolecule, θ is the complementary angle to the valent angle, η is the mean cosine of the angle of internal rotation. When internal rotation is completely free, $\eta = 0$ and the chain is particularly flexible; on the other hand, in a stiff chain η is close to unity. If rotation is not free, then

$$\eta = \overline{\cos \varphi} = \int_0^{2\pi} e^{-U(\varphi)/kT} \cos \varphi \, d\varphi \left/ \int_0^{2\pi} e^{-U(\varphi)/kT} \, d\varphi \right. \tag{4.16}$$

It is evident that, in order to calculate η and in order to compare the theoretical value of $\overline{h^2}$ in Eq. (4.15) with experimental values, it is necessary to know the form of the function $U(\varphi)$. In the case of all real polymers, there are still large difficulties which have not been overcome so far in the finding of $U(\varphi)$. In real macromolecules, $U(\varphi)$ never coincides with (4.10). However, as we shall see below, this problem may be examined in a simplified theoretical way.

It must be emphasized that this discussion was valid only for a thermodynamically reversible behavior of macromolecules. In other words, we were not interested in the time of stretching of the macromolecule, nor in the speed of the deformation. The kinetic flexibility of a polymeric chain, i.e. its ability to change shape with time, is a function of the velocity of internal rotation about single bonds. Consequently, the kinetic

‡This equation is normally called the *Taylor equation*. However, I have recently learned that it was first obtained, not by Taylor (1947), but by the Japanese physicist Oka [*Proc. Phys.-Math. Soc., Japan*, **24**:657 (1942)].

properties of macromolecules are determined by energy barriers which the links must overcome during rotation.

ROTATIONAL-ISOMERIC THEORY OF MACROMOLECULES

When internal rotations of 120 and 240° occur in an ethane molecule, conformations similar to the initial one are obtained. This is understandable: The methyl groups which form the ethane molecule are symmetrical. The situation is quite different in less symmetric molecules, for example, in n-butane, $H_3C-H_2C-CH_2-CH_3$. In this case, C_2H_5 groups rotate about a central $C-C$ bond. As a result, the potential curve of internal rotation is no longer described by the simple Eq. (4.10). The $U(\varphi)$ curve for n-butane has the shape shown in Fig. 55. This curve contains energy minima of different depths at $\varphi = 0°$ and approximately 120 and 240°. Evidently, the n-butane molecules exist in states of minimum energy, i.e., they exist most of the time in conformations which correspond to rotation angles of 0, 120, and 240°. These conformations of n-butane are shown in Fig. 56. The rotational isomers of n-butane are different, contrary to the case found in ethane. The first isomer is a trans isomer and has a center of symmetry. The other two differ from the trans isomer and are related to each other as a right and a left hand. Rotational isomers may be detected by a number of physical methods, but they cannot be resolved chemically.

The heights of the internal rotation barriers are of the order of several kilocalories per mole. The rate of transition over a barrier (the rate of rotational isomerization) is expressed by the Arrhenius equation. It can be calculated that with such height of barriers, rotations in small molecules occur 10^{10} times per second. It is obvious that rotational isomers cannot be separated into individual samples; they are constantly changing into each other.

How can we detect, then, the existence of rotational isomers? If we have N molecules of n-butane, then at any moment, part of these will be in the trans form and part in folded forms. The fractions of these forms (the

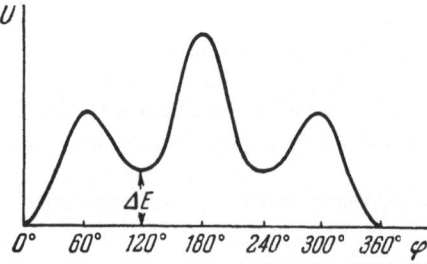

Fig. 55. Potential curve of internal rotation for n-butane.

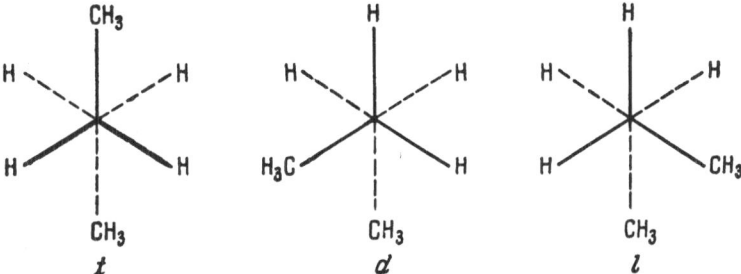

Fig. 56. Rotational isomers of *n*-butane. Projection on a plane perpendicular to the central C—C bond.

fractions of rotational isomers) are a function of temperature, in accordance with Eq. (1.14) and

$$N_t = N\frac{1}{1 + 2e^{-\Delta E/kT}} \qquad N_d = N_l = N\frac{e^{-\Delta E/kT}}{1 + 2e^{-\Delta E/kT}} \qquad (4.17)$$

and

$$\frac{N_d}{N_t} = \frac{N_l}{N_t} = e^{-\Delta E/kT} \qquad (4.18)$$

i.e., at the given temperature T, the fraction of coiled conformations will decrease with an increase of their relative energy ΔE. When $T \to \infty$, $N_t = N_d = N_l$; on the other hand, when $T \gg 0$, only trans isomers remain.

The existence of rotational isomers in such a dynamic system can be detected, for example, by infrared spectroscopy, which examines the frequencies of vibration of atoms within molecules. These frequencies are of an order of 10^{12} to 10^{13} vibrations per second; consequently, during the lifetime of a rotational isomer (10^{10} seconds), the molecule has time to vibrate 100 to 1,000 times, that is, it has time to emit its infrared spectrum. The vibrational spectra of rotational isomers differ from each other. Therefore, the spectrum of *n*-butane must be the superposition of the spectra of its rotational isomers. The ratio of the intensities of the spectra is described by Eq. (4.18), and the energy difference between the rotational isomers, ΔE,[1,2] can be determined from the change of this ratio with an increase or decrease in temperature. For *n*-butane, $\Delta E = 800$ cal/mole.

Let us turn now to macromolecules. A polyethylene chain is built in the same way as a molecule of *n*-butane. Rotations in this case are possible about each of the single C—C bonds, in each link of the chain. It is natural to suppose that, as a result of such rotations, discrete conformations will arise, i.e., rotational isomers similar to those of *n*-butane. In fact, the rotational isomers of polyethylene are quite similar to those of butane relative to each bond, with the one difference that instead of the CH_3

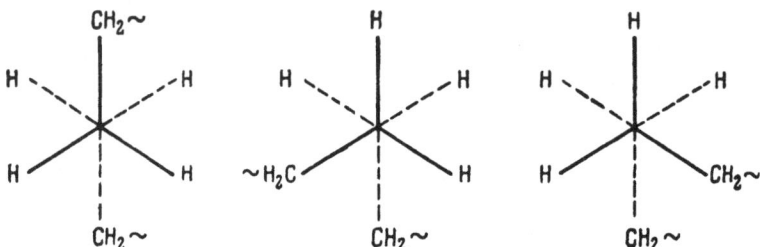

Fig. 57. Rotational isomers of polyethylene.

groups of butane, $CH_2\sim$ groups are involved in polyethylene ($\sim$ indicates the extension of the chain) (Fig. 57).

In this way, it is possible to regard a macromolecule in a state of thermodynamic equilibrium as a mixture of rotational isomers.[3,2] The fractions of rotational isomers in such a mixture, i.e., the fractions of links which exist in the corresponding discrete conformations, are expressed by Eq. (4.17), where $N = N_t + N_d + N_l$, that is, the total number of links (the degree of polymerization). In a macromolecule, the links exist, not in a state of disordered rotations, but distributed in discrete positions. In our examination, we quantize the macromolecule, by replacing a continuous curve of internal rotation with several discrete positions, which correspond to various values of the energy (Fig. 58). Similarly, in the calculation of $\eta = \cos \varphi$, integration of (4.16) is replaced with summation over the rotational isomers:

$$\eta = \frac{\sum\limits_{i=1}^{m} g_i \cos \varphi_i}{\sum\limits_{i=1}^{m} g_i} \qquad (4.19)$$

Here g_i is the statistical weight of the rotational isomers

$$g_i = e^{E_i/kT} \qquad (4.20)$$

and E_i is the energy of the ith rotational isomer, corresponding to the angle φ_i. We have a total of m rotational isomers for each link. In the case of the

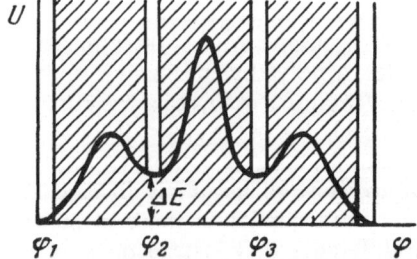

Fig. 58. Replacement of the continuous curve of internal rotation by discrete minima.

rotational isomers of polyethylene (Fig. 57), $m = 3$, and we obtain

Trans isomer: $\quad E_1 = 0 \qquad \varphi_1 = 0° \qquad \cos \varphi_1 = 1$

Right-folded: $\quad E_2 = \Delta E \qquad \varphi_2 = 120° \qquad \cos \varphi_2 = -0.5$

Left-folded: $\quad E_3 = \Delta E \qquad \varphi_3 = 240° \qquad \cos \varphi_3 = -0.5$

Introducing these values into (4.19), we find

$$\eta = \frac{1 - e^{-\Delta E/kT}}{1 + 2e^{-\Delta E/kT}} \tag{4.21}$$

When $\Delta E = 800$ cal/mole and $T = 0°C$, we find $\eta = 0.53$. The mean square of the chain length calculated according to Eq. (4.15) is equal to $\overline{h^2} = Nl^2 \times 1.63$ ($\pi - \theta = 109°28'$, the tetrahedral valence angles; $\cos \theta = -\frac{1}{3}$).

This result, however, is consistent with experiments only in order of magnitude. The reasons for lack of agreement will be examined below.

The thermodynamic flexibility of a macromolecule, i.e., the degree of coiling of the statistical coil, is greater, the smaller η is, i.e., the smaller ΔE is. If $\Delta E = 0$, $\eta = 0$, i.e., the macromolecule behaves as a chain with free internal rotation.

The rotational-isomeric theory permits the difficult integration (4.13) to be avoided and the conformations of the macromolecules to be represented in a clear and simple way. This is accompanied by the following questions: Is the rotational isomeric theory just some mathematical simplification? Are there really rotational isomers in macromolecules? Is it possible that the $U(\varphi)$ curve does not have strongly expressed minima, which would mean the presence of continuous rotations rather than discrete conformations of the links of the macromolecule? These questions can be answered only by experiment. The corresponding experimental investigations were carried out by infrared spectroscopy. It was found that, in fact, polyethylene has rotational isomers; ΔE is about 500 to 800 cal/mole, just as in n-butane.[4] Similar results were obtained for other polymers as well. The most direct evidence for rotational isomerism of macromolecules was given by studies of their elasticity.

Thus, the existence of rotational isomerism in macromolecules has been demonstrated. One may ask, however: Is the rotational isomerism, in fact, responsible for the flexibility of polymer chains, for the fact that the macromolecule can coil in a solution? There is also another possibility, related to the torsional oscillations of the links.

Let us assume that each link corresponds to only a single, for example, trans, isomer. This does not mean necessarily that the chain will be stiff and will not be able to coil. The thermal motion of the links, expressed through their torsional oscillations about single bonds, even though the amplitude may be small, will result in the coiling of a sufficiently long chain and in the absence of rotational isomerism. Theoretical calculations, however, of the dimensions of the coils and of other properties of macromolecules have shown that the hypothesis on the determining role of torsional oscillations is not consistent with experimental results. On the other hand, the rotational isomeric model is completely borne out.

Another very important problem is the determination of the true values of the angles φ_i, at which the minima of the potential energy $U(\varphi)$ occur, that is, the determination of the true conformation of the chain. We have examined polyethylene by analogy with n-butane, setting φ_i equal to 0, 120, and 240°. It is very important to find these values from an independent experiment, following experimental and not speculative data on the conformation of macromolecules. As we shall see below, it is possible to obtain these data by studying crystalline polymers.

Let us examine now the way in which the rotational isomeric theory explains a particular property of polymers, namely, the high elasticity of rubber.

When a macromolecule is stretched, the chain passes from a more coiled to a less coiled state. This means that a change in the set of conformations of the chain occurs as the result of the exertion of an external force. Such a change is carried out by internal rotation, by rotational isomerization. The maximally stretched state of the polyethylene chain is obtained when all the links are trans isomers (Fig. 59); the coiled state is given by a mixture of t-, d-, and l-rotational isomers. This can be simply explained with the aid of a unidimensional model of the macromolecule. Let us assume that each link of the chain is expressed by an arrow of length l, which may be turned to the right or to the left. One rotational isomer is described by two neighboring arrows turned in the same direction; let this be the trans isomer. A second rotational isomer is expressed by two adjacent arrows which are turned in opposite directions. The total length of the chain is expressed by the vector sum of all the arrows. In Fig. 60, a chain consisting of 10 links is shown. Of these, 5 are directed to the right and 5 to the left; consequently, the total length of the chain is $h = 0$. The numbers of rotational isomers are 5 of the first type, indicated by the letter t, and 5 of the second type, indicated by the letter s (in order for the total number of rotational isomers to be equal to the number of links, i.e., 10, we add mentally still another "minus first" link, indicated by a dotted

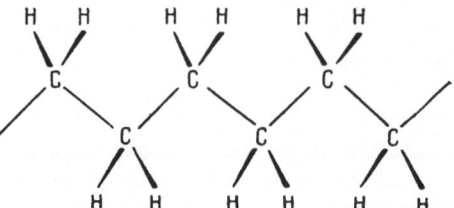

Fig. 59. Trans chain of polyethylene.

line). Let us stretch such a chain, i.e., let us exert on our links a force directed from the left to the right. Figure 60b depicts one of the stretched (but not the most stretched) states of the chain. Here the number of arrows turned to the right is 7, and those turned to the left is 3; consequently, the total length of the chain is $h = (7 - 3)l = 4l$. As far as the rotational isomers are concerned, their fractions have remained as before, namely, 5 of type t and 5 of type s. The chain has been elongated simply as a result of a redistribution of the rotational isomers, without changing their relative contents. Initially, the sequence of rotational isomers was *ttsststsst*; at the end, it was *tttsstssts*. Evidently such a change is not accompanied by a change in energy; it remains constant and equal to $5E_t + 5E_s$. This is evidently accompanied by an entropy change; it decreases as the ordering in the distribution of t and s links changes.

It is evident that it is not possible to attain complete extension of the chain by simple redistribution of t and s isomers. In our model, a completely extended state of the chain is given by a set of only t isomers (Fig. 60c and Fig. 59). When the macromolecule is extended, the redistribution of rotational isomers is accompanied by a transformation of the coiled isomers into the more extended trans isomers, i.e., transformations

s-isomer $\longrightarrow$ t-isomer

Such a transformation is accompanied not only by a change in entropy, but also by a change in the internal energy, since $10E_t \neq 5E_t + 5E_s$.

We have come to quite interesting and not at all trivial conclusions from the examination of a highly simplified model. The process of the stretching of the polymer consists in its rotational isomerization. The mechanism of high elasticity can be reduced to this process; thus, high elasticity receives a clear molecular interpretation. Moreover, the reverse resilient force in rubber must be not purely entropic, but partly energetic when the rotational isomers have different energies.[2,5] It is clear that these deductions from the theory are subject to experimental verification.

An investigation of the stretching of polymers by infrared spectroscopy has shown, in fact, that the relative contents of rotational isomers change during stretching.[2,6,7] In other words, stretching does mean a forced change of the conformation of the macromolecule.

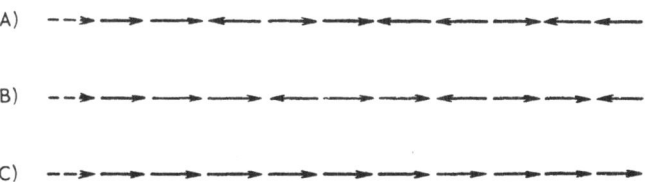

Fig. 60. Unidimensional model of the stretching of a macromolecule; (*a*) unstretched chain, (*b*) extended chain, (*c*) completely extended chain.

At the same time, a careful examination of the thermomechanical properties of rubberlike polymers, which has been carried out by Flory and coworkers on the basis of the theory just presented,[9,10] has shown that the purely entropic character of high elasticity, which has been discussed above, is a property of far from all polymers and that there is also an energetic force of resilience. The experimental data turned out to be not only in qualitative but also in quantitative agreement with theory. An energetic force of resilience, $f = \partial E/\partial L$, was determined during an examination of the thermomechanical properties of high elastic polyethylene; from this force, it was possible to calculate the energy difference of the rotational isomers, ΔE; this value coincided with that found in experiments.

In this manner, the rotational-isomeric theory of macromolecules was fully verified by experimental means. With this theory, it has been possible to calculate the physical characteristics of a macromolecule, namely its shape and dimensions, dipole moment and optical polarization, in good agreement with experiments.

So far, we have spoken of synthetic polymers and, in particular, of polyethylene. All that has been said, however, has a general significance. To understand the structure and physical properties of biological macromolecules, namely, of proteins, nucleic acids, and carbohydrates, it is necessary to invoke the same conformational principles, the same rotational isomeric theory.

MACROMOLECULES— COOPERATIVE SYSTEMS

A simplified qualitative description of the rotational-isomeric theory has been given above. To render a theory quantitative, and so result in quantitative agreement with experiments, it is necessary to take into account important properties of macromolecules which have not been discussed before.

In n-butane, individual molecules are linked only through intermolecular forces, which in general are much weaker than intramolecular forces. Both in the gas and in the liquid, butane molecules are essentially independent of each other and transformations of their rotational isomers take place essentially independently of each other. On the other hand, in a macromolecular chain, all the links are bonded to each other. As a result, rotations in a given link are not independent of the conformation of neighboring bonds. For example, the conformation of three consecutive polyethylene links shown on Fig. 61 is absolutely impossible, since in it the hydrogen atoms, indicated by open and full circles, approach each other too closely, with the result that strong repulsive forces arise between them. Consequently, the probability of finding a given link of a macromolecule

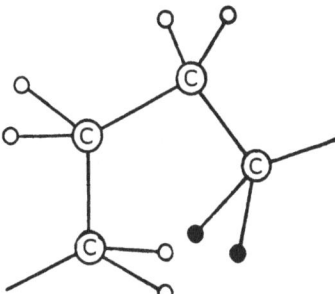

Fig. 61. One of the forbidden conformations of three consecutive links in a polyethylene chain.

in a given conformation is a function of the conformation of neighboring links. In this sense, a macromolecule chain is a Markov chain. The Russian mathematician Markov has developed the theory of chain probability processes. If in a totality of consecutive random events the probability of each subsequent event is a function of the preceding events, then such a totality forms a Markov chain. Let us give an example of such a chain.

Let us take one card at random out of the deck. The probability that the card will be of the given suit is 25%. If the card is returned to the deck, then the probability of the appearance of the same suit does not change when another card is selected. On the other hand, if the card is not returned to the deck, the probability of the appearance again of a card of the same suit is a function of the result of the previous drawing. Thus, if a spade had been pulled out, then the deck contains 13 cards of each of the other suits and only 12 spades. Consequently, the probability of picking a spade again is now 12/51, that is, less than 25%. To the contrary, if the selected card had been of another suit, then the probability of drawing a spade in the second drawing would be 13:51, that is, more than 25%. This is an example of a complex Markov chain, i.e., the probability of drawing a card of a given suit depends on the results of all the previous drawings and not just of the preceding one.

Another example belongs to genetics. Let us assume that we are crossing heterozygotic species with genotype Rr. The results of the crossing in the first generation may be expressed as

$$(\tfrac{1}{2}R + \tfrac{1}{2}r) \times (\tfrac{1}{2}R + \tfrac{1}{2}r) = \tfrac{1}{2}Rr + \tfrac{1}{4}RR + \tfrac{1}{4}rr$$

i.e., the probabilities of appearance of the species Rr, RR, and rr are equal $\tfrac{1}{2}$, $\tfrac{1}{4}$ and $\tfrac{1}{4}$, respectively. What is the probability of the appearance of a heterozygote Rr in any subsequent generation, obtained as a result of random crossing in a population which contains Rr, RR, and rr with the given probabilities? This probability, evidently, depends on the nature of the parent in the preceding generation, i.e., on the preceding event. This is again a Markov chain.

The mathematical theory of Markov chains is complicated, and we shall not discuss it here. From what has been said, however, it is evident that chain processes of such a type are very important. Markov chains have most important applications in the modern statistical physics, in meteorology (the probability of a given weather today depends on the weather on the preceding day), etc.

It is interesting to note that Markov himself applied his theory to the investigation of the order of vowels and consonants in the Russian literary language, having studied the texts of *Eugene Onegin* by Pushkin and *The Childhood Years of the Bagrov Grandson* by Aksakov. It is evident that the probability of the appearance of a consonant or vowel depends on the letter which preceded it.

In the investigation of the properties of macromolecules (their specificity is linked directly with conformational changes), it is necessary to take into account the interaction between links, the synchronization of their motion. In macromolecular physics, systems, the behavior of which is a significant function of the interaction between the constituent parts, are called cooperative processes which depend on such interactions are called cooperative processes. In this manner, a macromolecule is a rotational isomeric cooperative system.

How can we approach the examination of the properties of a cooperative system? Cooperative processes, namely phase transitions, have been studied both experimentally and theoretically long before the development of modern macromolecular physics. These encompass the crystallization of liquids and the melting of crystals, transitions of binary alloys (for example, of copper with zinc) from a disordered to an ordered state, transitions of ferromagnetic materials (for example, iron) from a ferromagnetic into a paramagnetic state, etc. It must be said that, even though physics has a general thermodynamic theory of phase transitions, it has not been able to solve up to now the problem of their complete molecular interpretation on the basis of statistical mechanics.[11,12] Up to now, for example, there is no quantitative molecular theory of the melting of crystals. How can we handle polymers? Their structure, after all, is much more complicated than the structure of a simple crystal, such as NaCl.

A simplifying circumstance is the linear structure of a polymeric chain. In a macromolecule, the strongest valent bonds are distributed in the form of a linear sequence. In this sense the chain is unidimensional. On the other hand, in a crystal of NaCl, six identical strong bonds emanate from each ion into all three space dimensions. The crystal is three-dimensional.

Very important successes have been attained in theoretical physics by finding a relation between phenomena which, at first glance, are quite remote from each other. For example, the relation between the theory of the heat capacity of solid bodies and the scattering of light by these bodies, developed by Mandel'shtam, played an important role in the development of molecular physics. It turns out that an understanding of the nature of the high elasticity of rubber enables us to understand why a chick will never develop from an already cooked egg. It enables us to understand what happens to a protein during its denaturation. These totally unrelated phenomena have, in a certain sense, the same general cause; they are determined by changes in the conformations of macromolecules.

An idea which was first proposed in the theory of ferromagnetism has been used in the development of the modern statistical theory of macromolecules.

The ability of iron to become magnetized is explained by the behavior of its electrons. Each electron can be regarded as a small magnet; it has a magnetic moment, a spin. As a result, it is possible to express an electron by a little arrow. When a magnetic field is acting, the electronic magnets turn; their spins become oriented in a definite direction. In a ferromagnetic material, such as iron, however, the electronic spins furthermore interact with each other in such a manner that it is advantageous to have a parallel distribution of neighboring spins, i.e., such a situation has a lower energy than an antiparallel distribution. As a result, iron may be magnetized even in the absence of a magnetic field because of the parallel orientation of the spins of the little magnet.

In 1925, Ising proposed a one-dimensional model of a ferromagnet, which consisted of a sequence of arrows directed to the right and to the left; furthermore, two neighboring parallel arrows have an energy different from that of two neighboring antiparallel arrows.[13] Such a model is absolutely similar to that shown in Fig. 60, namely, to the model of a unidimensional polymeric chain. It is evident that the Ising model is cooperative; the energy of each link depends on the orientation of neighboring links. In this manner, we are dealing with a particular Markov chain. It turns out that the statistical method for the solution of such problems can be applied with particular success to the physics of macromolecules.[2,10,14,15]

Up to now, we have examined the problem of high elasticity only qualitatively. Let us now carry out a quantitative calculation of the stretching of a one-dimensional chain by an external force.

Let the energy of two neighboring parallel links, i.e., the energy of the t conformation of the link, be equal to $-\varepsilon$, and the energy of two neighboring antiparallel links, i.e., the energy of the s conformation of the link, be equal to ε. The difference between the energies of the two rotational isomers is equal to

$$\Delta E = \varepsilon - (-\varepsilon) = 2\varepsilon \qquad \text{i.e., } \varepsilon = \frac{\Delta E}{2}$$

This condition may be written out in the form

$$E_{ij} = \begin{cases} -\varepsilon\sigma_i\sigma_j & \text{if } i = j + 1 \\ 0 & \text{if } i \neq j + 1 \end{cases} \tag{4.22}$$

Here $\sigma_i = \sigma_j = 1$ if the link is turned to the right, and $\sigma_i = \sigma_j = -1$ if the link is turned to the left; i and j are the numbers of the links. The energy is equal to zero when the links are not neighboring, i.e., when $i \neq j + 1$. If the system is under the influence of an external force f directed from the left to the right, then each arrow requires an additional energy

$$E_f = -lf\sigma_j \tag{4.23}$$

where l is the length of the arrow, i.e., the length of the link which is being rotated by the force. In fact, the potential energy of the arrow in the field by the external force is

$$E_f = -lf\cos(l,f)$$

where $\cos(l,f)$ is the cosine of the angle between the direction of the arrow and the direction of the force. In our case, the arrow may have only two directions, corresponding to the angles of $0°$ ($\sigma_j = 1$) and $180°$ ($\sigma_j = -1$). Rotation of each arrow in the model of Fig. 60 results in a change of state of its neighbors, i.e., in a cooperative change of their conformation. Let us turn, for example, the fifth arrow from the right in Fig. 60a. This will result, as well, in a change of the conformation of the fourth link; there will be a transition from $ttssttsst$ to $ttssttsst$, as a result of which the chain will become elongated.

Let us calculate now the partition function of this system,

$$Q = \sum_r e^{-E_r/kT} \tag{4.24}$$

Statistical physics show that any equilibrium properties of a system may be calculated if its partition function is known.[16] Thus, the energy of a system as a whole is expressed in the general case by the equation

$$\bar{E} = kT^2 \frac{\partial \ln Q}{\partial T} \tag{4.25}$$

while the length of the chain, i.e., the algebraic sum of the lengths of all the arrows, is

$$\bar{h} = kT \frac{\partial \ln Q}{\partial f} \tag{4.26}$$

In our case, the number of states of the system is equal to 2^N, since each of the N arrows may have two orientations, to the right and to the left. The partition function has the form

$$Q = \sum_{\sigma_1 = \pm 1} \sum_{\sigma_2 = \pm 1} \cdots \sum_{\sigma_N = \pm 1} \exp\left(\frac{\varepsilon}{kT} \sum_{j=1}^{N-1} \sigma_j \sigma_{j+1} + \frac{lf}{kT} \sum_{j=1}^{N} \sigma_j \right) \tag{4.27}$$

Each sum has two terms, corresponding to the right and left orientations of the arrow, $\sigma_j = \pm 1$. How can we calculate this expression for large values of N?

To carry out the calculation, it is convenient to introduce an additional interaction between the first and last arrows, $\varepsilon \sigma_N \sigma_1$ (i.e., $\sigma_{N+1} = \sigma_1$); this is physically equivalent to the closing of a linear chain into a loop. When $N \gg 1$, this introduces essentially no error. Then Q may be rewritten in the form

$$Q = \sum_{\sigma_1 = \pm 1} \sum_{\sigma_2 = \pm 1} \cdots \sum_{\sigma_N = \pm 1} \prod_{j=1}^{j=N} P(\sigma_j, \sigma_{j+1}) \tag{4.28}$$

where

$$P(\sigma_j, \sigma_{j+1}) = \exp\left(\frac{\varepsilon}{kT} \sigma_j \sigma_{j+1} + \frac{lf}{kT} \sigma_j \right) \tag{4.29}$$

The quantity $P(\sigma_j, \sigma_{j+1})$ may assume four values which correspond to four cases: $\sigma_j = \sigma_{j+1} = 1$; $\sigma_j = 1$ and $\sigma_{j+1} = -1$; $\sigma_j = -1$ and $\sigma_{j+1} = 1$; $\sigma_j = \sigma_{j+1} = -1$. Consequently, (4.29) may be regarded as elements of a second-order matrix

$$P = \begin{bmatrix} P(1,1) & P(1,-1) \\ P(-1,1) & P(-1,-1) \end{bmatrix} \tag{4.30}$$

It can be shown quite easily that the sum in (4.28) is equal to the trace of the matrix P raised to the Nth power. The trace of the matrix is the sum of its diagonal terms; it is expressed by the symbol Sp:

$$Q = \text{Sp}\,(P^N) \tag{4.31}$$

The trace of the matrix does not change during its diagonalization.[17] In such a form we have

$$P^N = \begin{pmatrix} \lambda_1 & 0 \\ 0 & \lambda_2 \end{pmatrix}^N = \begin{pmatrix} \lambda_1^N & 0 \\ 0 & \lambda_2^N \end{pmatrix} \tag{4.32}$$

and

$$\text{Sp}\, P^N = \lambda_1^N + \lambda_2^N \tag{4.33}$$

where λ_1 and λ_2 are the roots of the matrix, which are found by solving the equation

$$\begin{vmatrix} P(1,1) - \lambda & P(1,-1) \\ P(-1,1) & P(-1,-1) - \lambda \end{vmatrix} = 0 \tag{4.34}$$

or, in explicit form,

$$\begin{vmatrix} e^{a+b} - \lambda & e^{-a+b} \\ e^{-a-b} & e^{a-b} - \lambda \end{vmatrix} = 0 \tag{4.35}$$

where

$$a = \frac{\varepsilon}{kT} \qquad b = \frac{lf}{kT}$$

Equation (4.35) is a quadratic equation

$$\lambda^2 - \lambda(e^{a+b} + e^{a-b}) + e^{2a} - e^{-2a} = 0 \tag{4.36}$$

Its roots are

$$\begin{aligned} \lambda_1 &= e^a \cosh b + (e^{2a} \sinh^2 b + e^{-2a})^{1/2} \\ \lambda_2 &= e^a \cosh b - (e^{2a} \sinh^2 b + e^{-2a})^{1/2} \end{aligned} \tag{4.37}$$

Let us remind the reader that the hyperbolic cosine and sine are

$$\cosh b = \frac{e^b + e^{-b}}{2} \qquad \sinh b = \frac{e^b - e^{-b}}{2}$$

The root λ_1 is greater than λ_2. At large values of N, it is possible therefore to neglect λ_2^N relative to λ_1^N, and we obtain

$$Q \cong \lambda_1^N = [e^a \cosh b + (e^{2a} \sinh^2 b + e^{-2a})^{1/2}]^N \tag{4.38}$$

The length of our model is

$$\bar{h} = kT\frac{\partial \ln Q}{\partial f} = NkT\frac{\partial \ln \lambda_1}{\partial f} = N\frac{l \sinh b}{(\sinh^2 b + e^{-4a})^{1/2}} \qquad (4.39)$$

In this manner, it has been possible to solve completely by very elegant calculations the statistical problems of the cooperativity of a unidimensional system of links. The measure of cooperativity, i.e., the measure of interactions, is the quantity ε, i.e., the difference of energies of the rotational isomers ΔE. When $\varepsilon \to 0$, $\Delta E \to 0$, and consequently, $a \to 0$

$$\bar{h} \to Nl \tanh b = Nl \tanh \frac{fl}{kT} \qquad (4.40)$$

In other words, in the case of free internal rotation, in the absence of rotational isomerization, the length of the chain increases with an increase in the force and a decrease in the temperature. On the other hand, when the difference in the energies of the rotational isomers, ΔE, is very large, i.e., when the chain is quite stiff, corresponding to the condition $\Delta E \gg kT$ and, consequently, $a \gg 1$, it is possible to neglect the quantity e^{-4a} in (4.39), and we obtain

$$\bar{h} = Nl \qquad (4.41)$$

This is independent of the acting force. The chain is stiff and, therefore, extended. The same result is obtained with large forces, i.e., when $fl \gg kT$. In this case, as well, $\sinh^2 b \gg e^{-4a}$; a strong force extends the chain completely.

With weak forces, i.e., when $fl \ll kT$, $\sinh b$ can be expanded into the series

$$\sinh b = \frac{e^b - e^{-b}}{2} \cong \frac{1 + b + \cdots - 1 + b -}{2} \cong b \ll 1$$

and it follows from (4.39) that, with free rotation when $a = 0$,

$$\bar{h} \cong Nlb = Nl^2 \frac{f}{kT} \qquad (4.42)$$

where

$$f = kT\frac{\bar{h}}{Nl^2}$$

i.e., the length of the chain is proportional to the stretching force (Hooke's law); the modulus of elasticity is proportional to the absolute temperature, in agreement with the thermodynamic properties of rubber. The cooperativity, i.e., the difference of a, or ΔE, from 0, introduces a definite correction.

The unidimensional model of Ising turns out to be inadequate for ferromagnetism, the specific properties of which may be explained only on the basis of a three-dimensional model. The calculations for the latter meet with extreme mathematical difficulties; it is impossible to obtain Q in a strict analytical form.[2,14,15] To the contrary, when we study the cooperative properties of macromolecules, the method of calculation based on the unidimensional model has been found to be extremely fruitful. This method was applied for the first time to macromolecules of synthetic polymers[2,5,10] and to molecules of proteins and nucleic acids in studies described in Chapters 5 and 6. The point is that a macromolecule is, in fact, linear in the sense that the predominant interatomic interactions take place along the linear chain.

In three-dimensional space real macromolecules are, obviously, coiled. Also any sample of rubber which is subjected to stretching is three-dimensional. Nevertheless, the theory of elongation, which is based on the one-dimensional model by Ising and takes into consideration also the three-dimensional conformations of the links, gives good agreement with experiments. The same theoretical considerations make it possible to develop methods for the calculation of the physical characteristics of macromolecules in solution, namely, their dimensions, shapes, dipole moments, and anisotropic optical polarizations. The rotational-isomeric theory of macromolecules, which takes into account the cooperativity of the conformations and which makes use of a semiempirical method of determining conformations, gives excellent agreement with experiment and affords a physical interpretation for a number of phenomena.[2,10]

CRYSTALLINE POLYMERS

Synthetic polymers exist normally in an amorphous state; they may be glasslike (plastics) or plastic and highly elastic (rubber). There are, how-ever, also crystalline polymers, and this is very important. The point is that fabrics which are used for clothing are made of crystalline polymers, natural and, more recently, synthetic fibers. Technology develops, fashions change, but crystalline fiber-forming polymers are, as ever, irreplaceable. Wool and silk are natural protein fibers. Modern capron and nylon are synthetic fibers. The crystallinity of such materials is demonstrated by all their properties. They have a relatively narrow melting-temperature inter-val, they are anisotropic. The crystallinity of these materials has been confirmed by X-ray analysis.

Contrary to the usual low-molecular-weight substances, synthetic polymers, as a rule, do not crystallize entirely. The substance contains both amorphous and crystalline material. This is explained by two causes. First, as has been pointed out, synthetic polymers are not uniform; they contain macromolecules of various lengths, and of different stereochemical struc-tures, etc. It is natural that a system which consists of such heterogeneous objects cannot, as a whole, form a regular crystal lattice. Such crystal lattices will accommodate only similar macromolecules or their similar parts. Second, in order to form a crystal, the macromolecules must become reoriented. This requires a large number of turns in the links of the chain, i.e., these links must overcome the corresponding energy barriers of internal rotation. This requires time, which increases with a decrease in temperature. Therefore, the second cause of incomplete crystallization has a kinetic nature; at the temperature of crystallization, the chains just do not have the time to enter into the crystal lattice during the allotted time. In fact, heating the polymer or swelling it with a low-molecular-weight solvent facilitates crystallization and increases the degree of crystallinity.

The more regular the structure of the macromolecules, the more easily they become crystallized. Thus, isotactic polystyrene is crystalline and it forms fibers, while the normal atactic polystyrene is amorphous and is a glasslike plastic.

A crystalline structure is characteristic of fibers. The macromolecules in the fiber are arranged in regular order; they are oriented in a definite fashion, usually in such a manner that the chains lie parallel to the axis of the fiber. This explains the great anisotropy of fibers. A fiber is an oriented polymeric polycrystal. Furthermore, a number of polymers also form monocrystals during their crystallization from solution. Monocrystals of polyethylene, n-rubber, cellulose, etc., have been investigated.

Proteins are homogeneous; a sample of a given protein consists of identical molecules. Therefore, many proteins are crystallizable. A vast number of enzymes have been obtained in crystalline form.[18] However, the importance of crystalline polymers for molecular biophysics is determined not by this factor. The crystallization of polymers is the formation of a certain regular supramolecular structure which consists of macromolecules. Supramolecular structures in the cells, namely, in the membrane and the organoids, also have a regular structure. The difference between the usual polymeric crystals and the biological supramolecular structures consists in that the latter contain complexes of various biological macromolecules, namely, nucleoproteins in chromosomes, lipoproteins in cellular membranes, etc. Up to now, there is no theory of supramolecular structures; our understanding of these is far from satisfactory. It can be surmised, however, that the way to the investigation of these structures will pass through the theoretical and experimental study of crystalline polymers.

The structure of a number of polymeric crystals has been well investigated by X-ray and electron diffraction. In Fig. 62 is shown the structure of the crystal lattice of polyethylene. It is evident that, in a crystal, the macromolecules are no longer present in the form of an unordered twisted coil, but in a definite regular conformation. Thus, the macromolecules of polyethylene in a crystal are trans chains (Fig. 59). In other words, one particular rotational isomer becomes fixed during crystallization, while other conformations are excluded. In crystalline polyethylene, it is the trans isomer; however, one encounters more frequently coiled rotational isomers; in such a case, the chain crystallizes in the form of a helix. In Fig. 63 is shown the crystalline conformation of the chain of isotactic polystyrene. This is a triple helix with rotational isomers characterized by the angles $\varphi_1 = 0°$ and $\varphi_2 = 120°$ about the $CHR-CH_2$ and CH_2-CHR bonds, respectively. Later we shall see the great importance of the helical structures of biological macromolecules, namely, of proteins and nucleic acids.

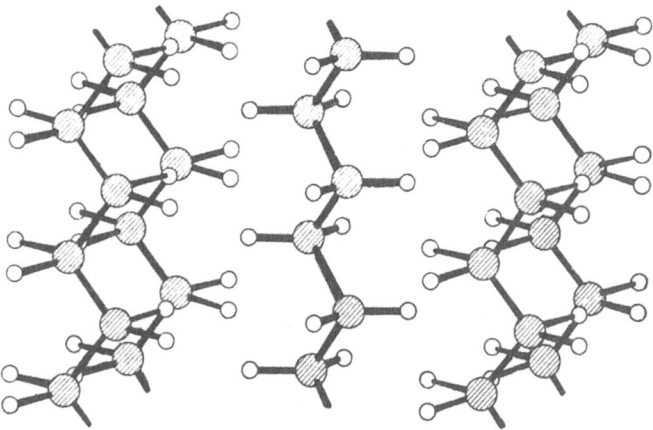

Fig. 62. Structure of the crystal lattice of polyethylene.

The fixation of a given rotational isomer in all the links of the chain in the crystal state leads to long-range order in the crystal. Knowledge of the position of the atoms of a given monomeric link enables one to indicate these positions for any link as far removed as possible, since the positions of the atoms are strictly periodic. On the other hand, there is evidently a definite order in the distribution of neighboring links, namely, short-range order. When the polymer is dissolved or melted, the long-range order vanishes as a result of rotational isomerization. On the contrary, the elements of short-range order must remain preserved. In this sense, the melting of a polymer crystal or its dissolution are similar to the melting of a usual low-molecular-weight crystal. Long-range order vanishes from

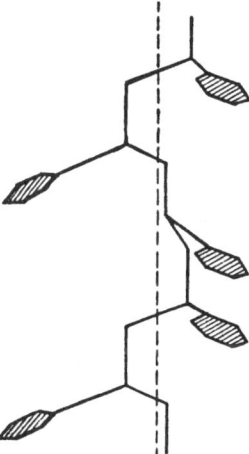

Fig. 63. Conformation of isotactic polystyrene in a crystal.

the liquid formed, but the atoms or molecules neighboring given atoms or molecules are distributed almost in the same manner as in the crystal; in a liquid there is short-range order.

The modern theory of liquids is based on such concepts of the quasi-crystallinity of their structures; this concept was first introduced by Frenkel.[19] Using the concept of short-range order in macromolecules, Ptitsin and Sharonov have proposed a method for determining the rotational isomers and their characteristic angles φ_i for individual macromolecules.[2,10,20] The method of X-ray analysis makes possible the determination of the conformation of the macromolecules in the crystal; it enables one to find the angles φ_i for the rotational isomers selected by crystallization.

In noncrystalline systems, this method cannot be used, and therefore, it becomes necessary to determine the conformations by other means.

Ptitsin and Sharonov have proposed that the most stable rotational isomer of a monomeric link in a polymer solution has the same set of angles as the crystal. It has been established that this is exactly the case for small molecules; for example, n-butane crystallizes in the form of a trans isomer and this is the isomer with the lowest energy in the gas or the liquid. This is particularly true of nonpolar substances in which the forces of intermolecular interaction are relatively weak. If the potential of internal rotation is symmetrical, then there are several rotational isomers with identical energy. Thus, in polymers of the type of

$$-CH_2-CR_2-CH_2-CR_2-$$

the potential is independent of whether the rotations about the single bonds occur clockwise or counterclockwise, i.e.,

$$U(\varphi_1,\varphi_2) = U(-\varphi_1,-\varphi_2)$$

where φ_i is the rotation angle about the CH_2-CR_2 bond, φ_2 is the rotation angle about the CR_2-CH_2 bond. Furthermore, we have the equalities

$$U(\varphi_1,\varphi_2) = U(\varphi_2,\varphi_1) = U(-\varphi_2,-\varphi_1)$$

since the two bonds which we are considering are identical. If the crystal contains the single isomer (φ_1,φ_2), then the solution must contain as well the isomers (φ_2,φ_1), $(-\varphi_1,-\varphi_2)$, $(-\varphi_2,-\varphi_1)$ with the same energy. The values of φ_1 and φ_2 have been found by X-ray methods. As a result, the entire set of rotational isomers is known and it is possible to calculate the properties of the macromolecules in solution. In this manner, excellent agreement with experiment has been obtained in the case of polyisobutylene in which $R = CH_3$. Here $\varphi_1 = \varphi_2 = 82°$. Both calculation and experiment give

$$(\overline{h^2})^{1/2} = 1.8(\overline{h_0^2})^{1/2}$$

where $\overline{h_0^2}$ is a quantity which corresponds to free internal rotation.

In this manner, the characteristics of a number of macromolecules, including stereoregular and iso- and syndiotactic ones, have been determined quite successfully.[2,10] As has been pointed out already, however, in many cases it is necessary to take into account the cooperative nature of the internal rotations, i.e., the correlation of the internal rotations in neighboring links. Detailed calculations carried out for small molecules, for example for n-butane, show that it is necessary to know, not the curve $U(\varphi)$, but the surface $U(\varphi_1, \varphi_2, \ldots)$.[21] n-Butane has three angles of internal rotation: φ is the angle of rotation about the CH_2—CH_2 bond, and $\varphi_1 = \varphi_2$ are the angles of rotation about the H_3C—CH_2— bonds. The function $U(\varphi, \varphi_1 = \varphi_2)$ has minima at $\varphi = 0°$, at $\varphi_1 = \varphi_2 = \pm 120°$, and at $\varphi = \varphi_1 = \varphi_2 = \pm 120°$.

Correlation of rotations in neighboring links changes the length of the chain. Calculation gives

$$\overline{h^2} = e^{-\Delta U/kT} (\overline{h^2})_{\Delta U = 0} \qquad (4.43)$$

where (in the simplest case of two equivalent isomers)

$$\Delta U = U[(\varphi,\varphi), (\varphi,\varphi)] - U[(\varphi,\varphi), (-\varphi,-\varphi)]$$

The first and second parentheses refer to two neighboring monomers.

In the case of chains which crystallize in helical conformations, a sequence of identical rotational isomers is more favorable than a sequence of different rotational isomers. In this case

$$U[(\varphi,\varphi), (\varphi,\varphi)] < U[(\varphi,\varphi), (-\varphi,-\varphi)]$$

and $\Delta U < 0$. Thus, correlation lengthens the chain, i.e., it renders it more rigid.

The quantities ΔU can be found by theoretical calculations. In the simplest polymer, polyethylene, we are dealing with two angles of rotation φ_1 and φ_2 in neighboring monomers. Calculations show that the following conformations are stable: $\varphi_1 = \varphi_2 = 0°$ (tt); $\varphi_1 = 0°$, $\varphi_2 = \pm 120°$ (ts); $\varphi_2 = 0°$, $\varphi_1 = \pm 120°$ (st); $\varphi_1 = \pm 120°$, $\varphi_2 = \pm 120°$ (ss). The conformation with $\varphi_1 = 120°$, $\varphi_2 = -120°$ ($s^+ s^-$) is energetically unfavorable.[22] In this case, the ts and st conformations contained nonequivalent rotational isomers, and it is a little more difficult to take correlation into account than with the use of Eq. (4.43). The values of ΔU for tt and st are equal to zero; for ts and ss they are equal to 0.5 kcal/mole; and for $s^+ s^-$ they are -2.3 to 2.7 kcal/mole. Calculation gives

$$\overline{h^2} = 3.2 \overline{h_0^2}$$

in excellent agreement with experiment.[10,23] Good results have been obtained as well in other cases, such as the calculation of the dimensions, dipole moments, and optical anisotropy of macromolecules.[10]

In this manner, the concept of the conservation of short-range crystalline order in a free polymer chain has been proven. The cooperativity of internal rotations is already expressed in the short-range order.

Let us return to polymeric crystals. The crystallization of a polymer is a conformational transition of the macromolecule; it is the selection of a definite rotational isomer. On the other hand, the melting of a polymeric crystal is rotational isomerization; namely, it is a transition from a single rotational isomer to a statistical sum of isomers, a transition from a regular conformation of the macromolecule as a whole to a coil. The general character of the conditions which determine the melting of a polymeric crystal may be found again from the theory of rotational isomerization. Thermodynamics give the following conditions of melting:

$$T_m = \frac{\Delta H}{\Delta S} \qquad (4.44)$$

i.e., the melting temperature is the greater, the greater the gain of energy (enthalpy) during crystallization and the less the loss of entropy. Evidently, ΔH is the heat of the melting of the crystal. This is a completely universal equation for any substances.

It is evident that the melting temperature of a crystalline polymer must be a function of the flexibility of the macromolecule, i.e., of the number of rotational isomers and of the differences of their energies. Since in a crystalline state, all chains exist in a single definite conformation which corresponds to, let us say, the trans isomer, the change of enthalpy during its melting is equal to its change as a result of the transition of a trans isomer to a mixture of rotational isomers, and in this manner, ΔH is related to ΔE, the energy difference between the rotational isomers.

The entropy ΔS increases during melting, since this results in the formation of a mixture of rotational isomers; we observe now the entropy of mixing and an additional entropy determined by the increase in the energy of the rotational isomers.

The greater the number of conformations which can be assumed by a link, i.e., the greater the flexibility of the macromolecule,[2] the lower is the temperature of the melting of the polymer. The rigorous theory developed by Flory takes into account, furthermore, an intermolecular entropy, i.e., the different distributions of the macromolecules in crystalline and melted polymers.[24] The principal physical result is not changed by this. The greater the flexibility of the macromolecules, the more favored the rotational isomerization, the less is the temperature of melting of the polymer. Intermolecular attraction is necessary for the crystallization of flexible macromolecules. On the other hand, stiff macromolecules must crystallize with the formation of ordered structures, even if the energy of their intermolecular interactions is equal to zero. In fact, let us try to fill a given volume with stiff rods. It is obvious that their most favorable position is one which corresponds to the greatest density, and it will be parallel; they will orient themselves like matches in a match box, notwithstanding the absence of any attractive forces between them. It is evident then also that incompletely stiff macromolecules, i.e., chains which have a limited flexibility, must distribute themselves not in a disordered manner, but in a manner in which they are coordinated with each other. Macromolecules with a limited stiffness form crystals at $T < T_m$, and at $T > T_m$ they form a partly ordered structure but they are amorphous as a whole. Kargin, Kitaigorodskii, and Slonimskii[25] have examined in a qualitative way such structures; they have called them stacks. Recently, DiMarzio investigated the problem of initiation of order in a totality of macromolecules on the basis of statistical theory.[26] He showed that the presence of intermolecular ordering and its changes during the stretching of the polymer must exert an effect on the dependence of stress upon deformation, a fact which apparently has been observed experimentally.[27] These problems are very important for molecular biochemistry. We see that macromolecules, by the very nature of their structure, as a consequence of the high molecularity of the chains, and of limitations of their flexibility, must form ordered supramolecular structures, such as crystals or stacks. Kargin has observed such structures in many polymers by means of electron microscopy.[28] In synthetic polymers, ordering is hindered by the heterogeneity of their dimensions and structures. But the cell "prints" its protein "texts" in many identical copies.

As we shall see, the macromolecules of proteins (and DNA) are sufficiently rigid, due to special reasons which will be discussed later. As a result, the formation of supramolecular highly ordered structures in organoids and membranes of the cell is determined, to a great extent, by the macromolecular structure of the biological molecules. So far, we do not have a rigorous quantitative theory which would permit us to predict the structure and properties of supramolecular structures from information on the structure of individual macromolecules. The development of such a theory which would explain all the thermodynamics and kinetics of formation of such structures is, at present, an important problem of the physics of polymers and of molecular biophysics. Bernal regards the automatic construction of the supramolecular structures from identical macromolecules as one of the necessary conditions for life.[29,30] An ordered structure is necessary for a regulated function of proteins, nucleic acids, carbohydrates, lipids, and their complexes. This is one of the reasons for the correct functioning of the cell, i.e., for its ability to synthesize a large number of identical molecules.

It is evident that the study of ordering in synthetic polymers, i.e., in compounds which in many respects are more stable than biological macromolecules, makes it possible to approach reasonably the study of biological supramolecular structures. For example, the elucidation of the role of conformation and the effect of the flexibility of macromolecules on ordering is an important step forward in the solution of these very difficult problems of molecular biophysics.

METHODS FOR STUDYING MACROMOLECULES IN SOLUTION

Information on the structure of individual macromolecules is obtained from studies of their properties in solution by physical methods. These methods can be used for any macromolecules, both synthetic and biological. What information do they yield? Let us list the principal physical characteristics which can be established by these methods: first, molecular weight and thus the degree of polymerization; second, the dimensions of the macromolecules, and thus their thermodynamic flexibility; third, the shapes of the macromolecules; fourth, the optical anisotropy of the macromolecules; fifth, the electric dipole moment; and sixth, the degree of ionization for those macromolecules which can bear electric charges.

This does not exhaust all the properties, but those listed are the most important ones. Here we shall describe briefly the physical methods and information which can be obtained with their help. More detailed information may be found in the monograph literature.[31-37]

The best method for determining the molecular weights of polymers is sedimentation in the ultracentrifuge. The basic principle of this method, which was developed by the Swedish scientist Svedberg, is simple.

A cell, containing polymer solution, is subjected to a very rapid rotation: in modern ultracentrifuges the speed of rotation is 60,000 rpm and higher. Molecules of the dissolved polymer are denser than those of the solvent and they become displaced in the direction from the axis of rotation toward the bottom of the cell. This displacement is registered by optical methods, by changes in the refractive index or absorption of light. The speed of displacement (the rate of sedimentation) is proportional to the speed of rotation and to the molecular weight M and inversely proportional to the frictional coefficient q of the macromolecule in the medium:

$$v = \frac{dx}{dt} = \frac{aM\omega^2 x}{q} \qquad (4.45)$$

where ω is the angular velocity of rotation, i.e., $2\pi v$; v is the number of rotations per second; x is the radius of rotation, i.e., the distance from the axis of rotation of the ultracentrifuge rotor to the moving boundary of the sedimenting polymer. The multiplier a is the buoyancy factor, which reflects the fact that the macromolecule is immersed in a liquid solution and loses from its mass the mass of the displaced solvent:

$$a = 1 - V\rho \qquad (4.46)$$

where ρ is the density of the solution, and $V = 1/\rho'$ is the partial volume of the polymer, i.e., a quantity reciprocal to its density. It is evident that expression (4.45) may be rewritten in the form

$$qv = aM\omega^2 x$$

This equation expresses the equality of the centrifugal force and the frictional force, i.e., the condition of a stationary state. The sedimentation coefficient is usually defined as

$$s = \frac{v}{\omega^2 x} = \frac{M(1 - V\rho)}{q} \qquad (4.47)$$

The dimension of s is time; s is usually expressed in special units, svedbergs; 1 svedberg $= 10^{-13}$ second.

The value of s may be determined from a knowledge of v, ω, and x; the molecular weight of the polymer, M, may be calculated from a knowledge of s, ρ, ρ', and q. The densities ρ and ρ' are measured independently by usual means. The frictional coefficient q is a function of the dimensions and shape of the macromolecule.

To find q it is essential to determine the diffusion coefficient of the polymer, D, in the same solvent. Diffusion follows Einstein's law

$$D = \frac{kT}{q} \qquad (4.48)$$

Consequently

$$M = \frac{SkT}{D(1 - V\rho)} \qquad (4.49)$$

In synthetic polymers, M is some average molecular weight; for proteins, it is the true molecular weight.

Other absolute methods of measuring molecular weight are osmometry and light-scattering. In ideal dilute solutions, the osmotic pressure is expressed by Van't Hoff's law

$$\pi = \frac{RT}{M} c \qquad (4.50)$$

where c is the weight concentration of the dissolved material. Solutions of polymers (and also of biopolymers) are always nonideal. In this case,

$$\pi = \frac{RT}{M} c + Bc^2 + Cc^3 + \cdots \qquad (4.51)$$

or

$$\frac{\pi}{c} = \frac{RT}{M} + Bc + Cc^2 + \cdots \qquad (4.52)$$

The coefficients B, C, ... (the second, third, etc., virial coefficients) are a function of the

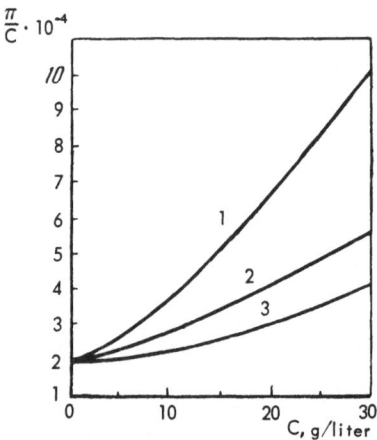

Fig. 64. Dependence of π/c on c for poly-methacrylate (1) in chloroform, (2) in tetra-hydrofurane, (3) in diethyl ketone.

interaction of the macromolecule with solvent molecules. Usually, for dilute solutions, $Cc^2 \ll Bc$. This means that when we plot π/c as a function of c, we obtain a line which extrapolates to a point equal to RT/M. The slope of the curve determines B (Fig. 64). From RT/M, we obtain M. The sedimentation method can be applied for polymers with high molecular weights, starting with 10,000. On the other hand, the precision of osmotic measurements falls with an increase in molecular weight. It is applicable, however, up to approximately 300,000.

The scattering of light is an extremely important phenomenon of molecular optics; its investigation gives very valuable and manifold information on the structure and properties of the molecules of a dissolved substance.[35] Light is scattered by a given volume of gas or liquid in all directions since any medium is nonuniform; in any medium, there are accidental regions of higher and lower density, statistical fluctuations of density, and also fluctuations of the orientations of molecules. In solutions, there are concentration fluctuations. Since light which passes through a material medium is scattered, its intensity decreases according to the law

$$I = I_0 e^{-\tau x} \tag{4.53}$$

where I_0 is the intensity of the incident beam; x is the length of the path which the light travels through the medium; and τ is the extinction coefficient, which measures the turbidity of the medium. For molecules with dimensions much smaller than the wavelength of the light, λ, theory gives

$$\tau = \frac{128\pi^5}{3\lambda^4} a^2 N_1 \tag{4.54}$$

where N_1 is the number of molecules in a unit volume, and a is the polarizability of the molecule. In a low-molecular-weight medium, the intensity of the scattering of light is inversely proportional to the fourth power of the wavelength. This is why the sky is blue; the atmosphere scatters predominantly short waves. The red color of the sun at dawn and dusk is explained by the same phenomenon; when the rays pass through a thick layer of the atmosphere, it is the shortwave blue and green frequencies which are weakened the most.

In a gas, polarizability a is simply related to the refractive index,

$$a = \frac{n-1}{2\pi N_1} \tag{4.55}$$

consequently, (4.54) may be rewritten in the form

$$h = \frac{32\pi^3}{3\lambda^4} \frac{(n-1)^2}{N_1} \tag{4.56}$$

If the refractive index n and the density ρ of the gas are known, then by changing τ we find N_1 and, consequently, M of the molecule, since

$$N_1 = \frac{\rho}{M} N_A$$

where N_A is Avogadro's number.

In solutions, scattering occurs as a result of concentration fluctuations. The expression for the extinction coefficient in this case has the form

$$\tau = \frac{32\pi^3}{3\lambda^4} \frac{n_0^2(n-n_0)^2}{N_1} \tag{4.57}$$

where n is the refractive index of the solution, n_0 is that of the solvent, N_1 is the number of dissolved particles per cubic centimeter. Introducing the weight concentration c of the solution, expressed in grams per cubic centimeter, we obtain

$$H\frac{c}{\tau} = \frac{1}{M} \tag{4.58}$$

where

$$H = \frac{32\pi^2}{3\lambda^4} \frac{n_0^2}{N_A} \left(\frac{n-n_0}{c}\right)^2 \tag{4.59}$$

since $N_1 M = cN_A$. This is valid for ideal dilute solutions. For nonideal solutions, just as in the case of osmotic pressure,

$$H\frac{c}{\tau} = \frac{1}{M} + 2\frac{B}{RT}c + \cdots \tag{4.60}$$

where B is the same second virial coefficient.

If we calculate H, using the determined values of n, n_0, and c and if we measure τ directly, we may plot the dependence of Hc/τ as a function of c and obtain $1/M$ and B in the same manner as in the osmotic method. The precision of the results obtained with light-scattering increases with molecular weight.

Finally, the simplest, but not absolute, method for determining M of a polymer is the change in the viscosity of the solution. At low concentrations, the specific viscosity of a solution is equal to

$$\eta_{sp} = \frac{\eta - \eta_0}{\eta_0} \tag{4.61}$$

where η is the viscosity of the solution, and η_0 is that of the solvent. The η_{sp} increases with an increase in c. The intrinsic viscosity is defined as

$$[\eta] = \lim_{c \to 0} \left(\frac{\eta_{sp}}{c} \right) \tag{4.62}$$

As has been shown by theory and experiment, it is a direct function of the molecular weight:

$$[\eta] = KM^a \tag{4.63}$$

The quantity a, which has values between 0.5 and 2.0, is a function of the shape of the macromolecule; $a = 0.5$ is valid for a solid sphere; $a = 2$ corresponds to a rigid rod. It is evident that, in order to determine M from a knowledge of $[\eta]$, we must know a and K. This quantity may be determined by independent measurements of the molecular weight of polymers of a given type by one of the absolute methods. If a and K are known, it is possible then to determine M for any other samples of the polymer of the same type in the same solvent. It is natural that viscosity is strongly dependent on the solvent. In poor solvents, interactions between links in a polymer chain which is twisted into a coil are greater than between the same links and the molecules of the solvent. As a result, the number of polymer–polymer contacts increases, and the coil becomes compact. In this case, $a = 0.5$. To the contrary, in good solvents, the polymer–solvent interaction is more favorable; as a result, the coil swells in the solvent and increases in dimensions. For highly swelled coils, $a = 1.0$. It is evident that, in the second case, the viscosity is greater than in the first. For the very same reason, it is evident that the viscosity must increase with an increase of the chain length, i.e., with an increase in polymer molecular weight.

In general, $[\eta]$ increases with an increase in the length and stiffness of the macromolecule. It is evident that a solution of rigid rods flows with a much higher friction than a solution of spherical particles.

The dimensions of the macromolecules, whether statistical coils or rigid rods, may also be determined by the method of light-scattering. In this case, however, it is not the absolute intensity of the scattering of light which is measured, but the dependence of the intensity of scattering on the angle of measurement.

For small molecules, the intensity of the light, I_θ, scattered at angles θ relative to the incident beam, is proportional to $1 + \cos^2 \theta$ and inversely proportional to the square of the distance from the scattering element to the point of observation, r^2. The angular dependence is expressed by the relation of Rayleigh

$$R_\theta^0 = r^2 \frac{J_\theta}{J_0} = \frac{3}{16\pi}(1 + \cos^2 \theta)h \tag{4.64}$$

For large molecules, the dimensions of which are of the order of λ, the dependence of the

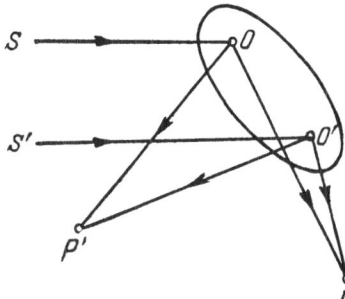

Fig. 65. The Mie effect. The difference in path lengths of rays SOP and $S'O'P$ at point P is less than the difference at point P' between the paths of the rays SOP' and $S'O'P'$.

intensity on θ is more complicated. Each point of the macromolecule scatters the light in a definite phase; the addition (interference) of the light rays scattered by various links of the macromolecule results in the scattering intensity in the forward direction being greater than that in the backward direction (Fig. 65). The exact expression for R_θ for macromolecules is a function of their shape. It has been shown by Debye that, in the case of statistical coils, the interference of rays scattered by its segments gives an expression for R_θ, relative to that for small molecules, expressed as

$$\frac{R_\theta}{R_\theta^0} = P(\theta) = \frac{2}{z^2 u^2}(e^{-zu} - 1 + zu) \tag{4.65}$$

where z is the number of statistical segments of length b, and u is equal to

$$u = \left(\frac{4\pi}{\lambda'} \sin \frac{\theta}{2}\right)^2 \frac{b^2}{6} \tag{4.66}$$

where λ' is the wavelength of the light in the solvent. For small molecules, as a result of (4.58), we have

$$h = HcM$$

$$\frac{Hc}{R_\theta^0} = \frac{1}{\dfrac{3}{16\pi}(1 + \cos^2 \theta) M} \tag{4.67}$$

For macromolecules, the similar relation has the form

$$H\frac{c}{R_\theta} = \frac{1}{MP(\theta)} + 2\frac{B}{RT}c + \cdots \tag{4.68}$$

The scattering factor $P(\theta)$ can be determined experimentally from the dependence of the intensity of scattering on θ. In turn $P(\theta)$ is a function of the dimensions of the macromolecule:

$$P(\theta) = 1 - \frac{\rho^2}{3}\left(\frac{4\pi \sin \theta/2}{\lambda'}\right)^2 + \cdots \tag{4.69}$$

where λ' is the wavelength of light in the solution and ρ is the radius of gyration of the macromolecule. For a rodlike molecule of length L, $\rho = L/\sqrt{12}$; for a statistical coil, $\rho = \sqrt{\overline{r^2}}/6$, where $\overline{r^2}$ is the mean square radius of the coil.

For sufficiently large macromolecules, it is also necessary to take into account the higher terms in expansion (4.68); in this case, it becomes possible to determine both the dimensions and the shape of the macromolecule. Zimm proposed a graphical method of determining simultaneously M and ρ. Hc/R_θ is plotted along the ordinate; $\sin^2(\theta/2) + $ constant $\times c$ is plotted along the abscissa. This results in two families of parallel straight lines [curves if it is impossible to ignore the higher terms in the expansion of (4.69)]. The straight lines of one family show the dependence of Hc/R_θ on c for given values of $\sin^2(\theta/2)$; the other family of lines shows the dependence of Hc/R_θ on $\sin^2(\theta/2)$ at given values of c. A Zimm plot for trinitrocellulose in acetone is shown in Fig. 66.[36] It gives $M = 400,000$ and $\sqrt{\overline{r^2}}$ = 1,500 Å. The macromolecule is found to be coiled more than fourfold. In fact, the monomer unit in this case has a molecular weight of 294 and a length of 5.15 Å. The degree of polymerization, i.e., the number of units in the chain, is equal to 400,000/294 = 1,350. The length of the extended chain would be 5.15 × 1,350 = 6,950 Å. The degree of coiling is 6,950/1,500 = 4.6. This is a relatively stiff chain.

From what has been said above about viscosity, a quantity which characterizes a hydrodynamic property of the molecule, it is evident that it is also a function of the dimensions and shape of the macromolecule; thus, if a is measured as a function of M, we obtain information on the structure of the macromolecule in the particular solvent.

A very fine and precise method for investigating the shape and optical anisotropy of macromolecules is the measurement of the double refraction (birefringence) of flow of the polymer solution. The theory and experimental methods are given in the work of Tsvetkov et al.[37] This phenomenon (the Maxwell effect[35]) consists in the following: If the solution of the polymer is caused to flow rapidly, the macromolecules become oriented in a certain way along the direction of flow. As a result, the liquid becomes anisotropic and, consequently, birefringent. From the value of the birefringence, it is possible to determine the intrinsic anisotropy of the macromolecule, which is related to the difference of its dimensions in different directions; a statistical coil always has the shape of a bean, and not that of a pea.

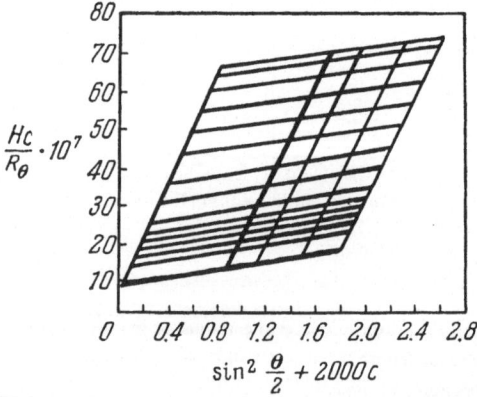

Fig. 66. Zimm plot for a solution of trinitrocellulose in acetone at 25°C.

Interesting information on the structure of oriented polymers, as well as on the dimensions and shapes of macromolecules in solution, may be obtained from a study of small-angle X-ray scattering (usually less than 2°). The application of this method seems particularly indicated in the case of biological macromolecules, the molecules of which maintain a specific conformation in solution.

Finally, large macromolecules may be examined directly in the electron microscope.

All these methods along with a number of others give valuable information on the macromolecules, in particular, on the macromolecules of proteins, nucleic acids, and other biopolymers. Molecular biophysics has inherited much from polymer physics and the corresponding genes have not yet been subjected to mutation.

POLYELECTROLYTES

Macromolecules containing atoms capable of ionization have special properties and are known as *polyelectrolytes*. They are particularly important for molecular biophysics, since both proteins and nucleic acids contain atomic groups which may carry positive or negative charges. Biopolymers are polyampholytes, since they contain both cationic and anionic groups. We have seen that there are both acidic and basic amino acid residues within proteins. In nucleic acids, the phosphate groups are anionic, while the nitrogen atoms in the nitrogen bases may gain a positive charge under proper conditions and play the role of cations. The biological function of biopolymers is closely related to their electrochemical properties.

Let us examine some simple polyelectrolytes, for example, polyacrylic acid. The ionized form of this polyacid has the form

There are considerable repulsive electrostatic forces between the singly charged links of the chain. As a result, the conformations of the macromolecule of a polyelectrolyte must be a strong function of its charge, i.e., of the acid or basic properties of the medium, determined by the concentration of hydrogen ions H^+ and the ionic strength of the medium, i.e., the concentration of any low-molecular-weight ions.

Polyelectrolytes, including proteins and nucleic acids, may be dissolved in water, and, in their charged state, they carry an electric current. Let us examine first how we can express the concentration of hydrogen ions. Water may be regarded both as acidic and basic. A rigorous physical-chemical definition of these concepts is that an acidic molecule is one that can release a proton, while a basic molecule is one that can bind a proton, being its receptor (Brönsted). In a reaction of the type

$$H_2O \; \rightleftharpoons \; H^+ + OH^-$$

water is an acid; in the reaction

$$H_2O + H^+ \; \rightleftharpoons \; H_3O^+ \quad \text{(oxonium ion)}$$

water acts as a base. Water, however, is a very weak electrolyte; its electric conductivity is very low. The constant for dissociation of water into the H^+ and OH^- ions, i.e., the equilibrium constant of the first reaction, is

$$K = \frac{[H^+][OH^-]}{[H_2O]} = 10^{-14} \text{ mole/liter} \qquad (4.70)$$

Since $[H^+] = [OH^-]$, the hydrogen ion concentration in water is $[H^+] = 10^{-7}$ mole/liter. It is inconvenient to work with such numbers; as a result, generally one uses the negative logarithm to the base 10 of the proton concentration, expressed as pH. For water,

$$pH = -\log[H^+] = 7$$

Taking water as a sample of an electrochemically neutral substance, we come to the conclusion that in an acid medium, containing an excess of protons relative to water, pH < 7, while in a basic medium, in which there are fewer protons than in water, $7 < pH \leqslant 14$. A polyelectrolyte, such as polyacrylic acid, is a weak electrolyte in aqueous solution, i.e., it is an electrolyte with a low degree of dissociation; the dissociation constant of the COOH group into COO^- and H^+ is equal to 10^{-5}. It is possible to increase the degree of dissociation of a polyacid by adding a base to the solution, for example, NaOH. The molecules of NaOH are almost completely dissociated in water into Na^+ and OH^-; the OH^- ions combine with H^+ ions to form water. As a result, $[H^+]$ decreases and the pH increases. This induces the displacement of the equilibrium of the acid dissociation reaction, and as a result, additional H^+ ions pass into solution, partly compensating for the increase in pH and increasing the degree of dissociation of the acid.

In other words, addition of a strong base to a weak acid results in the salt of a strong base and a weak acid, which is almost entirely dissociated. The process is that of neutralization of an acid by a base. By measuring a change of pH of solution during titration, we may find the dissociation constant of the acid

$$K = \frac{[H^+][A^-]}{[HA]} \qquad (4.71)$$

where HA is the acid and A^- is the anion. Considering that the salt of an alkali metal and a weak acid is completely dissociated, we may write

$$\frac{[A^-]}{[HA]} = \frac{\alpha}{1-\alpha} \qquad (4.72)$$

where α is the mole fraction of the added base, i.e., the degree of neutralization. Consequently,

$$K = [H^+]\frac{\alpha}{1-\alpha}$$

and, by taking the logarithm, we obtain

$$pH = pK + \log\frac{\alpha}{1-\alpha} \qquad (4.73)$$

where $pK = -\log K$. During titration, pH may be determined in a number of ways which we shall not discuss.[38] If we know α (from the amount of added base) and pH, we can find pK. For a weak base, we must replace the positive sign by a negative sign on the right-hand side of Eq. (4.73).

The simple Eq. (4.73) is valid for a univalent acid which contains a single ionizable group. We are dealing, however, with a polyacid. It is evident that a charged macromolecule induces around it a considerable electrostatic potential, which attracts H^+ ions in the case of polyacids and repels them in the case of polybases. This potential increases with an increase in α. Therefore, the increase in pH with an increase in α for polyacids and the decrease in pH with an increase in α for polybases are steeper than for univalent electrolytes. The titration curve, i.e., the dependence of pH on α, is determined by the interactions of neighboring groups in the chain and may be found theoretically, starting from an examination of the polyelectrolyte macromolecules as a cooperative system.

Changes of pH and ionic strengths may change the dimensions of the macromolecules of a polyelectrolyte fivefold and more. The theory of the dimensions of such molecules must also take into account the factors which determine the dimensions of usual macromolecules and Coulombic interactions between the charged groups, as well as the ionic atmosphere formed by the small counterions about each charged atom of the chain. The theory of the dimensions of polyelectrolytes has attracted the attention of many scientists, among whom the most prominent is Katchalsky.[39,40] Ptitsin has proposed a rigorous theory of the dimensions of polyelectrolyte molecules at low degrees of ionization[41] and an approximate theory valid for any degree of ionization.[42] Ptitsin's theory gave, for the first time, quantitative agreement with experiment and, thus, was able to explain it completely. It is impossible to present this theory here in detail; this would lead the reader too far afield.

One of the difficulties encountered with calculations on polyelectrolytes is related to the necessity of knowing the dielectric properties of the medium surrounding a macromolecule. In the equation for the Coulombic repulsion of univalent charges e_1 and e_2 immersed in some medium, we find the dielectric constant ε

$$f = \frac{e_1 e_2}{\varepsilon r^2}$$

For water at room temperature, $\varepsilon = 80$. Does this mean that in water the force of interaction between two charged groups on a macromolecule is 80 times weaker than in vacuum? With a knowledge of the expected shape of the theoretical dependence of the titration curve and the dimensions of the macromolecule on ε, it is possible to find an effective value of ε from experimental data. It is found that the local value of the dielectric constant which characterizes the interaction between close groups is much less than 80. For intergroup distances of about 4 Å, the local ε is 20; it becomes 80 only at $r = 8$ to 10 Å. In the region of 4 to 8 Å ε is proportional to r^{-2}, while at very small distances, less than 4 Å, it decreases rapidly as r^{-3}. The decrease of the local ε is explained by the fact that there are very few water molecules between neighboring groups.

A very interesting idiosyncracy of polyelectrolyte molecules follows from these considerations. The dimensions of these molecules may be changed by changing the pH of the medium. If, for example, the molecules form a fiber, then a change in the charge will be accompanied by an elongation or a shrinking of the fiber, and, consequently, the fiber may perform work. This work is performed at constant temperature and pressure. In this manner, chemical energy (or more exactly, electrochemical energy) may be transformed directly into mechanical energy, with no use of heat. Such mechanical processes will be discussed in Chapter 9. Looking ahead, however, let us point out an interesting idea which comes up at this point: Is it possible that muscle action may be reduced to the mechanochemistry of polyelectrolytes? This would be very pleasant since, in that case, one of the problems of biology would have a relatively easy solution!

Let us turn to proteins. They are constructed of amino acids, namely, of weak acids and weak bases. As we have pointed out already, an amino acid must be expressed, not by the usual formula

$$H_2N \cdot CR_1R_2 \cdot COOH$$

but by the formula of the dipolar ion

$$H_3N^+ \cdot CR_1R_2 \cdot COO^-$$

How is this proved? It is known that the heat of reaction for the ionization of organic acids

$$RCOOH \;\rightleftharpoons\; RCOO^- + H^+$$

is of the order of 1,000 cal/mole, while that of a substituted ammonium ion,

$$\overset{+}{R N H_3} \; \rightleftharpoons \; RNH_2 + H^+$$

is of the order of 12,000 cal/mole. Amino acids in acid solutions are characterized by heats of ionization of 1,300 to 2,100 cal/mole, and in alkaline solution by 10,000 to 13,000 cal/mole, i.e., the reactions are

$$\overset{+}{H_3 N} \cdot CR_1 R_2 \cdot COO^- + H^+ \; \rightleftharpoons \; \overset{+}{H_3 N} \cdot CR_1 R_2 \cdot COOH$$

and

$$\overset{+}{H_3 N} \cdot CR_1 R_2 \cdot COO^- + OH^- \; \rightleftharpoons \; H_2 N \cdot CR_1 R_2 \cdot COO^- + H_2 O$$

Other evidence of the dipolar state of amino acids may be found in the observations that, in their aqueous solution, the dielectric constant increases strongly and that solid amino acids have a high density and high T_m, since, in a crystal, the molecules are bound together by strong electrostatic forces between the positive and negative charges of the molecule.[43] Because of the amphoteric nature of amino acids, they are characterized by two dissociation constants, which correspond to the titration of the acid and alkaline groups. The corresponding values of pK are given in Table 7.

Table 7   Electrochemical Constants of Aliphatic Monoamino Acids

Amino acid	pK_1	pK_2	pH$_i$
Glycine	2.35	9.78	6.1
Alanine	2.34	9.87	6.1
Leucine	2.36	9.60	6.0
Serine	2.21	9.15	5.7

In Table 7, pH$_i$ indicates the pH at the isoelectric point. If the amino acid is charged positively, it migrates to the cathode; if negatively, it migrates to the anode. The isoelectric point corresponds to the neutral state of an amphoteric electrolyte; the amino acid remains immobile between the electrodes and does not participate in electron conductivity. We have

$$pH_i = \tfrac{1}{2}(pK_1 + pK_2)$$

The same naturally applies to proteins as well. Since they are polyampholytes, they are characterized by an isoelectric point, an important constant of the protein macromolecule. A small change in the amino acid composition of a protein is immediately reflected in its isoelectric point, on the ability of the protein molecule to migrate between electrodes immersed in the solution (electrophoresis). The difference in the electrophoretic properties of normal and mutant hemoglobins has been mentioned previously.

The titration curves of proteins and nucleic acids give valuable information on their properties. Of particular interest are the conformational transitions in proteins, nucleic acids, and their synthetic analogs, which are brought about by changes in pH. It is evident that the role of the ionic environment of biopolymers is very important and must affect their biological function. A number of examples of such phenomena will be found in the following chapters.

As a conclusion, let us point out an important practical significance of synthetic polyelectrolytes. Insoluble crosslinked polyelectrolytes, which swell in water and other liquids, are used as ion-exchange resins, called *ionites*.[44] Ionites are capable of adsorbing specific ions from solutions; this finds application in the purification and fractionation of various electrolytes and in the purification of nonionic substances from electrolytic impurities. An example of the application of ionites is the purification of antibiotics.[45] As a result of this procedure, it is possible to obtain penicillin, streptomycin, and other antibiotics free of harmful impurities.

It is evident that the questions of the ion exchange are important in biology, since cellular membranes in a number of cases are characterized by their special ability to adsorb ions. The same is apparently true of other supramolecular biological structures.

Having become acquainted with some aspects of the structure and properties of macromolecules, as well as with the methods of their theoretical and experimental investigation, we may return to biological molecules, and first of all to proteins.

Chapter 5

THE PHYSICS OF PROTEINS

SECONDARY STRUCTURE OF PROTEINS

We have already discussed the primary structure of proteins. This very term makes us think about the existence of secondary, tertiary, and higher structures. In fact, such a gradation of structure does exist in proteins.

In synthetic polymers, the sequence of monomer units (the macro-molecular text) is either a monotonic simple repetition of identical letters or, in copolymers, a random sequence of a small number of different letters (usually two). In most cases, there is no specificity of the primary structure, the macromolecule does not contain any information in this sense. On the other hand, the macromolecule possesses a certain structure as a whole, which, it is true, fluctuates, but which, on the average, is defined. The chain is twisted into a more or less stretched coil, as a result of the ability of the links to take on different conformations because of rotational isomerism.

In a protein the primary structure contains a vast amount of information[1] and is strictly determined. Furthermore, the protein differs from a simple polymer by the presence of specific groups in the amino acid residues which may form chemical, electrostatic, hydrogen, or Van der Waals bonds. Cystine residues combine with each other by disulfide bonds S—S. Such bridges, for example, link the two peptide chains A and B of the insulin molecule (Fig. 39). Positively and negatively charged groups enter into electrostatic interactions with each other. Van der Waals forces, i.e., the usual intermolecular forces, act between any atoms and groups of atoms, which approach each other; in this sense, a protein has no essential differences from other polymers. Hydrogen bonds play a particular role.

Hydrogen atoms, which enter into the composition of the hydroxyl group OH, the imino and amino groups NH and NH_2, and some other groups as well (but not CH, CH_2, CH_3), form specific noncovalent bonds with other atoms, such as O and N. These hydrogen bonds are indicated by dots

$$-O-H\cdots O\langle \qquad -O-H\cdots N\langle \qquad \rangle N-H\cdots O\langle \qquad \rangle N-H\cdots N\langle$$

The presence of hydrogen bonds is evidenced by the decrease of the interatomic distances between atoms of O and O, O and N, or N and N for these different groups. Thus, the smallest distance between two valently nonbonded atoms of oxygen is equal to the sum of their Van der Waals radii, that is, $1.40 \times 2 = 2.80$ Å. In the group $-O-H\cdots O-$, this distance must be greater; it must be equal to the sum of the length of the O—H bond (approximately 1 Å), the Van der Waals radius of hydrogen (from 1 to 1.4 Å), and the radius of oxygen (1.4 Å). This adds up to 3.4 to 3.8 Å. However, when a hydrogen bond is formed, this distance decreases to 2.7 Å, that is, the hydrogen attracts the two oxygen atoms. The energy of a hydrogen bond is equal to several kcal/mole. This is much less than the energy of the usual chemical bond (about 100 kcal/mole), but considerably greater than the energy of Van der Waals intermolecular interactions.

The presence of hydrogen bonds between molecules must be highly reflected in the property of the substance. Let us compare two isomeric substances, namely, ethyl alcohol C_2H_5OH and dimethyl ether $(CH_3)_2O$. The molecules of the first form hydrogen bonds

$$\begin{array}{c} H \\ | \\ O-H\cdots O-C_2H_5 \\ | \\ C_2H_5 \end{array}$$

The molecules of the second do not form hydrogen bonds. Consequently, the boiling temperature of alcohol is 78°C, much higher than that of ether (24°C). The heat of the evaporation of alcohol is 10.19 kcal/mole; that of ether is 4.45 kcal/mole. The high boiling temperature of water (with respect to, let us say, methane CH_4, the molecules of which contain as many electrons as those of H_2O), the high energy of evaporation, and the known anomalies of water may be interpreted in terms of the presence of hydrogen bonds which link together all the water molecules into a single network. The energy of the hydrogen bond for ice is 4.5 kcal/mole. Langmuir said that because of the presence of hydrogen bonds, the entire ocean consists of essentially a single molecule and the process of fishing is a process of dissociation.

The nature of the hydrogen bonds is complicated; it contains both electrostatic interactions and quantum-mechanical donor–acceptor bonds determined by the migration of an electron from the hydrogen atom to an atom which is not valently bonded with it. The theory of the hydrogen bond has been developed by Sokolov[2] and Coulson.[3]

Hydrogen bonds play an important role in tautomeric transitions. For example, the migration of the hydrogen in keto–enol tautomerism occurs through the momentarily formed hydrogen bond.

Hydrogen bonds can be easily detected by a number of methods, for example, by infrared spectroscopy. The absorption bands of the OH, NH, and NH_2 groups, which participate in hydrogen bonds, are displaced in the direction of longer wavelengths and are broadened. Spectroscopy makes it possible to determine the fraction of groups which participate in hydrogen bonds and to determine the character of these bonds, for example, to differentiate between the $O—H\cdots O$ and $O—H\cdots N$ bonds.

On the other hand, hydrogen bonds may be detected by X-ray diffraction and electron diffraction, by measuring changes in the interatomic distances. The presence of hydrogen bonds is reflected in the isotope exchange of hydrogen, deuterium, and tritium, slowing this exchange considerably. More detailed descriptions may be found in the literature.[4,5]

What is the role of hydrogen bonds in the structure of proteins? Each peptide group $—CO—NH—$ contains hydrogen atoms which are capable of donating a hydrogen bond and an oxygen atom which is capable of entering into a bond with such a hydrogen:

$$\rangle C{=}O\cdots H{-}N\langle$$

The structures and properties of proteins are determined to a great extent by hydrogen bonds. The same is true of other important biological molecules, namely, nucleic acids, carbohydrates, and lipids. Finally, life on earth requires water, the molecules of which interact with biological molecules by means of hydrogen bonds.

The importance of hydrogen bonds in nature is quite great. The first element of the periodic system, hydrogen, has related to it two extremely important phenomena, namely, the hydrogen bomb and the hydrogen bond. And if the hydrogen bomb is a menace to the very existence of mankind, hydrogen bonds may be called one of the sources of life.

Pauling and Corey[6,7] have applied general molecular-physics concepts to the structure of proteins, with the conclusion that they must have specific macromolecular structures; these structures are characterized by intramolecular hydrogen bonds. They assume that the peptide group $—CO—NH—$ is planar and that all the peptide groups in a macromolecule must be disposed in a similar manner forming two hydrogen bonds. Furthermore, in each $C{=}O\cdots H{-}N$ group, all four atoms must be located essentially along a single straight line. According to Pauling, the deviations from a linear disposition of $C{=}O$ and $H{-}N$ may not exceed 20°. From these starting precepts, Pauling and Corey proposed a protein chain structure in the form of what is called the α-*helix* (Fig. 67). Each turn of the helix contains 3.6 amino acid residues, i.e., the helical period contains 5 turns or 18 residues (the smallest whole number divisible by 3.6). The pitch of a single turn is 5.6 Å and the diameter of the helix is 10.1 Å. The helix is held together by hydrogen bonds directed parallel to its axis. These bonds link the first peptide group with the fifth, the fifth with the first and the ninth, etc.

The model of Pauling and Corey has been confirmed in a brilliant way by the direct X-ray diffraction studies of a number of proteins as well as of their analogs, synthetic polypeptides. This is a magnificent success of molecular biophysics. Another proof stems from a study of the isotope-exchange kinetics of the imino hydrogens of the protein for deuterium and tritium. A slow-down of the exchange relative to the low-molecular-weight peptides, found by Linderstrøm-Lang, has made it

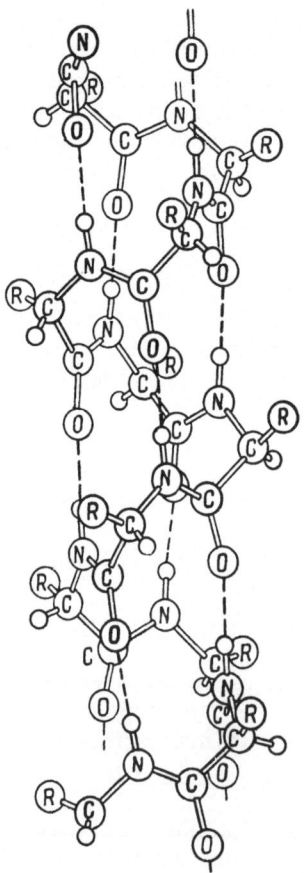

Fig. 67. The Pauling–Corey α-helix

possible to estimate in a rough way the number of peptide groups which participate in intramolecular hydrogen bonds.

In this manner, the protein macromolecules are found in a unique conformation, whose links have a 1 : 1 correspondence to definite rotational isomers; trans isomers are formed relative to rotation about the OC—NH bond, while rotational isomers, turned in a single direction by 120°, are formed relative to the NH—CHR and RHC—CO bonds.

The structure of the α-helix is similar to that of a polymer in the crystalline state. But while the succession of the rotational isomers in crystalline polymers takes place as a result of the intermolecular packing of the chains, in a protein-chain helix the rotational isomers are fixed by intramolecular hydrogen bonds which play the role of staples. For this reason, the structure of the α-helix of a given macromolecule is independent of the neighboring molecules and may remain intact in solution. A macromolecule in the form of a regular, periodic α-helix is similar to a crystal; the entire crystal in this case, however, consists of a single molecule.

The question comes up: What is the direction of rotation in the helix? Is the screw right-handed or left-handed? The right- and left-handed helices are related to each other as mirror images, as optical antipodes. These structures are asymmetric. If the amino acid residues had been symmetric by themselves, the right- and left-handed helices could have arisen with equal probability. All amino acids except glycine, however, are asymmetric and belong to a single stereochemical class; they are *l*-amino acids. It is possible to construct a *d*-helix from *l*-amino acids and an *l*-helix from *l*-amino acids. These two helices will no longer be mirror images of each other, but they will differ in reality from each other by the disposition and packing of their side chains. An overwhelming majority of the α-helices encountered are right-handed (*d*-helices); such a helix is shown in Fig. 67. Therefore, a protein α-helix possesses twofold asymmetry; it contains asymmetric amino acid residues and it is asymmetric as a whole, being a helical structure. This is reflected in the optical properties of proteins and polyamino acids.

The structure of the α-helix is adapted to the performance by the protein of its various functions. The interior of the α-helix contains the repeating group of the protein, namely, the peptide bonds. On the other hand, the different side-chain groups of the various amino acids of the protein are turned outward and may, as a result, interact with molecules in the surrounding medium.

Protein structure is not restricted to α-helices. Pauling and Corey proposed another variation of an ordered structure, namely, the β form. In this case, the protein chains are greatly extended, much more so than in the α-helix; they form a zigzag. Such a structure is stabilized now not by intramolecular hydrogen bonds but by intermolecular hydrogen bonds between parallel-extended antiparallel chains. The result is a pleated sheet structure, shown in Fig. 68. Since such a structure may be regularly linked

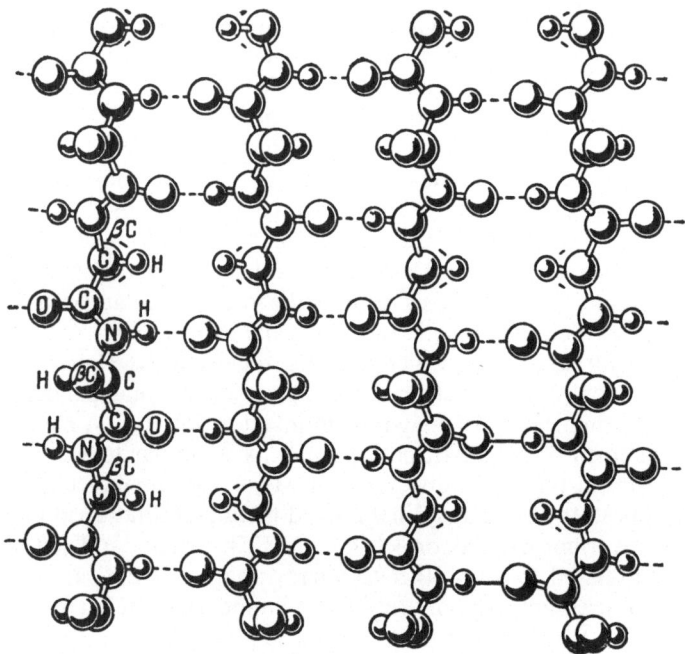

Fig. 68. The β form of polypeptide chains.

by hydrogen bonds only if the entire protein macromolecule is involved, it has much less chance of remaining intact in solution.

It should be noted that a structure similar to the β form of the protein is present in some synthetic crystalline polymers, for example, polyamides, such as

$$-CO-NH-(CH_2)_6-NH-CO-(CH_2)_4-CO-NH$$

Polyamides are widely found macromolecules; they are the materials from which the synthetic fibers, such as nylon and capron, are made. On the other hand, such a structure is also found in natural protein fibers, such as silk fibroin. The silk worm, *Bombyx mori*, makes an excellent highly structured fiber; it was only in the twentieth century that chemists learned to make materials that can compete with silk. The β form determines the important properties of silk, namely, its strength and elasticity.

Silk fibroin is an example of a fibrous protein. Other fibrous proteins are collagen, which enters into the composition of skin, and keratin, which is found in birds' feathers and the hair of mammals. Their structure is specific and it cannot always be reduced to the β form.

The *cross-β structure* of the macromolecule has been found in poly-O-acetyl-L-serine. In this case, perpendicular hydrogen bonds, similar to those found in normal β structures, link regularly located

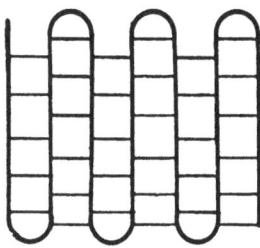

Fig. 69. Schematic representation of the cross-β form.

segments of a single chain (Fig. 69). The same form has been found in some proteins such as lysozyme, some globulins, etc.

The vast majority of biologically active proteins, namely, enzymes, antibodies, etc., exist in the form of globular compact structures. In a number of investigated globular proteins, the polypeptide chain contains α-helices.

In this manner, protein macromolecules may possess a highly ordered structure as a whole, a secondary structure. Later, we shall examine the conditions of formation and destruction of the secondary structure. Let us just remark at this point that in real proteins there is no complete secondary ordering; for example, along with a regular helical portion of the chain, there are also unordered portions, the links of which may be described by a whole number of conformations, of rotational isomers.

The secondary structure of a protein or synthetic polypeptide may be determined by a number of physical techniques. First, we have X-ray diffraction. Second, one may use infrared spectroscopy. Let us examine in greater detail now the infrared spectra of proteins.

The amide (peptide) group $-CO-NH-$ has a number of characteristic atomic vibrations, which are reflected by definite absorption bands in the infrared spectrum. Typical frequencies of such bands (expressed in wave numbers, i.e., cm^{-1}) are listed in Table 8 for liquid N-methylacetamide $H_3C-CO-NH-CH_3$.

Table 8   Characteristic Amide Bands[8]

Band	Frequency, cm^{-1}	Nature of vibration
A	3,280	Split, because of interaction with the amide II band
B	3,090	N—H vibration
I	1,653	C=O vibration
II	1,567	C—N—H in plane deformation and C—N stretching
III	1,299	Same
IV	627	O=C—N in plane deformation
V	725	C=N—H out of plane deformation
VI	600	O=C—N out of plane deformation
VII	206	Torsional oscillation about C—N

The amide I and II bands are characteristic for definite conformations of peptide chains. It is not only the frequency of the given vibration which is characteristic, but also its polarization, i.e., the direction of this vibration. If we study the absorption of plain polarized radiation in a stretched film of protein or polypeptide, we find the phenomenon of dichroism as the absorption bands split into parallel and perpendicular ones. In the first, the light polarized parallel to the axis of stretching is absorbed more strongly; in the second, it is the light polarized perpendicular to this axis which is absorbed more strongly. If macromolecules (in the α or β forms) are in themselves stretched along this axis, then the sign of the dichroism indicates the orientation of the vibrating bonds in the macromolecule. In an α-helix, the $C=O\cdots H-N$ bonds are located along the axis of the molecule; in the β form, they are located perpendicular to this axis. This is distinctly reflected in infrared dichroism.

The α form has characteristic frequencies of the amide I band close to 1650 to 1660 cm^{-1}, and an amide II band close to 1550 cm^{-1}. The first band is polarized parallel ($\parallel$), and the second band perpendicular ($\perp$). The β form has an amide I band close to 1630 cm^{-1} and band II close to 1520 cm^{-1}. Data for synthetic polypeptides are presented in Table 9.

The application of the amide I and II absorption bands to the determination of protein conformations is replete with difficulties. In the case of proteins, these bands are complex, and it is not easy to interpret them. It is also difficult to obtain globular proteins in an oriented state. Volchek, Purkina, and coworkers [10] have shown that the secondary structure of proteins may be studied with the aid of the amide V band; this band corresponds to the out-of-plane deformation vibrations of NH. The α form has a band at 620 cm^{-1}, the β form at 700 cm^{-1}, and the random coil at 650 cm^{-1}. These data show that it is possible to use this approach strictly in terms of frequencies, without applying dichroism, for the determination of the structure of a protein both in solution and in a lyophilized state in the form of a powder pressed in KBr. Further details on the infrared spectra of proteins and polypeptides may be found in the literature.[8,11]

The secondary structures may also be determined by other techniques, namely, by the use of optical activity, isotope exchange, measurements of viscosity, light-scattering, and dipole moments in solution of the materials under investigation. Evidently, the conformation of a macromolecule depends on the nature of the solution, on the composition of the solvent, on its pH, and on the dissolved ions. If the solution contains polypeptides in the α form, then its intrinsic viscosity, when expressed in the form of $[\eta] = KM^{a}$, has an exponent of 1.8; this is a value close to that which is expected for rigid rods. On the other hand, if the rigid conformation has been destroyed and the polypeptide has taken on the shape of a statistical coil, the exponent decreases to 0.9. From the laws of viscosity, it is possible to determine a ratio of the major to the minor axes of the α-helix, which is in good agreement with calculations. A similar agreement is obtained

Table 9   Amide I and II Bands in Synthetic Polypeptides[9]

Polypeptide	Deg. of polymerization	Mol. wt.	Amide I		Amide II		
			Freq., cm^{-1}	Dichroism	Freq., cm^{-1}	Dichroism	Structure
Poly-L-valine	190	19,000	1,638	⊥	1,545	=	β
Poly-L-leucine	35	3,900	1,650	=	1,543	⊥	α
Poly-S-methyl-L-cysteine	140	16,000	1,632	⊥	1,540, 1,525	=	β
Poly-S-benzyl-L-cysteine	365	77,000	1,632	⊥	1,542	=	β
Poly-L-methionine	135	18,000	1,648	=	1,540	⊥	α
Poly-O-acetyl-L-serine	130	16,800	1,637	⊥	1,520	=	β
Poly-L-serine	130	11,300	1,653	⋯	1,525–1,535	⋯	coil
Poly-β-benzyl-L-aspartate	190	39,000	1,668	=	1,563	⊥	α
Poly-γ-benzyl-L-glutamate	100	22,000	1,645	=	1,542	⊥	α

from experiments on the shapes of macromolecules by using the angular distribution of light-scattering and changes in dipole moments.[13] Finally, recently it has been possible to obtain directly photographs of the α form, as well as of the randomly coiled macromolecular peptide, by using an electron microscope.

The question of the dependence of the secondary structure of a polypeptide or protein on the amino acid composition and on the primary structure in the case of a protein is very important and interesting. This problem has not been solved entirely up to now; however, today we already possess such quantitative information on proteins and polypeptides that we may regard this question as close to solution.

Blout has made a hypothesis on conformations.[9] Formation of an α-helix may be hindered, first, by geometric (steric) factors and, second, by the formation of other hydrogen bonds which would interfere with the regular conformation. The formation of the α form is hindered in polypeptides which contain amino acid residues with bulky substituents on the β carbon atom. On the other hand, the α form is not formed if the residue contains hetero atoms O or S, which can enter into hydrogen bonds, different from the principal ones which stabilize the α-helix. Furthermore, the α-helix forms only at sufficiently high degrees of polymerization; some chains prefer a β form in the solid state. If a side-chain group can acquire charges, then the electrostatic interaction may interfere with the formation of the α-helix. This hampers what obviously must be a strong function of the medium. A conformational classification of amino acids is given in Table 10.

Guzzo has analyzed the primary and space structures of myoglobin and hemoglobin given by Kendrew and Perutz; he concluded that α-helical regions are stopped by the residues of Pro, Asp, Glu, and His.[13] Proline residues cannot enter into an α-helix since they have an imino and not an amino group. The Asp, Glu, and His residues carry charges, which induce an oriented shell of water molecules that interfere with the formation of an α-helix. With the knowledge of the primary structure of lysozyme, Guzzo predicted the distribution in it of α-helical regions along the chain; this turned out to be in reasonable agreement with experiment.

We encounter here a very serious conformational problem in the analysis of the rotational isomerism of cooperating monomeric units in proteins and polypeptide chains. The basis of such an analysis can and must be the rotational-isomerization theory of macromolecules which has been discussed in Chapter 4.

It follows from what has been said that synthetic polypeptides form the same secondary structures as proteins. Polypeptides consist of identical repeating amino acids. But, as long as the amino acids are linked by the peptide chains, all the Pauling and Corey concepts remain valid for a synthetic polypeptide as well. This is a fact which is very important for molecular biophysics. This makes it possible to have models of the

Table 10 Conformational Classification of α-Amino Acids[9]

α-Helix formers	α-Helix nonformers

α-Helix formers:

Alanine

$$-CH_3$$

Aspartic acid esters

$$-CH_2-\overset{\overset{\displaystyle O}{\|}}{C}-O-alkyl$$

Glutamic acid

$$-CH_2-CH_2-\overset{\overset{\displaystyle O}{\|}}{C}-OH$$

Glutamic acid esters

$$-CH_2-CH_2-\overset{\overset{\displaystyle O}{\|}}{C}-O-alkyl$$

Leucine

$$-CH_2-CH\overset{CH_3}{\underset{CH_3}{<}}$$

Lysine

$$-CH_2-CH_2-CH_2-CH_2-NH_2$$

Methionine

$$-CH_2-CH_2-S-CH_3$$

Phenylalanine

$$-CH_2-C_6H_5$$

Tyrosine

$$-CH_2-C_6H_4-OH$$

Norleucine

$$-CH_2-CH_2-CH_2-CH_3$$

Norvaline

$$-CH_2-CH_2-CH_3$$

O-Acetyl-homoserine

$$-CH_2-CH_2-O-\overset{\overset{\displaystyle O}{\|}}{C}-CH_3$$

α-Helix nonformers:

For steric reasons

Valine

$$-CH\overset{CH_3}{\underset{CH_3}{<}}$$

Isoleucine

$$-CH\overset{CH_3}{\underset{CH_2-CH_3}{<}}$$

For other reasons

O-Acetyl-serine

$$-CH_2-O-\overset{\overset{\displaystyle O}{\|}}{C}-CH_3$$

Serine

$$-CH_2-OH$$

S-methylcysteine

$$-CH_2-S-CH_3$$

Threonine

$$-CH\overset{OH}{\underset{CH_3}{<}}$$

O-Acetyl-threonine

$$-\underset{\underset{\displaystyle CH_3}{|}}{CH}-O-\overset{\overset{\displaystyle O}{\|}}{C}-CH_3$$

important properties of complex protein macromolecules in terms of the much simpler macromolecules of the synthetic polypeptides, which have a given structure.

Thus, most proteins and synthetic polypeptides may exist both in solution and as solid bodies, namely, as films, fibers, crystals. In cells, proteins exist either in solution or in organoids and membranes; in solution they have in general a globular shape in which they perform their enzymatic and other functions. In the second case, the proteins form complicated supramolecular structures, for example, fibers. The investigation of the nature of protein globules and protein supramolecular structures is one of the most important problems of molecular biophysics.

As a conclusion of this section, let us sum up in Table 11 some data on the higher levels of the structure of proteins and other biopolymers and synthetic polymers. We have listed their molecular weights M, the number of chains in a globule, n, and the number of disulfide bonds S—S.

TERTIARY AND SUPRAMOLECULAR STRUCTURES OF PROTEINS

How are the globules constructed? Is their structure random, i.e., is it similar to statistical coils, or does it have a regular character?

Globular proteins may be isolated in crystalline form. Just this fact leads one to think that the globules of a given protein are uniform and regular. The nature of these globules has been uncovered as a result of the study of the structure of several proteins by X-ray diffraction. X-ray diffraction studies of proteins were started in 1930 by Bernal, Fankuchen, Perutz, and Dorothy Crowfoot-Hodgkin. This important development of molecular biophysics has led to the major successes attained in only recent years. This may give an idea of the difficulty of this work. We cannot describe here the method of X-ray diffraction. The reader may become acquainted with it through the specialized literature.[15,16] Let us limit ourselves to the simplest information. X-rays differ from visible light in that they have very short wavelengths (shorter by a factor of 1,000). As is known, when light passes through small openings, the dimensions of which are of similar magnitude to the wavelength, we may see the phenomenon of diffraction; the light rays scatter on the other side of the opening. This is due to the fact that each point of the opening is now a source of secondary rays which reinforce and interfere with each other in different directions. The diffraction of light is used for its spectral decomposition; rays of different wavelengths deviate in different ways, and when white light passes through a small circular opening, it gives a diffraction pattern in the form of a number of concentric circles in spectral colors. To decompose light into its spectrum, we may use diffraction gratings, namely, a regular sequence of narrow slits. If we know the width of a slit, we may determine

Table 11[14]

Compact globular particles				Statistical coils		Rodlike particles	
Polymer	M	n	S—S	Polymer	M	Polymer	M
Polystyrene latex	10^9	1	0	Polystyrene in toluene		Fibrinogen	330,000
Insulin	5,800	2	3	Reduced ribonuclease	13,700	Collagen	345,000
Ribonuclease	13,700	1	4	Oxidized ribonuclease	14,100	Myosin	620,000
Lysozyme	14,400	1	5			DNA	5×10^6
Myoglobin	17,000	1	0	Ovalbumin in urea	44,000	TMV	4×10^7
Papain	20,900	1	3	Plasma albumin in urea	66,500		
Trypsin	23,800	1	6	Myosin in guanidine			
Chymotrypsin	24,500	3	5	hydrochloride	200,000		
Carboxypeptidase	34,300	1	0	RNA	1.5×10^6		
β-Lactoglobulin	35,000			Denatured DNA	5×10^6		
Hexokinase	45,000	2	0				
Ovalbumin	44,000						
Bovine plasma albumin	66,500	1	17				
Yeast enolase	67,000	1	0				
Hemoglobin	68,000	4	0				
Liver dehydrogenase	83,000	2	0				
Hemerythrin	107,000	8	0				
Aldolase	142,000	3	0				
γ-Globulin	160,000	4	25				
Yeast ribosomes	3.5×10^6						
Bushy stunt virus	8.9×10^6						

the wavelengths of the light by the declination of the diffracted rays; and
conversely we may determine the width of the slit from the wavelengths of
the light. For X-rays the role of the diffraction gratings is played by crystals
since they are regular gratings consisting of atoms and molecules which are
displaced from each other by distances of one to several angstroms, i.e.,
of the same order of magnitude as the wavelength of the X-radiation. An
X-ray which passes through a monocrystal gives a pattern of regularly
positioned spots on photographic film; the rays scattered by the crystal
disperse into strictly determined directions. In Fig. 70 we have shown the
X-ray diagram obtained with a crystal of sperm-whale myoglobin. If the
wavelength of the X-ray is known, the positions of the atoms in the crystal
may be determined from the positions of the diffraction spots.

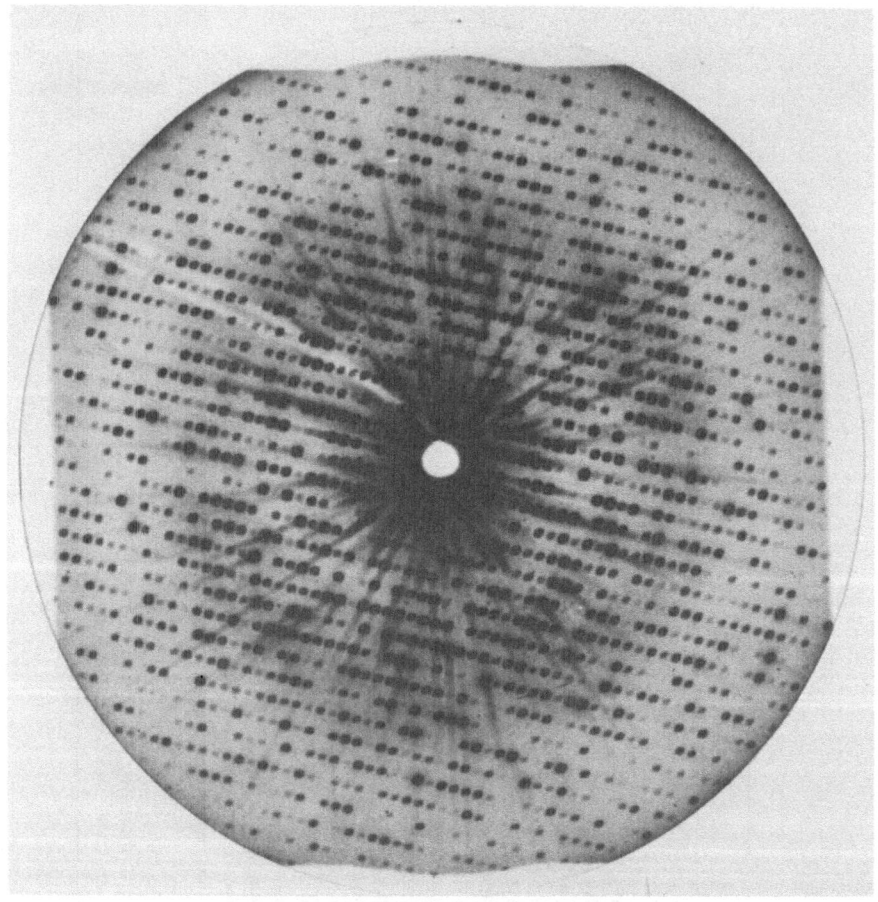

Fig. 70. X-ray diffraction diagram of crystalline sperm-whale myoglobin.

If the molecules which form the crystal are complicated, i.e., if they contain many atoms, the problem of such a determination becomes extremely difficult. Special highly refined methods have been developed for this purpose. A major achievement was the determination of the complete X-ray structure of vitamin B_{12} obtained by Crowfoot-Hodgkin. A molecule of vitamin B_{12} contains about 90 atoms, not counting the hydrogen atom. A protein molecule, however, contains many more atoms!

A serious difficulty of the X-ray analysis stems from the fact that the light atoms of which proteins are constituted scatter X-rays very weakly. In this way, the hydrogen atoms become completely unnoticeable, and their presence and distribution is determined only by special indirect methods.

In 1935 Perutz developed a method for the investigation of the structure of hemoglobin; this was based on the preparation of derivatives of the protein which contain heavy atoms, namely, silver and mercury. It was possible to prepare derivatives which crystallized in an isomorphous fashion with hemoglobin. The presence of the silver and mercury atoms (which replaced the hydrogen atoms in the sulfhydryl group) made it possible to obtain a much sharper diffraction pattern. Using this method, Perutz studied the structure of hemoglobin, while Kendrew investigated that of a simpler but related protein, myoglobin. Hemoglobin is contained in the erythrocytes of blood, while myoglobin exists in muscle. Both proteins contain the same porphyrin groups, at the center of which there is an iron atom. Myoglobin contains one heme group, while hemoglobin contains four such groups.

A model of myoglobin, constructed from a rigorous X-ray analysis is shown in Fig. 71.[16–18] The disk seen on edge in the lower left-hand part is the heme group; the small spheres are the heavy atoms linked to the myoglobin. The pretzellike structures are α-helices. The X-ray analysis is

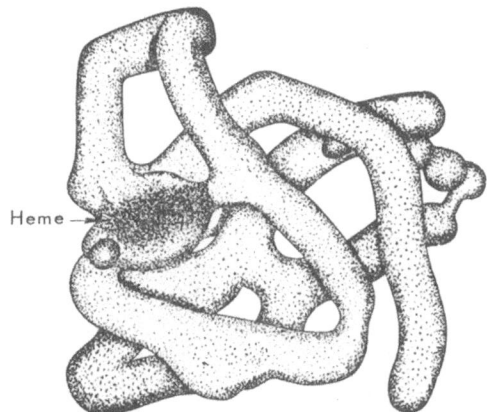

Heme

Fig. 71. Structure of myoglobin.

in good agreement with isotope-exchange and optical-rotation investigations; it shows that the helices constitute approximately 75 % of the amino acid contents of this protein. The other 25 % are contained in unordered regions. This complicated globular structure, the tertiary structure of myoglobin, is stabilized by chemical bonds, namely, disulfide bonds, hydrophobic and hydrogen bonds, and electrostatic interactions. The X-ray investigation makes it possible to decipher almost completely even the primary structure of this macromolecule, which contains 153 amino acid residues and has a molecular weight of 18,000.

In this manner, the α-helix within the globule is, in turn, twisted into a specific tertiary structure. The biological role of a tertiary structure is quite important as we shall see in Chapter 8.

Hemoglobin, which has been studied by Perutz, has been found to contain four polypeptide chains, four subunits, each of which is similar to myoglobin. The four chains are not identical; two of them are of type α, two of type β. The α and β chains differ slightly in their amino acid compositions and primary structures. The α chain contains 141 amino acid residues, the β chain 146 residues. Thus, hemoglobin contains a total of 574 residues. The four subunits are located approximately at the apices of a tetrahedron. Here we encounter already the quaternary structure; the four globular subunits form a regular system. It is also possible to cite other examples of quaternary structure. Thus, the protein of the tobacco mosaic virus particle consists of 2,000 identical subunits, each of which has dimensions similar to those of myoglobin. These subunits are arranged in the form of a hollow helix, inside of which is located the RNA of the virus. In the case of spherical viruses, the subunits are located symmetrically on the surface of regular or almost regular polyhedra.

Bernal had examined the general problem of the architectural organization of supramolecular structures in proteins, from the point of view both of the geometric conditions of such an organization and of the forces which act between the molecules.[19] He established two basic principles: first, the cell synthesizes a large number of similar macromolecules; second, the probability of the formation of a structure of great complexity from its elements increases, and the number of possible ways of forming it decreases, if the structure under examination may be decomposed into a finite series of substructures, which are systematically included in each other. Bernal showed that there is such an hierarchy of biological structures similar to cosmic structures; stars form galaxies, galaxies are part of metagalaxies, etc. In Fig. 72 are shown six consecutive orders of magnification of muscle cells. From right to left we pass, first, from a fragment of the protein chain to an α-helix, from an α-helix to a supramolecular structure which consists of parallel chains of protein molecules (myosin) in myofibrils, from such chains to the myofibril, from the myofibril to a parallel aggregation of myofibrils in the fibril, and finally, to a section of the cell of cross-striated muscle, as seen in the microscope.

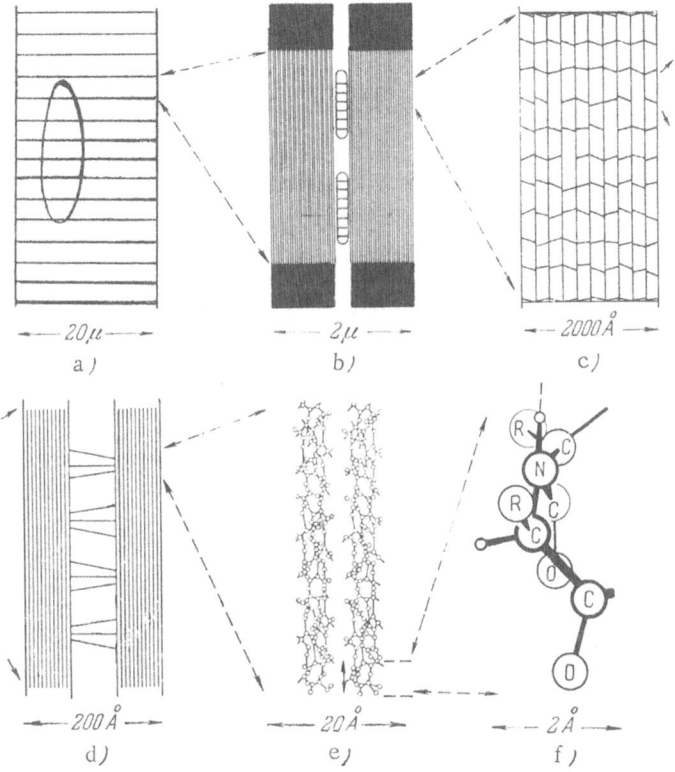

Fig. 72. Six consecutive orders of magnification of the muscle cell.

The nature of the structure on each level of organization is determined by the geometric properties of the structure of the preceding level and by forces which act between them. This "supercrystallography" which was initiated by Bernal has a general scientific significance. The automatic manner of the formation of absolutely definite structures from preceding ones is very important. The problem which has been clearly stated by Bernal has not been solved yet; it consists in seeking the rigorous conditions for organization of all the levels of structure. Ideally, having solved this problem, we should be able to predict the microscopic structure of the muscle from a knowledge of the chemical structure of its proteins.

Let us turn now to the conformations of proteins and polypeptides in a solid state.[20,21] We have already talked about the β form of silk. The same structure is found in the keratin of hair. Polyglycine gives a β form. It is interesting to note that glycine constitutes 44% of the amino acid residues present in fibroin. The β form means that there are flat layers of macromolecules linked by hydrogen bonds. These layers, in turn, are connected by intermolecular forces.

Along with the β form, there is also the cross-β structure which has been found in the insulin fibrils and epidermin. In this case, the chain axes are located perpendicular to the axis of the fiber. The same form also exists in the chain of the silk made by the fly *Chrysopa flava*. When this silk is stretched, the cross-β form is transformed into the usual β form.

Stretched α-helices give specific structures. The α-helices wind about each other, forming cablelike structures. In this superstructure, the helices are packed in a hexagonal array.

Poly-β-benzyl-L-aspartate, when cast as a film from chloroform and heated in vacuum, possesses a new so-called "ω structure" in which the α-helix has been distorted; it has now a fourfold rather than threefold symmetry. The projection of the ω form of polypeptides on a plane perpendicular to the axis of the helix is shown in Fig. 73.

The conformation of poly-L-proline is quite interesting. Poly-L-proline is an imino, and not an amino acid; it does not contain NH_2 groups. Consequently, it is impossible to form $N-H\cdots O$ hydrogen bonds. For steric reasons, polyproline cannot be completely extended; this is hindered by the cyclic groups of the monomer. It has been established that polyproline crystals have a triple helix with a repeat distance of 9.38 Å. The supramolecular and secondary structures of polyproline have not been completely established yet. It should be noted that this field of investigation is only in its infancy and actually we know very little about the supramolecular structures of proteins.

One of the well-studied proteins is collagen, the principal structural element of skin and connective tissue of animals of all types. Its function is structural. Collagen contains more than 30% glycine, a large number of pyrrolidine rings, and about 24% proline and hydroxyproline. The structure of collagen is similar to that of polyglycine in one of its crystal forms. Polyglycine II, studied by Crick and Rich,[20] has a hexagonal

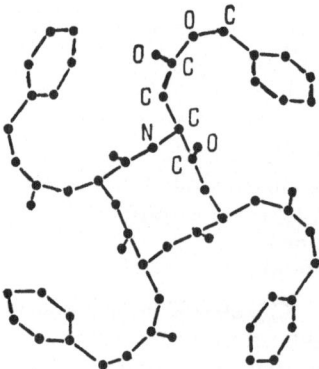

Fig. 73. The ω form of poly-β-benzyl-L-aspartate; projection on a plane perpendicular to the axis of the helix.

structure; all the polypeptide chains are parallel to each other and each of them is the triple helix. Each chain is connected to six others by hydrogen bonds, which have different directions and are not located in a single plane. This structure is shown in Fig. 74. The collagen molecule is obtained from a triplet of polyglycine chains which have been twisted into a single cable; this is a structure with a twisted helix (Fig. 75). The simultaneous twisting of three polypeptide chains can be explained by the steric interaction of the nonglycine residues, namely, of the prolyl and hydroxyprolyl groups. Since collagen is a protein with a relatively simple primary structure, it becomes possible to examine how this structure determines the secondary and tertiary structures of collagen. When collagen is partially hydrolyzed, Gly–Pro–HPro tripeptides are formed.

The problem of the synthesis of a polypeptide which would be similar to natural collagen has been approached by Andreeva et al. They synthesized a polymer (Gly–Pro–HPro)$_n$ with a molecular weight of several thousand. This polypeptide turned out to be similar to collagen, as shown

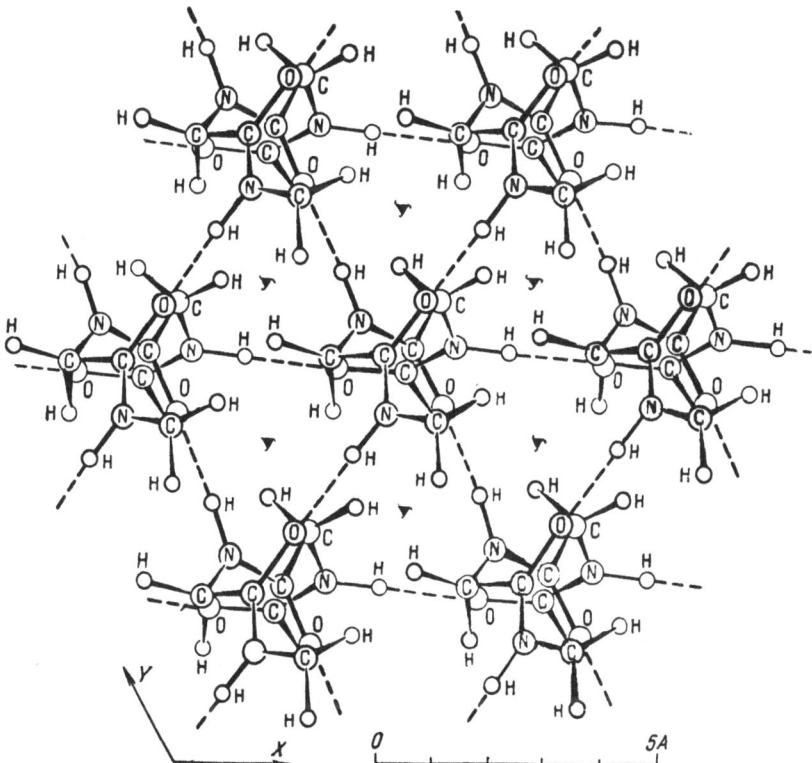

Fig. 74. Structure of polyglycine II.

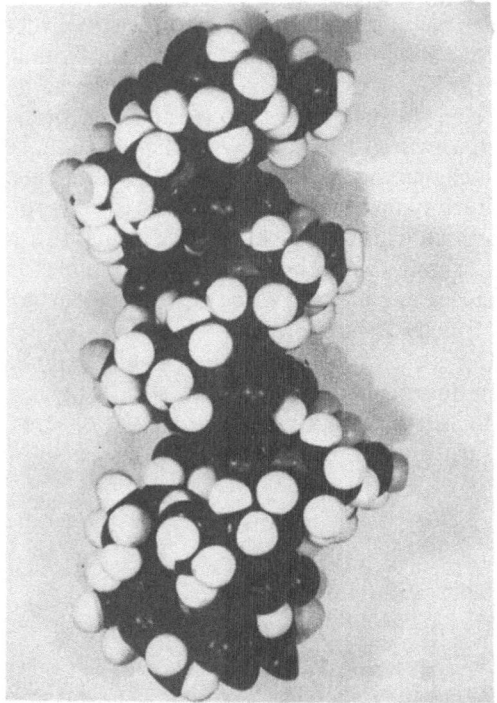

Fig. 75. Structure of collagen.

by X-ray, infrared spectroscopic, and polarimetric investigations.[22] This, of course, does not represent the synthesis of a natural protein. But the results of this study are cited, since this rather simple example has been used as a direct experimental verification for the relations between the primary and higher levels of protein structure.

There are good reasons to believe that the primary structure always determines the secondary, tertiary, and quarternary structures of a protein. In other words, we have that very automatism in the formation of a hierarchy of structures which was discussed by Bernal. The primary structure of a protein is genetically coded. Since the higher levels of structure are uniquely determined by the primary structure, they are also coded. Thus, the structure of the genetic material, the nature of the matrix, which records the protein text, dictates as well the molecular and supramolecular structures of biological materials. Further investigations in this field will lead to the solution of the problem related to the structures of cells and organisms and entire supramolecular systems with differentiation of cells and tissues.

INTERACTIONS WHICH DETERMINE THE THREE-DIMENSIONAL STRUCTURE OF PROTEINS

A native protein, i.e., a protein which exists in the organism and performs its biological activity, is a structurally organized system. This is not contradicted by the fact that proteins are not fully helical, as, for example, hemoglobin and myoglobin. The presence of unordered regions may be regarded as an element of a definite order, since the number of links is fixed in this region.

The space structure of the globule is stabilized by a number of different interaction forces: chemical S—S bonds; hydrogen bonds; ionic (salt) bonds; and ion–dipole, dipole–dipole, and dispersion forces. This variety is the result of the many functionalities of the 20 types of amino acid residues. A particularly important role is played by the so-called "hydrophobic interactions" which will be discussed later.

Since the secondary and higher structures of the protein are maintained by this entire ensemble of forces, the protein macromolecules maintain their structure in aqueous solution and, furthermore, this structure is stabilized by the aqueous medium. Had this not been so, then the dissolution of a protein in cytoplasm, for example, the dissolution of enzymes, would lead to a loss of their role. The three-dimensional structure of the protein macromolecule determines its biological function.

Proteins are functional not only in a dissolved form. They enter into cellular organoids, into biological structures, by themselves and also in the form of supramolecular complexes with lipids (for example, in mitochondrial membranes), with DNA (for example, in chromosomes), etc. We do not yet know well the factors which are responsible for the formation of supramolecular protein structures. In developing a future theory, it will be necessary to take into account the general properties of the macromolecules which have been investigated by Flory.[23]

In dilute solution, a macromolecule behaves as an independent entity. To the contrary, in highly concentrated solutions, ordered systems, such as liquid crystals, may and must be formed. We have seen that the crystallization of a polymer is related to the flexibility or rather to the stiffness of the macromolecule which cannot be packed in any manner other than an ordered fashion if its flexibility and the available volume are limited. It is evident that the situation may also be found in solution. Consequently, supramolecular structures may also originate without the action of specific interaction forces, as a result of the entropic factor.

In a concentrated solution, the polymer chains may not take on definite different conformations independently of each other. Only those conformations are possible in which each link of a given chain occupies a position in space which is not filled by another link of the same molecule or by links of other macromolecules. Let ω be the number of possible conformations for each link with the corresponding probabilities; the total number of conformations of a given chain which contains N links, is ω^N. The total number of conformations n of chains that are independent of each other is

$$Q = (\omega^N)^n = \omega^{Nn} \qquad (5.1)$$

Let us increase the concentration up to the point where the volume fraction of the polymer is equal to v. Flory has shown that, in this case, the number of conformations which may be assumed by each macromolecule is diminished by a factor of e^{-vN} and, consequently, Q becomes

$$Q = (\omega e^{-v})^{Nn} \qquad (5.2)$$

If a polymer is not diluted, $v = 1$ and

$$Q = \left(\frac{\omega}{e}\right)^{Nn} \qquad (5.3)$$

where ω is evidently a measure of the flexibility of the chain. If the flexibility is so low that ω approaches e, then Q tends toward unity. Since, for the unordered state, $Q > 1$, that is, since it may be realized in a large number of ways, the macromolecules must become ordered. When $v = 1$, crystallization occurs, while in solution ($v < 1$), an isotropic liquid crystalline phase develops. These theoretical considerations are consistent with experiment. In this manner, the behavior of a macromolecule, and in particular of biological macromolecules, in solution is a function of their stiffness, of the concentration of the solution, and also of the temperature, pH, and nature of the solvent. In Fig. 76 are shown the conformational transitions

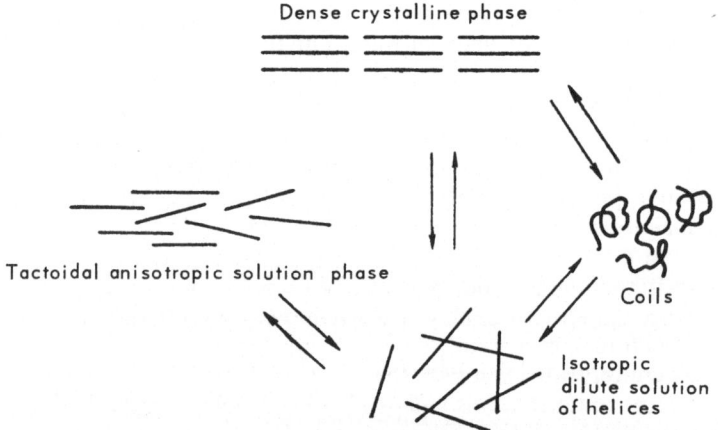

Fig. 76. Conformational transformations of a protein during dissolution.

which occur upon dissolution of a polymer and its conformational transformation in solution.

Let us note that a study of the conformation of polymers in solution is
very important for the understanding of the technological processes of
fiber formation. Synthetic fibers are obtained by forcing concentrated
solutions of polymers through spinnerets, i.e., very narrow openings.
The solvent evaporates and the fiber remains. The ordered crystalline
structure of the fiber exists already to some extent in the solution before
the evaporation of the solvent. The physical investigation of the process of
formation must evidently be based on conformational theory. One should
remark that living organisms make their fibers in the same manner. Both
a spider, in secreting its web, and the silk worm, in making an extremely
thin silk thread, force concentrated protein solutions through extremely
narrow openings. We see that at this point the problems of molecular
biophysics approach those of technology and it is quite evident that the
study of the biological process of structure formation of fibrous proteins
can be of great help to industrial production.

If we wish to understand the factors which control the three-
dimensional structure of proteins, we must first take into consideration
the effect of the surrounding aqueous medium on this structure. Those
proteins which are not part of supramolecular structures that form the
cellular organoids exist in an aqueous solution which contains various
organic molecules and inorganic ions. Water exerts the decisive effect on
the structure of a protein. This effect is due to particular properties of
water, a highly associated and structurized liquid.

An important factor in the determination of the secondary and
tertiary structures of a protein is the competition between the $NH\cdots OC$
hydrogen bonds inside the protein and the hydrogen bonds between these
groups and water molecules, $NH\cdots OH_2$ and $CO\cdots H_2O$. Quantitative
investigations of the thermodynamics of such interactions have shown that
the energy of formation of peptide hydrogen bonds is very small in an
aqueous medium and, consequently, the protein structure must be stabilized by other factors.

Bresler and Talmud[24] were among the first to point to hydrophobic
interactions as the principal driving force in the formation of protein
globules. Hydrocarbon nonpolar amino acid residues are essentially in
contact with each other and not with water; on the other hand, polar
residues interact with water. As a result, the protein chain folds into a
globule; i.e., the nonpolar hydrophobic residues are found inside the
globule in contact with each other, while the polar hydrophilic residues are
located on the surface of the globule in contact with water.

This pattern follows from a number of well-known factors. Nonpolar
hydrocarbon substances are poorly soluble in water. Langmuir has shown
in classical experiments on monomolecular layers of fatty acids cast on
the water surface that the polar carboxylic groups of the molecules of the

layer are immersed in water, while the nonpolar radicals are turned out-side. Similar factors determine the structures of aqueous colloidal solutions of soap; the hydrophobic residues are located inside the micelles, while the hydrophilic ones are on the surface.

The quantitative examination of these effects and the physical concepts of hydrophobic interactions were developed by Kauzmann.[25]

The physical causes of these interactions are found in the structure of water. Contacts between nonpolar groups and water are thermodynamically unfavorable, since they correspond to an increase in free energy. It has been shown by experiments that, when hydrocarbons are dissolved in water, the free energy increases, not because of an increase in enthalpy, but because of a decrease in entropy. Thus, for a solution of methane in water $\Delta F = 3.0$ kcal/mole, $\Delta H = -2.6$ kcal/mole, $\Delta S = -18$ cal mole^{-1} deg^{-1}. Similar quantities for benzene are 4.3 kcal/mole, 0 to 0.6 kcal/mole, -14 cal/mole^{-1} deg^{-1}. The decrease in entropy during the dissolution of hydrocarbons in water is due to its relatively loose structure. Ice is characterized by a low density because of the small coordination number in the crystal lattice; this number is 4. Bernal and Fowler have shown that the loose icelike structure is maintained in liquid water. The existence of a density maximum in water at 4°C makes it possible to regard water as a system consisting of two structures, namely, a loose icelike structure and a denser structure devoid of order. For the first structure, the values of the enthalpy, entropy, density, mobility, and coordination number are lower than the corresponding values of the second structure. The hydrophobic interactions may be regarded as a result of the displacement of the equilibrium in the direction of the more ordered structure; namely, an increase in order results in a decrease in entropy.

In recent years, a number of attempts have been made at the quantitative development of a theory of the structure of water and hydrophobic interactions. The theory by Némethy, Scheraga, and Poland[26,27] has received the greatest attention. These authors have examined five possible states of the water molecule. These are a state unbound by hydrogen bonds and states which possess one, two, three, and four hydrogen bonds. The last state is ice. The coordination number and the degree of packing of the molecules decrease in the opposite order. The introduction of a hydrophobic group or molecule into water increases the coordination number of the molecule in the icelike structure and, consequently, lowers the energy of the molecule with four hydrogen bonds by a quantity ΔE_1; to the contrary, the energy of all the other states is increased by an identical quantity, ΔE_2.

According to the estimate of Némethy and Scheraga, for aliphatic groups, $\Delta E_1 = -0.03$ and $\Delta E_2 = 0.31$ kcal/mole, whereas for aromatic groups, these values are 0.16 and 0.18 kcal/mole. By using these two parameters, it has been possible to obtain a reasonable agreement between calculated and measured constants of aqueous solutions of hydrocarbons.

The theory of Némethy and Scheraga is not all-encompassing. It does not explain a number of properties of water, such as the phase transition at 0°C. This theory, however, results in the reasonable quantitative estimates of hydrophobic interactions.

Independent of any theory, the concept of hydrophobic interactions may be regarded as well founded in experimental work. We should remember that we are talking here, not of some special bonds between the nonpolar groups, but of nonspecific interactions, which are the result of changes in the structure of water. As a result, the free energy of the system

$$-CH_3 \cdots H_2O$$

$$H_2O \cdots H_3C-$$

is greater than that of the system

$$-CH_3 \cdots H_3C-$$

$$H_2O \cdots H_2O$$

Up to now, we do not have a quantitative characterization of the degree of hydrophobicity of amino acid residues. It is evident that the data on the aqueous solutions of the monomeric acids are not adequate. The amino acids contain highly hydrophilic groups, namely, carboxyl and amino groups which are absent from the polypeptide chains. However, following the concept of hydrophobic interactions, it has become possible to reach quite important conclusions on the structure of proteins. Fisher has shown that such an approach gives quite valuable results.[28]

Instead of the 20 different elements of an ensemble, i.e., the different amino acid residues, let us examine just two types of elements, namely, the polar and nonpolar residues, p and n. This is obviously a very rough and approximate examination; it is, however, quite instructive.

Let us consider that all the residues have approximately equal volumes. The hydrophobic interactions force the n residues to be located inside the globule and the p residues to be located on its surface. The shape of a globule follows from a quite elementary mathematical examination; the most complicated formula used by Fisher is the equation of the volume of a sphere. Thus, the outside layer of the globule is a monomolecular layer consisting of the p residues and having a constant thickness d. If the globule is spherical, then the volume of the outside layer is

$$V_e = \frac{4\pi}{3}[r^3 - (r - d)^3] \tag{5.4}$$

where r is the radius of the globule. The internal volume of the globule is filled by the n residues and is

$$V_i = \frac{4\pi}{3}(r - d)^3 \tag{5.5}$$

while the total volume is

$$V_t = V_e + V_i \tag{5.6}$$

The ratio of the values of p and n is equal to

$$p = \frac{V_e}{V_i} \tag{5.7}$$

and for a sphere

$$p_s = \frac{r^3}{(r-d)^3} - 1 \tag{5.8}$$

From (5.6) and (5.7), it follows that

$$V_e = V_t \frac{p}{p+1} \tag{5.9}$$

Furthermore,

$$V_e = Ad \tag{5.10}$$

where A is the surface of the nonpolar nucleus of the globule. Fisher takes $d = 4$ Å. From (5.9) and (5.10), we obtain

$$p = \frac{A}{V_t/4 - A} \qquad A \text{ in } Å^2 \tag{5.11}$$

The smaller the value of V_t, that is, the molecular weight of the protein, the greater must be the relative polarity p. In fact, with small values of V_t, the nonpolar residues have no place to go. In the limiting case of a sphere with a radius of $r = d$, $p_s \to \infty$.

The theoretical curve of Eq. (5.11) for a sphere and the experimental points for a number of proteins are shown in Fig. 77. They fall close to a curve but do not follow it in general. The globule may be spherical only if $p = p_s$. Usually $p > p_s$, and the globule has a shape with a nonminimal surface-to-volume ratio. As a result, the majority of the points fall above the experimental curve. When $p \gg p_s$, the molecule is highly elongated and the protein is not globular, but fibrillar (tropomyosin, fibrinogen). To the contrary, when $p < p_s$, the p residues do not cover the n residues. One will find n residues on the surface of the globule and their hydrophobic interactions may lead to aggregation of the globules, i.e., to the formation of quaternary protein structures. These cases are shown by triangles in Fig. 77; aggregation displaces these points to the right.

In this manner, it has been possible to obtain a rough estimate of the shape of protein macromolecules in aqueous solution from very simple considerations. For this we need to know two quantities, p and V_t.

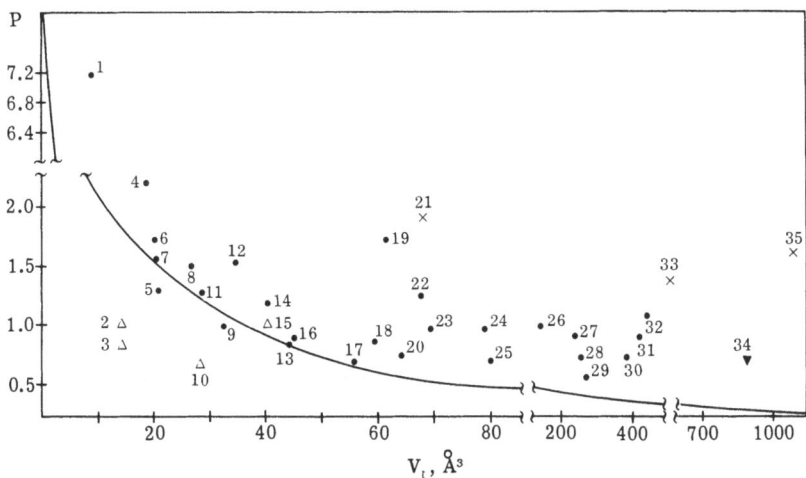

Fig, 77. Fisher's curve and experimental points: Δ, aggregating proteins; ×, proteins with an elongated rodlike structure. 1, Salmine; 2, insulin; 3, insulin; 4, ribonuclease; 5, lysozyme; 6, whale myoglobin; 7, horse myoglobin; 8, papain; 9, chymotrypsinogen; 10, structural protein; 11, corticotropin; 12, ovomucoid; 13, pepsin; 14, carboxypeptidase; 15, prothrombin; 16, β-lactoglobulin; 17, pepsinogen; 18, ovalbumin; 19, edestin; 20, α-amylase; 21, tropomysin; 22, human plasma albumin; 23, bovine plasma albumin; 24, avidin; 25, horse hemoglobin; 26, conalbumin; 27, human globulin; 28, aldolase; 29, phosphate dehydrogenase; 30, leucine aminopeptidase; 31, phosphorylase; 32, glutamate dehydrogenase; 33, fibrinogen; 34, β-galactosidase; 35, myosin.

Fisher has further shown that this regularity is also a function of the hydration of the protein molecules.[29] The hydrophilic amino acids on the surface of the globule interact with a monomolecular layer of water. These results and the concept of hydrophobic interactions as a whole contradict the widespread point of view that the water which surrounds the protein molecules is structured and as a result has special properties which permit the migration of energy (Szent–Györgyi). Hydrophobic interactions occur just because the structuring of water is thermodynamically unfavorable. In fact, there is no evidence whatever for such a structuring. To the contrary, the concept of hydrophobic interactions has received a number of confirmations.

The detailed X-ray diffraction investigations of the three-dimensional structures of myoglobin and hemoglobin, carried out by Kendrew and Perutz[30,31] has shown that the amino acid residues located on the side of the α-helix which is turned toward the inside of the globule are always nonpolar. The structure of myoglobin is shown in Fig. 78. The number of internal residues is 33; they are nonpolar in myoglobin and hemoglobin in all different types of vertebrates from the simplest organisms to man. On the other hand, one finds both polar and nonpolar residues on the surface of the globule. The rough rule defined by Fisher is not followed

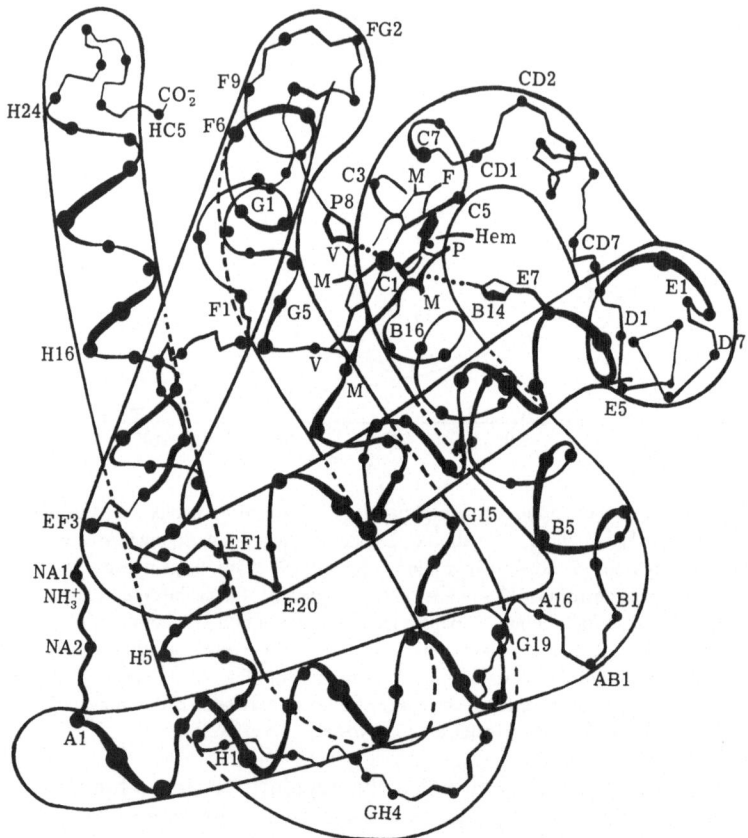

Fig. 78. Myoglobin structure, according to Perutz. *A, AB, B, C, CD, D, E, EF, F, FG, G, GH, H,* and *HC* are α-helical regions.

here; this should cause no surprise, since Fisher actually examined mixtures of independent amino acid residues and not polypeptide chains. Still, as a zero-order approximation, the classification of the 20 residues into two types, polar and nonpolar, is quite useful for an understanding of the three-dimensional structure of the protein. Furthermore, as we shall see, this concept based on hydrophobic interactions has made it possible to uncover the principal law of the genetic code and to understand how globular structure is formed.

The new studies by Levinthal[32] are quite important for the understanding of the three-dimensional structures of proteins. The author started from the following considerations. The equilibrium three-dimensional structure of the protein macromolecule in aqueous medium, which corresponds to minimal free energy, is uniquely determined by the primary structure of the polypeptide chain. The protein chain is synthesized from the amino end toward the carboxyl end, and its parts become folded into equilibrium secondary and tertiary structures independently of each other. Taking into account all the presently known quantitative characteristics of the electrostatic, Van der Waals, and hydrophic interactions, as well as hydrogen bonds, Levinthal carried out calculations on the equilibrium conformations of a number of proteins, using an electronic calculator. The problem consisted in the identification of the conformations and the rotational isomers which would correspond to a minimal free energy. The calculator gave the structure obtained as the model of the molecule directly on the screen of an oscilloscope. These calculations have given excellent results: structural segments of cytochrome c, myoglobin, and DNA, obtained in this manner, have turned out to be in good agreement with the X-ray data. The success of this interesting study proves the correctness of modern concepts on the factors which control the three-dimensional structure of proteins and testifies to the great possibilities opened to molecular biophysics by the use of electronic calculators.

CONFORMATIONAL TRANSITIONS OF PROTEINS

We shall speak now of the native biologically active structure of a protein. Various perturbing effects, such as heating, changes in pH, and the effect of various substances which compete with peptide groups in the formation of hydrogen bonds (for example, urea), may destroy the native structure and annihilate the biological activity of the protein. Processes of such a type, which occur without the breaking of peptide bonds and with the maintenance of the primary structure, are called *protein denaturation*. A denatured protein is biologically nonfunctional, but it can obviously serve as a source of essential amino acids by entering into the organism in the form of food. Boiled or broiled food contains denatured proteins. Denaturation occurs both in the solid state and in solution. It may be reduced to considerable conformational changes of the macromolecule.

A study of the process of destruction of the secondary and tertiary structures of proteins is quite important for biophysics, since it gives us information on the nature of the forces which maintain the structures and their quantitative characteristics. Denaturation means an order → disorder transition, a transition which is, in general, similar to transitions studied in molecular physics. We may compare denaturation with the melting of a crystal. The transition from a native to a denatured form of the protein is accompanied by a free-energy change (chemical potential) ΔF which sums up a number of effects that have been examined in great detail by Scheraga:[33]

$$\Delta F = \Delta F_\alpha + \Delta F_H + \Delta F_{Np} + \Delta F_e + \Delta F_m + \Delta F_x + \cdots \quad (5.12)$$

where ΔF_α is a free-energy change related to the unwinding of the α-helix, i.e., to the breaking of intramolecular hydrogen bonds and an increase in the number of conformations during the transition from helix to coil (see the next section); ΔF_H is the free-energy change due to the breaking of hydrogen bonds between side-chain residues of neighboring macromolecules; ΔF_{Np} is the free-energy change involved in hydrophobic interactions; ΔF_e is the difference in the electrostatic forces which act in helical and disordered forms; ΔF_m is the change in the free energy (entropy) from mixing the denatured molecule with a solvent; ΔF_x is determined by the destruction of cross links between the helices which exist in the crystal phase.

The values of ΔF_α and ΔF_x are independent of pH, since they are not related to processes which include the participation of ionizable groups. All the other terms of ΔF are dependent on pH. The quantitative expression for the changes listed above is

$$\Delta F_\alpha = (N - 4)\Delta H_0 - (N - 1)T\Delta S_0 \qquad (5.13)$$

where N is the number of links in a chain, ΔH_0 and ΔS_0 are the enthalpy and entropy changes per link of an infinitely long chain during the helix–coil transition. A chain made up of N amino acids has $N - 4$ hydrogen bonds; the $(N - 1)$st amino acid residue is fixed in space, i.e., it is devoid of conformational mobility. It follows immediately, from expression (5.13), that the helix is stable, that is, $\Delta F_\alpha > 0$, only if N is sufficiently large. In fact, (5.13) may be rewritten in the form

$$\Delta F_\alpha = (N - 4)(\Delta H_0 - T\Delta S_0) - 3T\Delta S_0$$

If we consider ΔH_0 and ΔS_0 as positive, we see that, for small values of N (of the order of 4), $\Delta F_\alpha < 0$.

$$\Delta F_H = -kT\sum_{i,j} \ln(1 - x_{ij}) \qquad (5.14)$$

where x_{ij} is the fraction of molecules which have a side hydrogen bond between the given groups i and j.

$$\Delta F_m = -2pkT\ln(1 + Kc) \qquad (5.15)$$

where K is the equilibrium constant for the interaction of the NH or CO group of the protein with a small molecule in solution. Thus, urea $(NH_2)_2CO$ forms hydrogen bonds with proteins and aids denaturation. In this case, c is the molar concentration of urea, and p is the number of residues of the crystalline chain, which, after destruction of the helix, may interact with urea. Water molecules certainly also interact with the protein; but this interaction has already been taken into account by ΔF_α, ΔF_H, and ΔF_p^N. At small values of c,

$$\Delta F_m = -2pkTKc \qquad (5.16)$$

The term ΔF_x is purely entropic in nature:

$$\Delta F_x = -T\Delta S_x = kT\frac{3v}{4}(\ln n + 3) \qquad (5.17)$$

where v is the number of helices linked by cross bonds, and n is the number of links between the cross bonds. A quantitative expression of ΔF_{Np} and ΔF_e encompasses three difficulties. By ignoring these terms (there are reasons to believe that they are relatively small), it is possible to write

$$\Delta F = (N - 4)\Delta H_0 - (N - 1)T\Delta S_0 + \Delta F_H - T\Delta S_x \qquad (5.18)$$

In the presence of urea we must also add ΔF_m.

Schellman[34] has estimated that $\Delta H_0 = 1500 \, \text{cal/mole}$, $\Delta S_0 = 4.2 \, \text{cal/mole}^{-1} \, \text{deg}^{-1}$. The magnitude of ΔH_0 is considerably smaller than the usual energies of hydrogen bonds, which are of the order of 4 to 6 kcal/mole. This is due to the fact that ΔH_0 is not the energy difference of peptide groups linked and unlinked by hydrogen bonds, but the difference in energies between groups linked by intramolecular bonds and groups linked by intermolecular hydrogen bonds with water molecules.

It is difficult to measure ΔH_0 directly, because of its low value. Privalov has developed an exact calorimetric method, which has resulted in values close to Schellman's estimate.[35]

A particularly important circumstance must be pointed out. ΔH_0 and ΔS_0 are not small, but ΔF_α is a small quantity, since ΔH_0 and ΔS_0 compensate for each other. In fact, intramolecular hydrogen bonds are stronger than the bonds with water, and their destruction means a loss in enthalpy. On the other hand, during denaturation the entropy increases, since the free chain, i.e., the coil, has many more conformations than the helix. At large values of N

$$\Delta F_\alpha \cong N(\Delta H_0 - T\Delta S_0) \qquad (5.19)$$

and when $T = 300°\text{K}$, $\Delta F_\alpha \cong N(1,500-1,260) = 240 \times N \, \text{cal/mole}$ of links. The value of 240 cal/mole of links is less than $kT = 600 \, \text{cal/mole}$. For a chain length of $N = 14$, $\Delta F_\alpha = -1,380 \, \text{cal/mole}$. ΔF_H is a strong function of pH. At high and low pH's, ΔF_H tends toward zero. An estimate of ΔS_x gives, for $n = 14$ and $v = 1$, $\Delta S_x = -8.5 \, \text{cal/mole}^{-1}$ deg^{-1}, that is, $-T\Delta S_k = 2,550 \, \text{cal/mole}$. In the presence of urea, ΔF_m is negative, i.e., this term decreases the stability of the native protein. To the contrary, ΔF_{Np} increases its stability. The transition temperature is found from the condition

$$\Delta F = 0 \qquad (5.20)$$

If the protein does not contain such hydrophobic bonds, (5.20) means

$$(N - 4)\Delta H_0 - (N - 1)T_m\Delta S_0 - T_m\Delta S_x = 0$$

and

$$T_m = \frac{(N - 4)\Delta H_0}{(N - 1)\Delta S_0 + \Delta S_x} \qquad (5.21)$$

The quantity ΔS_x is also proportional to the number of links in the chain N. Thus, the longer the chain, the greater the number of cross bonds that can be formed in it. This means that $v = \beta N$. If $N \gg 1$, then T_m is independent of N, since N is a multiplier both in the numerator and in the denominator in (5.21) and is cancelled out. It is evident that if ΔH_0 and ΔS_0 are known, then it is possible to determine ΔS_x from a knowledge of T_m. Such a determination is consistent with the estimate cited above.

If ΔF is known as a function of temperature, pH, and so forth, then the fraction of denatured material, x, may be found with the use of the expression

$$x = \frac{e^{-\Delta F/kT}}{1 + e^{-\Delta F/kT}} = \frac{1}{1 + e^{\Delta F/kT}} \qquad (5.22)$$

At the point of transition, $\Delta F = 0$ and obviously $x = \frac{1}{2}$. A detailed analysis of the hydrogen and hydrophobic bonds in the water–protein system has been carried out by Némethy, Steinberg, and Scheraga.[36] Quantitative estimates of the values of ΔF, ΔH, and ΔS are given. These estimates are consistent with what has been stated; the total free energy of formation of a simple structure, which includes both the energy of hydrogen and hydrophobic bonds, is found to be within the limits of -0.5 to 2.4 kcal/mole, the value of ΔH is between -1.5 and -0.5 kcal/mole, and that of ΔS is between -3 and $+7$ cal mole^{-1} deg^{-1}. It has been shown that hydrophobic bonds generally reinforce hydrogen bonds. The behavior of hydrogen bond-forming groups, which are enclosed within the nonpolar region of the molecule, has been examined. Such groups are capable of forming additional hydrogen bonds with water during the conformational transitions of the proteins.

Schachman[14] has estimated the free-energy change during the denaturation of a protein; the conformational entropy changes F by -1.5 kcal/mole of peptide bonds; the interpeptide hydrogen bonds make a contribution of 0.5 kcal/mole (assuming that 50% of the protein is in the form of an α-helix); the contribution of the hydrophobic bonds is 1.0 kcal/mole. The side-chain hydrogen bonds and ionic bonds make small contributions; thus, the total free energy of denaturation of a protein is close to zero.

ΔF may be obtained from the potentiometric titration of polar amino acids and proteins, as well as from calorimetric changes. As an example of such a study, let us cite the values obtained by Hermans for polyglutamic acid (PGA). At $25°C$, $\Delta F = 100$ cal/mole of monomer. Such a small value is the result of the compensation of the entropic and enthalpic terms. In fact, in such a case, $\Delta H = 1000$ cal/mole of monomer and $\Delta S = 3$ cal deg^{-1} mole^{-1} of monomer. The value of $\Delta F = 16$ kcal/mole has been obtained in a reversible denaturation of myoglobin at pH 9 and $25°C$; this value is of the same order as that for PGA, since myoglobin contains 150 monomeric units, i.e., amino acid residues.

The denaturation of globular proteins is due first to the destruction of the compact globular structure of the molecule. Denaturation proceeds over a narrow temperature interval or a narrow pH interval, similar to a phase transition. Consequently, denaturation processes have a cooperative character.

The highly cooperative nature of protein denaturation has been interpreted recently by Ptitsyn and Eizner[37] by analogy between the denaturation process and processes of evaporation of a drop of liquid. They have examined the intermolecular interactions in a number of macromolecules and have shown that, at sufficiently low temperature, the dependence of the volume of the macromolecule on temperature forms a loop similar to that found in a Van der Waals dependence of the pressure of a real gas on volume. The presence of such a loop indicates the existence of a phase transition of the first type, similar to the gas-liquid transition. At some temperature, the molecule must undergo an intermolecular phase transition from a noncondensed coillike state into a condensed globular state. Normally such a transition is maximized by the simultaneous intermolecular aggregation and precipitation of the polymer. If the macromolecules, however, can be kept in solution with the aid of other groups which do not participate in the aggregation, such as happens in the case of proteins, it is possible to form a globular state of the macro-molecule in solution.

According to this study,[37] the denaturation of globular proteins is a globule–coil transition similar to a liquid–gas transition in simple substances.

In this manner, the denaturation of a protein becomes a function of many factors, the study of which is faced with difficulties. They are related to the difference between the amino acid residues, that is, to the complexity of the primary structure of a protein. The study of macromolecules, which are capable of transition from an ordered to a disordered form, of melting by the breaking of hydrogen bonds, but which do not have a simple structure, may simplify the situation quite considerably. The problem of the investigation of the ability of synthetic polypeptides in solution to exist in regular structures, for example, in an α-helix, has been investigated by Doty and coworkers. These investigators have synthesized several poly-peptides and have shown by several methods that these macromolecules can exist in solution in two forms, namely, in an ordered, probably α-helical state and in a statistical coil. At some definite conditions a sharp helix–coil transition occurs, i.e., the polypeptide melts.

Poly-γ-benzyl-L-glutamate (PBG)

$$RNH-[CO-CH-NH]_n-$$
$$\underset{|}{}CH_2$$
$$\underset{|}{}CH_2$$
$$\underset{|}{}CO$$
$$\underset{|}{}O$$
$$\underset{|}{}CH_2$$
$$C_6H_5$$

was studied in detail, and also polyglutamic acid (PGA) which was obtained from PBG by means of hydrolysis. Since it is soluble in a number of organic solvents, PBG is convenient to study; PGA is soluble in water. Further, PBG has no ionizable groups. Its conformations in various solvents are different. In solvents which are not capable of forming hydrogen bonds, for example, in benzene, the macromolecules are strongly associated. In dichloroethane, chloroform, and formamide, PBG exists in the form of rigid rods. A study of these solutions by the methods of light-scattering and viscosity and also direct photography of the particles in the electron microscope give results which are compatible with a helical structure. The same is true of PGA in aqueous solutions at pH's below 5. On the other hand, if PBG is dissolved in solvents capable of forming hydrogen bonds, such as trichloroacetic and trifluoracetic acids, the macromolecules exist in the form of statistical coils. Such solutions have, for example, very low flow birefringence. PGA undergoes the transition into the coiled state at pH's above 5.

The dependence of ionization, intrinsic velocity (η), and specific rotation (α) of PGA on the pH of the medium is shown in Fig. 79. In regions close to pH 6, a helix–coil transition takes place; this is accompanied by an increase in the ionization and a sharp drop in the viscosity and in the specific rotation. The reasons for the last observations will be discussed later. Here we shall point out only the basic facts. It has been known for a long time that the optical activity of a protein decreases during denaturation. The optical activity of the α-helix is the sum of the optical activities of the asymmetric amino acid residues plus the optical activity of the helix as a whole. Melting results in an unordered coil, the optical activity of which is due only to the first factor. The helix–coil transition is accompanied also by a change in the absorption of ultraviolet light in the region of 190, 197, and 205 mμ. The absorptivity in a coil is greater than that in the helix.

The Doty school obtained similar results for polylysine, copolymers of lysine and glutamic acid, etc. It is evident that, from studies with synthetic polypeptides, it is possible to obtain information on the conformational transitions of proteins. It is possible to estimate the degree of helicity of a protein from values of (α) and the absorption of ultraviolet light, as well as from the H–D isotope exchange. It is noteworthy that the different methods give consistent results (Table 12).

The helix–coil transitions in polypeptides and proteins are an extremely interesting physical phenomenon. Here the extremely sharp transition is of interest. In Fig. 80, we have shown for PBG the dependence of (α) on solvent composition in the binary mixture of chloroform with dichloroacetic acid. At 80% of the latter, an extremely sharp drop occurs in the optical activity.

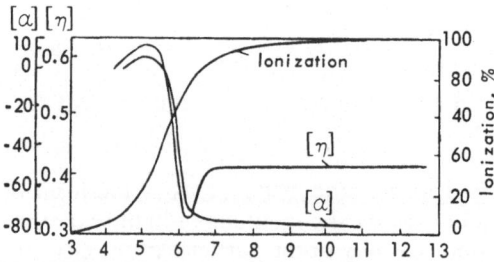

Fig. 79. Helix–coil transition in PGA.

Table 12   Degree of Protein Helicity

In percent

Protein	From UV absorption	From deuterium exchange	From X-ray diffraction	From optical rotation
Paramyosin	100	..	..	100
Myoglobin	82	> 70	77	75
Hemoglobin	..	..	75	78
Insulin......................	66	60	..	38
Ribonuclease	40	35	..	16
β-Lactoglobulin..............	30 (pH = 6.4)	25	..	30
	16 (pH = 8.7)			

The iodiosyncracy of this transition is that, being similar to melting, it takes place in an individual molecule and not in a crystal. This is intramolecular melting, i.e., the destruction of ordered secondary structure.

THEORY OF HELIX-COIL TRANSITIONS

The sharpness of the transition during heating may be calculated from expression (5.22) for the fraction of denatured material. Let us differentiate x with respect to T. We obtain

$$\frac{dx}{dT} = -\frac{e^{\Delta F/kT}}{1 + e^{\Delta F/kT}}\left(-\frac{\Delta F}{kT^2} + \frac{1}{kT}\frac{d\Delta F}{dT}\right)$$

and since ΔH and ΔS are practically independent of temperature,

$$\frac{d\Delta F}{dT} = -\Delta S = -\frac{\Delta H - \Delta F}{T}$$

When $T = T_m$, $\Delta F = 0$ and

$$\left(\frac{dx}{dT}\right)_{T_m} = \frac{\Delta H}{4kT_m^2} \tag{5.23}$$

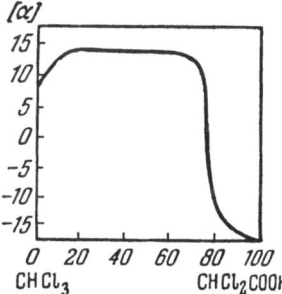

Fig. 80. Optical activity of PBG as a function of solvent composition ($M = 350,000$).

Therefore, the slope of the curve $x(T)$, that is, the sharpness of the transition, is the greater, the greater ΔH is. We have seen that the transition temperature T_m is independent of the number of elements which take part in it. To the contrary, ΔH must be a function of the number of elements. If during the transition the state of r links of the chain changes simultaneously, then

$$\Delta H \cong r \, \Delta H_0$$

and the sharpness of the transition is proportional to r.

The very sharp transitions which have been observed are a manifestation of the fact that they involve a large number of links, i.e., they are cooperative phenomena in which many elements participate. This is understandable. In an α-helix, each link is connected by hydrogen bonds with two others which are removed from it by four links along the chain. This is illustrated by the scheme shown in Fig. 81. To liberate a given link and give it the possibility of assuming any random conformation, it is evidently insufficient to break two hydrogen bonds; it is necessary to break several such bonds in a row. Therefore, a transition is possible only if a series of links participate in it.

The theory of this phenomenon is built on the very same principles as the theory of cooperative processes which was applied with such great success in the investigation of the properties of synthetic macromolecules, in the stretching of rubber, etc. This theory is based on the unidimensional Ising model. A simplifying condition is the linearity of the amino acid sequence in the protein.

It is interesting to see how the theory of the conformations of macromolecules was developed, as well as the statistics of polymeric chains. Russian scientists have developed the rotational-isomeric theory of macromolecules based on an investigation of these substances as cooperative systems, with the help of the one-dimensional Ising model.[40,41] Somewhat later, the same idea was applied independently by a number of American scientists to the development of the theory of helix–coil transitions.[42–46] Here we shall present the version of this theory by Zimm and Bragg,[42] which has the advantage of great simplicity and lucidity. A much more detailed description of this theory can be found in the monograph by Birshtein and Ptitsyn.[41] Finally, the same idea of the one-dimensional

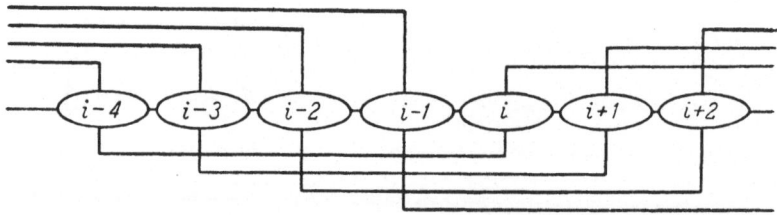

Fig. 81. Unidimensional model of the α-helix.

cooperative system was applied in the development of the theory of the matrix reduplication in the synthesis of DNA (see Chapter 6).

And so, let us examine a polypeptide chain which can have hydrogen bonds as depicted in Fig. 81. As we have stated already, the formation of a hydrogen bond between the ith and $(i - 4)$th monomer links results in the restriction of conformation of links i, $i - 1$, $i - 2$, and $i - 3$. In order to find the conditions of a helix–coil transition and the characteristic state of the chain, it is necessary to calculate the partition function of the chain, while taking into account all possible distributions of hydrogen bonds between the links. We shall distinguish between two states of each link, $\mu_i = 0$ and $\mu_i = 1$. The first state corresponds to the absence of a hydrogen bond in the link, the second to its presence; i.e., when $\mu_i = 0$ the monomeric link is found in a free state, while when $\mu_i = 1$, it is in a bound state. It is evident that the free energy of the chain is given by the sum

$$F(\mu_1, \mu_2, \ldots, \mu_N) = \sum_{i=1}^{N} F(\mu_{i-3}, \mu_{i-2}, \mu_{i-1}, \mu_i) \tag{5.24}$$

where N is the number of links in the chain. If $N \gg 1$, it is possible to neglect the states of the end links. The terms found within the sum may be expressed in the form

$$F(\mu_{i-3}, \mu_{i-2}, \mu_{i-1}, \mu_i) = F(\mu_i) + \tilde{F}(\mu_{i-3}, \mu_{i-2}, \mu_{i-1}, \mu_i) \tag{5.25}$$

Here $F(\mu_i)$ is the intrinsic free energy of the ith link, including also the energy of the hydrogen bonds, and $\tilde{F}$ takes into account the dependence of the free energy of the ith link on the state of the three preceding links. If $\mu_i = 0$, then

$$F(\mu_{i-3}, \mu_{i-2}, \mu_{i-1}, 0) = F_0 = F_{\text{free}} \tag{5.26}$$

since the state of a free link is independent of the state of the preceding links. If the ith link is bound, then its state depends on these other states. If the $(i - 1)$st link is also bound, then the conformation of the $(i - 1)$st and $(i - 2)$nd and $(i - 3)$rd links are drastically restricted independent of whether they are present in a bound or free state. Consequently, if a bound link $i - 1$ is followed by a bound link i, the free energy of the latter is

$$F(\mu_{i-3}, \mu_{i-2}, 1, 1) = F(1) + F_{\text{bound}} \tag{5.27}$$

The free-energy change in the liberation of the ith link is

$$\Delta F = \Delta H - T \, \Delta S = F_{\text{bound}} - F_{\text{free}} \tag{5.28}$$

$\Delta H < 0$ is the energy difference (enthalpy) of the intramolecular hydrogen bond and the hydrogen bond of the free link with solvent molecules; $\Delta S < 0$ is the decrease in the entropy as a result of the binding of the link, which follows the already bound links.

When a hydrogen bond is formed in the link which follows the free links, it is necessary to use up some free energy. In this case, four links are fixed at once; they are forced to assume a rigid helical conformation. Consequently, an entire helical region is formed, and the entropy drops at the same time,

$$\tilde{F}(0,0,0,1) = F_{\text{init}} = -2T \, \Delta S \tag{5.29}$$

where F_{init} is the free energy of initiation of a helical region. If we assume that, in the rotation of a free link about each bond, three rotational isomers are possible, then

$$F_{\text{init}} \cong 2 \times 300 \times 4 \ln \tfrac{1}{2} \cong 2.5 \, \text{kcal/mole}$$

i.e., F_{init} has a significant value.

The appearance of one or two free links between bonded ones is accompanied by a large increase in free energy and is therefore almost impossible. After all, hydrogen bonds may be broken and the links kept in a rigid helical conformation, i.e., without compensating for the used-up energy by a gain in entropy. Thus

$$\tilde{F}(\mu_{i-3}, 1, 0, 1) \to \infty$$
$$\tilde{F}(1, 0, 0, 1) \to \infty$$

(5.30)

The partition function of the chain is

$$Q = \sum_{\{\mu_i\}} e^{-F\{\mu_i\}/kT}$$

(5.31)

Summation is carried out over all the states of the chain, i.e., over values of $\mu_i = 0.1$ for each link. If we take the free energy of the free link as zero, then the contribution of a given state to Q is given by a product of the following multipliers:

1. Multiplier 1 for each link in a free state ($\mu_i = 0$).
2. Multiplier s for each link in a bonded state ($\mu_i = 1$), which is equal to

$$s = e^{-\Delta F/kT} = e^{-(\Delta H - T\Delta S)/kT}$$

(5.32)

where s is the equilibrium constant for the reaction of the formation of a hydrogen bond in the link which follows a bonded link, i.e., an equilibrium constant for the increase of a row of bonded links by one link, at the expense of the neighboring region of free links.

3. Multiplier σ for each link in a bonded state ($\mu_i = 1$), which follows three or more free links ($\mu_{i-1} = \mu_{i-2} = \mu_{i-3} = 0$); it is evident that

$$\sigma = e^{-F_{\text{init}}/kT} = e^{2\Delta S/k} \cong 10^{-2}$$

(5.33)

4. Multiplier zero for each bonded link ($\mu_i = 1$) which follows free links, if their number is less than three ($\mu_{i-1} = \mu_{i-2} = 1$, or $\mu_{i-1} = \mu_{i-2} = 0$, and $\mu_{i-3} = 1$). This zero corresponds to Eq. (5.30). Consequently,

$$Q = \sum_{\{\mu_i\}} \prod_{i=1}^{N} s^{\mu_i} \sigma^{\mu_i(1-\mu_i)} [1 - \delta_{\mu_i,1} \delta_{\mu_{i-1},0}(1 - \delta_{\mu_{i-2},0} \delta_{\mu_{i-3},0})]$$

(5.34)

Here

$$\delta_{ab} = \begin{cases} 1 & \text{if } a = b \\ 0 & \text{if } a \neq b \end{cases}$$

As we have seen, Q, which is expressed as a sum of products, may be expressed as a trace of a certain matrix taken to the power N:

$$Q = \text{Sp}(P^N)$$

(5.35)

In this case, the matrix is of the order $2^3 = 8$, since it is necessary to take into account the states of three consecutive links, each of which may be bonded or free. Zimm and Bragg have carried out a complete analysis of this problem, and they have shown at the same time that almost the same result may also be obtained with a simplified treatment which takes into account only the state of two, and not four, consecutive links. In fact, the relation between the state of the ith link and the state of the $(i - 2)$nd and $(i + 3)$rd links is given only by condition (5.30). If we neglect this condition, then instead of (5.24), it is possible to write out

$$F(\mu_1, \mu_2, \cdots, \mu_N) = \sum_{i=1}^{N} F(\mu_{i-1}, \mu_i) \qquad (5.36)$$

and instead of (5.25), equation

$$F(\mu_{i-1}, \mu_i) = F(\mu_i) + \tilde{F}(\mu_{i-1}, \mu_i) \qquad (5.37)$$

Thus, in agreement with (5.26), (5.27), and (5.28), we obtain

$$F(0) = F_{\text{free}}$$
$$F(1) = F_{\text{bound}}$$
$$\tilde{F}(0,1) = F_{\text{init}}$$
$$\tilde{F}(0,0) = \tilde{F}(1,0) = \tilde{F}(1,1) = 0$$
$$\qquad (5.38)$$

Each bonded link introduces a multiplier s into the partition function; each free link, the multiplier unity; and the first of the bonded links which follows one or more free ones, the multiplier σ. In this way,

$$Q = \sum_{\{\mu_i\}} \prod_{i=1}^{N} s^{\mu_i} \sigma^{\mu_i(1-\mu_{i-1})} = \text{Sp}(P^N) \qquad (5.39)$$

The matrix P in this case is of second order, since the situation is completely analogous to the examination of the simplest model of a unimolecular rotational-isomeric chain:

$$P = \begin{pmatrix} \mu_i/\mu_{i-1} & 0 & 1 \\ \hline 0 & 1 & \sigma s \\ 1 & 1 & s \end{pmatrix} \qquad (5.40)$$

The characteristic equation of the matrix is

$$(\lambda - 1)(\lambda - s) = \sigma s \qquad (5.41)$$

Let us examine the two extreme cases, $\sigma = 1$ and $\sigma = 0$. If $\sigma = 1$, there is no cooperativity, since this means that $F_{\text{init}} = 0$ [compare Eq. (5.33)]. Furthermore, the roots of Eq. (5.41) are equal to $1 + s$ and 0 and

$$Q = (1 + s)^N$$

i.e.,

$$Q = q^N \qquad (5.42)$$

where $q = 1 + s$ is the partition function of a single link. Consequently, the partition function of the chain is equal to the product of the partition functions of the links which are independent of each other.

Let us find the fraction of the links which are bonded by hydrogen bonds. It is equal to

$$x = \frac{1}{N} \frac{\partial \ln Q}{\partial \ln s} = \frac{s}{1+s} = \frac{e^{-\Delta F/kT}}{1 + e^{-\Delta F/kT}} \qquad (5.43)$$

i.e., it corresponds to the normal noncooperative equilibrium. The term x is a monotone function of temperature and, consequently, the case of $\sigma = 1$ corresponds to a gradual

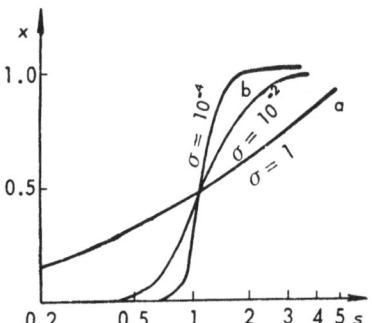

Fig. 82. Helix–coil transition, (a) noncooperative
and (b) cooperative.

transition from a helix to a coil during heating (Fig. 82, curve a). This case is not consistent with experiment; the experimentally observed transitions are extremely sharp.

When $\sigma = 0$, the roots of the equation are equal to 1 and s, and, consequently,

$$Q = 1 + s^N \tag{5.44}$$

i.e., the chain may exist only in two states, either totally helical, $Q_{\text{helix}} = s^N$, or totally free in the form of a statistical coil, $Q_{\text{coil}} = 1$. When $s > 1$, then $s^N \gg 1$ and $Q = s^N$, that is, the chain is completely helical. To the contrary, when $s < 1$, then $s^N \ll 1$ and $Q = 1$, that is, the chain is totally in the form of a random coil. When $s = 1$, a sharp cooperative helix–coil transition takes place. The fraction of bonded links is

$$x = \frac{1}{N} \frac{\partial \ln Q}{\partial \ln s} = \frac{s^N}{1 + s^N} \tag{5.45}$$

so that, when $N \gg 1$,

$$x \cong \begin{cases} 1 & s > 1 \\ 0 & s < 1 \end{cases} \tag{5.46}$$

This sharp transition, which occurs in the all-or-none form, is shown by curves b of Fig. 82. During a transition, the energy, just as x, changes in steps

$$E = kT^2 \frac{\partial \ln Q}{\partial T} = NkT \frac{s^N}{1 + s^N} \frac{d \ln s}{d \ln T} \cong \begin{cases} NkT \dfrac{d \ln s}{d \ln T} & s > 1 \\ 0 & s < 1 \end{cases} \tag{5.47}$$

The case of $\sigma = 0$ corresponds to $F_{\text{init}} \to \infty$, i.e., to total cooperativity.

As we have seen, while σ is small, it is nevertheless different from zero. In the general case of $0 < \sigma < 1$, the roots of Eq. (5.41) are equal to

$$\lambda_{1,2} = \frac{1 + s}{2} \pm \sqrt{\frac{(1 - s)^2}{4} + \sigma s} \tag{5.48}$$

The transition is less sharp for higher values of σ, and it takes place over a change of s from $1 + \sqrt{\sigma}$ to $1 - \sqrt{\sigma}$. Calculations have shown that, when $\sigma \ll 1$, the temperature interval for melting an α-helix, ΔT, is

$$\Delta T = \sqrt{\sigma} \frac{kT_m^2}{|\Delta H_m|} \tag{5.49}$$

When the finite number of links is taken into account, the interval of the transition is increased and it is displaced toward higher values of T. Comparison of the theory with the experiments of Doty and coworkers on PBG in solutions of mixtures of dichloroacetic acid and dichloroethylene results in good agreement between theory and experiment, with $|\Delta H| = 900$ to $1,100$ cal/mole and $\sigma = 2 \times 10^{-4}$. The σ turns out to be very small, i.e., the degree of cooperativity is quite large; it corresponds to the simultaneous transition of 10^2 links. This value of ΔH is close to the estimates quoted above.

In connection with this theory there is an essential physical problem which had been examined only in abstract form up to the discovery of the helix–coil transition. Is the helix–coil transition a melting, i.e., a phase transition of the first type? In other words, is it possible to regard the simultaneous existence of helical and unordered regions in the chain in the region of transition as an equilibrium between two phases, i.e., an equilibrium between two segments of the chain of any size, which touch each other at a point?

Landau and Lifshitz[47] have shown that such an equilibrium is impossible in a unidimensional system. The free energy of the system is equal to the free energies of the total quantities of the two phases plus the free energy of the regions of contact between the phases. If the number of such regions of contact is m, then

$$F = NxF_{bound} + N(1 - x)F_{free} + mkT \ln \frac{m}{eN} + m\psi \tag{5.50}$$

The logarithmic term is the entropy of mixing of the points of contact between the links of the chain, and ψ is the energy of the surface tension at the boundary points. From Eq. (5.50), it follows that

$$\frac{\partial F}{\partial m} = kT \ln \frac{m}{N} + \psi \tag{5.51}$$

F tends to a minimum, which corresponds to the condition $\partial F/\partial m$, i.e.,

$$\frac{m}{N} = e^{-\psi/kT} \tag{5.52}$$

Consequently, the two phases will tend to mix with each other until they become divided into little finite segments which satisfy condition (5.52). Therefore, there is no phase equilibrium in a unidimensional system; phase transitions are equally impossible.

The statistical theory enables one to calculate the average fraction of links which are present at points of contact between bonded and free regions. This fraction is equal to[41]

$$\frac{m}{N} = \sqrt{\sigma} = e^{-F_{init}/2kT} \tag{5.53}$$

Comparison of Eqs. (5.52) and (5.53) shows that the role of the energy of surface tension is played by $F_{init}/2$. In the limiting value of F_{init} and, consequently, when $\sigma < 1$, the transition is sharp, but not infinitely sharp, and thus it is not a true phase transition, not a true melting. A phase transition exists only when $\sigma = 0$, that is, when $F_{init} \to \infty$ and $\psi \to \infty$, which is impossible. In practice, however, as long as $\sigma \ll 1$, the transition is very sharp and may be regarded as a quasi melting of a unidimensional chain.

The theory of helix–coil transitions caused by a change in solvent or in the pH, and not by an increase in temperature, is based on the same principles.[41] In the first case, it is necessary to take into account the fact that each link may exist, not in two, but in three states: without a hydrogen bond, with intramolecular hydrogen bonds, and with molecular bonds with the molecules of the solvent. The intermolecular bonding may be regarded as noncooperative, and each nonbonded link present inside the molecule makes a contribution to Q which is not 1, but $1 + e^{\Delta\mu/kT}$, where $\Delta\mu$ is the difference between the chemical potentials of the link in a state in which it has an intramolecular hydrogen bond and the free state, i.e., the change of the free energy of a monomeric link in the formation of a hydrogen bond.[43,46] Similarly the condition for transition is a function of $\Delta\mu$. Just as before, the sharpness of the transition is a function of σ. When the concentration of the specifically linked solvent molecule increases, T_m decreases. If, in the absence of an intermolecular interaction, the melting temperature is equal to T_m^0, then, in the presence of the solvent[46] it is

$$T_m = \frac{T_m^0}{1 + (kT_m^0/|\Delta H|) \ln (1 + e^{\Delta\mu/kT})} \tag{5.54}$$

When $\Delta\mu \to -\infty$, that is, when a state with an intermolecular hydrogen bond is practically existent, $T_m = T_m^0$. The greater $\Delta\mu$, the smaller T_m. This theory describes well the results obtained for a number of synthetic polypeptides.

Polypeptides which are constituted of ionizable amino acid residues (for example, polyaspartic acid, polyglutamic acid, polylysine, etc,) undergo helix–coil transitions when the pH is changed. The transition in this case may be followed not only by measuring it viscometrically or polarimetrically, but also by the method of potentiometric titration, which gives the degree of ionization of the macromolecule, α. The theory of transition in polypeptides with ionizable groups has been developed by Peller[46] and Zimm and Rice.[48] The last theory is, as before, based on the Ising method.

In a chain with ionizable groups, each link must be characterized not only by the parameter $\mu_i = 0.1$ for a free and a bonded link, but also by another parameter $\eta_i = 0.1$ for an uncharged and a charged ionizable group. The partition function of the chain has the form

$$Q = \sum_{\{\mu_i\}} e^{-F\{\mu_i\}/kT} \sum_{\{\eta_i\}} \prod_{i=1}^{N} a^{\eta_i} e^{-F_{\{\mu_i\}}^{(e)}(\{\eta_i\})/kT} \tag{5.55}$$

Here $F\{\mu_i\}$ is the free energy of the uncharged chain for a given set of values of μ_i, which is the same free energy that we have already discussed; a is the ratio of the activities of the charged and uncharged links, related to pH by

$$\log a = \pm(\text{pH} - \text{pK}) \tag{5.56}$$

The plus sign is for acidic and the minus sign for basic groups. The term $F_{\{\mu_i\}}^{(e)}(\{\eta_i\})$ is the free energy of the electrostatic interaction between the charged groups for given sets of μ_i and η_i. The expression for Q takes into account in this manner the mutual repulsion of charged groups; the magnitude of this repulsion is evidently a function of the conformation of the links, i.e., of the set of μ_i. Calculations show that, in the helical conformation, repulsion increases more rapidly than in the coiled conformation, and, consequently, the charge of the chain must favor the helix–coil transition. If the partition function is found by the Ising method, it becomes possible to determine quantitatively the decrease in T with an increase in the degree of ionization of the chain, α, and also the titration curve of the polypeptide (Fig. 83). The last displays a quite distinct break which corresponds to the helix–coil transition. Theory is in good agreement with experiment.[41,48]

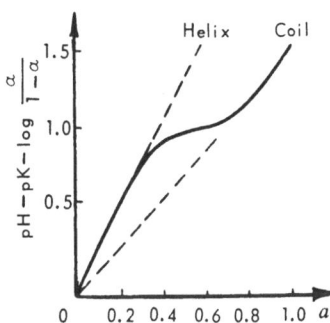

Fig. 83. Potentiometric titration curve for a poly-peptide.

Thus, the nature of helix–coil transitions in polypeptides is clear. It is a cooperative conformational transition which is not a phase transition of the first order, but which differs little from it and which may, therefore, be regarded as a quasi melting of the macromolecule.

Helix–coil transitions in proteins have the same character. The essential difference of a protein from a synthetic polypeptide consists of the presence in the protein of a specific primary structure and, as a result, of a tertiary structure. This is expressed in the incomplete helicity of a protein, which is due to the variety of the amino acid residues and their specific distribution and also, in part, to the presence of cystine disulfide bridges. As a result, the transition turns out to be less sharp than in a polyamino acid. The estimates of Scheraga have been discussed above. The development of a rigorous theory for a protein is a problem that has not been solved yet and which has met with serious difficulties. So far, we have no statistical theory which takes into account effectively the individual properties of the different components of an ensemble; and in the protein there are 20 such components which corresponds to 20 types of amino acids. Furthermore, the properties of a protein are a function of the sequence of amino acids; a protein is an aperiodic crystal. As a result, the analysis by Fisher presents a particular interest; it shows that it is possible to solve such problems approximately, by dividing the 20 amino acids into two groups, polar and nonpolar, and by making use of the concept of hydrophobic interactions.

OPTICAL PROPERTIES OF PROTEINS AND POLYPEPTIDES

It has been pointed out already that the optical rotatory power changes sharply during helix–coil transitions. It is possible to study the transition by following this change, since it is quite characteristic. Furthermore, the transition is accompanied by a change in the intensity of ultraviolet absorption.

Optical methods have a particular significance in the physics of molecules. What factor expresses directly the structure of a system consisting of atomic nuclei and electrons, of atoms and molecules? It is possible to calculate all the basic constants which characterize the properties of atoms and molecules from the spectrum of such a system if the energy levels and the probabilities of transitions between them are known. The distances between the levels are expressed in the frequencies, or wavelengths, of the spectrum; the probabilities of the transitions, in the intensities of the corresponding lines and bands. The structure of the atom was determined from its spectrum; this was the problem examined in the theory of Bohr. The spectroscopy of molecules is one of the most important sources of information about their structure. It is natural, therefore, that the physics of proteins as well is turning to spectra for the solution of problems related to the structure of macromolecules, and in particular to their secondary structure. We have seen already that infrared spectra make it possible to differentiate between the α and β forms of proteins and polypeptides. Infrared absorption is due to transitions between vibrational energy levels of the molecules or, more simply, to the vibration of its atoms. On the other hand, the ultraviolet spectra are determined by transitions between different states of the electrons in the molecule. In polypeptides which do not contain aromatic rings, the nearest ultraviolet bands are present in a rather far region of the spectrum, namely, close to 200 mμ. These bands are determined by the changes in the states of the electrons in the planar amide group. It has been pointed out above that the π electrons of such a system are partly conjugated and possess greater mobility. The scheme of the electronic levels of the amide group is shown in Fig. 84, while Fig. 85 shows the directions of polarization of the corresponding transitions according to theoretical calculations.[49] These are the directions of what are called the *transition dipole moments* μ or, if we use a classical simple model, the directions of the *vibrations of the electronic oscillators.*

First we must take into account the fact that these oscillators are under the influence of the environment. The amide group under investigation is the chromophoric group which is responsible for the absorption of light; it is under the influence of neighboring molecules and other elements of the chain. These particles which surround the group, in their turn, possess

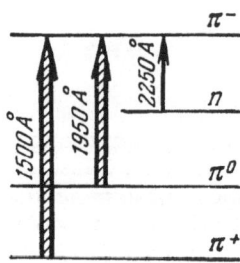

Fig. 84. Energy levels of the electrons of an amide group.

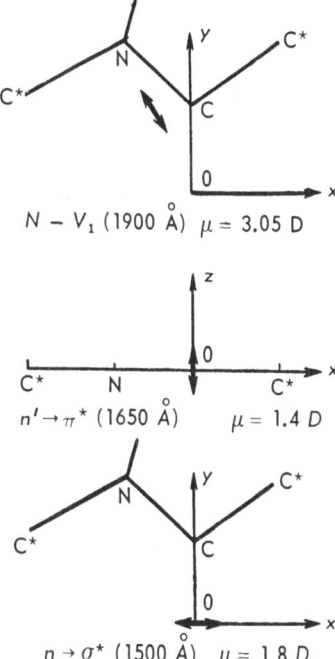

$N - V_1$ (1900 Å) $\mu = 3.05$ D

$n' \rightarrow \pi^*$ (1650 Å) $\mu = 1.4$ D

$n \rightarrow \sigma^*$ (1500 Å) $\mu = 1.8$ D

Fig. 85. Directions and magnitudes of the transition moments in an amide group.

charges, permanent dipole moments, and polarization. The electrostatic interactions of the chromophoric groups with the medium result first of all in a change in the frequencies of absorption of the groups relative to these frequencies in vacuum. A detailed experimental and theoretical investigation of the influence of intermolecular forces on the electronic spectra of absorption of complex molecules has been carried out by Neporent and Bakhshiev.[50] A schematic representation of what occurs is given in Fig. 86.[49] The greater the polarizability and the dipole moment of the solvent molecule, the greater is the displacement of the absorption frequency in the "red direction," in the direction of lower frequencies, or longer wavelengths. As a result, denaturation may cause a displacement of frequencies in the opposite direction; it may increase them and decrease the wavelengths. During denaturation, the amide groups may shift from a medium with a higher polarizability in the native proteins to a medium with a lower polarizability, to the denatured form in the solvent. This is indeed observed.[51]

The presence of a regular ensemble of a large number of identical groups leads to another no less significant effect. Identical groups have identical energy levels. Consequently, a quantum of light absorbed by one group may migrate into the neighboring group, etc., if these groups have a regular fixed distribution. Each group taken separately has its own energy levels which correspond to the fundamental and excited states.

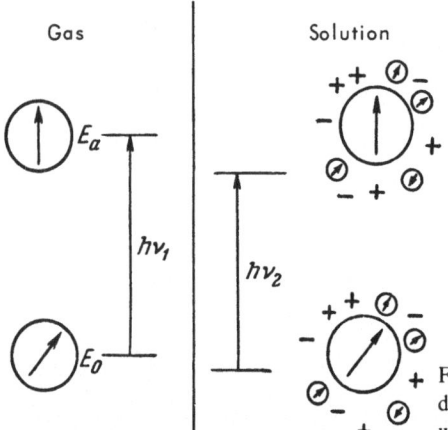

Fig. 86. Schematic representation of the displacement of the energy of a molecule when it is placed in a solvent.

Let us assume that there is a fundamental and a single excited state. Transition of the electrons from the first into the second occurs as a result of the absorption of light of a definite frequency and with a definite wavelength. If we have a large number of identical and regularly distributed groups, the absorbed energy may pass from the first group to the second, etc. An excited wave, called an *exciton*, spreads through the system. Simultaneously, as a result of the resonance interactions between the groups, each energy level splits into an almost continuous zone. The band in the absorption spectrum splits into two or more components. This has been established in the studies of Davydov,[52] who developed the theory of absorption of light by molecular crystals.[53] The term *Davydov splitting* is now universally accepted.

An α-helix also has a regular ensemble of a large number of identical amide groups. Once again we come to the strong analogy between the properties of a regular stiff macromolecule and a crystal. Calculations show that the peptide band at 190 mμ splits in an α-helix into two bands at 198 mμ and 189 mμ, as a result of the Davydov splitting. The direction of the oscillator in the first of these bands is parallel to the axis of the helix, while in the second band, it is perpendicular to this axis.[49,54] This is in agreement with experiment.

The effect of conformational changes on the absorption intensity may also be explained theoretically. The intensity of absorption in any given band, which corresponds to the transition $0 \rightarrow i$, is expressed by the oscillator strength

$$f_{0i} = \frac{8\pi^2 m}{3he^2} v_{0i}\mu_{0i}^2 \qquad (5.57)$$

where e and m are the charge and mass of the electron; h is Planck's constant; v_{0i} is the frequency of the transition, equal to the energy difference between the levels divided by h,

$$v_{0i} = \frac{1}{h}(E_i - E_0) \qquad (5.58)$$

and μ_{0i} is the dipole moment of the transition, called the *matrix element of the dipole moment*,

$$\mu_{0i} = \int \psi_0 \mu \psi_i^* \, d\tau \qquad (5.59)$$

where ψ_0 and ψ_i^* are the wave functions of the states of the molecule between which the transition takes place.

The oscillator strength is related to the observed absorptivity of the band, ε_{0i}, by the simple relation

$$f_{0i} = 4.32 \times 10^{-9} \int \varepsilon_{0i} \, dv'$$

The integral is taken over the entire absorption band, v' is the frequency expressed in cm^{-1} ($v' = v/c$, where c is the velocity of light).

It can be shown quantum mechanically that the sum of the oscillator strengths for all the electronic transitions in a given electronic system is a constant quantity which is equal to the number of the electrons. Therefore, intensity change in individual bands may occur only as a result of a redistribution of the intensities. It occurs thanks to the interaction of the transition moments. One band steals intensity from another.

We are interested in the behavior of the highest wavelength bands during the conformational transition, i.e., during the helix–coil transition. If the interacting transition moments of two bands are perpendicular to each other, the intensity does not change at all. If the moments are colinear, that is, if they are directed along a single axis, the intensity of higher wavelength bands increases (at the expense of the short-wavelength bands), i.e., a hyperchromic effect occurs. To the contrary, if the moments are parallel to each other, the intensity of the longer-wavelength band decreases and we have a hypochromic effect. In an α-helix, the amide groups are parallel to each other; the corresponding transition moments are also parallel. Therefore, the highest wavelength amide band has a lower intensity in an α-helix than in a statistical coil, and we observe hypochromism.

The experimental results for polyglutamic acid are shown in Fig. 87.[49] At pH 4.9, the polypeptide is in the form of an α-helix; at pH 8, it is in the form of a coil. We observe two bands which have arisen as a result of Davydov splitting and a high degree of hypochromism; the absorption intensity is much higher in the case of the coil than in the case of the helix. The ratio of the oscillator strength f of the helix to that of the coil is 0.7; this is in good agreement with the result of theoretical calculations. During denaturation there is a small shift of the maximum in the red direction.

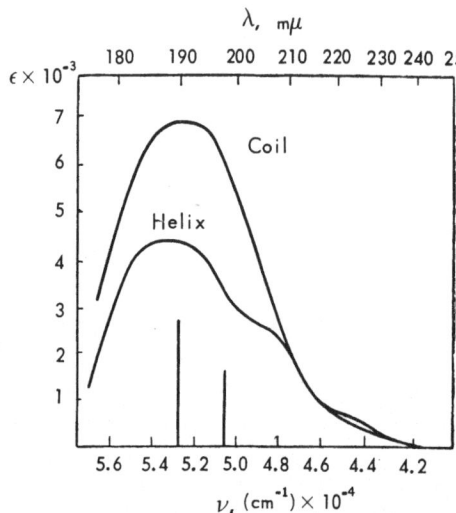

Fig. 87. Absorption curves for polyglutamic acid. The vertical lines indicate the calculated positions of the absorption bands of the helix.

There is also observed a small hypochromism in the region of 215 mμ, the reasons for which are not clear. A detailed presentation of this theory has been given by DeVoe.[55]

From what has been said, it follows that the ultraviolet spectra may give extremely valuable information on protein structures. Here we encounter a vast number of experimental and theoretical problems. More or less complete calculations have been carried out only for the simplest polypeptides. There are very serious difficulties connected with studies in the far ultraviolet spectral region. However, the study is worth the effort since future investigations promise much. In relation to this, it is necessary to point out the great need for the development of a general theory of the electronic spectra of polymers which would take into account the cooperative nature of the conformations of neighboring elements even in statistical coils.

Let us pass now to optical activity. Just as all other optical phenomena, it is closely related to electronic spectra.[53,56] The optical properties of molecules are determined first of all by their polarizability, i.e., by the ability of the electron clouds of the molecules to become displaced under the action of the electric field of a light wave. In its turn, the polarizability is expressed through the frequencies of the absorption bands and their intensities, i.e., through the positions of the energy levels and the probabilities of transition between them. At frequencies removed from the intrinsic absorption, the expression for the polarizability has the form

$$a = \frac{e^2}{4\pi^2 m} \sum_i \frac{f_{0i}}{v_{0i}^2 - v^2} = \frac{e^2 \lambda^2}{4\pi^2 mc^2} \sum_i \frac{f_{0i}\lambda_{0i}^2}{\lambda^2 - \lambda_{0i}^2} \tag{5.60}$$

where e and m are the charge and mass of the electron, and c is the velocity of light. In the numerator of this equation we find the oscillator strength; the resonance denominator contains the difference between the squares of the frequencies of the corresponding electronic transition, v_{0i}, and the frequencies of the incident light, v. Introducing Eq. (5.57) into (5.60), we obtain

$$a = \frac{2}{3h} \sum_i \frac{v_{0i}\mu_{0i}^2}{v_{0i}^2 - v^2} = \frac{2\lambda^2}{3hc} \sum_i \frac{\lambda_{0i}\mu_{0i}^2}{\lambda^2 - \lambda_{0i}^2} \qquad (5.61)$$

The refractive index of a gas consisting of molecules with polarizability a is equal to

$$n = 1 + 4\pi N_1 a = 1 + \frac{8\pi N_1 \lambda^2}{3hc} \sum_i \frac{\lambda_{0i}\mu_{0i}^2}{\lambda^2 - \lambda_{0i}^2} \qquad (5.62)$$

In this manner, the difference of the refractive index from unity and its dependence on frequency or, what is the same, on the wavelengths of the incident light, is determined by the presence of absorption bands in the spectrum of the molecule. For water, $n = 1.35$ in yellow light, because it absorbs light in the far ultraviolet region of the spectrum. As we approach the absorption bands, n increases and we observe the dispersion of light. In the region of absorption, $n - 1$ changes its sign and we observe anomalous dispersion. Equations (5.60) to (5.62) are valid only for normal dispersion, that is, in the spectral region removed from the absorption bands of the substance.

Natural optical activity, i.e., the rotation of the plane of polarization of light by a medium consisting of asymmetric molecules, is also related to the positions of the energy levels and the probability of transition between them. As has been shown by Fresnel, optical activity is a manifestation of the circular birefringence of light. In other words, the rotation of the plane of polarization in an active medium is determined by the difference in the velocities of propagation of light polarized circularly to the right and to the left. In fact, the electric vector in linearly polarized light may be represented as a superposition of two waves, one circularly polarized to the right and the other circularly polarized to the left. When the velocity of propagation of these two waves is different, there arises between them a phase difference and the resulting vector of the linearly polarized wave will be rotated by an angle φ' given by

$$\varphi' = \frac{\pi}{\lambda}(n_L - n_R)l \qquad (5.63)$$

where n_L and n_R are the refractive indices for the left and right polarized waves, λ is the wavelength, and l is the pathlength of the light in the active medium. For a unit pathlength (usually expressed in decimeters), the rotation angle is

$$\varphi = \frac{\pi}{\lambda}(n_L - n_R) \qquad \text{deg/dm} \qquad (5.64)$$

Optically active substances are characterized by a specific rotation $[\alpha]$;

$$[\alpha] = \frac{\varphi}{c'} \frac{1{,}800}{\pi} \qquad \frac{\text{deg} \cdot \text{cm}^3}{\text{dm} \cdot \text{g}} \tag{5.65}$$

where c' is the concentration of the dissolved substance in grams per cubic centimeter. The molar rotation is

$$[\Omega] = [\alpha]\frac{M}{100} = \varphi \frac{18}{\pi} \frac{M}{c'} \tag{5.66}$$

where M is the molecular weight.

Why is it that a medium consisting of asymmetric molecules possesses circular dichroism? In a qualitative way this question has been answered by electronic theory (the studies of Drude, Born, and Oseen) and in a strictly quantitative form by quantum mechanics (the studies of Rosenfeld). In the molecular theory of the propagation of light within a substance (in the theory of polarizability), the dimensions of the molecule are taken as vanishingly small relative to the wavelength of the light. In fact, the order of magnitude of the dimensions of molecules, but not macromolecules, is 10^{-8} to 10^{-7} cm, while the wavelengths of visible light are of the order of 10^{-5} cm. This means that the theories disregard the difference in phases of the light waves within the different points of the molecule. Such a theory explains very well the refraction and dispersion of light, but it is insufficient in principle for the interpretation of optical activity. Optical activity is an effect of the order of the ratio d/λ (d is the dimension of the molecule); this ratio is defined by the very difference in phases of the light waves in different parts of the molecules or, so to say, by intramolecular interference. Calculations show that, if we have within the molecule a plane or center of symmetry, there will be no optical activity. The simplest model of an optically active molecule has been proposed by Kuhn; it consists of two oscillators (two electrons) which are present at some distance from each other and which vibrate in different planes (Fig. 88). Here both conditions are respected; there is a distance d which cannot be neglected, and the model is devoid of a plane or center of symmetry. However, in order to have optical activity, there is still a third condition which must be fulfilled; it is always satisfied in molecules, namely, the vibrations of the electrons must be related to each other. Rigorous quantum theory expresses optical

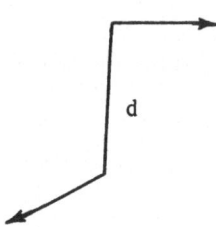

Fig. 88. The two-oscillator model by Kuhn.

activity through transition moments, namely, the electric moment μ_{0i} and the magnetic moment m_{i0}:

$$m_{i0} = \int \psi_i m \psi_0^* \, d\tau \tag{5.67}$$

where m is the vector of the magnetic moment. We have

$$\varphi = \frac{16\pi^3 N_1 v^2}{c^2} \beta \tag{5.68}$$

where β is the molecular parameter of optical activity given by

$$\beta = \frac{c}{3\pi h} \sum_i \frac{\mathrm{Im}\{\mu_{0i} m_{i0}\}}{v_{0i}^2 - v^2} \tag{5.69}$$

The symbol Im means that we must take only the imaginary part of the complex expression $\mu_{0i} m_{0i}$.

Let us compare Eq. (5.69) and (5.61). They are quite similar; they are dispersion equations. In the numerator of each term of the sum, the polarizability is expressed by the dipolar strength

$$\mu_{0i}^2 = \frac{3he^2}{8\pi^2 m v_{0i}} f_{0i} \tag{5.70}$$

while in the numerator of each term of the sum in (5.69), we have the rotational strength

$$F_{0i} = \mathrm{Im}\,(\mu_{0i} m_{i0}) \tag{5.71}$$

The dipolar force (i.e., the oscillator strength) can be found indirectly from the dispersion of the refractive index and directly from the intensity of absorption. The rotational strength which corresponds to the given transition $0 \rightarrow i$ can be found indirectly from the optical rotatory dispersion and directly from the circular dichroism within the absorption band.

Just as polarizability is related to the usual absorption of light, optical activity is related to the Cotton effect and to circular dichroism. The circular dichroism, i.e., the different absorptions of right and left circularly polarized light, is expressed in the fact that light of a wavelength that corresponds to an absorption band will acquire not only rotation of the plane of polarization after passing through an optically active medium, but will also become elliptically polarized. The measure of ellipticity is

$$\theta = \frac{\pi}{\lambda}(\kappa_L - \kappa_R) \tag{5.72}$$

where κ_R and κ_L are the absorption coefficients for right and left polarized waves. If a light with intensity I_0 passes through a medium with an absorptivity κ over a path l, its intensity decreases to

$$I = I_0 e^{-(4\pi\kappa/\lambda)l} \tag{5.73}$$

The coefficient κ is related to the molecular absorptivity coefficient ε by

$$\kappa = \frac{\lambda c}{4\pi}\varepsilon \tag{5.74}$$

where c is the concentration of the absorbing substance in moles per liter.

The rotational strength of a given band is expressed through the ellipticity in a manner similar to the way that the dipolar strength or oscillator strength is expressed through the absorption coefficient

$$R_i = \frac{3hc}{8\pi^3 N_1} \int \frac{\theta_{0i}}{v'} dv' \tag{5.75}$$

The integral is taken over the entire absorption band. Thus, a direct measurement of the Cotton effect gives a value of the principal constants, namely, of the rotational strengths.

The theory makes it possible to recalculate the data of the anomalous dispersion of optical activity into circular dichroism and *vice versa*, with the help of the *Kronig–Kramers relations*. In Fig. 89 we have shown schematically the relation between the anomalous dispersion of the refractive index and the absorption of light and the similar relationship between the anomalous optical rotatory dispersion and circular dichroism.

It is evident that the dispersion Eqs. (5.60) and (5.69) cannot be used to describe anomalous dispersion, since at $v = v_{0i}$ they go to infinity. The

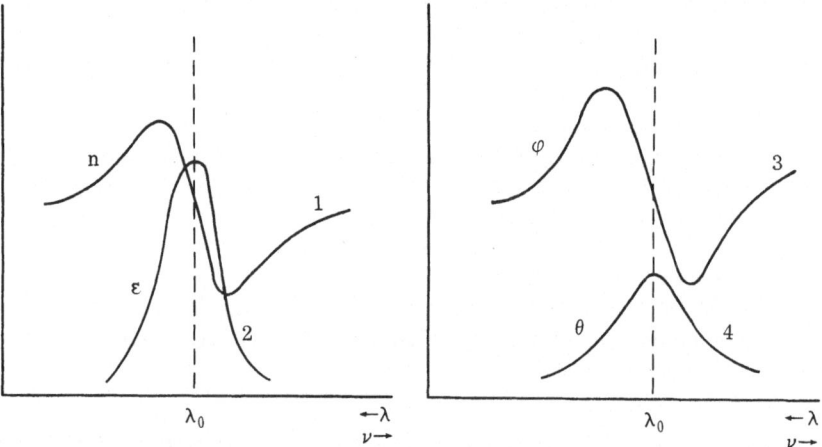

Fig. 89. Curves showing (1) the anomalous dispersion of the refractive index, (2) absorption of light, (3) anomalous dispersion of optical activity, (4) circular dichroism.

point is that, in the development of these equations, the extinction of the oscillators related to absorption was not taken into account. In the region of the intrinsic absorption, instead of Eq. (5.69) the valid equation is

$$\beta = \frac{c}{3\pi h} \sum_i \frac{R_{0i}}{v_i^2 - v^2 + iv\Gamma_i} \tag{5.76}$$

where i in the denominator is $\sqrt{-1}$, and Γ_i is the absorption parameter, which is approximately equal to the half-width of the absorption band expressed in frequency units.

Theory shows that, while the sum of all the dipolar forces is a constant equal to the total number of electrons in the molecule, the sum of the rotational strengths is equal to zero:

$$\sum_i R_i = \sum_i \text{Im} \,(\boldsymbol{\mu}_{0i} \boldsymbol{m}_{i0}) = 0 \tag{5.77}$$

According to the rules of matrix calculations, it is true that

$$\sum_i \boldsymbol{\mu}_{0i} \boldsymbol{m}_{i0} = \int \psi_0 \boldsymbol{\mu} \boldsymbol{m} \psi_0^* \, d\tau = \int \psi_0 \boldsymbol{\mu} \psi_0^* \, d\tau \int \psi_0 \boldsymbol{m} \psi_0^* \, d\tau$$

Such an expression is finite, and its imaginary part is identically equal to zero. From this expression, it follows that an optically active substance must be characterized by at least two absorption bands which have different values of R_i or θ_i. The model of Kuhn satisfies this condition; in the simplest case, when both oscillators are identical but vibrate, since they are linked with each other, the model has two normal vibrations, in phase and out of phase, with opposite signs of θ. For further details on the theory of optical activity, see Refs. 53 and 56.

In such a manner, the optical activity has a specific dependence on v (or λ). Far from the intrinsic absorption band, the dispersion of optical activity of simple molecules may be approximately expressed by the Drude equation

$$[\alpha] = \frac{A}{\lambda^2 - \lambda_0^2} \tag{5.78}$$

Here λ_0 is some effective wavelength corresponding to the absorption of light. The Drude equation corresponds to a single term of Eq. (5.69). In fact,

$$[\alpha] = \frac{\varphi}{c'} \frac{1{,}800}{\pi} = \frac{28{,}800}{c'} \frac{\pi^2 N_1}{c^2} v^2 \beta$$

$$= \frac{28{,}800 N_1}{c'} \frac{\pi}{3hc} v^2 \sum_i \frac{R_i}{v_{0i}^2 - v^2} = \frac{28{,}800}{c'} \frac{\pi N_1}{3hc} \sum_i \frac{R_i \lambda_{0i}^2}{\lambda^2 - \lambda_{0i}^2} \tag{5.79}$$

And if we limit ourselves to a single term of this equation, setting λ_{0i} equal to λ_0, we obtain Eq. (5.78), where

$$A = \frac{28{,}800}{c'} \frac{\pi N_1}{3hc} R_0 \lambda_0^2 \qquad (5.80)$$

In the calculations it is also necessary to take into account the refractive index of the medium (the Lorentz correction). Thus, it is necessary to introduce the multiplier $(n^2 + 2)/3$ into the expression for φ.

Let us turn now to the optical activity of proteins and synthetic polypeptides. Let us examine the principal experimental data and their theoretical interpretation.

OPTICAL ACTIVITY OF POLYPEPTIDE CHAINS

A helix–coil transition is accompanied by a large change in optical activity. This change has turned out to be of the same order of magnitude in a number of cases. Polybenzyl glutamate in dioxane (helix) has $[\alpha]_D^{20°} = +9$; and in trifluoroacetic acid (coil), $-46°$. The change is 55°. As has been stated above, this is due to the disappearance of the rotating ability of the helix as a whole.

To verify this natural assumption, Doty and coworkers studied the optical activity of copolymers of L- and D-γ-benzyl glutamate.[57] As the number of D residues increased in the chain, the specific rotation of the helix increased up to large positive values, while in a coil, it fell proportionally to zero. Doty and Lundberg[58] obtained a number of copolymers by initiating mixtures of D- and L-N-carboxyanhydrides of benzyl glutamate by pure L-polypeptides which already had a helical conformation. If these initiating polypeptides had helices with a single direction of rotation, this should have been maintained also during the growth of the chain by the addition of new D and L groups. After the addition of 25 such residues to each chain, the optical activity was measured and the optical activity of the initiating chain was subtracted from the value obtained. The dependence of $[\alpha]$ obtained in this way on the percent content of D groups shows that the specific rotation grows linearly with an increase of D groups in the content up to 30%. At D-group contents higher than 30%, the curve turns downward, which is explained by the refolding of the left helix into a right helix when there are sequences of four D groups.[59] At a concentration of less than 30%, the presence of such sequences has a small probability. The straight line obtained was extrapolated to 50% D groups (racemic mixture of amino acid within the polypeptide), and Doty and Lundberg obtained $[\alpha] = 54°$. This value characterizes the optical activity of the helix itself and is found to be in excellent agreement with the experimentally found difference of 55°. This can be expressed very well by the following relations:

$$[\alpha]_{\text{helix}} = [\alpha]_{\text{helix,total}} + [\alpha]_{\text{residues}} = +9°$$

$$[\alpha]_{\text{coil}} = [\alpha]_{\text{residues}} = -46°$$

Consequently,

$$[\alpha]_{\text{helix,total}} = 9° - (-46°) = 55°$$

On the other hand, for a copolymer,

$$[\alpha]_{\text{helix}} = [\alpha]_{\text{helix,total}} + (x_L - x_D)[\alpha]_{\text{residues}}$$

where x_L and x_D are the fractions of L and D residues. At 50% L and D

$$[\alpha]_{\text{helix}} = [\alpha]_{\text{helix,total}} = 54°$$

Consequently, during the helix–coil transition there is, in fact, a melting of the helix and its optical activity disappears just as the optical activity of crystalline quartz when it is melted. Once again we find the same analogy with a crystal!

There is another idiosyncracy of the optical activity of a helix. The optical rotatory dispersion of small molecules or individual groups can be expressed by the Drude equation, which we shall rewrite in the form

$$[\Omega] = \frac{3}{n^2 + 2} \frac{M}{100}[\alpha] = a\frac{\lambda_0^2}{\lambda^2 - \lambda_0^2} \tag{5.81}$$

It turned out that the dispersion of the helix can be described by a more complicated relation due to Moffitt:[60]

$$[\Omega] = (a + a_{\text{helix}})\frac{\lambda_0^2}{\lambda^2 - \lambda_0^2} + b\left(\frac{\lambda_0^2}{\lambda^2 - \lambda_0^2}\right)^2 \tag{5.82}$$

where a, a_{helix}, and b are constants. For α-helical synthetic polypeptides, $a_{\text{helix}} \cong 650$, $b \cong -630$, and $\lambda_0 = 212\ \text{m}\mu$. The a_{helix} and b constants are determined independently by using the variation of the dispersion. It should be pointed out that the optical rotatory dispersion of crystalline quartz is also described, not by a simple Drude equation, but by a quadratic equation of the type of (5.82).[61] How do we explain the specific dispersion of optical activity by helices?

The general equation of the rigorous theory of optical activity (5.79) does not yet make it possible for us to carry out practical calculations which could be applied to concrete problems. For this, we should need to know the complete set of wave functions of the molecule, ψ_i. As always in quantum chemistry, it becomes necessary to seek approximations, i.e., simplified methods which would permit the circumvention of this difficulty. Until recently, such methods in the theory of optical activity were quite unsatisfactory; they permitted the calculation of the rotatory strength at best as an order of magnitude. It is this alone which explains the insufficient application of optical activity to the study of the structure of molecules. After all, optical activity is a phenomenon which quite possibly is the most sensitive to any changes in the structures of molecules and their environment. However, Djerassi[62] has shown that there are theoretical concepts which make it possible to extract important information on the structure of small molecules and, in particular, on their conformations, from optical rotatory dispersion (earlier the aim was to calculate the optical activity for a single wavelength, usually the sodium D line, and this led to poor results). Second, the specificity of structure of linear polymers also introduces some definite simplifications which permit the interpretation of the optical rotatory dispersion of helices and coils.

The approximate theory of the optical activity of molecules stems from the possibility of breaking molecules down into separate groups between which there is no exchange of electrons. For a protein, such groups are the individual peptide groups and the amino acid residues. In such a case, the parameter of optical activity β may be expressed in the form of the sum

$$\beta = \beta_1 + \beta_2 + \beta_3 + \beta_4 \tag{5.83}$$

Here β_1 is the contribution of an individual group, equal to zero if the group is symmetrical; β_2 is the "one-electron" term, determined by the behavior of the electrons of the chromophoric groups of atoms which are responsible for the highest wavelength absorption of light by the molecule in an asymmetric field of the other group; and β_3 stems from the interaction of the magnetic dipolar transition in a single group with the electric dipolar transition in another. Finally, β_4 is determined by the interaction of the electric dipoles, induced in the different groups of the asymmetric molecule by the electric field of the light waves. The theory of β_4 may be called the *theory of polarizability*. Each group of atoms within the molecule has an anisotropic polarizability. Thus, the C—H bond has a polarizability $\alpha_1 = 7.9 \times 10^{-25}$ cm^3 along the bond and $\alpha_2 = 5.8 \times 10^{-25}$ cm^3 in the perpendicular direction. Consideration of the anisotropic polarizabilities of the bonds permits the investigation of a number of the optical properties of the molecules based on what is called the *valence optical* scheme.[53] In a theory of optical activity it is necessary to examine the interaction of dipoles induced by the electric field of the light waves in each of the bonds or groups of the molecule. This results in an approximate expression for the optical activity through the polarizability of the groups and bonds, α, and their relative positions within the molecule

$$\beta = \frac{1}{6}\sum_{k,l}\frac{1}{r_{kl}^3}(\alpha_{k1} - \alpha_{k2})(\alpha_{l1} - \alpha_{l2})(r_{kl}[kl])\left\{kl - 3\frac{(kr_{kl})(lr_{kl})}{r_{kl}^2}\right\} \tag{5.84}$$

where r_{kl} is the vector connecting group k and l, k and l are unit vectors directed along the axes of symmetry of these groups, and α_{k1} and α_{k2} are the parallel and perpendicular polarizabilities. It is easy to show that β is equal to zero when there is a center or plane of symmetry. Equation (5.84) was derived by Kirkwood by quantum-mechanical methods.[63] Later it was possible to obtain the same expression on a purely classical basis and to take into account further approximations.[64]

The theory of β_4, or the theory of polarizability, is inapplicable to the examination of weak absorption bands. In fact, the magnitudes of the anisotropic polarizabilities are determined by dipolar forces (oscillator strength), i.e., they increase with an increase in the intensity of the given band, with an increase in μ_{0i}. Moreover, the rotational strength may be great for weak rather than strong bands, i.e., for bands with small μ_{0i} but large m_{i0}.

Another limitation of the theory of polarizability is that it cannot be applied to molecules with conjugated bonds. First, bonds cannot be characterized by autonomous ellipsoids of polarizability; it becomes necessary to examine the entire system of conjugated bonds; for example, the benzene ring must be regarded as a single group. This limitation of the valence optical scheme is not so important in our case; proteins and polypeptides are not conjugated systems. This is what makes it possible to develop the theory of the optical properties of α-helices (and also of the β structure) by analogy with the theory of molecular crystals, on the assumption that amide groups are independent of each other in the first approximation.

Furthermore, it should be said that when the distances between the atoms within the molecules are small, the assumption of dipole–dipole interactions is not rigorous. As a result, the theory of polarizability is much more applicable to the investigation of the effect of intermolecular interactions on optical activity.[61]

The relations between the various components of β [Eq. (5.83)] have been analyzed,[65] and a semiempirical method of calculation has been proposed.[66,67] In those cases in which it is possible to carry out sufficiently complete calculations, the dispersion of optical activity can be described sufficiently well by a sum of the polarization and one electron terms. Using the approximate theory, Djerassi and others have examined the structures and conformations of a number of quite complicated organic compounds, such as terpenes and steroids.[62]

Let us return to the Moffitt equation (5.82). Does it have any basis other than empirical?

It is always possible to express $[M]$ in a spectral region removed from the absorption band in the form

$$[M] = \sum_i \frac{\alpha_i \lambda_i^2}{\lambda^2 - \lambda_i^2} \tag{5.85}$$

Expanding this equation in a Taylor series in $\lambda_i^2 - \lambda_0^2$ ($\lambda_i^2, \lambda_0^2 < \lambda^2$), we obtain

$$[M] = \sum_i \frac{a_i \lambda_0^2}{\lambda^2 - \lambda_0^2} + \sum_i \frac{a_i \lambda^2 (\lambda_i^2 - \lambda_0^2)}{(\lambda^2 - \lambda_0^2)^2} + 0 \frac{1}{(\lambda^2 - \lambda_0^2)^3} \qquad (5.86)$$

Rearranging the second term of this expression, we obtain

$$\frac{\lambda^2 (\lambda_i^2 - \lambda_0^2)}{(\lambda^2 - \lambda_0^2)^2} \equiv \frac{(\lambda^2 - \lambda_0^2 + \lambda_0^2)(\lambda_i^2 - \lambda_0^2)}{(\lambda^2 - \lambda_0^2)^2}$$

$$\equiv \frac{\lambda_i^2}{\lambda^2 - \lambda_0^2} - \frac{\lambda_0^2}{\lambda^2 - \lambda_0^2} + \frac{\lambda_0^2 (\lambda_i^2 - \lambda_0^2)}{(\lambda^2 - \lambda_0^2)^2}$$

Summing up the first and second terms, we obtain

$$[M] = \sum_i \frac{a_i \lambda_i^2}{\lambda^2 - \lambda_0^2} + \sum_i \frac{a_i \lambda_0^2 (\lambda_i^2 - \lambda_0^2)}{(\lambda^2 - \lambda_0^2)^2} + 0 \frac{1}{(\lambda^2 - \lambda_0^2)^3} \qquad (5.87)$$

And setting

$$a_0 = \sum_i a_i \frac{\lambda_i^2}{\lambda_0^2} \qquad b_0 = \sum_i a_i \left(\frac{\lambda_i^2}{\lambda_0^2} - 1 \right)$$

we obtain

$$[M] \cong \frac{a_0 \lambda_0^2}{\lambda^2 - \lambda_0^2} + \frac{b_0 \lambda_0^4}{(\lambda^2 - \lambda_0^2)^2} \qquad (5.88)$$

This expression is similar to the Moffitt equation. The sign of b_0 is different from that of a_0 if $\lambda_0 > \lambda_i$. It is evident that a simple calculation points to the usefulness of the Moffitt equation, but it certainly does not prove its validity for the α-helix.

Moffitt obtained Eq. (5.82) by a theoretical approach. He started from the Davydov theory of exciton splitting and he took into consideration the dipole–dipole interactions between the peptide groups of the α-helix.[64] Moffitt calculated the exciton splitting of the $\pi_0 \pi_-$ band, which is close to 190 mμ. According to Moffitt's estimate, the splitting amounts to $\Delta v = v_{\parallel} - v_{\perp} = 1{,}500 \ \text{cm}^{-1}$; this is actually close to the splitting which is observed experimentally. Using the theory of polarizability, i.e., the theory of β_4, Moffitt obtained his equation.

Later on, Moffitt, Fitts, and Kirkwood established that these calculations included an error related to an incorrect setting of the boundary conditions of the problem. The complete solution of the problem of the absorption spectrum and optical activity of α-helices is given in the papers of Tinoco;[49,68,69] it is shown that the theoretical expression for the dispersion of optical activity of α-helices is more complicated than Moffitt's equation; it also contains other terms. A simplified and quite interesting variation of this theory was given by McLachlan and Ball,[70] which was improved later by Harris.[71]

Thus, the Moffitt equation must be regarded as semiempirical. This does not decrease its value for the experimental determination of the degree of α-helicity of proteins.

In these theoretical investigations, only the $\pi\pi$ transitions in the peptide bonds of α-helices were examined. However, the weak $n\pi$ transitions may also make a considerable contribution to the optical activity, if they correspond to large magnetic transition moments m_{01}. Schellmann and Oriel, using the one-electron model, have shown that the moment m_{0i} for $n\pi$ transitions is indeed large if the peptide groups form a helix.[72] The moment is directed along the axis of the helix. The same study included an experimental investigation of the optical rotatory dispersion in the absorption bands of polyglutamic acid. The $n\pi$ transitions make a significant contribution to the optical activity, but they are considerably smaller than the contributions of the $\pi\pi$ transitions. The optical rotatory dispersion of PGA is shown in Fig. 90 in the α-helical and the denatured, coiled forms.

Tinoco has shown that if, within the absorption spectrum of the α-helix, the $\pi\pi$ band is split into two components, polarized parallel ‖ and

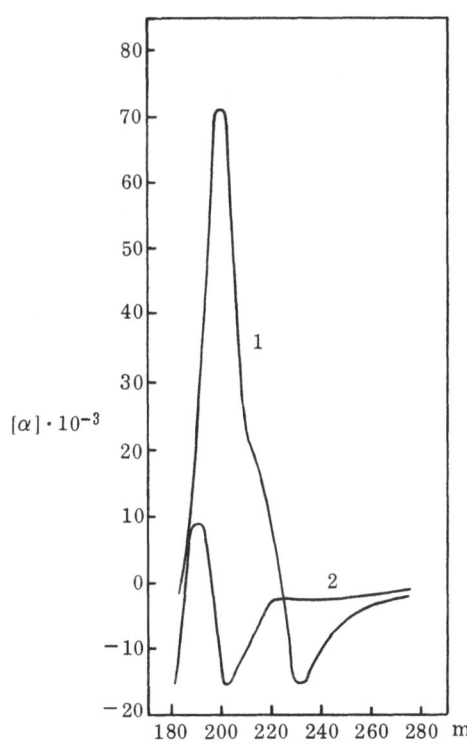

Fig. 90. The anomalous dispersion of optical rotation of polyglutamic acid: (1) helix and (2) coil.

perpendicular $\perp$ to the axis of the helix, then there must be a double optical-rotatory-dispersion curve in the region of the $\perp$ band; this band is split into two. The experimental data and theoretical values of the wavelengths, the oscillator strengths, and the rotational strengths for the α-helix of PGA are given in Table 13.[73] The measurements of R were carried out with the help of circular dichroism. The agreement between theory and experiment is far from satisfactory; however, theory gives the correct signs of R, the sequence of their values, and the orders of magnitude.

The Moffitt equation may be used to determine the degree of α-helicity of proteins. Let us express the equation in the form

$$[M] = (a + fa_{hel}) \frac{\lambda_0^2}{\lambda^2 - \lambda_0^2} + bf \frac{\lambda_0^4}{(\lambda^2 - \lambda_0^2)^2} \tag{5.89}$$

where f may be determined from the slope of the straight line in the plot of which the ordinate is $[M](\lambda^2/\lambda_0^2 - 1)$, and the abscissa is $(\lambda^2/\lambda_0^2 - 1)^{-1}$. This gives a determination of $b_0 = bf$. Independently, f is determined from the intercept of this straight line across the ordinate if a is known from data for the denatured protein. This is a determination of $a_0 = a + fa_{hel}$. At small values of b_0, the determination of f is not reliable, but if $b_0 > 200$, the values obtained turn out to be in good agreement with the result obtained by other methods, for example, by X-ray diffraction (for hemoglobin and myoglobin).

In empirical determinations of f, it is also possible to use the one-term Drude equation (5.85). In this case, the value of λ_c is not constant, but is a function of f. For 40% helicity, $\lambda_c = 269$ mμ; for a coil, $\lambda_c = 212$ mμ. If the dependence of λ_c on f is regarded as linear, it is possible to determine f from the dispersion of the optical activity.

Blout and Schechter have proposed determining f with the aid of a two-term Drude equation.[74] Taking into account the experimentally observed maxima $\lambda_1 = 193$ mμ and $\lambda_2 = 225$ mμ, we have

$$[\alpha] = \frac{A_1 \lambda_1^2}{\lambda^2 - \lambda_1^2} + \frac{A_2 \lambda_2^2}{\lambda^2 - \lambda_2^2} \tag{5.90}$$

where the constants $A_1 > 0$ and $A_2 < 0$ are related linearly. In aqueous solutions of polyamino acids (dielectric constant $\varepsilon > 30$),

$$A_2 = -0.55A_1 - 280$$

in organic solvents ($\varepsilon < 30$),

$$A_2 = 0.55A_1 - 430$$

In the coiled state, $A_2 = 0$, $A_1 = -600$ (1,1-copoly-L-methyonyl-L-methyl-S-cysteine in a 1:1 mixture of dichloroacetic and trifluoracetic acids); for the α-helix, $A_1 = 3{,}020$, $A_2 = -1{,}900$ (poly-L-methyonine in

Table 13   Optical Properties of the α-Helix

| | | Absorption | | | | | | Rotation | | |
| | | λ, mμ | | Oscillator strength f | | λ, mμ | | Rotatory strength, $R \times 10^{40}$ erg-cm³-rad | |
Transition	Character	Theory	Exp	Theory	Exp	Theory	Exp	Theory	Exp
$n_1-\pi^-$	In absorption	210–230	222	0.0001	0.007	210–230	222	−3.4	−22
$\pi^0-\pi^- \parallel$	In absorption	198	206	0.09	0.03	198	206	−126	−29
$\pi^0-\pi^- \perp$	In absorption	188	189	0.16	0.10	191	190	+242	+81
$\pi^0-\pi^-$	Only in rotation	...	...	...	...	185	...	−115	

chloroform). From this, the degree of helicity is

$$f_{193} = \frac{A_1 + 600}{36.2} \qquad f_{225} = -\frac{A_2}{19.0}$$

It is easier to use the combined empirical equation

$$f = \frac{A_1 - A_2 + 600}{55.8}$$

The method by Blout and Shechter makes it possible not only to determine f but also to characterize the environment of the amino acid residues by determining whether the points fall on a curve for $\varepsilon > 30$ or $\varepsilon < 30$. This is important in studies of the partial or complete uncoiling of globular structures, during which the residues change their amino acid environment for an aqueous medium.

All the listed methods of the empirical determination of the degree of α-helicity of proteins from optical rotatory dispersion outside of the absorption band give results which are in fair agreement with each other. The direct investigations of optical rotatory dispersion and circular dichroism permit the degree of α-helicity to be determined with great reliability. In recent years, spectropolarimeters and dichrographs have become available for studies in the far ultraviolet region, in the region of wavelengths close to 200 mμ and below.

In relation to this, one should point out the inaccuracy of the use in the literature of the term *Cotton effect*. Generally it is used to refer to optical rotatory dispersion. In fact, Cotton discovered circular dichroism, whereas optical rotatory dispersion was known even before that time. They are certainly closely related, but it is just as senseless to call optical rotatory dispersion a Cotton effect as to call the absorption of light the anomalous dispersion of the refraction of light.

The second regular form of a polypeptide chain is the β form; it is characterized by the usual Drude dispersion of optical rotation. Both theory and experiment show that in this case b_0 is close to zero. For a completely planar β form, b_0 must approach zero; a deviation of b_0 from zero means a folding of the β structure.

Blout and coworkers[75] discovered an important way of studying the anomalous dispersion of the optical rotation of biopolymers, not in the ultraviolet, but in the visible range of the spectrum. They studied the optical activity of polypeptides to which dye molecules had been adsorbed. These molecules are symmetrical, they are planar and therefore they have no optical activity of their own. However, complexes of the helical molecules with acridine dyes possess an anomalous dispersion of optical rotation within the absorption bands of the dye. Similar results are obtained for proteins and for enzymes containing prosthetic groups, with absorption bands at long wavelengths, e.g., for myoglobin and hemoglobin.[76] During

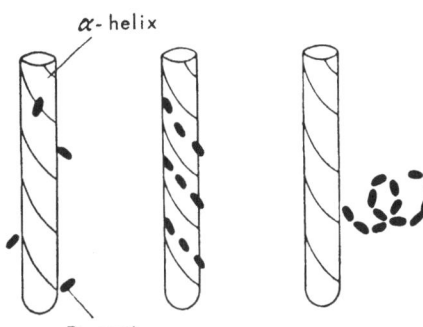

Fig. 91. Possible models of the interaction of dye molecules with an α-helix.

the denaturation of a biopolymer this effect vanishes; there remains no anomalous optical rotatory dispersion within the visible range. How can this be explained? Three alternatives are possible: Asymmetry is induced in each molecule of the dye by the asymmetric surroundings, the molecules of the dye form helices on the α-helix of the polypeptide, or the molecules of the dye become polymerized under the influence of the α-helix and form their own helix (Fig. 91). It has turned out that only the first assumption is possible. The effect is observed even with a very small relative content of dye molecules, which excludes the formation of secondary helices with sufficiently close dye molecules.[77] The induced anomalous optical rotatory dispersion is quite important in the study of enzymes. This phenomenon is observed also in complexes of nucleic acids with dyes.

In this manner, the investigation of optical activity makes it possible to answer important questions of molecular biophysics, to investigate the conformations of biopolymers and their changes. We must be grateful to nature for this gift, the asymmetry of biological molecules.

A number of problems in this realm have not been solved yet. Further development of both theory and experiment is necessary. It would seem quite useful to continue direct measurements of Cotton effects; a quantitative interpretation of the anomalous dispersion of optical rotation requires a knowledge of the ellipticity. Measurements of anomalous optical rotatory dispersion are necessary within the absorption bands of biopolymers, in the far ultraviolet region of the spectrum. Similar to polarizability, the quantity β and the rotational strength are tensors; optical activity is different in different directions along the molecule. Calculations show that optical activity along and perpendicular to the helix may have different signs. There is no doubt that determination of the anisotropic optical activity promises new valuable information on the structure of biopolymers. For such determinations it is necessary to orient macromolecules. Here unfortunately it is difficult to use the orienting effects of an electric field, since biopolymers are polyelectrolytes and subject to electrophoresis within the field. Up to the present it has been possible to carry out experiments in only one case, for polybenzylglutamate, which is not a polyelectrolytic polyamino acid.[78] However, one may hope that these difficulties will be surmounted.

ANOMALOUS DISPERSION OF MAGNETO-OPTICAL ROTATION

As investigations have shown in recent years, along with the natural rotation of the plane of polarization, optical activity, valuable information on the structure of molecules, of protein molecules among these, may be given by the Faraday effect, namely, the magnetic rotation of the plane of polarization. In 1846 Faraday discovered that when polarized light passes along the direction of a magnetic field through any substance, the plane of polarization of the light wave is rotated by an angle proportional to the thickness of the layer of the substance, l, and the strength of the magnetic field H.

$$\varphi = VlH \tag{5.91}$$

where V is the constant of magnetic rotation, the Verdet constant. Faraday wrote "... finally, I was able to magnetize and electrize a lightwave and to illuminate a magnetic line of force." The essence of the Faraday effect consists in the action of the magnetic field on the substance, as the result of which the substance acquires circular dichroism and circular birefringence. Contrary to optical activity, determined by the asymmetry of molecules and crystals, the magnetic rotation is determined by the asymmetry of the magnetic field and may be observed in any substances.

The Faraday effect is closely related to the Zeeman effect, the splitting of spectral lines in a magnetic field due to the splitting of degenerate energy levels. It is easy to explain this in terms of the classical model of the electron, namely, a harmonic oscillator. If, in the absence of the magnetic field, the oscillator vibrates with an angular frequency $\omega = 2\pi v$, then, in the presence of a field parallel to the ray of light, a longitudinal Zeeman effect takes place; the ω band splits into two bands, which are circularly

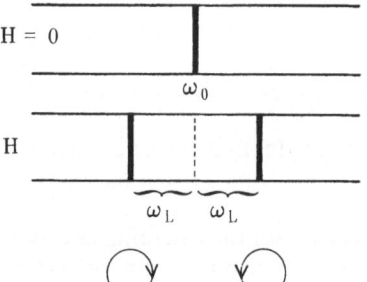

Fig. 92. Schematic representation of the longitudinal Zeeman effect.

polarized to the right and to the left (Fig. 92). The magnitude of the splitting is equal to $2|\omega_L|$, where ω_L is the frequency of the Larmor precession

$$\omega_L = -\frac{eH}{2mc} \tag{5.92}$$

As a result, light polarized to the left is absorbed with frequency $\omega + \omega_L$, while light polarized to the right is absorbed with frequency $\omega - \omega_L$. Circular birefringence [see Eq. (5.63)] is expressed by

$$\Delta n = \left(\frac{\partial n}{\partial \omega}\right)_{\omega_L = 0} 2\omega_L = -\frac{eH}{mc}\left(\frac{\partial H}{\partial \omega}\right)_{\omega_L = 0} \tag{5.93}$$

or

$$\Delta n = \lambda^2 \left(\frac{\partial n}{\partial \lambda}\right)_{H = 0} \frac{eH}{2\pi mc^2} \tag{5.94}$$

Introducing these into (5.63), we obtain

$$\varphi = \frac{eH}{2mc^2}\lambda \left(\frac{\partial n}{\partial \lambda}\right)_{H = 0} l \tag{5.95}$$

From here

$$V = \frac{e\lambda}{2mc^2} \left(\frac{\partial n}{\partial \lambda}\right)_{H = 0} \tag{5.96}$$

This is the Becquerel equation. The dispersion of the Faraday effect is expressed by the derivative of the refractive-index dispersion with respect to the wavelength. Consequently, anomalous dispersion of magnetic rotation (ADMR) is no longer expressed by an asymmetric curve such as is shown in Fig. 89, but by a symmetric one (Fig. 93). This is easy to understand. The curves of the dispersion of n, which correspond to the two components of Zeeman splitting, are identical but displaced from each other by $2\omega_L$ (Fig. 92). The quantity φ is expressed by the difference of these curves, and one obtains a symmetric curve. The first observation of ADMR of such a type was made by Macaluso and Corbino in 1898[53] on sodium vapors. It would seem only correct to call the symmetrical ADMR the *Macaluso–Corbino effect*. This effect may be expressed quite sharply and must be quite sensitive both to the positions and intensities of spectral bands; in fact, what is observed is the difference interference effect inside a given absorption band.

We have presented the most elementary variant of the theory, which has only limited applicability. Rigorous theory of magnetic rotation developed from the perturbation theory of quantum mechanics[79–81] leads

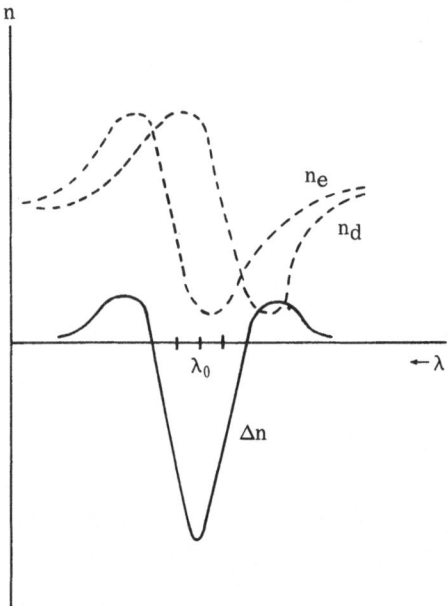

Fig. 93. The Macaluso–Corbino effect.

to a more complicated expression for V. For diamagnetic substances, V has the form

$$V = V_A + V_B \qquad (5.97)$$

where (far from the absorption line)

$$V_A = \frac{16\pi^2 N_1 v^2}{3h^2} \sum_{(\xi,\eta,\zeta)} \sum_i \frac{v_i[(m_\zeta)_{ii} - (m_\zeta)_{00}] \operatorname{Im}\left[(\mu_\xi)_{0i}(\mu_\eta)_{i0}\right]}{(v_i^2 - v^2)^2}$$

$$V_B = \frac{8\pi^2 N_1 v^2}{3h^2} \sum_{(\xi,\eta,\zeta)} \sum_i \sum_k \left\{ \frac{\operatorname{Im}\left[(m_\zeta)_{ik}(\mu_\xi)_{0i}(\mu_\eta)_{k0} + (m_\zeta)_{ki}(\mu_\xi)_{0k}(\mu_\eta)_{i0}\right]}{v_{ki}(v_i^2 - v^2)} \right.$$
$$\left. + \frac{\operatorname{Im}\left[(m_\zeta)_{k0}(\mu_\xi)_{0i}(\mu_\eta)_{ik} + (m_\zeta)_{0k}(\mu_\xi)_{ki}(\mu_\eta)_{i0}\right]}{v_k(v_i^2 - v^2)} \right\}$$

Summation is carried out over the circular transformations of the coordinates ξ, η, ζ, to a system fixed within the molecule.

The Verdet constant term V_A characterizes the Macaluso–Corbino effect; it is related to the Zeeman splitting of spectral levels. To the contrary, the term V_B is a function of $v_i^2 - v^2$ and is asymmetric. Classical theory does not give V_B; this quantity arises as a result of the quantum-mechanical examination. It is related to what is called the *Van Vleck paramagnetism*, which does not depend on temperature.

The quantity V_A is different from zero only if the molecule has a magnetic moment different from zero in the fundamental or excited state, if the quantity m_{00} or m_{ii} is not zero. For diamagnetic molecules $m_{00} = 0$. Consequently, the Macaluso–Corbino effect can be observed in this case only when $m_{ii} \neq 0$. As a rule, the absorption bands of complex molecules are related to singlet–singlet transitions; the spin moment of the molecule is equal to zero in both the fundamental and excited states. Consequently, the difference of m_{ii} from zero may be related only to orbital magnetism. This is a characteristic of molecules which have axial symmetry, i.e., such

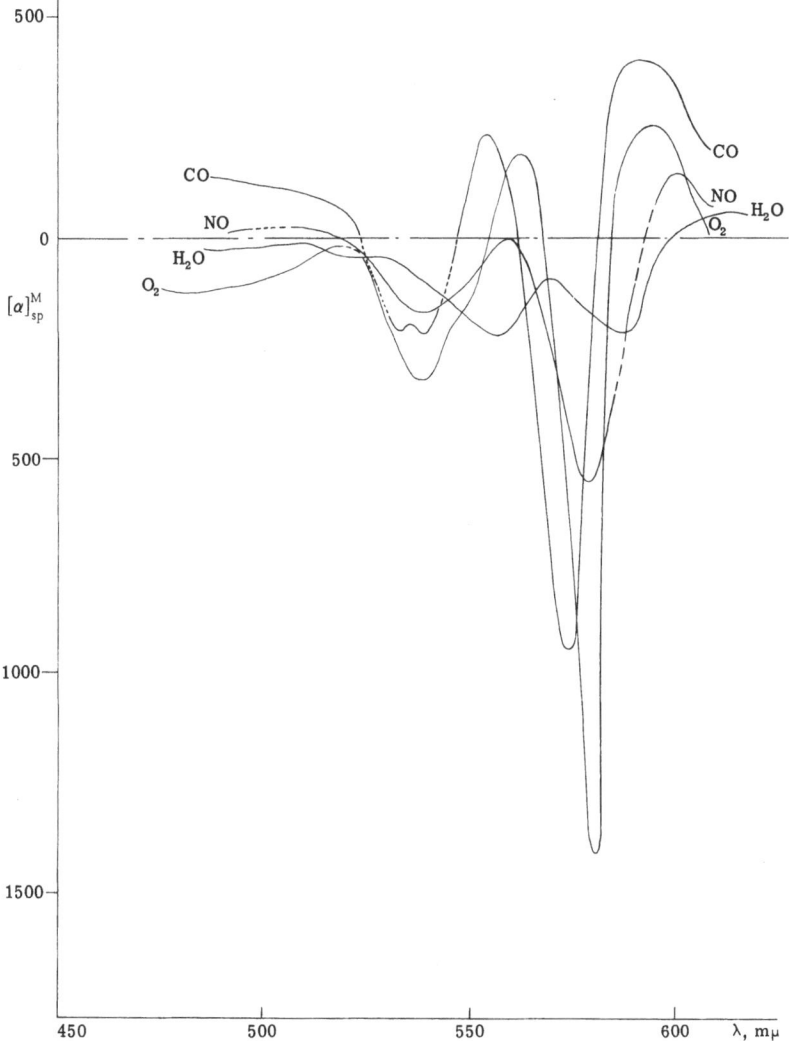

Fig. 94. The anomalous dispersion of magneto-optical rotation in the α absorption band of ferromyoglobin with various ligands.

molecules as have an axis of symmetry of third order or above. The Macaluso–Corbino effect must be absent from asymmetric diamagnetic molecules.

If the molecules are paramagnetic in the fundamental state, that is, $m_{00} \neq 0$, then the expression for V contains as well a third term V_C, which is a function of temperature,

$$V_C = \frac{8\pi^2 N_1 v^2}{3hkT} \sum_{(\xi,\eta,\zeta)} \sum_i \frac{(m_\zeta)_{00} \, \text{Im} \, [(\mu_\xi)_{0i}(\mu_\eta)_{i0}]}{v_i^2 - v^2} \tag{5.98}$$

The dispersion of V_C is the same as that of V_B, that is, it is asymmetric.

Magnetic rotation dispersion (ADMR) may give important information on the structures of molecules and, in particular, on the conformations of macromolecules. The numerators in the dispersion equations for V_A, V_B, and V_C contain different products of three vectors, one magnetic and two electric vectors. The directions of these vectors are fixed within the molecule and must change with changes in conformation. As a result, ADMV must be quite sensitive to such changes. This has been noted by Tinoco and coworkers, who have carried out calculations of orientation for biopolymers.[69]

It is quite evident that the information which may be obtained from ADMV is independent of that which is obtainable from other optical methods.

At present (the end of 1966), there are only a few data on ADMV. There are three papers devoted to proteins, all of which are related to ADMV within the absorption band of the heme. Shashova has studied the oxidation of cytochrome c.[82] The high sensitivity of ADMV to changes in the state of the cytochrome has been demonstrated, and as a result, a new method has been proposed for the quantitative investigation of the kinetics of oxidation. The oxygenation of hemoglobin and myoglobin has also been investigated.[83] Some of these results will be described later. Finally, ADMV has been studied in various myoglobin derivatives.[84] Let us cite as an example a few facts which indicate the high potential of this method.

In Fig. 94 we have shown the ADMV for the α absorption band of ferromyoglobin with four different ligands, namely, H_2O (reduced myoglobin), NO, CO, and O_2. The curves are quite different from each other. The most interesting observation is that the depths of the minima in the ADMV curves correlate quite well with the characteristics of the conformational stability and cooperativity of the protein. It follows, therefore, that a change in the electronic properties of the heme brought about by the ligand is reflected directly in the conformational properties of myoglobin.

Chapter 6

THE PHYSICS OF NUCLEIC ACIDS

BIOLOGICAL ROLE OF NUCLEIC ACIDS

Prior to speaking of the structure and physical properties of nucleic acids, it would seem desirable to examine their biological role. In this book it has been remarked already several times that DNA is the fundamental genetic substance and nucleic acids are responsible for the synthesis of proteins. It would seem that the time has come to bring out the evidence for these statements.

As we have said already, DNA is localized in the chromosomes, i.e., in the organoids which are responsible for transferring the hereditary information. The contents of DNA within chromosomes are quite characteristic. First, the DNA contents of diploid (somatic) cells of different tissues of the same species are quite constant. Second, in the haploid (sex) cells (spermatozoids), the amount of DNA is half that found within somatic cells. This is illustrated in Table 14.

These facts in themselves are quite convincing evidence about the genetic role of DNA. It is quite characteristic that the contents of other biologically functional substances which are found in chromosomes and in cell nuclei (ribonucleoproteins and proteins), are not constant but vary from one tissue to another.

The DNA contents within cells double during mitosis, i.e., during the doubling of the number of chromosomes.

If DNA is a genetic material, it must be, on one hand, particularly stable toward external interactions, and, on the other hand, it must be capable of changing under the action of mutagenic agents. And in fact, DNA is a substance which possesses a particular metabolic stability, a

Table 14[1]   Average DNA Contents in Cellular Nuclei

In picograms $(10^{-12}$ g) per nucleus

Organ	Rats	Chicken	Bovine	Frog	Newt	Carp
Liver	6.4	2.6	6.4	15.7	...	3.3
Kidney	6.7	2.3	6.3			
Spleen	6.5	2.6				
Lung	6.7					
Leucocytes	6.6					
Erythrocytes	...	2.6	...	15.0	7.3	3.5
Heart	6.5	2.5				
Pancreas	7.3	2.7				
Brain	...	2.3‡				
Muscle	...	2.5‡				
Sperm	...	1.3	2.8	...	3.7	1.6

‡Embryo chicks.

fact which is proven by direct experiments with labeled atoms. Labeled adenine (containing ^{32}P) is not incorporated in the DNA of nonduplicating cells. If the cells are duplicating (experiments with *E. coli*) in a medium which contains labeled adenine, they incorporate this material in the newly formed DNA and this material remains within the DNA practically without participating in metabolism. DNA synthesis is an irreversible process. It is interesting that, within the cells of malignant tumors (experiments with mice), the metabolic stability of DNA becomes violated and there is observed a significant incorporation of ^{32}P and ^{14}C. As a whole, a vast number of experiments show that DNA is not an absolutely inert material in metabolism, but it is much more stable than other components of the cell. The reasons for the enhanced stability of native DNA lie within the specificity of its secondary structure. The question on the amazing stability of the genetic material, which was first raised by Schrödinger,[2] has been solved by modern science.

Furthermore, the participation of DNA in mutagenesis has been established. The action spectrum of ultraviolet rays, which brings about mutations, coincides with the absorption spectrum of nucleic acids (absorption maximum at 260 mμ). This fact, in itself, is not an absolute proof, since the mutagenic activity of shortwave radiation is quite complicated; the light quanta may be absorbed by the DNA, but they may exert their influence on some other substance by a chemical reaction. There is, however, evidence that changes occur within the DNA as a result of the radiation. Samples of DNA have been sent into space on one of the Soviet space ships; this was done to study the effect of cosmic rays on DNA in order to understand their mutagenic activity.

Much more definite information may be obtained from chemical mutagenesis. Zamenhof and coworkers[3] have shown that when 5-bromouracil is incorporated in DNA, the number of mutations increases; 5-bromouracil is a substance analogous to thymine. The mutagenic action of nitrous acid may be followed chemically. It consists in the replacement of an amino group by a hydroxyl, in a deamination

$$-NH_2 + NHO_2 \longrightarrow -OH + N_2 + H_2O$$

In this manner, nitrous acid changes cytosine into uracil, guanine into xanthine, and adenine into hypoxanthine. Here we are talking about quite definite changes of the DNA which are reflected in mutations.

An extremely interesting proof of the genetic role of DNA is found in the transformation of bacteria. Already in 1928 Griffith discovered the transformation of pneumococci (*Diplococcus pneumoniae*). The mutant strains of the pneumococci differ in the character of their polysaccharide cell walls. These differences can be detected by immunological methods and seen directly by observation of the colony. Strain S forms "smooth" shiny colonies, i.e., the cells have strong capsulelike walls. Strain R consists of cells which have lost their ability to produce a polysaccharide, which is essential for such capsules; the colonies are "furry." If a living culture of type R was injected into a mouse, together with killed pneumococci of type S, the mouse became infected with pneumonia and a culture of the bacteria, which had multiplied within it, gave a living culture of virulent pneumococci of type S. It follows from this that dead S bacteria contain a certain transforming factor which can transform strain R into strain S. Later it was shown that bacterial transformation may be carried out not only *in vivo*, i.e., not only in the organism of a mouse or another animal, but also *in vitro*. Transformations have been established in a number of bacteria within *Bacillus subtilis*, within *E. coli*, etc. It is possible to carry out transformations of resistance to various antibiotics and sulfonamides. At first, the nature of the transforming factor was not clear. But finally, as a result of quite complicated and difficult work, it was possible not only to isolate this factor in pure form but also to characterize it completely.

It turned out that the transforming factor is DNA. This was shown for the first time by Avery, McLeod, and McCarty.[4] If DNA is extracted from the S type of cells and added to a medium containing a ground culture of R, R becomes transformed into S; type S multiplies without limit after this. In these and other experiments, it was shown that it is pure DNA which performs the transformation; additions of proteins and other substances were eliminated. Furthermore, deoxyribonuclease (DN'ase), an enzyme which splits DNA, stops transformation.

The transformation phenomenon has been investigated in particularly great detail by Hotchkiss.[5,6,7] In particular, he studied the ability of pneumococci to induce the synthesis of an enzyme, namely, mannitol-dehydrogenase. There is a strain which can use mannitol [a six-carbon alcohol $CH_2OH(CHOH)_4CH_2OH$] as a carbon source, i.e., this strain produces the corresponding enzymes; furthermore, the presence of mannitol stimulates the production of the enzyme; in its absence, the cells are devoid of the necessary enzymatic activity. The mutant strain is devoid of such an ability. The DNA, extracted from the first strain, transforms the mutant cells into cells which are able to manufacture the enzyme with the inducing activity of mannitol.

DNA may carry several genetic indicators at once. Thus, for example, one form of DNA may transform to pneumococci the stability toward penicillin and streptomycin, and also the ability to form capsules. Among the transformed cells, 98% acquire one of the three characteristics, 2% acquire two of the three, and only 0.01% acquires all three. This is quite understandable; the probability of such transformations should be regarded as independent, and, as a result, the probability of a double transformation is equal to the product of the probabilities of the two single ones. This points to the introduction into the transformed cells of only part of the material contained within the DNA. This is illustrated in Table 15.

Table 15[8]   Transformation of Resistance to Streptomycin and Penicillin and Ability to Use Mannitol in Pneumococci

Properties		Properties of transformers	
Donor DNA	Acceptor cells	Single	Double
MS	ms	Ms, ms	Much MS
Ms + mS	ms	Ms, mS	Less than average MS
Ms	mS	MS (ms)	Much MS
ms	Ms	MS (ms)	Much mS
MsP	msp	Msp, msP	Less than average MsP

What happens during transformation? This is illustrated by the scheme in Fig. 95. In the starting culture of pneumococci, mutants arise during growth; let us assume that these mutants are stable toward penicillin. When penicillin is added to such a culture, all the cells perish, except the mutant ones; the last ones form culture P after multiplication. We extract a DNA from it. When this DNA is added to the starting culture, we find many more P mutants within it than in the absence of this DNA, i.e., of the transforming factor. The increase in the number of mutants is very large. If the number of spontaneous mutants is 1 in 10^7 cells, then, as a result of this transformation, this is an increase to 1 in 10^2 to 10^3 cells, that is, by 10^4 to 10^5 times!

These results prove unequivocally the transfer of genetic information by DNA molecules.

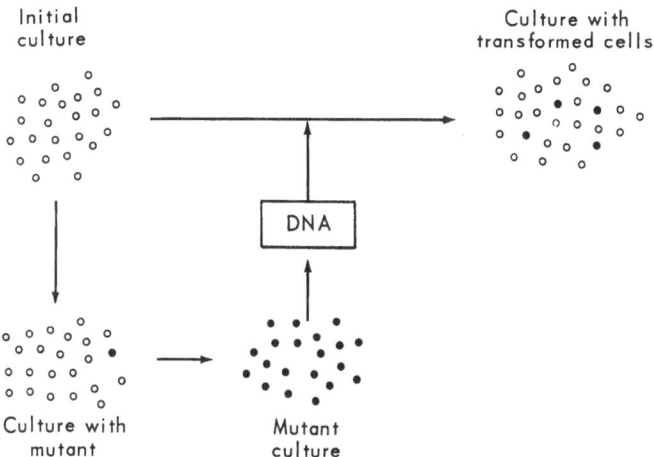

Fig. 95. Schematic diagram which explains the transformation of bacteria
by DNA.

Another set of proofs of the genetic role of nucleic acid was obtained
in studies of viruses and phages. In Chapter 2, we discussed the phenomenon of lysogeny. It turns out that some phage strains can carry the
genetic material from a donor bacterium within which the phage multiplied
to a receptor bacterium which is becoming infected by this phage. This is
the phenomenon of transduction. Various genetic characteristics of a bacterium are transferred by transduction; the phage particle carries over
both its own genetic material and some of the genetic material of the donor
bacterium. This genetic material is again DNA.[9,10]

Experiments with bacteriophages, and in particular with T2, labeled
with ^{35}S and ^{32}P, show that when a bacterial cell is infected with a phage,
practically no protein (labeled with sulfur) penetrates into it; however, the
major part of the DNA (labeled with phosphorus) does penetrate. The
particle of the T2 phage consists of DNA surrounded by a protein envelope.
It behaves like a syringe, i.e., it injects its DNA into the bacterium. This is
shown schematically in Fig. 16. Since afterward phage particles multiply
within the cells, it is evident that the DNA organizes both the synthesis of
its copies and a synthesis of the protein envelopes. In other words, DNA
is the genetic material of the phage. Direct experiments with labeled atoms
have shown that, during the multiplication of phage particles within the
cell, its own DNA is destroyed, while the nucleotides formed in this manner
are used to synthesize a DNA similar to that of the phage.[11,12]

In the case of plant viruses, the role of the DNA is played by RNA.
Fraenkel-Conrat[13] and Gierer and Schramm[14] have shown that RNA
extracted from TMV virus has infectious activity. This means that the
molecules of the virus RNA are able to organize the synthesis of new virus

particles when they are introduced within plant cells. The results are of the same nature as found with the bacteriophages; it proves the genetic role of the virus RNA.

Fraenkel-Conrat was able to reconstitute entire particles of TMV by mixing the TMV RNA with its protein. These reconstituted particles were found to have the biological properties of the starting virus. Hybrid viruses were also obtained by combining the RNA of one strain of TMV with the protein of another strain. The protein of the progeny of the hybrid virus contains methionine and histidine, which are characteristic for the virus donor protein, but are absent from the protein of the second strain. These results again manifest the genetic role of the viral RNA.[15]

The total of all these facts proves beyond any doubt that the DNA of cells, as well as of phages, and the RNA of viruses are genetic materials, i.e., that these nucleic acids contain the information about the synthesis of a protein. We shall see later how these substances perform their work.

A natural question comes up right away: What is the matrix for the DNA? At present, we have no direct data on the primary structure of DNA. However, the total of all results obtained in molecular biology and in chemical genetics prove that the primary structure of DNA (and of viral RNA) is fixed. And what kind of matrix would it be without fixation of the text? Well, what directs the primary structure of the genetic substance? Perhaps it is necessary to have some other substance which serves as a matrix for the synthesis of DNA, and so on, up to infinity. The answer to this question has been found as a result of studies on the secondary structure of DNA. It is found that, contrary to proteins, DNA is capable of duplicating itself, i.e., it serves as a matrix for self-copying.

We have examined the biological roles of DNA and viral RNA. What is the function of three other types of RNA, i.e., ribosomal, transfer, and matrix? Here we shall limit ourselves to a brief description; a detailed description will be presented in the following chapter. Matrix, or information, RNA is synthesized in the nucleus and is transferred to the ribosomes. It copies the code in which the genetic information is inscribed within the DNA of the chromosomes and, in this way, it is a transfer matrix. In this sense, informational RNA is similar to the matrices of newspapers which are daily carried by airplanes from one city to another. Ribosomal RNA serves apparently as a support for messenger or matrix RNA. The role of transport, low-molecular-weight RNA, consists in interaction with activated amino acids. The number of types of transport RNA is equal to the number of the different amino acids, i.e., 20 or a little more. It is not the amino acids which are assembled on the matrix, but molecules of transport RNA to which the amino acids have been attached. As a result of such an adsorption, the amino acids become joined to each other, and protein synthesis occurs.

Thus, three types of RNA participate in the synthesis of the protein according to a program dictated by the DNA.

SECONDARY STRUCTURE OF DNA, RNA, AND POLYNUCLEOTIDES

The key to the biological properties of DNA and to its genetic functions is found within the structure of the molecule. The primary structure of DNA is not yet known, but its secondary structure has been well established.

The X-ray diffraction investigation of the structure of DNA had an important significance for molecular biophysics and molecular biology. Wilkins examined oriented fibers of the Li salt of native DNA; he obtained excellent X-ray pictures which contained up to 100 diffraction spots.[16,17] It was found that the DNA molecules may crystallize in various forms. At a relatively low humidity (about 70 %), a monoclinic A form is obtained with lattice parameters $a = 22.2$ Å, $b = 40.0$ Å, $c = 28.1$ Å, and $\beta = 97.1°$. At high humidity, a paracrystalline B form is formed; it corresponds to orthorhombic symmetry with lattice parameters $a = 22.7$ Å, $b = 31.3$ Å, and $c = 33.6$ Å. Later on, still another form, C, was found, while recently Mokulskii *et al.* have discovered a fourth crystalline modification of DNA obtained from T-even phages.

On the basis of these X-ray data, Watson and Crick deciphered the molecular structure of DNA.[18,19]

The observed diffraction patterns could be explained completely by a double helical structure of native DNA. According to Crick,[20] the DNA molecule consists of two polynucleotide chains, antiparallel and wound about each other. These two chains are held together by means of hydrogen bonds between the nitrogen bases; furthermore, each base of one chain is linked to its partner. The pairing of the bases is specific: Adenine A is paired only with Thymine T while guanine G is paired only with cytosine C.

The structure of the DNA double helix is shown in Fig. 96. The diameter of the helix is approximately 20 Å, a single turn contains 10 monomer units and 2 nitrogen bases, and 2 nucleotides are found per each 3.46 Å of the helix length. Such a structure of DNA is characteristic for material obtained from absolutely different sources, from calf thymus to bacteriophage. Furthermore, in the X-ray diffraction investigations of heads of spermatozoa, the same picture is obtained, i.e., the Watson–Crick double helix is observed directly within living cells.

As has been said already, there is a 1:1 correspondence between the nitrogen bases of the two chains. In Fig. 97, we have shown the hydrogen bonds of the paired bases of purines and pyrimidines. The planes of the bases are parallel to each other; these planes and the hydrogen bonds are located perpendicular to the axis of the helix. It is noteworthy that A and T are linked by two hydrogen bonds, while G and C are linked by three. Thus, if the primary structure of one of the chains is known, then the

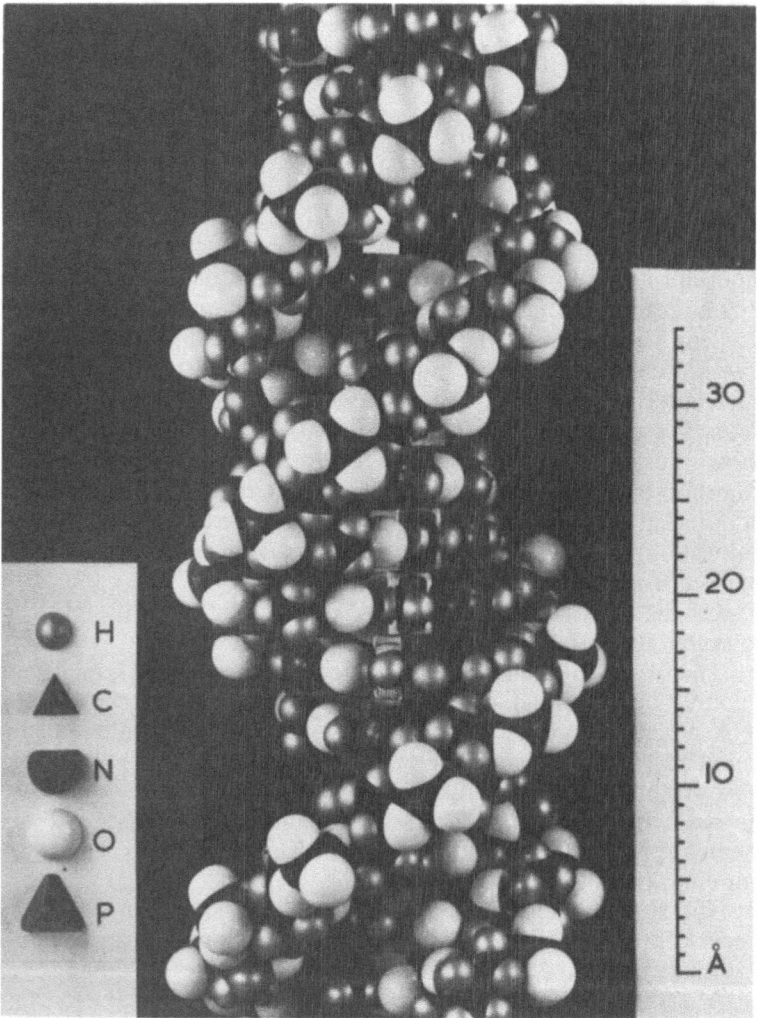

Fig. 96. Double helix of DNA.

structure of the second one is uniquely determined, since it is a complement of the first one. The indicated 1 : 1 pairing of the bases explains the rule of Chargaff that the contents of A within DNA are always equal to those of T and those of G are always equal to those of C.

Two chains may be either antiparallel or parallel. The two variants are shown in Fig. 98. The first is in better agreement with the X-ray diffraction data. Furthermore, there is a direct proof of the antiparallel nature of the chains obtained by biochemical means.[21] DNA can be synthesized

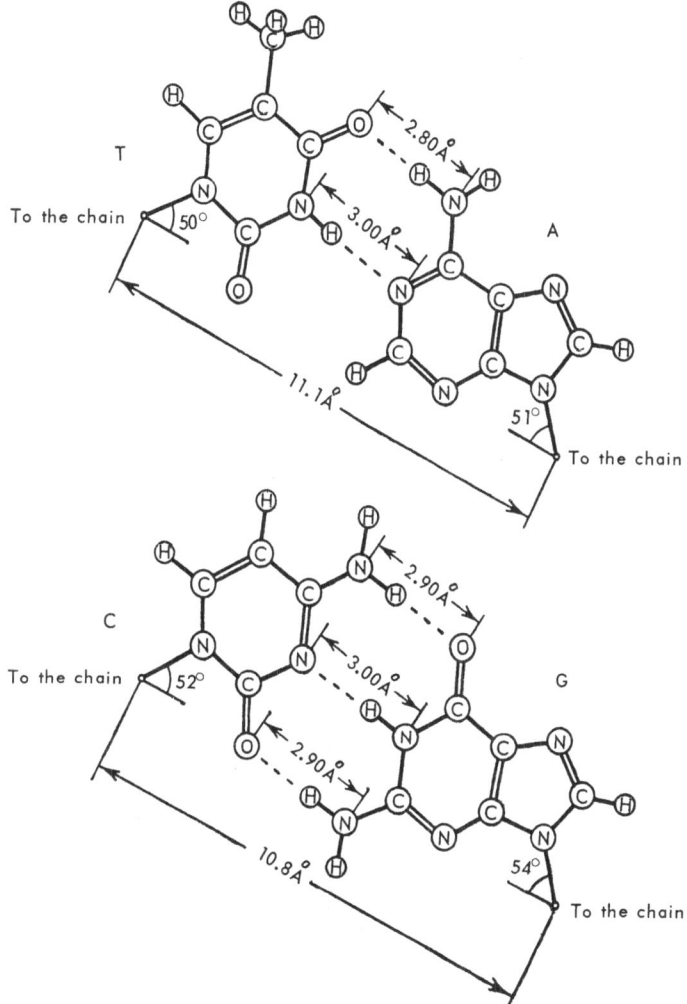

Fig. 97. Nitrogen-base pairs linked by hydrogen bonds.

in vitro. Such a synthesis takes place if we introduce into the mixture nucleoside triphosphates of all four types, necessary enzymes, some ions, and also native DNA as an initiator.

Experiments were carried out in which one of the four triphosphates contained labeled ^{32}P in the phosphate linked to the sugar (deoxyribose) (Fig. 99). The DNA obtained was broken down to 3'-deoxynucleotides with the use of enzymes (micrococcal DN'ase and phosphodiesterase from calf spleen); thus, the labeled P atoms were found in the 3' position of the nearest nucleotide (Fig. 99). Four different 3'-deoxynucleotides were

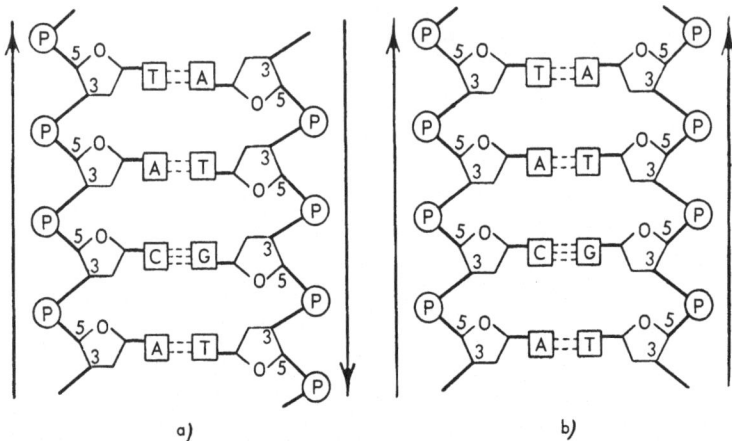

Fig. 98. The DNA double helix (a) from antiparallel chains and (b) from parallel chains.

obtained. These were separated by electrophoresis on paper. It is important that the starting nucleotide was labeled in the 3′ and not the 5′ position. Thus, by determining the radioactivity of the obtained monomers, it was possible to establish to what extent ^{32}P was linked to other nucleotides. If only one of the four starting triphosphates had been labeled, then four combinations could have been possible (let us assume that A was labeled; then the possible combinations are APA, APT, APG, APC, the P indicating ^{32}P phosphate). These experiments were repeated with all four nucleotides and, as a result, all 16 possible combinations were investigated. The results obtained are shown in Table 16. The rule of Chargaff is, of course, obeyed. The same Roman numerals indicate the combinations which must be identically equal if the chains are antiparallel; the same

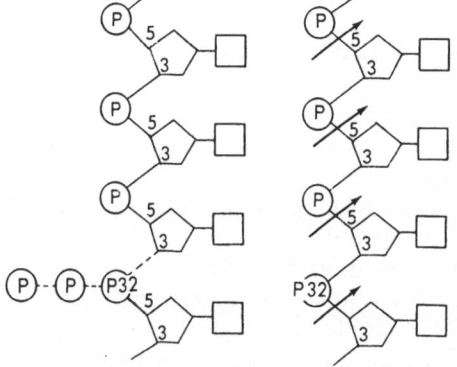

Fig. 99. Synthesis of ^{32}P-labeled DNA and its subsequent enzymatic breakdown to 3′-deoxyribonucleotide.

Table 16[21]   **Relative Contents of Various Combinations of Neighboring Nucleotides in a DNA Preparation Obtained by Using Initiator Native Calf Thymus DNA**

Labeled triphosphate	Isolated 3'-deoxyribonucleotides			
	TP	AP	CP	GP
DATP-^{32}P	*a* TPA 0.053 I	*b* APA 0.089 II	*c* CPA 0.080 III	*d* GPA 0.064
DTTP-^{32}P	*b* TPT 0.087 I	*a* APT 0.073	*d* CPT 0.067 IV	*c* GPT 0.056 V
DGTP-^{32}P	*e* TPG 0.076 II	*f* APG 0.072 IV	*g* CPG 0.016	*h* GPG 0.050 VI
DCTP-^{32}P	*f* TPC 0.067 III	*e* APC 0.052 V	*h* CPC 0.054 VI	*g* GPC 0.044
Total	0.238	0.286	0.217	0.214

Latin letters indicate those which must be identically equal if the chains are parallel. In fact, in the first case (Fig. 98) the contents of the combinations APC and GPT must be equal, for example, while in the second case, the combinations APC and TPG must be equal, since the bonds with ^{32}P which are being cut by the enzyme are distributed differently in the two cases under examination. Table 16 is the absolute proof of the antiparallel nature of the two DNA chains which take part in the Watson–Crick double helix.

What are the properties of native DNA in solution? The molecular weight of DNA, which has been determined by the methods of sedimentation and light scattering, is very high. For usual preparations, it has values of the order of 10^6. However, it seems that the DNA obtained from cells and phages is partly broken down during the extraction process. A more careful extraction of DNA leads to an increase of the molecular weight. The genes of *E. coli* and bacteriophage T2 have a single chromosome. One may think that the entire information in these cases is for a single molecule of DNA. In fact, molecules with a length of 40 μ have been extracted from T2 phage, while molecules of lengths of up to 400 μ have been extracted from *E. coli*. This corresponds to a molecular weight of the order of 10^9. Molecules as large as DNA can be easily observed in the electron microscope in the form of long filaments (Fig. 100; see also Fig. 31).

It would seem necessary to relate the very interesting experiments of Levinthal and Thomas.[22] They studied the T2 phage. They infected with

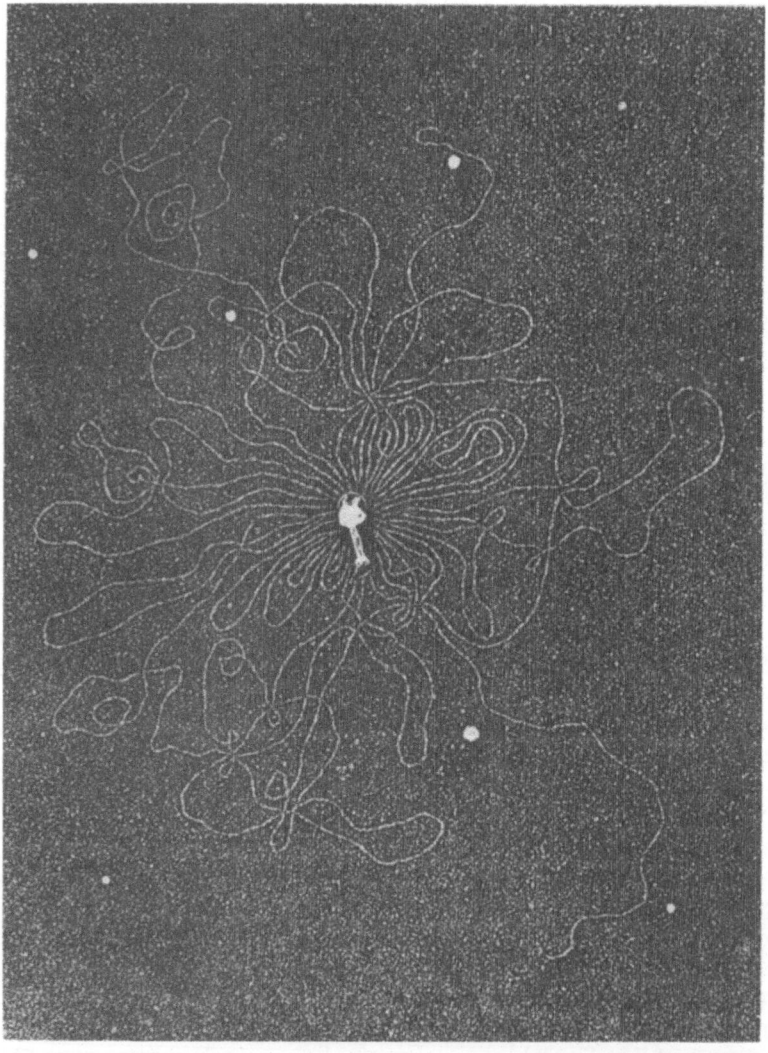

Fig. 100. Native DNA as seen by the electron microscope.

phage cells of *E. coli* grown in a medium with a high content of radioactive phosphorus, i.e., in a hot medium. This phosphorus was incorporated into the DNA of the multiplying phage particles. The phage population was extracted, a drop of a suspension of phage particles was placed on a layer of photographic emulsion containing a large amount of AgBr and thus sensitive to electrons, and this was covered on top by a similar photographic emulsion. After one week, this was developed. During this time, one-quarter of the ^{32}P atoms decomposed and the electrons which were emitted during the decomposition left their traces on the plate. These traces, or tracks, are seen under a microscope; each particle of the phage gives its star, containing 10 to 15 rays. The number of rays determines the number of ^{32}P atoms in each phage particle. Since the specific activity of the phosphorus is known, i.e., the fraction of ^{32}P atoms is known, the molecular weight of the phage DNA is determined from the stars. It turned out to be 1.4×10^8, or much greater than that of DNA studied in solution. It is evident that this can be explained by the fragility of the DNA molecules; during extraction they are torn to pieces; furthermore, DNA molecules undergo destruction even during solution flow.[23] The entire DNA of a phage particle is a single molecule.

In this manner, the method of thick-layer photographic emulsions, developed in nuclear physics, has been successfully applied in the studies on DNA.

The structure of DNA in solution has been studied by all the possible methods which had been developed earlier for synthetic polymers (see Chapter 4). The Watson and Crick model was proposed originally for the interpretation of X-ray data obtained with crystalline DNA. Luzzati, Nicolaieff, and Masson[24] have applied to this problem the method of small-angle X-ray scattering; they showed that the mass per unit length of DNA in solution is consistent with the Watson–Crick structure. This, as well as other data, prove that the DNA double helix is maintained in solution. Solutions of native DNA are very viscous. This does not mean, however, that the DNA macromolecules exist in solution in the form of rigid rods.

Data obtained by the methods of light scattering, sedimentation, viscosity, and flow birefringence have shown that the DNA double helix is twisted in solution into a large solvent penetrable coil. The linear dimensions of a DNA macromolecule with a molecular weight of the order of 10^6 are as high as $0.5\ \mu$. If the macromolecule had been extended, its length would have been $5\ \mu$.[25] Thus, the penetrable coil is folded tenfold.

The intrinsic viscosity $[\eta]$ of DNA solutions is proportional to the first power of M, while that of a rigid rod is proportional to M^2. This corresponds to a penetrable coil. Its structure may be characterized by the value of what is called the *persistence length L* of the macromolecule, i.e., of the length of a linear segment of the coil. A value of $L = 360\ \text{Å}$

was found by studies of the dependence of the sedimentation coefficient of DNA on its molecular weight.[26,27]

Sadron[28] has carried out a detailed critical analysis of the methods used to investigate DNA and RNA; he has found that the Zimm method of treating light scattering is not applicable to such large molecules; it turns out that it is impossible to establish the initial slopes of such curves. Ptitsyn and Fedorov have proposed a method for determining the persistence length from scattering at large angles and have found values of L of the order of 300 Å in agreement with those cited above.[29] Contrary to the opinion of Sadron, available experimental information does not give a unique possibility of characterizing the penetrable coils which we are discussing. It is impossible to say whether the macromolecule is in the form of a zigzag, consisting of strictly linear segments with breaks or has a wormlike structure with a continuous curvature (under such circumstances the significance of L is relative). The situation is further complicated by the heterogeneity of the DNA molecules.

To what extent are the DNA molecules heterogeneous? It is more difficult to answer this question than similar ones for a protein, since the isolation of DNA is accompanied by its partial degradation. However, the application of the recently developed very interesting method of sedimentation of macromolecules has made it possible to reach some rather definite conclusions. We are discussing here the method of sedimentation in a density gradient, proposed by Meselson, Stahl, and Vinograd.[30,31] The sedimentation of DNA is carried out in a rather concentrated aqueous solution of CsCl. A density gradient of the solution is established as a result of the force of sedimentation; it increases with distance from the axis of rotation toward the periphery. Since the solution densities lie in the same range as the density of DNA (of the order of 1.7), the DNA molecules of a given density become concentrated in a given layer. During the photography of the obtained distribution in the ultraviolet light (the DNA absorption maximum is at 260 mμ), a density spectrum is obtained, containing bands which correspond to DNA's of different densities. Such a spectrum of a mixture of seven samples of DNA of different sources is shown in Fig. 101. It is quite characteristic that the bands are quite narrow; this points to the high degree of homogeneity of the DNA. The most homogeneous DNA is that prepared from bacteriophage; the least homogeneous is that isolated from the cells of higher organisms. A rough estimate shows that the number of different sorts of DNA molecules, with the same molecular weight of the order of 10^7, is 10^6 in a mammalian cell, while in a bacterial cell, it is 100, and in the bacteriophage particle, it is 10.

The studies by Doty have shown that the density of DNA increased linearly with an increase in contents of G + C. This is quite understandable; the molecular weight of A + T = 247 and G + C = 273. These pairs, however, occupy approximately identical volumes within the double

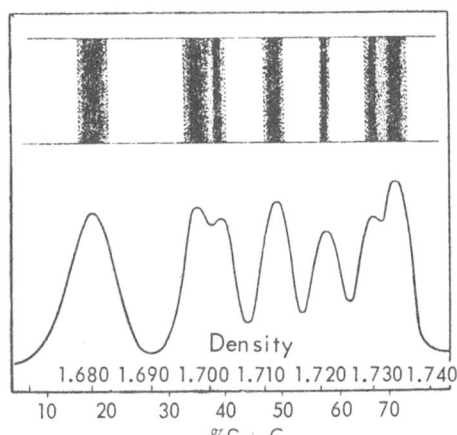

Fig. 101. Density spectrum of a mixture
of different samples of DNA.

helix. For each percentage unit of increase of contents of G + C, the density increases by 0.00103 g/cm^3. The method of sedimentation within a density gradient is so sensitive that it permits the determination of molecules labeled with ^{15}N nitrogen instead of ^{14}N.

The second conformation which DNA may assume in solution is the statistical coil; this is the conformation of denatured DNA. If native DNA is heated or if the pH is changed, etc., the double helix separates into two chains, each one of which assumes the shape of a coil. The denaturation of DNA is discussed in the next section.

Luzzati *et al.* have reported, in 1964, on the existence of a third form of DNA, intermediate between native and denatured. It was shown by studies using the small-angle X-ray scattering technique that, when DNA is heated, first the general native form is kept, while the mass per unit length and the cross-sectional axial radius decrease. The molecule becomes loosened. The helix–coil transition occurs afterward. These results are very interesting, but this question requires further investigation.

We have seen that, to understand the secondary structure of a protein, very important information has been obtained from studies of synthetic polypeptides. A similar situation exists in the case of nucleic acids. An investigation of synthetic polynucleotides (polyribonucleotides) has made it possible to understand better the structures of DNA and RNA; it has also given an extremely valuable result for the solution of the problems of the genetic code (Chapter 7).

Grünberg-Manago and Ochoa have isolated an enzyme from bacterial cells, polynucleotide phosphorylase; this enzyme has the property of joining various nucleoside diphosphates, with the formation of a synthetic polyribonucleotide.[32] The reactions proceed according to the scheme

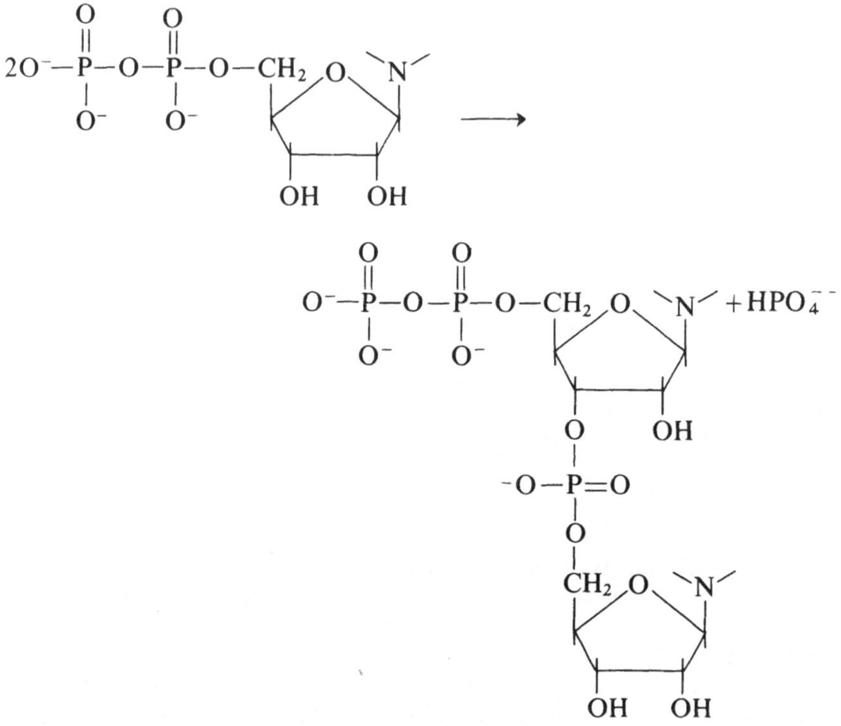

$$2O^- - \overset{\overset{O}{\|}}{\underset{\underset{O^-}{|}}{P}} - O - \overset{\overset{O}{\|}}{\underset{\underset{O^-}{|}}{P}} - O - CH_2 \quad O \quad N \qquad \longrightarrow$$

OH   OH

$$O^- - \overset{\overset{O}{\|}}{\underset{\underset{O^-}{|}}{P}} - O - \overset{\overset{O}{\|}}{\underset{\underset{O^-}{|}}{P}} - O - CH_2 \quad O \quad N \quad + HPO_4^{--}$$

O   OH

$$-O - P = O$$

O

$$CH_2 \quad O \quad N$$

OH   OH

In this manner, it was possible to synthesize low-molecular-weight poly-nucleotides, such as polyuridylic acid (poly U), polyadenylic acid (poly A), polyinosinic acid (poly I), which contains hypoxanthine, and even the unusual polyribonucleotide, namely, poly T (RNA does not contain thi-mine). Similarly, it has been possible to synthesize copolymers of these nucleotides which contain definite ratios of the monomers; of course, the sequence is unknown and random.

Just as synthetic polypeptides frequently form α-helices, the synthetic polynucleotides form Watson–Crick double helices, and triple helices. It has been shown by the method of birefringence and X-ray diffraction, for example, that two separate chains of poly A and poly U combine to form a double helix. Independent evidence is obtained from the change of the ultraviolet absorption at 260 mμ. A double helix characteristically has a hypochromous effect in this spectral region, i.e., a decrease of the inten-sity of the absorption relative to the denatured form. If poly A and poly U solutions are mixed in a medium of 0.1 M NaCl, the highest hypochromism is observed at a composition of 1:1, with the simultaneous formation of the double helix.[33] If divalent metal ions are introduced into the solution, such as 1.2×10^{-2} M MgCl$_2$, the maximal hypochromism is observed at a poly-U–poly-A ratio of 2:1 (Fig. 102). A triple helix is formed; this is

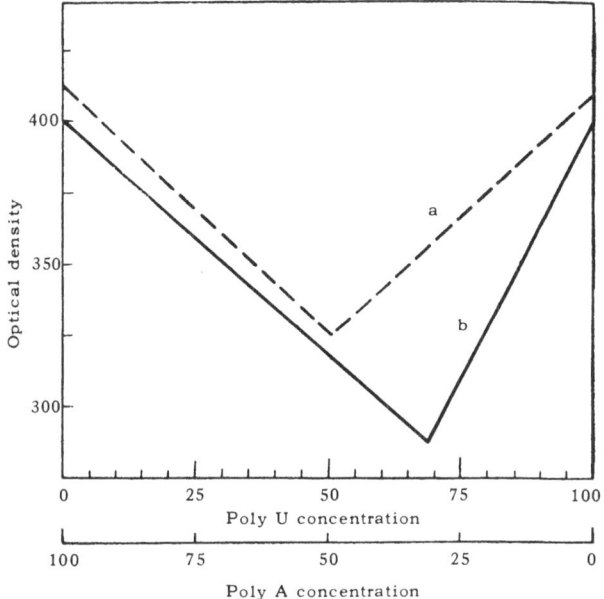

Fig. 102. Hypochromism in a poly-U–poly-A mixture: (a) formation
of the 1:1 complex; (b) formation of the 2:1 complex.

confirmed by measurements of the sedimentation coefficient. A triple helix
is formed as well by polyinosinic acid, in the absence of any other polynucleotide.

The role of metal ions in the formation of double and triple helices
consists probably in the neutralization of the charged phosphate groups
and the decrease in the mutual repulsion of the chains. To form a triple
helix, in the opinion of Rich, it is essential to overcome a large electrostatic
barrier, which is done as a result of the specific binding of Mg^{++} ions
(or Mn^{++}) by the phosphate groups.[31] Another possibility consists in the
binding of these ions by the nitrogen atoms of the heterocyclic ring, similar
to the way that this occurs in chlorophyll. The ions penetrate into the
double helix and bind to it. This question cannot be considered finally
solved; but the importance is quite evident, since magnesium ions play
an important role in the process of protein synthesis in the living cell.

It is characteristic that neither poly U nor poly A by themselves form
double or triple helices; A complements with U, just as, in DNA, A comple-
ments with T. In the double and triple helices, A combines with U by
hydrogen bonds.

Fresco and Alberts have studied the hypochromism of mixtures of
solutions of poly U and copolymers of poly A with U of different compositions. They obtained very interesting results.[33] The hypochromic effect
shows that it is possible to obtain double and triple helices from poly U

and poly A (poly AU + U and poly AU + poly U + poly U). However, as long as A and U are bound to each other, the helices must be defective. Here one may have two possibilities shown in Fig. 103. However, there are no hydrogen bonds in those positions where U faces U, the structure remaining helical (case a); the same is true in those regions where the U groups are outside the helix and loops are formed (case b). These two cases differ in the percentage of the composition of poly U and poly AU of identical molecular weight in the double helix; these systems have a low hypochromism. In the helical fragments shown in Fig. 103, the molecular ratios of poly AU to poly U are the following: In case a the ratio is 10:10, that is, 1:1; in case b it is 6:10, that is, 3:5. It is, thus, possible to establish experimentally which case exists in fact. The results obtained are shown in Table 17.

Table 17[11]   Molar Percentage of Poly U in Mixtures with Maximal Hypochromism

Mixture	Double helix			Triple helix		
	Found	Calc case a	Calc case b	Found	Calc case a	Calc case b
Poly A + Poly U	50	50	...	67	67	
Poly AU 90 + Poly U (90% A).	...	...	...	65 ± 1	67	64.5
Poly AU 66 + Poly U	38 ± 1	50	40	54 ± 1	67	57
Poly AU 53 + Poly U	34 ± 1	50	34.5			
Poly AU 62 + Poly U	...	...	...	57	67	55
Poly AU 82 + Poly U	...	...	...	64	67	62
Poly AU 37 + Poly U	38 ± 3	50	39			

These results show that loops are actually formed. The examination of such a defective structure with the aid of molecular models has shown that the presence of loops does not interfere with the formation of the Watson–Crick double helices if we neglect the relatively small convergence of phosphate groups, the shortest distance between which decreases from 7 to 6 Å; this is not very significant. This makes it possible to think that an unfitting base may be accidentally included in DNA; A, G, or C, instead of T, facing an A causes the formation of a loop.

These data are for synthetic polynucleotides; they enable us, however, to make assumptions on the secondary structure of RNA. Contrary to DNA, it is not possible to obtain a viral or ribosomal RNA in crystalline form. As a result, it is not possible to form a sufficiently clear picture in X-ray diffraction experiments of such RNA molecules. In solution, samples of native RNA have a much lower viscosity than those of DNA and

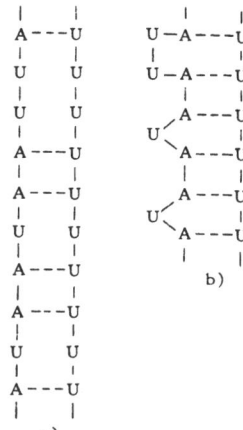

```
        |    |           |        |
        A---U        U—A---U
        |    |         |        |
        U    U        U—A---U
        |    |             |    |
        U    U           ,A---U
        |    |        U<
        A---U           `A---U
        |    |           |    |
        A---U          ,A---U
        |    |       U<
        U    U          `A---U
        |    |           |    |
        A---U              b)
        |    |
        A---U
        |    |
        U    U
        |    |
        A---U
        |    |
             a)
```

Fig. 103. Two possibilities of the formation of double helices from poly U and poly AU.

a greater velocity in sedimentation; this points to their more compact coiled-up shape. The intrinsic viscosity of native RNA is proportional to the square root of the molecular weight; this means that the shape of the molecule is coiled. During denaturation, the viscosity of RNA does not decrease, as DNA, but increases, i.e., the coil becomes looser.

The double refraction of flow of RNA is considerably less than that of DNA. The persistence length is no greater than 100 Å. At low ionic strength, the RNA coils swell and increase in size; when the ionic strength is increased, they become more compact. As has been pointed out already, ribosomal RNA does not obey the law of Chargaff.

The ability of RNA to denature is expressed, for example, in the increase in the intensity of absorption at 260 mμ and in the greater compactness of the coil; this led Doty to conclude that RNA is partly helical. It is possible that segments of the molecule are partly helical and that these come back on themselves like a hairpin. Uracil combines by hydrogen bonds with A, T with C. It is significant that the stability of RNA relative to heat denaturation increases considerably with an increase in the contents of G + C (let us remember that, contrary to the T + A pairs in DNA and the U + A pairs in RNA, G + C forms, not two, but three hydrogen bonds). The melting temperature of the helical segments is linearly dependent on the contents of G + C. According to the estimate of Doty and coworkers, the smallest helical segment of RNA without loops consists of four pairs of nucleotides. A defective helix in RNA may contain loops. A helical segment with loops is evidently less stable than a completely structured segment; the loops, however, increase the number of bases which are linked by hydrogen bonds. As a whole, the secondary structure of RNA, according to Doty, is similar to that of a sea star (Fig. 104). Such a structure does not contradict the available experimental data and does explain the observed high percentage of helical segments, up to 77% (ribosomal RNA) and even 88% (RNA in TMV[34]).

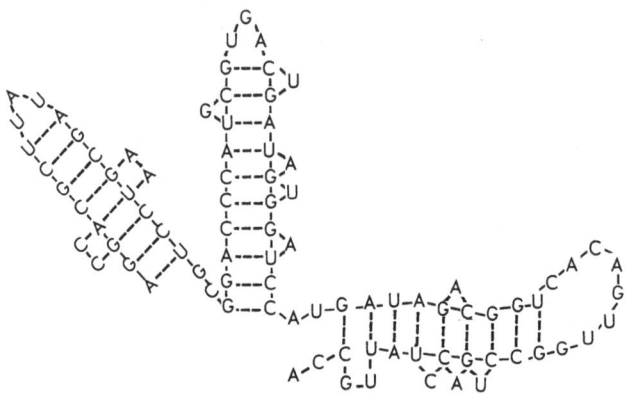

Fig. 104. Secondary structure of RNA according to Doty.

Spirin, following the same concept of the partial double helicity of the RNA structure, has proposed another model (Fig. 105). On the basis of electron microscopy and other data, he drew the following conclusions: The RNA molecule is a single chain; at high ionic strength, RNA forms a tight coil; and at low ionic strength, it is a compact rod.[35] During heating, acidification, or complete deionization of solution, RNA denatures, which results in a nonhelical loose coil.

There are no great differences between the models by Doty and Spirin. Neither model contradicts experimental data. However, these data do not permit a unique determination to be made of the secondary structure of RNA in solution. It is not possible to exclude the possibility of an unusual pairing of RNA nucleotides as a result of tautomeric transitions. Thus, for

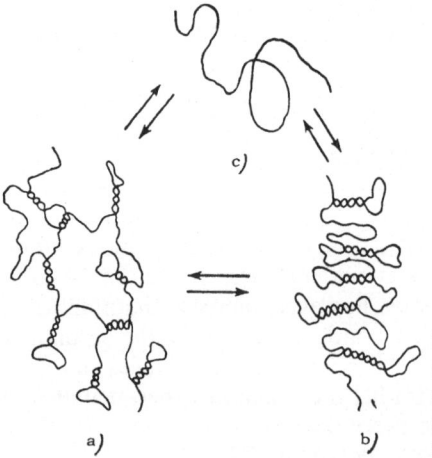

Fig. 105. Secondary structure of RNA according to Spirin (a) at high ionic strength; (b) at low ionic strength; and (c) after heating, acidification, or complete deionization of the solution.

example, uracil is usually present in the keto form

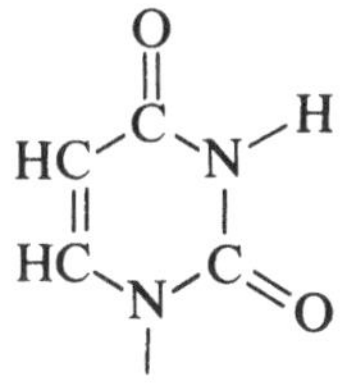

In this form, it gives a Watson–Crick pair with A. However, the enol form is also possible:

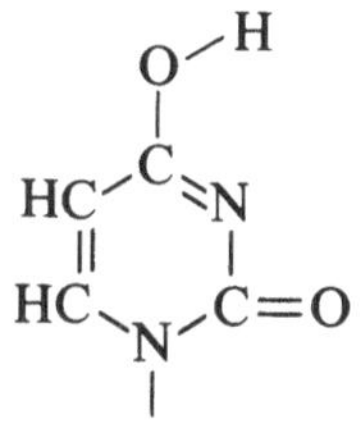

In this form, U may combine by three hydrogen bonds with G. The tautomerism of pyrimidines and purines has not been studied in sufficient detail up to now. One cannot exclude the possibility that such transformations occur and that the resulting hydrogen bonds stabilize the tautomers. The important role of tautomerism in mutagenesis is quite well known. The problems relating to polymeric chains, which include tautomerism along with rotational isomerism, have not been examined up to now. If this possibility actually occurs in RNA, then tautomerism may be the cause of the high contents of helical regions within RNA molecules which have no regular primary structure.

Finally, non-Watson–Crick pairs may be formed with a decreased number of hydrogen bonds, one instead of two or two instead of three. Such pairs are bonded less strongly, but they nevertheless contribute to the stabilization of the helix. It is necessary to emphasize that the interaction between the nitrogen bases which participate in the complementary chains is not restricted to hydrogen bonds only. We also have the Van der Waals forces; in particular there must be a significant interaction of parallelly organized bases of a single chain, which participate in the double helix.

The question is thus still open. In order to solve it, it is necessary to continue studies on the structure of RNA and to follow experimentally and theoretically the problem of tautomerism in nucleotides, plus a detailed analysis of possible conformations of the chains which contain tautomers and form hydrogen bonds. This is a problem of molecular biophysics; it is quite important, since RNA plays a very important role in protein synthesis.

In the following chapter, we shall discuss the result of a direct X-ray investigation of paired interactions between nucleotides.

The secondary structure of T-RNA has not been established directly, although there are interesting results obtained by the X-ray[37] and dynamic birefringence[37a] methods. It has been possible, however, to establish the primary structure of several types of T-RNA. With the concept of the formation of Watson–Crick UA and CG pairs, it is possible to construct several different models of the secondary structure of T-RNA. One of these models is for alanyl T-RNA; it is shown in Fig. 106.[38] This model has the shape of a clover leaf; it seems to be the most plausible and it does not contradict experimental data.[37,37a] The T-RNA anticodon is indicated in the upper leaf.

DENATURATION OF DNA

The information discussed above on the secondary structure of nucleic acids is the result both of direct investigations of native DNA and RNA in solution and of an analysis of the patterns of their denaturation.

The most extensive data on the denaturation of DNA have been obtained by Doty et al.[39] When a solution of DNA is heated, or if it is treated with acid or alkali under mild conditions, sharp changes are

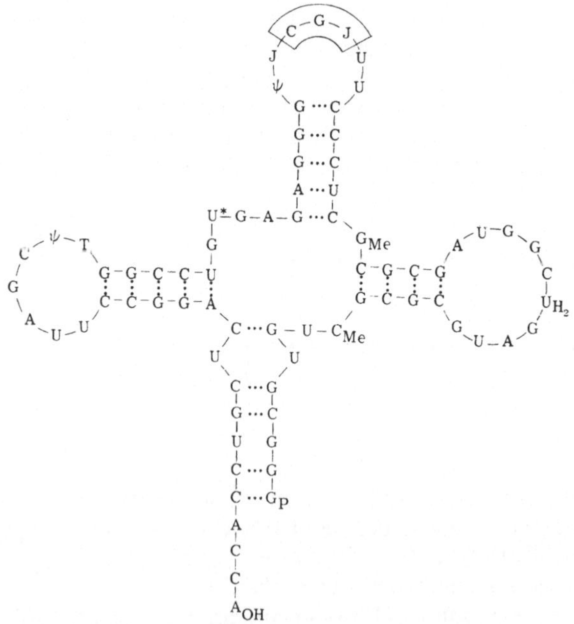

Fig. 106. One of the possible structures of alanine T-RNA (according to Holley et al.).

observed in a number of properties, such as in hydrodynamic and optical properties. The viscosity drops considerably, the intensity of absorption of light at 260 mμ increases (i.e., the hypochromic effect vanishes), and the optical activity decreases. It has been demonstrated that all these phenomena are determined by the division of the DNA double helix into two independent chains, which curl into compact statistical coils. After denaturation, the dependence of the intrinsic viscosity on the molecular weight at high ionic strengths (about 0.15) is typical for a statistical coil:

$$[\eta] = 7.5 \times 10^{-4} M^{0.52}$$

The division of the double helix into two chains is proved by the centrifugation of DNA labeled with ^{15}N, in a density gradient of CsCl. Cells of *E. coli* grown in a medium of ^{15}N were transferred into a medium containing normal nitrogen ^{14}N. When the cells divided, duplicated double helices were formed; one chain contained ^{15}N and the other ^{14}N. Such native DNA was heated and studied by the method of Meselson. Before denaturation, one could see a single peak with a density of 1.717 g/cm^2, that is, the density of native DNA with ^{15}N + ^{14}N. After denaturation, two peaks appeared, one at 1.740, corresponding to a single-chain coil with ^{15}N, the other at 1.724, corresponding to a coil with ^{14}N. The density of denatured DNA is greater than that of the native molecule because of the compactness of the coil.[31,40] Direct determinations of the molecular weight of DNA show that, during denaturation, it decreases by a factor of 2.[39,41,42]

The formation of coils during denaturation may be observed directly in the electron microscope. Such a photograph is shown in Fig. 107. It should be compared with Fig. 100 taken with native DNA. The difference is quite striking! At low ionic strength, the macromolecules of denatured and, thus, single-chain DNA display polyelectrolytic behavior. The chains are charged, the charges are not screened by counterions, and, as a result, strong repulsive forces act between the links of the chain. Under such conditions (ionic strength of 0.01), the macromolecule is no longer in the shape of a coil, but is considerably stretched. As a result,

$$[\eta] = 2.7 \times 10^{-5} M^{0.93}$$

The denaturation of native DNA is a typical helix–coil transition, similar to that described by Doty in polypeptides. Contrary to the case found with proteins, hydrogen bonds are broken here during a transition. The process may be considered an intramolecular melting; its sharpness testifies to its cooperative nature. Consequently, just as in the case of polypeptides, this transition may be regarded as the melting of the helix. Similar transitions are observed also in the double helices of synthetic polynucleotides, such as poly A + poly U. This model double helix melts at 65°C (in a solution of 0.15 M NaCl at neutral pH). The intensity of light

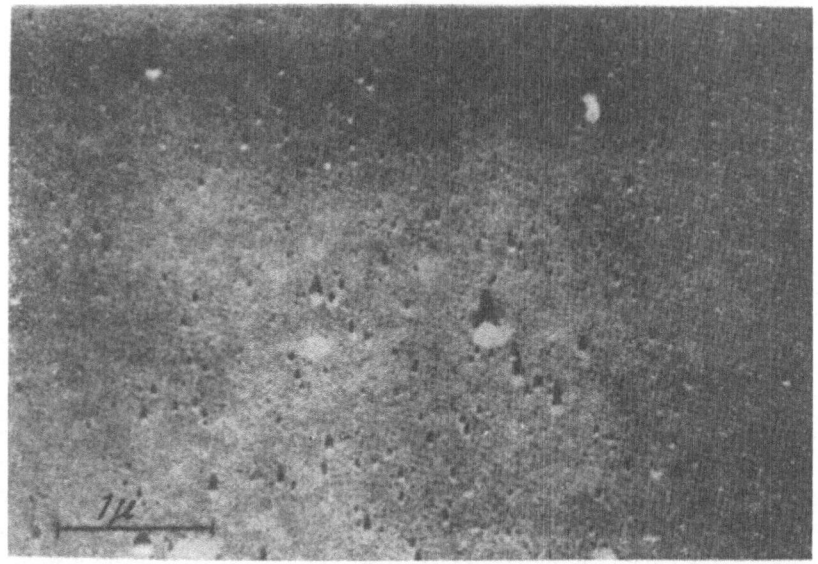

Fig. 107. Electron micrograph of denatured DNA.

absorption at 260 mμ increases by 34% during melting, while the specific rotation decreases by 275°.

The T_m of DNA, as could be expected, depends on the relative contents of G + C; these bases are mutually bonded more strongly than A + T. The dependence on the contents of G + C is linear.[43] Extrapolation of this straight line gives a limiting value of $T_m = 69°$ for poly-d (A—T) and $T_m = 110°$ for poly-d (G—C). The experimental values obtained by Kornberg for such synthetic copolymers are 65 and 104°. The agreement is quite good.

The T_m of DNA is a strong function of ionic strength, increasing approximately in a logarithmic form with a concentration of the cations. The T_m for DNA from *Diplococcus pneumoniae* in 0.01 M KCl is 68°C, while in 0.6 M KCl, it is 90°C. The temperature of boiling water is sufficient to denature any samples of DNA, since the G + C contents never exceed 75–80%. Total denaturation during boiling occurs over approximately 10 minutes.

Just as in the case of polypeptides, the theory of transitions in DNA has been examined by many authors.[44-49] The most perfect and simplest variant of this theory is that proposed by Zimm;[48] in this theory the one-dimensional model of Ising has been applied.

A somewhat idealized double helix of DNA is examined; the fact that the strength of the bonding of A with T and G with C are different is neglected. It is necessary to calculate the partition function. Each pair of bases may be in two states, namely, bonded ($\mu_i = 1$) and nonbonded ($\mu_i = 0$) by hydrogen bonds. The state of a given pair of bases depends on the states of the previous pairs. If j preceding pairs of bases are nonbonded, then the formation of a hydrogen bond in the $(j + 1)$st pair means the formation of a ring from j pairs of unbonded bases. This places a limitation on the conformations of the chains of the links which enter into the ring, and this determines the cooperativity of melting of the double helix. It is evident that, if the molecular links which precede the base pair under examination are nonbonded, a change of free energy of formation of the hydrogen bond includes a decrease in the entropy, as a result of the decrease in the number of conformations of the linked chains.

In this manner, the cooperativity of the properties of DNA is the result of the fact that the destruction of bonding between individual nucleotides within the double helix is unfavorable since, in this process, the links are not freed and their entropy does not increase. As a result, such a break is immediately repaired. An irreversible break may occur only over a sufficiently long sequence of hydrogen and Van der Waals bonds between the bases; in this case the loss of enthalpy is compensated by a gain in entropy.

The contribution of each state of the two chains to the partition function consists in the product of the following multipliers: (a) One for state 0 of the pair of links. This means that the free energy of such a pair is taken as zero, $F_{free} = 0$. (b) The multiplier

$$s = e^{-\Delta F/kT} \tag{6.1}$$

for the bonded state of the pair. Here s is the equilibrium constant of the process of increasing the helical segment by one pair of monomers at the expense of the neighboring segment of nonbonded units. (c) The multiplier

$$\frac{u}{V} = e^{-F_{init}/kT} \tag{6.2}$$

for the first pair of monomer units in state 1. Here V is the volume of the system, u is the volume available for the center of gravity of one of the bonded chains when the center of gravity of the other chain is fixed. (d) The multiplier

$$\sigma_j = e^{-(\bar{F}_j)_{init}/kT} \tag{6.3}$$

for a ring from j independent monomers of each chain.

The constant s depends evidently on the ionic strength of the solution. The energies F_{init} and $\bar{F}_{j,init}$ determine the degree of cooperativity. If these energies are equal to zero, then $u/V = 1$, $\sigma_j = 1$, and there is no cooperativity. To the contrary, at values of F_{init} and $\bar{F}_{j,init} \gg kTu/V = 0$, $\sigma_j = 0$ and cooperativity is total. In this case, when $s = 1$, that is, $\Delta F = \Delta H - T\,\Delta S = 0$, a sharp jumplike transition must take place from the state of the double helix in which all the pairs of bases are bonded ($s > 1$, the fraction of bonded bases is $x = 1$) to a state of two free coils ($s < 0$ and $x = 0$). This is a cooperative transition which takes place according to the all-or-none principle.

Real transitions are different from the described limiting cases. To examine real transitions, it is necessary to take into account all the possible states in the partition function. The solution of this problem for double helices is much more difficult than for polypeptides, in particular because of the specificity factor, σ_j, that is, states with rings. It is not possible to obtain an analytical expression for the partition function and it is necessary to use graphic methods of solution.[46,48] In Fig. 108, we have shown a dependence of the fraction of bonded bases, x, on s, calculated by Zimm for two values of the cooperativity parameter σ_0 (0.1 and 0.01), related to σ_j by the expression

$$\sigma_0 = \sigma_j(j + 1)^{3/2} \qquad (6.4)$$

The helix–coil transition occurs over a narrow temperature interval close to the value

$$s = \frac{1}{1 + 1.612\sigma_0} \cong 1 \qquad (6.5)$$

The theory makes it possible to determine also as a function of s and temperature the mean fraction of base pairs present at helix–coil junctions. At $\sigma_0 = 0.1$ close to the transition temperature, this fraction is close to 1 %. Experiment is in agreement with theory if we assume that $\sigma_0 = 10^{-1}$ to 10^{-2}. The value of the heat of melting for DNA is $\Delta H = 5$ kcal/mole of a pair of bases; this is found experimentally at pH 2.5, a temperature of 25°C, and an ionic strength of 0.1.[50] In a neutral medium, this quantity increases by the heat of ionization of the purines and pyrimidines of the bases and amounts to about 8 kcal/mole. One should note that the values of ΔH are large, and they are considerably greater than the energies of hydrogen bonds. The forces which stabilize the double helix are not just hydrogen bonds.

Helix–coil transitions in DNA may be caused not only by heating. They can also occur as a result of changes in pH. The native form is stable at room temperature at pH's between 3.5 and 4.5 and between 9 and 11.5 (depending on the ionic strength). The fact is that the dissociation constants of the NH_2 and $NH–CO$ groups of the nitrogen bases which can ionize when hydrogen bonds are broken are different. The value of pK_{NH_2} is 2.35 to 2.9 for G, 3.7 to 3.85 for A, and 4.5 to 4.85 for C; $pK_{NH–CO}$ is 9.5 to 11.4 for G, T, and U.[50] The NH_2 group is charged at pH $< pK_{NH_2}$ and becomes NH_3^+, while the $NH–CO$ group at pH $> pK_{NH–CO}$ becomes $N^- – CO$. The helical structure is stable within the region of $pK_{NH_2} < pH < pK_{NH–CO}$, when the groups are uncharged and there are no repulsive forces between them.

Birshtein has carried out a theoretical examination of the effect of ionization of the bases on the helix–coil transition and the effect of these transitions on titration curves.[51,48] The condition for transition has the form

$$s = e^{-(\Delta H - T_m\Delta S)/kT_m} = 1 + a_m \qquad (6.6)$$

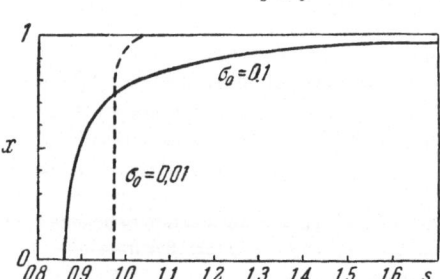

Fig. 108. Dependence of the fraction of bonded bases on the parameter s.

where a is the ratio of activities of charged and uncharged groups; it is a function of pH and pK. Therefore, by measuring the dependence of the melting temperatures of the double helix, T_m, on pH$_m$ (i.e., on a_m), it is possible to determine ΔH and ΔS. Such a determination gives, for a poly-U–poly-A helix, $\Delta H = 8 \pm 1$ kcal/mole and $\Delta S = 25$ cal mole^{-1} deg^{-1}. The theory is in good agreement with experiment; ln $(1 + a_m)$ is actually a linear function of $1/T_m$.

Similar determinations have been carried out for DNA;[52] it was found that $\Delta H = 10$ to 11 cal/mole and $\Delta S = 30$ cal mole^{-1} deg^{-1}. The theory also accounts very well for the shape of the titration curves of native and denatured DNA.

A more elementary version of the theory of DNA denaturation is presented below.

From a consideration of the general theory of polyelectrolytic macromolecules,[53] Ptitsyn has shown that charged coils must be highly swelled because of electrostatic interactions between links that are removed from each other. This is how the macromolecules of denatured DNA behave. These long-range interactions must interfere with the helix–coil transition and, correspondingly, helices with higher molecular weights must melt at higher temperatures than short helices. This has been confirmed experimentally on poly A–poly U. It follows, furthermore, that for high-molecular-weight DNA, there must be a temperature interval in which the melting of short segments is favored, while that of long segments is unfavorable. Thus, the denaturation transition occurs in two stages; first, the short segments must melt and a partly denatured state must be formed with many large loops, and only then does the complete melting of the double helix take place. The first stage must evidently be reversible; the second, irreversible. These theoretical conclusions are confirmed by the data of Doty, who has shown that the denaturation of calf-thymus DNA is reversible over a certain temperature interval and becomes irreversible afterwards.[54] Partial denaturation of DNA has been observed directly with the help of the electron microscope.[55]

As has been pointed out already, the denaturation of DNA (and also RNA) may be followed most easily by a disappearance of the hypochromic effect; the denatured single-chain DNA absorbs ultraviolet light at a wavelength of 260 mμ to a much greater extent than the native DNA. The theory of the hypochromic effect in double-helical polynucleotides has been developed by Tinoco.[56,57] It starts from the same concepts as the analogous theory for polypeptides. Absorption at 260 mμ is determined by an electronic transition within the plane of the nucleotide nitrogen bases or, speaking in classical terms, the electronic oscillator vibrates within this plane. As we have seen, when such oscillators are regularly organized, a Davydov splitting of the electron levels may occur, with a resulting splitting of the spectral bands. If the oscillators, i.e., the transition moments, are located colinearly to each other, the intensity of the high wavelength bands increases (the hyperchromic effect); if the moments are parallel, they decrease (the hypochromic effect). The nitrogen bases in polynucleotides and nucleic acids lie parallel to each other like a pile of coins and, consequently, we must observe a hypochromic effect (just as in polypeptides). Calculation gives quantitative agreement with experiment. The oscillator strengths for the band at $260\,\mu$ of free nucleotides are $f_A = 0.30$, $f_G = 0.40$, $f_C = 0.17$, and $f_T = 0.21$. Depending on the angle of rotation within the helix of the next pair of bases with respect to the previous one, the oscillator strength decreases by 10 to 40%. The theory gives the dependence of hypochromism on the length of the helix (it grows with the length of the helix, and tends toward a limit); this is confirmed by experiments.[39] Recently, more detailed and rigorous calculations have been carried out.[58]

Studies of complexes of DNA with acridine compounds and other dyes are interesting from many points of view, in particular, from the point of view of the denaturation of such complexes. Let us present some studies in this realm.

The interaction of DNA with the dye acridin orange

$$(CH_3)_2N \underset{\displaystyle \overset{H}{N^+}}{\underbrace{\hspace{4cm}}} N(CH_3)_2 \tag{6.7}$$

has been studied.[59,60] The polarization of the fluorescence of the complex was determined. The phenomenon of fluorescence consists in that the molecule absorbs light of a given wavelength and then, after a short but finite time τ, it reemits light of another wavelength (usually lower). In the terminology of classical physics, we may say that the electromagnetic waves of a given frequency are absorbed by some electronic oscillators within the molecules which oscillate in a fixed direction. The absorbed energy is transmitted to another oscillator which oscillates with another frequency in another direction. It reemits the light. Since the two oscillators have fixed directions within the molecule and the angle between them depends on the molecular structure, the emitted light is polarized if the molecules are in rather slow motion. However, if the fluorescent molecule can turn during the time τ, the polarization vanishes. Consequently, the degree of polarization P is a function of mobility of the molecules. Theory results in the equation[61,62]

$$\frac{1}{P} = \frac{1}{P_0} + \left(\frac{1}{P_0} - \tfrac{1}{3}\right)\frac{RT}{v\eta}\tau \tag{6.8}$$

where P_0 is the degree of polarization of the radiation in the absence of rotational depolarization, (i.e., for $T \to 0$, $\eta \to \infty$), v is the molecular volume, and η is the viscosity of the medium.

It has been found that the polarization of the native complex falls slowly with an increase of T/η (and conversely the reciprocal quantity $1/P$ increases slowly) up to the melting temperature of the double helix, i.e., 75°C. At this temperature, $1/P$ increases sharply up to a value which corresponds to the free dye molecules. For complexes with denatured DNA, the curve of dependence of $1/P$ on T/η has no sharp break. These results may be interpreted in terms of the intercalation of the dye molecules into the space inside of the double helix, i.e., the formation of what is called a *clathrate*.[63] During denaturation, the clathrate becomes destroyed and the dye molecule is liberated. Luzzati has come to the same conclusion after studying these complexes by the method of small-angle X-ray scattering.[64] The two-stage nature of the denaturation has also been established,[59] i.e., reversible and irreversible, in agreement with other experimental and theoretical results.

Permogorov and Lazurkin have investigated the complexes of DNA with dyes by spectropolarimetry.[65-67] The dye which is intercalated into the double helix acquires an induced anomalous dispersion of the optical activity. Pinocyanol becomes intercalated into native DNA in the form of dimers; this can be proved by the symmetric shape of the optical-rotatory-dispersion curve, similar to a Macaluso–Corbino curve shown on Fig. 93. In fact, the appearance of such a curve may be explained by invoking the existence of two Kuhn models. The magnitude of the effect, the amplitude of the optical rotatory-dispersion curve, is a function of temperature. This dependence can be easily explained by the fact that the dye molecules are more easily bound to the native regions of DNA than to the denatured ones. During the melting, the dye molecules are redistributed from the melted DNA segments to those that have remained native.

An important result has been derived from these studies. It has been shown that dye molecules intercalated within the DNA stabilize the double helix, forming so-called staples between the nucleotides of the complementary chains. This is expressed in an increase of the temperature and heat of melting of the double helix. Furthermore, the presence of a hinge has also an effect on the biological properties of DNA. Thus, actinomycin decreases the ability of DNA to cause the genetic transformation in *B. subtilis*, probably because of the hindrance to the separation of the parent chains during the introduction into the genome of the receptor cells. Furthermore, actinomycin increases the temperature at which a full inactivation of transforming DNA takes place.

Frank-Kamenetskii has developed the theory of melting of a DNA with staples.[68] This theory leads to the following equations (neglecting a term that contains the cooperativity factor, the magnitude of which is of the order of experimental error):

The difference between the melting temperatures of DNA with staples and of pure DNA is

$$\delta T_m = 2\frac{p-1}{p+1}\frac{T_0{}^2}{H}c \tag{6.9}$$

where $p = \exp(\Delta/T)$, Δ is the additional free energy of interaction introduced by the hinge, T_0 is the melting temperature of pure DNA, H is the average energy of the bond per pair of bases, and c is the molar concentration of the bound dye relative to the molar concentration of nucleotide pairs; the temperature is expressed in energy units.

The difference in the widths of the interval of melting of DNA with hinges and pure DNA is

$$\delta(\Delta T_m) = 4\left(\frac{p-1}{p+1}\right)^2\frac{T_0{}^2}{H}c \tag{6.10}$$

From experimental data it follows that $\Delta \sim 1$ cal/mole. Theory is in good agreement with experiment. It is evident that the theory may be used to account for the heterogeneous structure of DNA in the absence of dye; this has been carried out by the author.

These studies are of further great interest, since dyes of this type are mutagens and carcinogens.

Now we have become acquainted with a convincing statistical thermodynamic theory of denaturation of DNA. What is the path over which denaturations occur? Statistics and thermodynamics do not give answers to this question, since they ignore completely the time factor. On the other hand, we may not be completely disinterested in the rate of the process of the unwinding of a double helix. Here we encounter difficulties immediately.

Let us assume that a change in the medium or temperature has occurred which favors thermodynamically the breaking of the hydrogen bonds that link the two DNA chains. The links are broken. To separate the chains, however, it is necessary to unwind a large number of turns. This question was examined first by Kuhn.[69] He took a DNA macromolecule with a molecular weight of 3×10^6 (in fact, the molecular weight may be much larger). The length of the double helix in this case is $L = 3 \times 10^{-4}$ cm; it contains approximately $v = 900$ turns. What is the time necessary for the separation of the two helices by the consecutive unwinding of its turns as a result of thermal motion? It is necessary to unwind 450 turns, since separation may occur simultaneously from two ends. During the unwinding of a single turn, the chain follows a path

$$s = \sqrt{q^2 + (2\pi r)^2} \tag{6.11}$$

where $q = 3.4 \times 10^{-7}$ cm is the length of a turn along the helical axis, and $r = 10^{-7}$ cm is the radius of the helix. If $j - 1$ turns have already been unwound, then in the unwinding of the jth turn, the end of the chain follows the path $l = 2\pi j s/\sqrt{2}$ (Fig. 109). The time of the diffusional Brownian motion at the jth turn is

$$\tau_j = \frac{l^2}{2D} = \frac{(2\pi j s/\sqrt{2})^2}{2D} \tag{6.12}$$

Kuhn assumes that the diffusion coefficient D is equal to one-fourth that of a free chain of length $js/4$:

$$D = \frac{2kT}{3\pi\eta_0} \frac{4}{js} \tag{6.13}$$

where η_0 is the viscosity of the surrounding medium. Consequently,

$$\tau_j = \frac{3\pi^3}{8} \eta_0 s^3 \frac{j^3}{kT} \tag{6.14}$$

The total time of unwinding is

$$\tau = (\sqrt{\tau_1} + \sqrt{\tau_2} + \cdots + \sqrt{\tau_{v/2}})^2 \tag{6.15}$$

since all the turns unwind independently of each other. At $T = 300°K$ and $\eta_0 = 10^{-2}$, calculation gives $\tau = 1.3 \times 10^7 = 150$ days!

Thus, diffusional unwinding must proceed very slowly. In order to denature DNA, much less time is actually necessary; one needs minutes only. Kuhn finds a very clever solution. He has shown that the partial unwinding of the double helix may result in its separation into two chains. If we force the end of the helix to undergo a certain number of turns $\theta v (\theta < 1)$ in a direction opposite to the turning of the helix, so that the distance between the center of the helix and the end (and consequently q) remains unchanged, a displacement by a segment q along the axis of the helix does not require a full turn but only a fraction equal to $1 - \theta$. In fact, after θv such turns, the radius of the cylinder has increased since

$$2\pi(1 - \theta)r' = 2\pi r$$

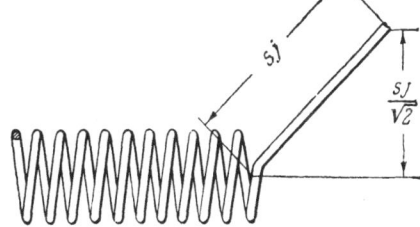

Fig. 109. Scheme for the calculation of the duration of the unwinding of a DNA double helix.

and

$$r' = \frac{r}{1 - \theta}$$

If, for example, $\theta = 0.3$, it is possible to separate the double helix after only 150 turns from its end, and not after 450 turns, just by pulling one single-chain helix from the other, i.e., by simple translational displacement. In this case the time of helix separation is

$$\tau = (\sqrt{\tau_1} + \sqrt{\tau_2})^2 \tag{6.16}$$

where τ_1 is the time of rotation and τ_2 is the time of translational diffusional motion. A calculation based on the same reasonable values of molecular constants gives $\tau_1 = 3$ seconds and $\tau_2 = 35$ seconds and $\tau = 58$ seconds, at $\theta = 0.3$, a quantity quite in agreement with experiment. Thus, Kuhn has come to the conclusion that the separation of the double helix is the result of simultaneous rotational and translational motion.

Longuet-Higgins and Zimm[70] give another interpretation of the unwinding. The separated chain ends have a higher degree of freedom than similar chain segments bonded within the double helix. Consequently, unwinding is accompanied by an increase in entropy or a decrease of the free energy, $-\Delta F = T \Delta S$. Thus, in addition to a purely diffusional Brownian motion, there must be a rotation caused by the rotational moment of an entropic nature. We have already talked about entropic forces. One may also speak of the entropic moment of the force, P, equal to

$$P = \frac{\Delta F}{\varphi} = \frac{57.3}{36} |\Delta F| = \frac{57.3}{36} T \Delta S \tag{6.17}$$

Here $\varphi = 37°/57.3°$ is an angle expressed in radians by which the chain end must be rotated in order to free one pair of nitrogen bases. The rate of the unwinding is

$$\omega = \frac{P}{K} \tag{6.18}$$

where K is the fractional coefficient. Calculation gives values of τ of the order of seconds for M of the order of 10^6. A free energy ΔF of about 10 cal/mole of links is required.

Fixman has calculated the time of unwinding of the double helix, assuming that this process proceeds from one end as a result both of diffusion and of the rotational moment due to the gain of free energy, or entropy.[71] The principal result of the calculation is the following: If the decrease in free energy during the unwinding of one turn is

$$\Delta F = akT \tag{6.19}$$

and if

$$aN \gg 1 \tag{6.20}$$

where N is the total number of turns within the molecule, then the time of unwinding is

$$\tau \cong 3.6 \times 10^{-9}a^{-1}N^2 \tag{6.21}$$

This time is much smaller than that given by Longuet-Higgins and Zimm, whose calculations are less exact. If $a = 0$, i.e., the mechanism is purely diffusional, then

$$\tau \cong 1.2 \times 10^{-9}N^3 \tag{6.22}$$

which gives $\tau = 20$ minutes for $N = 10^4 (M = 10^6)$. If we take[70] $a = 17.5$, then it follows from Equation (6.21) that $\tau \sim 0.02$ second. If condition (6.20) is maintained, then diffusion plays practically no role during the unwinding.

RENATURATION OF DNA AND FORCES WHICH STABILIZE THE DOUBLE HELIX

Is it possible to obtain a native double helix from completely separated chains of denatured DNA? We have always spoken of reversible and irreversible denaturations, and we have established that the complete separation of chains means irreversibility. It would seem then that the answer to this question must be negative.

On the other hand, however, why should the two chains, separated at high temperature, not recombine during cooling? After all, at low temperature the double helix is more stable than the separated coils; it corresponds to a state of lower free energy. It is evident that for such recombination or renaturation of the chains to occur, it is necessary only for the chains, which are mutually complementary, to encounter each other in solution and bind to each other, at one of the specific loci, by hydrogen bonds. The pair formed could then serve as a nucleus for the subsequent crystallization; the gain of free energy during the formation of the hydrogen bonds would give the force to unwind the coils and to place the chains in a regular double helix. In fact, experiments with denatured synthetic polynucleotides show that this process is reversible; after the cooling of melted binary complexes, they reform again.

Thus, an assertion on the irreversibility of denaturation must have a relative character. It is possible to find conditions under which the irreversible denaturation, i.e., the complete separation of the chains, turns out to be a reversible process.

The renaturation of DNA was discovered and investigated by Doty, Marmur, *et al.*[72,73] When DNA from *Diplococcus pneumoniae* is heated at 100° for 10 minutes and is then cooled at the same rate, there is no renaturation; the coils remain stable even at room temperature. They seem to be in a thermodynamically nonequilibrium annealed state. This is manifested both by the absence of the hypochromic effect and by the loss of biological activity from DNA. The fact is that biological functionality is a property of native double helical, and not denatured, DNA, as evidenced by the transforming ability. The explanation for this is unclear; it may be the result of the ability of penetration into the bacterial cell or a change in the matrix genetic function. Probably both factors play a role.

During slow cooling of heat-denatured DNA, its biological activity is restored by 30%. In other cases, it has been possible to obtain 60%. The hypochromic effect is also restored. This is shown in Fig. 110. It is not surprising that the transforming activity is not fully restored; the DNA molecules are very fragile and probably must undergo partial destruction during heating, i.e., they must be broken into pieces. It is significant, however, that renaturation does occur. Renaturation can occur also after rapid cooling of denatured DNA if the frozen sample is again heated to 67°C, i.e., to a temperature less than the temperature of melting.

Renaturation may be followed also from a change in the density by Meselson's method. Density peaks for native annealed and melted DNA are shown in Fig. 111. The left peak at 1.725 g/cm^3 is a mark for calculation. Native DNA has a density of 1.704, denatured and annealed DNAs have a density of 1.716, while the renatured form has a density of 1.700. The

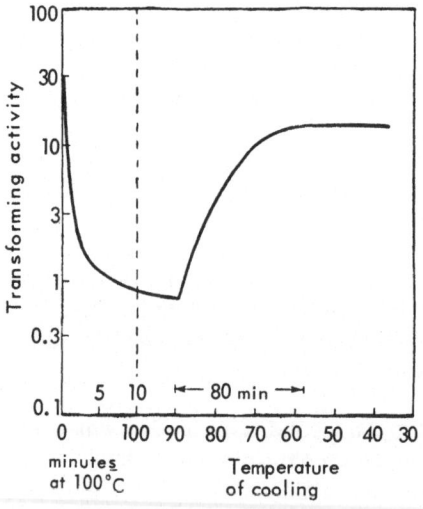

Fig. 110. Renaturation of DNA.

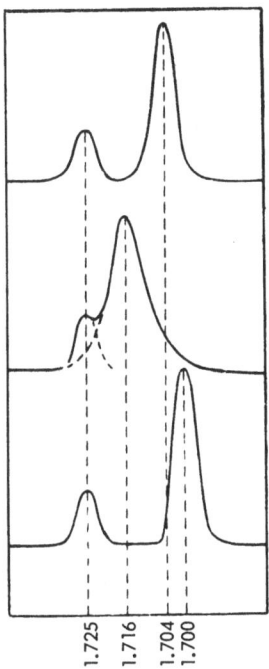

Fig. 111. Density peaks of DNA.

density is restored by 75 % during renaturation. The fact of renaturation has been observed also by electron microscopy.

Renaturation requires, on one hand, temperatures which are sufficiently high to permit the coils to unwind, and on the other hand, temperatures which are sufficiently low for the double helix to be considerably more stable than the separated chains. One can expect, therefore, a maximum in the curve of the rate of renaturation as a function of temperature; renaturation must proceed most rapidly at some optimal intermediate temperature. This has been found to be so; at 67°C, DNA from *Diplococcus pneumoniae* is renatured 2.5 times faster than at 80 or 50°C.

Renaturation cannot be achieved in all samples of DNA. It is achieved in experiments with DNA extracted from bacteria, and particularly easily with phage DNA; but it has not been achieved with DNA prepared from multicellular organisms, in particular, from mammalian cells. This is easy to understand. If the samples of DNA obtained from calf thymus, a typical bacterium, and bacteriophage, all had molecular weights of the order of 10^7, then the number of different DNA molecules per cell (or phage particle) would be 10^6, 100, and 10, respectively. Consequently, at an identical weight concentration, these three samples have completely different concentrations of chains which are mutually complementary. Compared with bacterial DNA, the concentration of complementary chains in calf-thymus DNA would be 10^{-4}, while that in phage would be

10. It is evident that the probability of the encounter with each other of two complementary chains that resulted from the denaturation of a single molecule is very low in a sample prepared from mammalian cells.

Renatured DNA is nevertheless somewhat poorer than the starting native DNA. Its thermal stability is somewhat lower and its hypochromous effect is also lower (by 15%). The double helix of the renatured DNA is somewhat spoiled.

These experiments are interesting in themselves, since they permit better understanding of the physical nature of DNA and its conformational transitions.

After the discovery of the renaturation of DNA, it was natural that there arose an idea of the production of hybrid DNA which consisted of renatured double helices from two chains which had not been paired previously. In this manner, it was possible to obtain a hybrid helix from a DNA chain containing labeled N^{15} atoms and deuterium from an unlabeled chain of DNA obtained from *B. subtilis*. Three peaks were obtained in density-gradient sedimentation corresponding to densities of 1.744, 1.725, and 1.704 g/cm^3 (Fig. 112). The first peak is a double helix in which the two chains are labeled; the second is a helix in which one chain is labeled and heavy and the second one is light, i.e., the hybrid chain; and

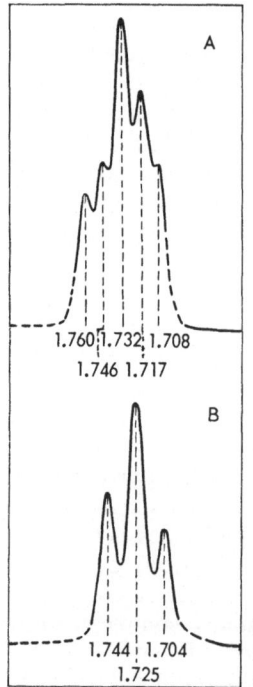

Fig. 112. Density peaks of nonhybrid and hybrid double helices of DNA from *B. subtilis*: (*a*) before treatment with enzyme; (*b*) after treatment with enzyme.

the third peak corresponds to two unlabeled chains. Since the area under each peak is proportional to the relative contents of DNA, it is easy to see that the number of hybrid molecules is about twice as large as that of each nonhybrid component.

In these experiments it is very easy to mistake, for the formation of the hybrid, the formation of nonspecific amorphous aggregates; the DNA chains simply intertwine with each other and give peaks with corresponding densities in CsCl gradient sedimentation. Such aggregates, however, are destroyed by the enzyme phosphodiesterase, while true double helices are not touched by it. This has made it possible to differentiate between hybrids and aggregates.[73]

Marmur and Doty have naturally attempted to prepare hybrid DNAs from the DNA of different types of bacteria. It is evident that, to pair two chains of DNA, it is necessary that they be complementary to each other. The difference in the types of bacteria must be related fundamentally to differences in the nucleotide text of their DNAs. It is evident that only very close strains may give two hybrid DNAs. Hybridization is quite successful with DNA molecules prepared from different mutant strains of *E. coli*. One must assume that loops are formed in these double helices in those places which have been mutated. Such hybrid DNAs may result in previously unknown mutants during the transformation of bacteria. This presents a very interesting possibility for the artificial preparation of new mutants of bacteria, if not of new types.

It has been possible to obtain also two interspecies hybrids in those cases in which the two types have the same nucleotide composition within the DNA and are linked by the fact that one form may transform the other. The density curves for the renaturation of a mixture of DNAs are shown in Fig. 113; the light one is from *Bacillus natto*, and a labeled heavy one is from *B. subtilis*.[73]

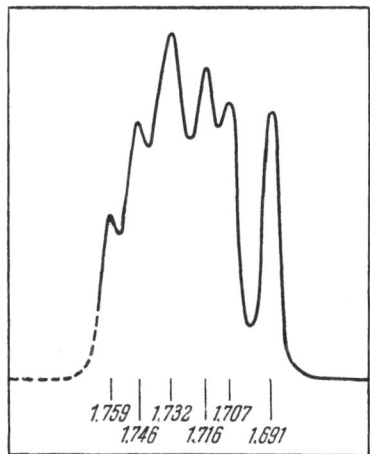

Fig. 113. Density peaks of DNA during the formation of interspecies hybrids: 1.691, standard marker (DNA from *Cl. perfringens*); 1.707 and 1.716, renatured and denatured DNA from *B. natto*; 1.732, hybrid *B. natto–B. subtilis*; 1.746 and 1.759, renatured and denatured ^{15}N deuterated DNA from *B. subtilis*. After treatment with phosphodiesterase, the peaks at 1.716 and 1.759 vanish.

1.759 1.732 1.707
1.746 1.716 1.691

DNA hybrids have been obtained also from *E. coli–Schigella dysenteriae* and from *E. coli* and different strains and forms of Salmonella.

DNA hybridization uncovers also internal genetic links between organisms. These links were noticed by Belozerskii and Spirin, who examined the composition of the DNA of a number of species.[74] We see that those organisms which have similar DNA molecules are genetically related, while the genetic relations are different from taxonomic ones, i.e., they differ from those determined by external traits. The formation of molecular hybrids also gives an excellent possibility of studying the genetic and evolutionary interactions of organisms (so far unicellular) along with transformation and transduction. Here, molecular biology penetrates directly to the fundamental biological problems.

Let us return to physics. Experiments on renaturation have shown quite clearly the forces which link the nitrogen bases of the two DNA chains in a Watson–Crick double helix. We have assumed all along that these forces are hydrogen bonds, G and C being linked by three such bonds and A and T by two. These facts may be regarded as established. On the other hand, hydrogen bonds are not the only interactions between the nitrogen bases in the double helix.

DeVoe and Tinoco,[75] using the methods of quantum chemistry, have calculated the energies of the usual intermolecular interactions of neighboring nitrogen bases, both along the chain and between pairs linked by hydrogen bonds. By usual interactions, we mean those interactions which define the difference of a real gas from an ideal one. These are the physical interactions between molecules in which the chemical bonds have been saturated. There are three types of such interactions. First, dipole–dipole forces. Two electric dipoles are attracted to each other if the positive pole of one is close to the negative pole of the other, and they repel each other if poles of the same sign are close to each other. The energy of interaction of two dipoles is equal to

$$E_{\mu\mu} = - \frac{\mu_1\mu_2}{\varepsilon R^3}(\cos\theta - 3\cos\varphi_1\cos\varphi_2) \qquad (6.23)$$

where μ_1 and μ_2 are interacting dipole moments, R is the distance between them, ε is the dielectric constant of the medium, θ is the angle between the dipoles, and φ_1 and φ_2 are the angles formed by the first and second dipoles with vector R. Secondly, the dipole of one molecule induces in the second molecule a dipole collinear with it, and, consequently, it creates induced interactions which increase with an increase in the polarizability α of the molecules. The energy of induced interaction in the case of a molecule which is isotropically polarizable is

$$E_{\mu\alpha} = - \frac{\mu_1^2\alpha_2}{\varepsilon R^6} - \frac{\mu_2^2\alpha_1}{\varepsilon R^6}$$

Since, in actuality, the polarizations of molecules are anisotropic, i.e., they have different signs in different directions, the rigorous expression for $E_{\mu\alpha}$ is more complicated; it contains the anisotropic polarizabilities and the geometric factors which are a function of the relative position of the molecule. Molecules without dipoles also interact. After all, gases which consist of spherically symmetric atoms, such as noble gases, may be liquified. This means that there is a certain type of intermolecular interactions which can be examined only in terms of quantum mechanics. The *dispersion* or *London forces* are due to the interactions of momentary dipoles which arise from the motion of electrons within the atoms and molecules. The energy of these forces is a function only of the polarizability of the molecule. It may be expressed by

$$E_L = -\frac{3}{2}\frac{J_1 J_2}{J_1 + J_2}\frac{\alpha_1 \alpha_2}{\varepsilon R^6} \tag{6.25}$$

where J_1 and J_2 are the energies necessary for the ionization of the first and second molecules. When the polarizabilities are anisotropic, the expression for E_L is again more complicated. For further details, see the monograph literature.[76]

DeVoe and Tinoco have calculated the dipole moments and isotropic polarizabilities of the nitrogen bases of DNA. In Fig. 114 are shown the calculated dipole moments in the A + T and G + C pairs. It is noteworthy that, in the A + T pairs, these are small and repulsive, while, in the G + C pairs, they are large and attractive. Calculations result in the energies shown in Table 18.

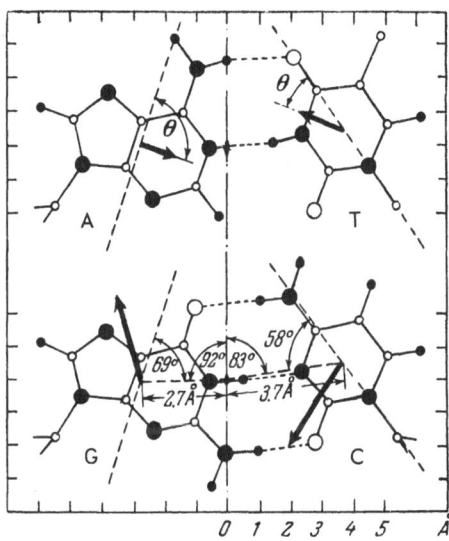

Fig. 114. Dipole moments in the nitrogen-base pairs A + T and G + C.

Table 18[75]   Energies of Base Interactions

In kcal/mole

Base pair	$E_{\mu\mu}$	$E_{\mu\alpha}$	E_L	Sum
G + C	−3.1	−0.3	−0.5	−3.9
A + T	0.8	−0.1	−0.5	0.2

The G is attracted by C, while A is repelled by T. We must add to these the energies of the hydrogen bonds. The energies in Table 18 are calculated for $\varepsilon = 1$; thus, the actual energies must be smaller. We must introduce, of course, not $\varepsilon = 81$ (with water at 25°C), but $\varepsilon = 2$ to 5, the local value of ε during interactions at short distances.

In a double helix, however, dipole–dipole, induced, and London forces act, not only between the paired bases, but also between other neighbors, such as those along a single chain. The London interaction between planar bases located one on top of the other is quite strong. Calculations (again for $\varepsilon = 1$) lead to interesting results, cited in Table 19.

To obtain the total energies of interactions between two pairs, it is necessary to add up the energies of Tables 18 and 19. For example, for ↑ AT, GC ↓ we obtain − 7.4 + 0.2 − 3.9 = − 11.1.

DeVoe and Tinoco consider that these energies are considerably larger than the energies of hydrogen bonds and that, therefore, the formation of the double helix is determined principally by the usual molecular interactions. It is difficult to agree with this; had this been so, it would not be possible to understand why G is paired with C and A with T.

One should also not rely on the numbers found by DeVoe and Tinoco. Quantum-mechanical calculations of such complicated molecules are still not reliable. The local values of ε are unknown. However, these results are important from a qualitative point of view. They indicate strongly that the GC pairs are linked more strongly than the AT pairs and that the interaction energies must be different for different sequences along a given chain. Thus, the thermodynamic stability of different nucleotide texts is different and it is possible that this places some limits on the genetic information which may be included in the DNA. To verify the last assertion, it is necessary to know the text, i.e., the nucleotide sequence. As we have pointed out already, the nucleotide texts have not been read so far.

In an examination of the Van der Waals forces, DeVoe and Tinoco do not take into consideration the role of water, which cannot be reduced just to a dielectric constant. It is necessary to take into account the energy of interaction with water during its exclusion from the double helix. Up to now, the role of hydrophobic interactions remains absolutely unresolved.

Table 19[75]   Half Sum of Four Interactions between Unpaired Bases

In kcal/2 moles of bases

Adjacent pairs‡	$E_{\mu\mu}$	$E_{\mu\alpha}$	E_L	Sum
↑CG ↓ GC	− 5.8	− 4.1	− 6.0	− 19.8
GC GC	2.1	− 3.3	− 6.8	− 11.9
TA CG	− 0.5	− 2.0	− 6.8	− 11.2
AT CG	− 0.9	− 1.4	− 3.6	− 7.8
AT GC	3.3	− 2.0	− 6.8	− 7.4
TA GC	3.1	− 2.4	− 6.0	− 7.2
GC CG	4.2	− 2.4	− 3.6	− 5.7
TA AT	1.3	− 0.7	− 6.0	− 5.2
AT AT	2.2	− 0.6	− 6.8	− 5.0
AT ↑TA ↓	2.2	− 0.4	− 3.6	− 1.6

‡The arrows indicate the antiparallel nature of the chains; for example, TA means T-sugar-3′-phosphate-5′-sugar-A.

Hydrogen bonds in nucleotide pairs indicate in any case the structural correspondence of the two complementary chains. It is evident that the energy of denaturation, of which we have talked before, sums up all these effects. In what follows, we shall continue to speak of hydrogen bonds, but we shall have in mind what has been discussed above.

DNA REDUPLICATION

We have become acquainted with the particular structure of the DNA molecules and with some properties of these molecules. The biological function of DNA is directly linked to its structure and its very interesting chemical properties.

The principal way of transferring the genetic information from cell to cell consists in the reduplication of the chromosomes during mitosis and meiosis in cell separation. A detailed study of these processes testifies that the progeny chromosome is a copy of the parent one, built anew from material present within the dividing cell. Since the principal material within chromosomes is DNA, it is quite natural to assume that the doubling of the supramolecular structures, chromosomes, is determined on the molecular level of organization by the doubling (reduplication) of the DNA macromolecules. As has been stated already, DNA serves as the matrix for protein synthesis. If, in fact, it is reduplicated, then the question of some other matrix necessary for the synthesis of DNA is removed; it serves as its own matrix.

The Watson–Crick model yields a qualitative explanation of DNA reduplication. Let us suppose that the original DNA double helix, consisting of complementary chains, has separated as a result of the breaking of hydrogen and Van der Waals bonds and has become unwound into single chains (Fig. 115a and b). Each of the individual chains may serve as a matrix for the assembly of a new chain out of monomers (nucleoside triphosphates). The nucleotides combine with the matrix; this combination is in such a way that Watson–Crick pairs are formed, i.e., A combines with T, T combines with A, G combines with C, and C combines with G (Fig. 115c). At the same time, polycondensation of nucleotides occurs and, as a result, instead of a single parent helix, we have two helices identical with the starting one. Such a reduplication is covariant; in other words, if in some place there had been an unsuitable nucleotide which had formed a loop in the starting helix or any other defect, then in the further duplication, this nucleotide would have found its legitimate place and the defect would vanish. Such a fixation of a structural defect is a good model for mutation.

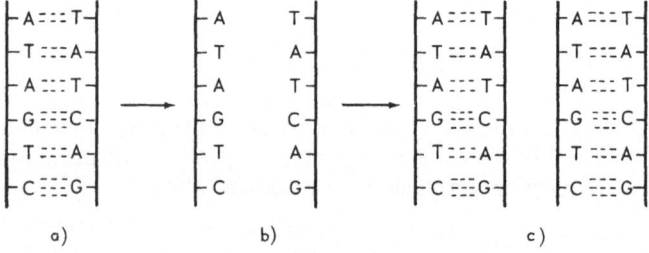

Fig. 115. Model of DNA reduplication: (a) parent double helix; (b) two separated chains; and (c) two progeny double helices identical with the starting one.

In principle, we may have three different mechanisms of reduplication.[77] In the conservative mechanism, the matrix double helix is maintained and a new progeny helix is formed with its help (Fig. 116). A semiconservative mechanism is shown in Figs. 115 and 116; in the new double helices, one is old, the other one is new. Finally, in the dispersion mechanism, we have a uniform distribution of parent material between the four chains of the two-progeny double helices (Fig. 116c). The question of the nature of the mechanism is subject to experimental examination. A semiconservative mechanism is the most attractive one, since it guarantees covariance without any difficulties, i.e., during reduplication a defect stops being a defect. In fact, experimental evidence favors the semiconservative mechanism.

Meselson and Stahl have studied the reduplication of DNA during the division of *E. coli* cells with the help of tagged atoms and density-gradient sedimentation.[78] The bacteria were grown in a medium containing only labeled nitrogen ^{15}N. Thus, a cell population was obtained which was completely labeled with ^{15}N. During the interphase period, the cells were transformed into a new medium which contained only normal nitrogen ^{14}N. The cells duplicated in this medium. The starting population, its children, and grandchildren were examined. The DNA was extracted from the starting cells and from the cells of the progeny of the first and second generations; this DNA was studied in the ultracentrifuge, and its density and isotope effect were determined. It was found that the starting DNA had the highest density; it was completely labeled with ^{15}N. The DNA of the children was labeled to one half; i.e., its density was equal to the arithmetic mean between the densities of ^{15}N DNA and ^{14}N DNA, which had been isolated from cells grown in unlabeled medium. Finally, the DNA of the grandchildren was separated in a density gradient into two zones, one corresponding to ^{14}N DNA and the other to DNA labeled to 50 %. The results obtained are shown in Fig. 117. This is exactly the pattern which must be obtained with semiconservative reduplication according to the schemes of Figs. 115 or 116b.

The questions of the structure of the chromosome as a whole and of the distribution in it of the DNA double helices are at present outside the realm of molecular biophysics. A chromosome is a complicated supramolecular structure which is not yet completely understood. The complication is related in part to the fact that the chromosomes are not composed of just DNA, but of complexes of DNA with proteins (histones, protamines),

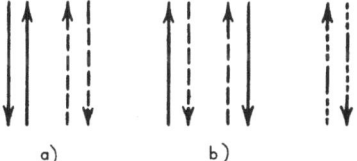

Fig. 116. Possible mechanisms of DNA reduplication: (*a*) conservative (*b*) semiconservative, and (*c*) dispersed.

a)   b)   c)

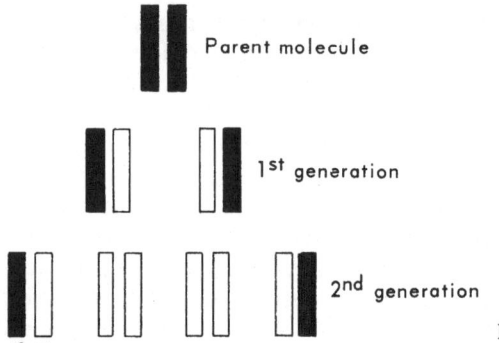

Fig. 117. Results of the Meselson
and Stahl experiment.

i.e., of nucleoproteins, of which very little is known at present. Chromosomes also contain RNA. Up to now, molecular biophysics has attained a great deal of success in the investigation of the structure and properties of individual biological molecules, but not of complicated supramolecular structures.

However, we know quite a lot even about the structure of chromosomes. The chromosomes apparently consist of a number of filaments, each of which has a thickness of about 200 Å. These filaments are in principle parallel to each other but they are wound about each other in the regions of disks, which are observed in chromosomes. An entire hierarchy of supramolecular helical structures has been observed.[79-81] If the first-order unit of structural organization is a filament with a diameter of about 200 Å, then the structural unit of the next order must be a thicker double filament with a diameter of 500 Å. According to Ris, in this way the chromosome is similar to a virus particle; its core consists of nucleic acids, and its membrane, of protein.[81]

Taylor has shown that the reduplication of a chromosome as a whole also proceeds in a semiconservative manner just as the reduplication of the DNA macromolecules from which the chromosomes are constructed.[82] This was examined by the method of radioautography. The chromosomes of the cells of some plants (*Vicia faba, Bellevalia, Crepis, Allium*) contained tritium-labeled thimidine. The medium surrounding the chromosomes did not contain tritium; thus, during the duplication according to the semiconservative mechanism, the progeny chromosomes had to be nonradioactive. This scheme is shown in Fig. 118. Since the tritium is β radioactive, it photographs itself, i.e., it gives spots on a photographic emulsion. In fact, the results obtained were in agreement with this scheme.

Taylor assumed that the model of chromosome reduplication must follow as a result of DNA reduplication (Fig. 119). This model is in accord with a number of experimental facts.[82] In the study of bacterial and

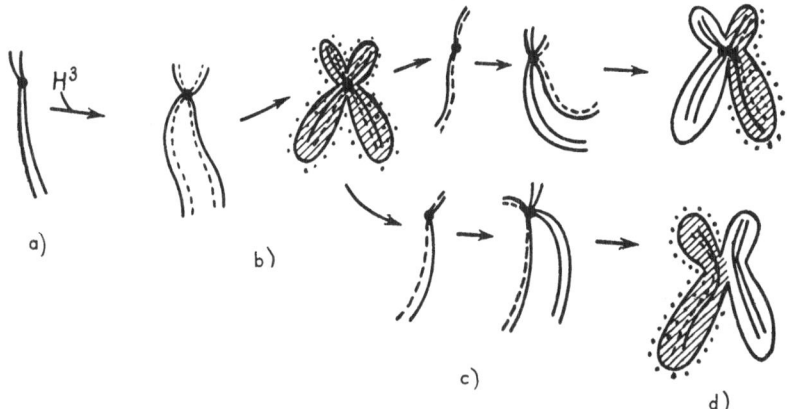

Fig. 118. Scheme of DNA distribution during chromosome duplication. The dotted lines are tagged DNA; the dots indicate grains on the radioautograph. (a) Duplication in the presence of H³-thimidine; (b) the first metaphase after the introduction of the label; (c) duplication in the absence of the label; (d) the second metaphase. (According to Taylor.)

phage genetics, one does not meet the difficulties involved in the investigation of complicated supramolecular structures. In these cases, the chromosome is just a macromolecule of DNA, i.e., a double helix wound into a closed loop. A few known exceptions are the single-chain DNA of phage φ X174[83] and also of phage S13 and φ R.[84]

The same method of radioautography was used to study the reduplication of the ring chromosomes of *E. coli*. The labeling was done with tritium. The results obtained are shown schematically in Fig. 120.[85] It follows from these results that the inclusion of the labels is in sequence from nucleotide to nucleotide. The chromosome undergoes reduplication starting with some locus. The chemical evidence of the DNA duplication is obtained from the studies by Kornberg on the synthesis of DNA *in vitro*.[86,87]

DNA can be synthesized, if the incubation mixture contains triphosphates of thimidine, deoxycytosine, deoxyguanosine, and deoxyadenosine (TTP, CTP, GTP, and ATP), Mg^{++} ions ($6 \times 10^{-3} M$), pH 7.5, the enzyme polymerase extracted from *E. coli*, and native DNA as an initiator. For this, calf-thymus DNA was used. If a system is complete, then DNA synthesis proceeds in it, and the nucleotide composition of the product is completely similar to that of the initiator DNA. The reaction does not proceed if any one of the components of the mixture is absent, or if the initiator DNA is treated first with deoxyribonuclease, an enzyme which destroys DNA. The results are illustrated in Table 20.

The inclusion of nucleotides in the new DNA is followed by means of ³²P-labeled nucleotides. As each nucleotide is included in the chain, inorganic diphosphate, or pyrophosphate PP (that is, $H_4P_2O_7$), is

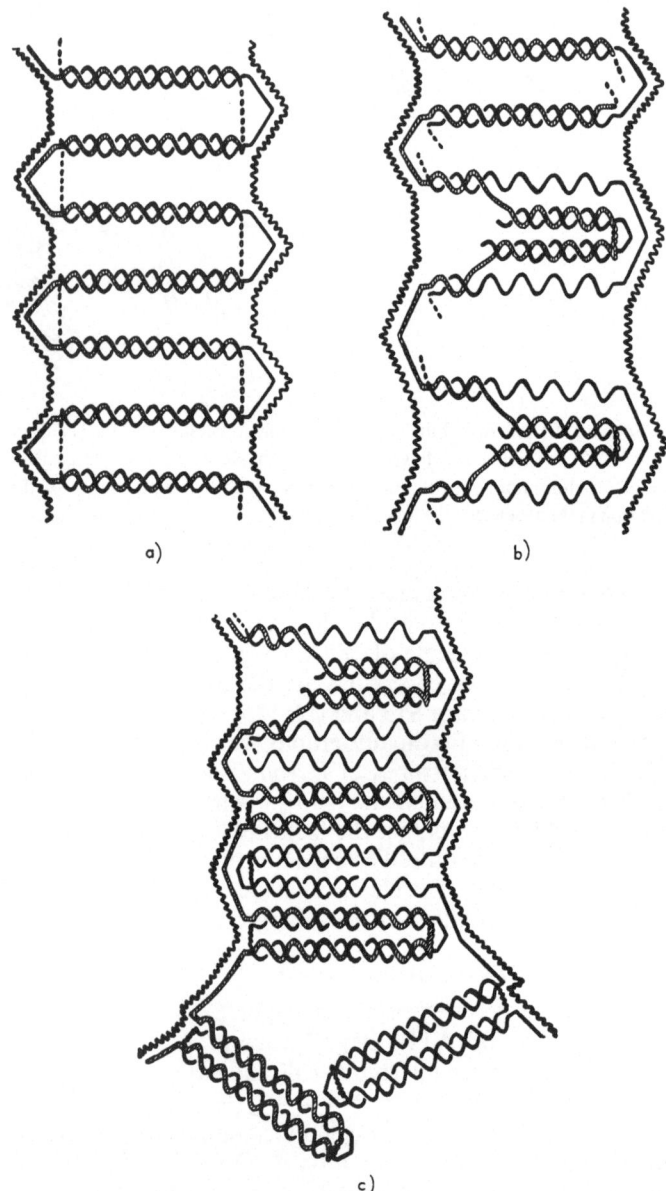

Fig. 119. Scheme of the reduplication of chromosomes: (*a*) the starting
state; (*b*) the opening of the 3′ links by the corresponding enzymes;
(*c*) the beginning of reduplication. (According to Taylor.)

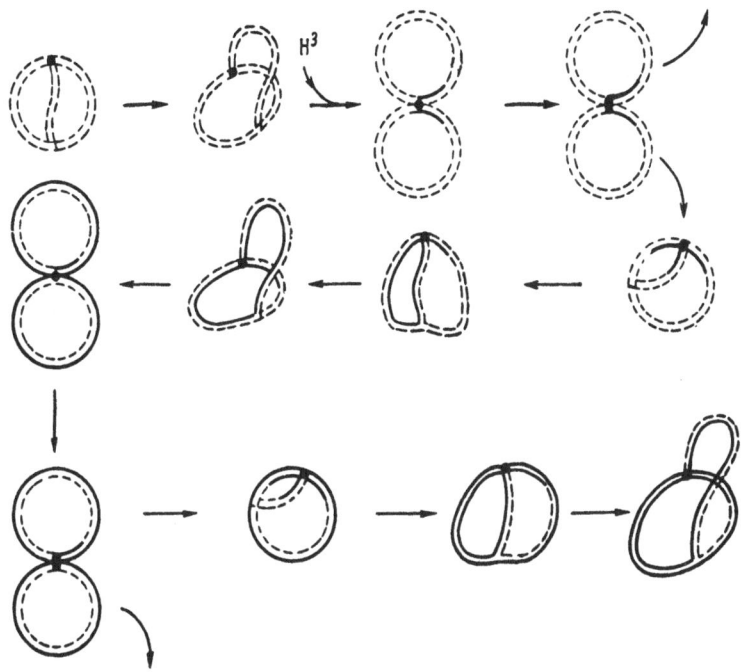

Fig. 120. Reduplication of the cyclic chromosome of *E. coli*: solid line, labeled DNA; filled circle, hinge. (According to Cairns.)

released. The reaction proceeds according to the scheme

$$
\begin{matrix} n\text{TPPP} \\ n\text{GPPP} \\ n\text{APPP} \\ n\text{CPPP} \end{matrix} + \text{DNA} \rightleftharpoons \text{DNA} - \begin{bmatrix} \text{TP} \\ \text{GP} \\ \text{AP} \\ \text{CP} \end{bmatrix}_n + 4n\text{PP}
$$

This is a polycondensation reaction, and not a polymerization. In accordance with the law of mass action, the reaction may be slowed down and reversed by increasing the concentration of pyrophosphate. In fact, if pyrophosphate is present in an amount 100 times greater than the total amount of deoxynucleoside triphosphates, the reaction is slowed down by 50%.

Unfortunately, there are no more detailed experimental investigations of the thermodynamics and kinetics of synthesis of DNA *in vitro*, not to speak of the synthesis *in vivo*. The studies of Kornberg, however, are very important achievements, since they make it possible to approach an extremely important phenomenon, namely, the reduplication of DNA, as a simple physical-chemical process.

Table 20[86]   **Requirements for Various Components during the Incorporation of Deoxynucleotides in DNA**

System	Amount of DNA, mμM
Entire system (5 mμM of each of the four triphosphates, 1 μM MgCl$_2$, 10 μg DNA, 3 μg enzyme, pH 9.2)	0.50
Same, without CTP, GTP, ATP	<0.01
Same, without CTP	<0.01
Same, without GTP	<0.01
Same, without ATP	<0.01
Same, without Mg^{++}	<0.01
Same, without DNA	<0.01
DNA pretreated with deoxyribonuclease	<0.01

DNA molecules, in which some of the nitrogen bases have been replaced, have been obtained by this same method. The base T may be replaced by U (the rate of the synthesis drops considerably during this) and by 5-Br–U (this is not accompanied by a change in the rate of synthesis), C may be replaced by 5-Br–C and 5-CH$_3$–C, and G may be replaced by hypoxanthine. The Kornberg enzyme works also if the mixture does not contain all the monomers; this requires a long incubation period. A mixture of ATP and TTP results in a copolymer of regular periodic structure with double helices of the type

$$-A-T-A-T-A-T-A-T-$$
$$-T-A-T-A-T-A-T-A-$$

A mixture of GTP and CTP gives homo, and not copolymers, with the formation of double helices

$$-C-C-C-C-C-C-C-C-$$
$$-G-G-G-G-G-G-G-G-$$

In these cases, first a matrix is slowly polymerized and then matrix synthesis occurs.

The matrix synthesis of polynucleotides can take place also with short oligomer matrices. This was shown by Kornberg *et al.*[88] in the study of the synthesis of poly-*d*–AT using oligomers of AT. In this, just as in the polymer matrix, chains with a length much greater than n are formed. The process takes place with a considerable lag, which decreases rapidly with an increase in n.

It is evident that the reduplicating synthesis of DNA may and must be the object of thermodynamic examination. As always in problems of molecular physics and physical chemistry, a thermodynamic analysis precedes a detailed theoretical investigation of the process. Thermodynamics yields the general conditions under which a process takes place; it says nothing of its velocity, if we talk of equilibrium conditions. The statistical thermodynamic examination of DNA reduplication has been the subject of several studies.[89,90]

We shall examine an initial state of a system formed of a DNA double helix and an aqueous solution of nucleoside triphosphates (NTP) which surrounds it and a final state which contains, instead of a single starting double helix, two double helices identical with the starting one, while the solution has acquired the corresponding amount of pyrophosphate PP (Fig. 121). Let us calculate the difference in the chemical potentials of the final and initial states

$$\Delta F + p\,\Delta V \cong \Delta F \tag{6.26}$$

The change of free energy may be expressed in the form

$$\Delta F = -N(2E_1 + \bar{E}_2) - T\,\Delta S \tag{6.27}$$

where N is the number of nucleotides in each of the chains of the helix, E_1 is the energy released during the inclusion of one nucleotide into the chain, $\bar{E}_2$ is the mean energy gain during the formation of the hydrogen bonds between the nitrogen bases in the Watson–Crick pair, ΔS is the entropy change. It is evident that $\bar{E}_2$ must be a function of the relative

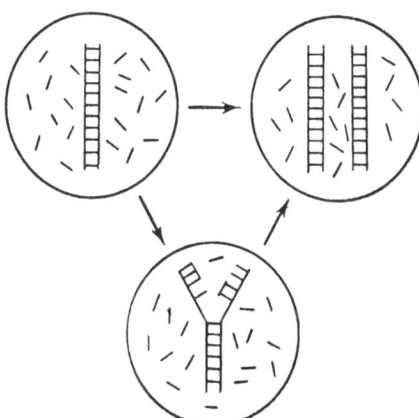

Fig. 121. Initial, intermediate, and final thermodynamic states of DNA reduplication.

contents of G + C, since the AT bonding is weaker than GC. Therefore, we may write

$$\bar{E}_2 = \alpha E_{GC} + (1 - \alpha) E_{AT} \tag{6.28}$$

where α is the fraction of GC in DNA, and

$$\alpha = \frac{[G + C]}{[A + T] + [G + C]} \tag{6.29}$$

How can we represent the entropic term? The entropy change during the introduction of the nucleotide into the chain is determined by three factors. First, the replacement of a nucleoside triphosphate by pyrophosphate in a solution is expressed by the difference of their intrinsic entropies in the free state; these molecules are of different structures, masses, and shapes. Second, a nucleotide which had been introduced into a double helix is deprived of freedom of translational and rotational motion. Third, during the introduction of a nucleotide, the entropy of the mixing of the solution changes. Expressing the first two factors by the single term $k \ln q$, which expresses the change in intrinsic entropy per single nucleotide, we may write

$$\Delta S = 2Nk \ln \left(q \frac{\bar{x}_{NTP}}{x_{PP}} \right) \tag{6.30}$$

where x_{PP} is the concentration of PP in the solution, and $\bar{x}_{NTP}$ is the average concentration of NTP and is equal to

$$\bar{x}_{NTP} = (x_G^\alpha x_C^\alpha x_A^{1-\alpha} x_T^{1-\alpha} x_T^{1-\alpha})^{1/2} \tag{6.31}$$

where x_G, x_C, x_A, and x_T are the concentrations of the corresponding NTPs. If these are equal to each other, then

$$x_G = x_C = x_A = x_T = \frac{x_{NTP}}{4}$$

and

$$\bar{x}_{NTP} = \frac{x_{NTP}}{4} \tag{6.32}$$

The thermodynamic condition of reduplication consists in the equality of the free energies of the final and initial states. The condition of reduplication is

$$\Delta F \leqslant 0 \tag{6.33}$$

i.e.,

$$2E_1 + \bar{E}_2 + 2kT \ln \left(q \frac{\bar{x}_{NTP}}{x_{PP}} \right) > 0 \tag{6.34}$$

In other words, the larger the gain in energy, $2E_1 + \bar{E}_2$, the more favorable the process is. The role of the entropic term depends on its sign. If, during reduplication the entropy increases, then it acts in the same direction. If the entropy decreases, then there is a competition between the energetic and entropic factors.

At first glance it may seem that an increase in the ordering of a substance during the formation of the second helix must be accompanied by a decrease in the entropy, and that such a competition will take place. This would be true if the DNA synthesis had occurred by polymerization and not polycondensation. The release of PP into the solution results in a dependence of the sign of ΔS on the ratio $\bar{x}_{NTP}/x_{PP}$ and, evidently, on the quantity q.

An estimate of the quantities in Eq. (6.34) is not very rigorous but it may be useful for orientation. A reasonable value of E_1 is of the order of 4 kcal/mole. Let us take a value of $\bar{E}_2$ of about 5 kcal/mole ($\bar{E}_2$ may be even greater). The quantity q has the value of

$$q = \frac{q_1' q_2}{q_1 q_2'} \tag{6.35}$$

where q_1 and q_1' are the intrinsic partition functions of NTP in solution and in the state of attachment to the matrix chain by hydrogen bonds, and q_2 and q_2' are the intrinsic partition functions of free PP in solution and of the PP incorporated into the NTP attached to the matrix by hydrogen bonds. Considering that, in solution, the molecules will have independent degree of freedom of translational and rotational motions, which are lost during the fixation of the molecules on the matrix, we find $q \sim 10^{-2}$ by application of the usual equations of statistical mechanics and taking into consideration the masses and dimensions of the molecules. Thus, at 300°K, the condition of reduplication has the form

$$13 + 1.2 \ln \left(10^{-2} \frac{\bar{x}_{\text{NTP}}}{x_{\text{PP}}} \right) > 0 \tag{6.36}$$

This condition is fulfilled as long as the ratio $x_{\text{PP}}/\bar{x}_{\text{NTP}}$ is less than $1 : 400$. In cells (and also in the experiments of Kornberg) this ratio is much greater. Thus, both the energetic and entropic factors work in the same direction, and the thermodynamic condition of reduplication is obeyed. The actual reasons which hinder the reduplication during certain stages of cellular development must probably be, not thermodynamic, but kinetic.[92] They may be due to the absence from the cell of some necessary enzyme, which is synthesized in a late state of mitosis. It is also possible that it is, not the relative, but the absolute concentration of NTP in the interphase which is insufficient for the synthesis of DNA. Reduplication is certainly controlled by complicated regulatory laws, which cannot be explained by the present model. These calculations should be compared, not with the process *in vivo*, but with the experiments of Kornberg. Theory does not contradict these experiments, but it will be possible to speak of complete agreement only after an investigation of the concentration and temperature dependence of DNA synthesis.

The denaturation of DNA may be described within the realms of the same theory; it may be regarded as a breaking of hydrogen and Van der Waals bonds between the nitrogen bases of complementary chains. The free-energy difference of the separated chains on the double helix may be expressed in the form

$$\Delta F' = N\bar{E}_2 - 2NkT \ln q' \tag{6.37}$$

where q' is the change of the intrinsic entropy of the chain per nucleotide. The condition for denaturation has the form

$$\Delta F' \leqslant 0$$

As a first approximation, it is possible to consider that the change of entropy of the chains during their denaturation is determined by their flexibility. Then $q' \cong z$, where z is the number of conformations of a link of a single polynucleotide chain. A link contains five single bonds and each bond corresponds to two to three rotational isomers. Because of steric limitations, it is evident that not all possible conformations may be realized and Z must be a quantity of the order of 10 to 100.

From (6.35), it follows that the denaturation temperature is

$$T_m = \frac{\bar{E}_2}{2k \ln q'} = \frac{\alpha E_{\text{GC}} + (1 - \alpha) E_{\text{AT}}}{2k \ln q'} = \frac{E_{\text{AT}}}{2k \ln q'} + \alpha \frac{E_{\text{GC}} - E_{\text{AT}}}{2k \ln q'} \tag{6.38}$$

that is, T_m has, in fact, a linear dependence on α. The numerical agreement with the dependence of T_m on the contents of G + C is the same as found by Marmur and Doty,[72]

$$T_m = 342°K + 41\alpha° \; K$$

This is obtained at $E_{AT} = 4.6$ kcal/mole, $E_{GC} = 5.3$ kcal/mole, and $z = 32$.

According to thermodynamic theory, the fate of the DNA double helix depends on the ratio between the values of ΔF and $\Delta F'$ corresponding to the differences of the free energies. If

$$\Delta F < \Delta F' \quad \text{and} \quad \Delta F \leqslant 0$$

then reduplication takes place. If

$$\Delta F' < \Delta F \quad \text{and} \quad \Delta F' \leqslant 0$$

denaturation occurs. If

$$\Delta F = \Delta F' \leqslant 0$$

both processes are thermodynamically of equal probability.

If

$$\Delta F > 0 \quad \text{and} \quad \Delta F' > 0$$

the double helix remains without any changes.

It should be pointed out that the thermodynamic theory is based on the reversibility of the processes of reduplication and denaturation. The inclusion of labeled PP in the experiments of Kornberg in the free NTP indicated that the intermediate process is reversible. Denaturation may be reversible at a given stage. As a whole, however, processes which occur *in vivo* are accompanied by irreversible reduplication of DNA. The total denaturation of DNA is also irreversible. To examine these phenomena rigorously, it is necessary to apply both the kinetics and thermodynamics of irreversible processes. The equilibrium theory presented here can serve as a model and is approximate. Its results are useful, while the theory itself does not pretend to describe the process as a whole.

This theory makes it possible to estimate the thermodynamic probability of mutations regarded as loops. To do this, it is possible to use the condition of duplication [Eq. (6.34)].[90,91] In the formation of a loop, the energy of hydrogen bonds is lost. The probability of mutation in any given link which contains one of the bases (A, T, C, or G) may be expressed by

$$W = e^{-\Delta F_k/kT} \qquad (6.39)$$

where ΔF_k is the free-energy difference between the mutated and nonmutated states of the link containing the given base within the DNA chain. Calculation gives different probabilities for different types of loops shown in Fig. 122. If one takes the values of the energies and entropies cited above, the results in Table 21 are obtained.

These probabilities are considerably greater than those observed experimentally. In bacteria, the number of mutations per gene does not exceed 10^{-5} to $10^{-7}\%$ during a single generation. It is evident that if the number of nucleotides in the loop is increased, the probability of its forma-

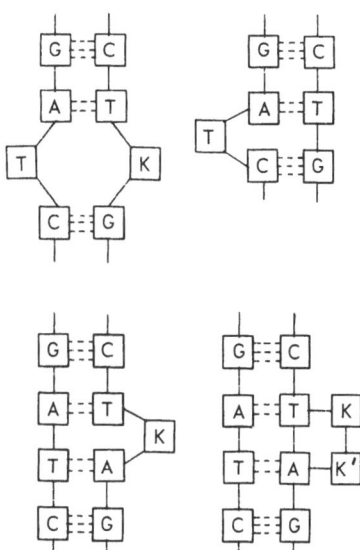

Fig. 122. Different types of loops which may serve
as models for point mutations.

tion decreases rapidly, approximately by a factor of 100 for each nucleotide. The situation cannot be corrected in this way, however. The deciphering of the DNA code shows that point mutations are linked with a replacement of just a single nucleotide. If the tautomerism of nucleotides is taken into account, the situation may only become worse, since the ability of an unsuitable nucleotide in a tautomeric form to form the necessary hydrogen bonds will lead to an increase, and not a decrease, in the probability of mutations. It is possible to consider that if the loops are correct models for point mutations, their formation is limited, not by thermodynamic, but by kinetic factors, i.e., by the presence of a high activation barrier which results in a sharp decrease of the probability of loop formation.[92] On the other hand, if a larger value of the energy $\bar{E}_2$ is taken, such as that which may be deduced from experimental results,[52] the theoretical probability decreases considerably and it is quite possible to obtain agreement with experimental

Table 21   Probability of Loop Formation

	In the normal state, the ring contains	
Type of loop	G or C	A or T
1. Loops in both chains	1.4×10^{-4}	4×10^{-4}
2. Loops in the parent chain	10^{-2}	2.5×10^{-2}
3. Loops in the progeny chain	1.5×10^{-2}	1.5×10^{-2}
4. Double loop in the progeny chain	2.0×10^{-4}	2.0×10^{-4}

data. With further development of the theory of spontaneous mutations, it is necessary also to take into account the calculations of DeVoe and Tinoco.[58] It is evident that the number of different types of loops and, thus, of different types of mutations with different probabilities increases sharply during this.

However, even at this stage, thermodynamic calculations lead to interesting consequences. First we obtain a natural result, i.e., replacement of AT by GC is thermodynamically more probable than the reverse, since GC is a stronger bond. It does not follow from here with certainty that the processes of evolution must result in an increase in contents of GC in DNA. Thermodynamics tell us nothing on the natural selection of mutations. In fact, in higher organisms, the GC contents in the DNA are stabilized at a level of 40 to 45%, while in bacteria, it may reach as high as 80%. These facts are explained by molecular biology.

Second, it follows from the calculations that the probability of mutations depends on the relative concentrations of the different NTPs in solution.

The numbers given in Table 21 were obtained for an equimolar mixture of all four NTPs. In order to have a constant level of spontaneous mutations, the concentration ratio of NTP in solution must remain constant. Any disturbance of the balance leads to a change in the number and types of mutations. In principle, this situation may be checked experimentally. It is apparently valid even when the mutation has a kinetic, not a thermodynamic, origin. One of the side effects of radiation on mutations may consist in the disturbance of the balance of nucleotides within the cytoplasm. It is known that short-wavelength radiation acts differently on different nucleotides. Here local interactions of quanta with nucleotides are important, in particular if the nucleotides are located in the immediate domain of the duplicating DNA.

STATISTICAL PHYSICS AND KINETICS OF DNA REDUPLICATION

The thermodynamic analysis of the reduplication process certainly does not give any information on the mechanism of this process. The conditions of the transition are examined, but no light is shed either on the nature of this transition or on its sharpness. The situation is quite similar to what exists in the theory of crystallization of liquids or the melting of crystals. It has been shown above that the examination of the macromolecular chain as a cooperative system on the basis of the one-dimensional Ising model makes it possible to explain all the basic physical properties of the macromolecules, the phenomena of helix–coil transitions, and the denaturations of proteins and nucleic acids. Let us assume that the process of reduplication, i.e., the synthesis of a new chain on a single-chain matrix, is also cooperative. The one-dimensional model of Ising is also applicable to this question.[89,90]

Let us assume that new nucleotides are added on the liberated links of the chain, i.e., on those links whose hydrogen bonds have been broken. Let us consider that the separation of the double helix proceeds from one end and does not occur in the middle (the calculation practically does not change if the separation occurs from two ends). In this manner we examine not only the initial and final states of the system, but also any intermediate in which a certain number of links r per chain, are separated and freely bind NTP from solution (Fig. 121).

The total partition function of the system is

$$Q = \sum_{r=0}^{N} e^{-rE_2/kT} Q_{2r} \qquad (6.40)$$

The first multiplier under the summation sign expresses the breaking of the interchain bonds into r Watson–Crick pairs, which requires the expenditure of rE_2 energy. The second multiplier is the partition function of $2r$ free DNA links. To calculate Q_{2r}, let us use the Ising method.

We shall assume that each of the $2r$ links may exist in three states: (1) A suitable NTP is bonded out of solution to the monomer by hydrogen bonds (A to T, T to A, G to C, or C to G); (2) an unsuitable NTP is bonded out of solution (G, C, or T to A, etc.); (3) the link has remained free, i.e., it may bind water molecules. This is obviously a rough model, since the energy differences of interaction in the Watson–Crick model are not taken into account in it, as well as such interactions between nucleotides which are not in Watson–Crick pairs. States 1, 2, and 3 are described by the following free energies:

$$F_1 = -\bar{E}_2 - kT \ln \left(\bar{x}_{NTP} \frac{q_1'}{q_1} \right)$$

$$F_1' = -\bar{E}_2 - kT \ln \left(\bar{x}_{NTP}' \frac{q_1'}{q_1} \right) \qquad (6.41)$$

$$F_2 = -kT \ln z$$

$\bar{x}_{NTP}'$ is different from $\bar{x}_{NTP}$, since each link of the chain may bind one suitable and three unsuitable NTPs. We have

$$\bar{x}_{NTP}' = [(x - x_G)^\alpha (x - x_C)^\alpha (x - x_A)^{1-\alpha} (x - x_T)^{1-\alpha}]^{1/2} \qquad (6.42)$$

where $x = x_{NTP}$. When $x_A = x_T = x_C = x_G$,

$$\bar{x}_{NTP}' = \tfrac{3}{4}x = 3\bar{x}_{NTP} \qquad (6.43)$$

The principal hypothesis is that polycondensation, i.e., the formation of phosphoester bonds between two nucleotides with a simultaneous liberation into solution of one molecule of PP, may occur only when the neighboring links on the chain are bonded to suitable nucleotides. If two neighboring links are in state 1, then an additional free-energy change occurs,

$$F_3 = -E_1 - kT \ln \left(x_{PP} \frac{q_2'}{q_2} \right) \qquad (6.44)$$

This is the nature of the cooperative process.

In accordance with the theory of the one-dimensional Ising model, the partition function Q_2 may be expressed as a trace of the probability matrix, of order $2r$. Here the matrix has the form

$$P = \begin{pmatrix} P(1,1) & P(1,2) & P(1,3) \\ P(2,1) & P(2,2) & P(2,3) \\ P(3,1) & P(3,2) & P(3,3) \end{pmatrix} \tag{6.45}$$

where $P(ij)$ is the thermodynamic probability that any given link exists in state 1 if the link following immediately is in state j: $i, j = 1, 2, 3$, since each link may exist in three states.
Setting

$$e^{-F_1/kT} = a_1 \qquad e^{-F_1'/kt} = a_1' \qquad e^{-F_2/kT} = a_2 \qquad e^{-F_1-F_3/kT} = a_3 \tag{6.46}$$

we obtain in our case

$$P = \begin{pmatrix} a_3 & a_1 & a_1 \\ a_1' & a_1' & a_1' \\ a_2 & a_2 & a_2 \end{pmatrix}$$

In order to find the trace of the matrix P, we must solve its characteristic equation

$$\begin{vmatrix} a_3 - \lambda & a_1 & a_1 \\ a_1' & a_1' - \lambda & a_1' \\ a_2 & a_2 & a_2 - \lambda \end{vmatrix} = 0 \tag{6.48}$$

or

$$\lambda^3 - (a_1' + a_2 + a_3)\lambda^2 + (a_1'a_3 - a_1a_2 - a_1a_1' + a_2a_3)\lambda = 0 \tag{6.49}$$

The maximal root is

$$\lambda_{max} = \frac{a_1' + a_2 + a_3}{2} + \left[\frac{(a_1' + a_2 - a_3)^2}{4} + a_1a_1' + a_1a_2 \right]^{1/2} \tag{6.50}$$

The partition function Q_{2r} is

$$Q_{2r} = \text{Spur}\,(P^{2r}) = \lambda_1^{2r} + \lambda_2^{2r} + \lambda_3^{2r} \tag{6.51}$$

and when $2r \gg 1$,

$$Q_{2r} \cong \lambda_{max}^{2r} \tag{6.52}$$

If the values of E_1, $\bar{E}_2$, etc., cited above, are taken over the entire range of concentration and temperature of interest, a_1 and a_1' are considerably larger than a_2 and a_3. In other words, the probability of the bonding of NTP by hydrogen bonds without polycondensation is very small. In fact, when $\bar{x}_{NTP} \sim 10^{-6}\,M$ (the concentration in the Kornberg experiments), a_1 and a_1' have values of the order of 10^{-5}, where a_2 at room temperature is approximately 30, and a_3 at the point of reduplication is approximately 10^2. Consequently, we obtain with a high degree of accuracy

$$\lambda_{max} = \frac{a_2 + a_3}{2} + \left[\frac{(a_2 - a_3)^2}{4} \right]^{1/2} \tag{6.53}$$

If $a_2 - a_3 > 0$, then $\lambda_{max} \cong a_2$. If $a_2 - a_3 < 0$, then $\lambda_{max} \cong a_3$. If $a_2 - a_3 = 0$, then $\lambda_{max} = (a_2 + a_3)/2$.

With a fixed value of r, the mean number of polymerized links is

$$\bar{l}_r = \frac{\partial \ln Q_{2r}}{\partial \ln a_3} \cong 2r \frac{\partial \ln \lambda_{max}}{\partial \ln a_3} \tag{6.54}$$

When $\lambda_{max} = a_2$, $\bar{l}_r = 0$, that is, there is no reduplication. When $\lambda_{max} \cong a_3$, $\bar{l}_r = 2r$, that is, new chains are synthesized on the free chains, and reduplication occurs. Finally, when $\lambda_{max} = (a_2 + a_3)/2$, $\bar{l}_r = 2ra_3/(a_2 + a_3) = r$, since $a_2 = a_3$. This corresponds to duplication which had proceeded to the half-way point.

Now we know Q_{2r} and we may write out the complete partition function

$$Q = \sum_{r=0}^{N} e^{-r\bar{E}_2/kT} \lambda_{max}^{2r} \tag{6.55}$$

Summation is carried out over all states, from the nonseparated double helix ($r = 0$) to a completely separated double helix ($r = N$). Substitution of Q_{2r} by λ_{max} is valid for large values of r. It is possible to show, however, that a more rigorous solution which takes also the second root into account in Q_{2r} is

$$Q_{2r} = \lambda_1^{2r} + \lambda_2^{2r} \tag{6.56}$$

(the third root $\lambda_3 = 0$ does not change the result).

Expression (6.55) is a geometric progression. Carrying out the summation, we obtain

$$Q = \frac{b^{N+1} - 1}{b - 1} \tag{6.57}$$

where

$$b = \lambda_{max}^2 e^{-\bar{E}_2/kt} \tag{6.58}$$

We have obtained a simple analytical expression for the complete partition function. Using Q, we may calculate all the thermodynamic functions of the system. We find the mean number of separated links

$$\bar{r} = \frac{1}{Q} \sum_{r=0}^{N} r b^r \tag{6.59}$$

This is an arithmetic-geometric progression. Carrying out the summation, we obtain

$$\bar{r} = \frac{Nb^{N+2} - (N+1)b^{N+1} + b}{b^{N+2} - b^{N+1} - b + 1} \tag{6.60}$$

When $N \gg 1$, $\bar{r} \cong N$ when $n > 1$, and $\bar{r} \cong 0$ when $b < 1$.

The case of $b < 1$ corresponds to the conservation of the initial double helix; the case of $b > 1$ corresponds to complete chain separation. When $b > 1$ and $a_2 - a_3 > 0$, the state is limited to denaturation; when $b > 1$ and $a_3 - a_2 > 0$, reduplication takes place. The points of denaturation and reduplication are determined by the condition $b = 1$. These conditions coincide with the conditions (6.34) and (6.37), which are obtained from purely thermodynamic considerations.

The dependence of $\bar{l}_r$ on $\bar{x}_{NTP}/\bar{x}_{PP}$ for reduplication and the dependence of $\bar{r}$ on temperature for denaturation are shown in Figs. 123 and 124. Already when $N = 1,000$,

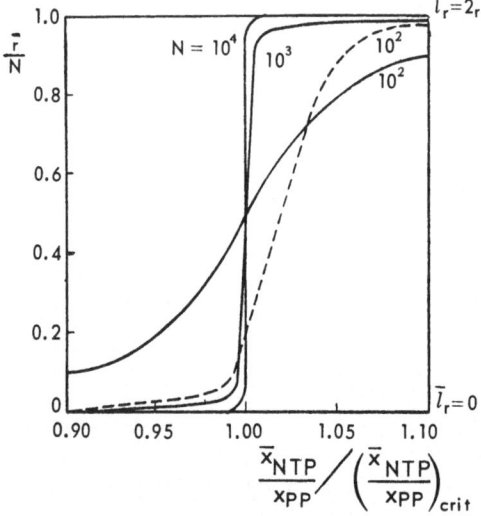

Fig. 123. Dependence of the degree of reduplication on the concentration of nucleoside triphosphates.

the transition which occurs is absolutely sharp according to this theory and it is similar to a phase transition. The denaturation theory which is contained within these calculations is evidently rougher than that of Zimm.

Let us examine now a second model which may be called antiparallel. The chain separation proceeds from one (or two) ends; both chains act as matrices to the same degree. In actuality, however, the two chains are not equivalent in each given segment of the double helix, since they are antiparallel. There are data which indicate that the synthesis of the new DNA chain occurs by the sequential reaction of NTP with 3'-nucleoside at the end of the growing chain.[93] Then reduplication occurs simultaneously at the lower end of one chain and the upper end of the other one. The two other ends remain temporarily free. The linking proceeds link by link without the NTP's being adsorbed randomly on the liberated part of the chain. Calculation may be carried out with the help of the model shown in Fig. 125.[94]

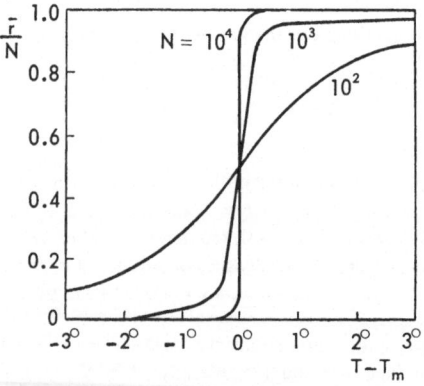

Fig. 124. Dependence of the degree of chain separation on temperature

Fig. 125. Antiparallel model of DNA duplication.

The partition function in this case has the form

$$Q = \sum_{r=0}^{N-1} \sum_{i=0}^{r} \sum_{p=0}^{N-1-r} \sum_{l=0}^{p} b_1^{r+p} b_2^{i+l} + b_1^N \sum_{r'=0}^{N} \sum_{r''=0}^{N} b_2^{r'+r''} \qquad (6.61)$$

where

$$b_1 = e^{-\bar{E}_2/kt + 2\ln z} \qquad b_2 = \frac{a_3}{a_2}$$

The first term of the summation corresponds to the state in which the DNA chains have not separated yet and are held together at least by single hydrogen bonds. For the random intermediate state, we have taken the state in which r bonds between the bases have been broken from one end and a new chain from i nucleotides has grown, while from the other end, p bonds have broken and a chain of l nucleotides has grown. The second term in the summation refers to a state in which all the hydrogen bonds are broken and synthesis proceeds on single DNA chains. As an intermediate state, we have taken the state in which there has grown, on one chain, a new one of r' and, on the other one of r'' nucleotides. After rather cumbersome summations, we have obtained

$$Q = \frac{1}{(1-b_2)^2}\left[\frac{(1-b_2)^2}{(1-b_1)^2(1-b_1b_2)^2} + \frac{b_1^N(b_2-1)}{(1-b_1)^2(1-b_1b_2)} - \frac{b_1^N b_2^{N+1}(b_2-1)}{(1-b_1)(1-b_1b_2)^2}\right.$$

$$\left. - \frac{Nb_1^N}{1-b_1} - \frac{b_1^N b_2(b_2^N-1)(1-2b_1b_2+b_2)}{(1-b_2)(1-b_1)(1-b_1b_2)} - \frac{Nb_1^N b_2^{N+2}}{1-b_1b_2} + b_1^N - 2b_1^N b_2^{N+1} + b_1^N b_2^{N+2}\right]$$

$$(6.62)$$

This complicated expression may be simplified. At moderate temperatures, when denaturation is thermodynamically unfavorable, $b_1 < 1$. Reduplication takes place only when $b_1 b_2^2 \geqslant 1$. Thus, Q interests us only at the values $b_1 < 1$ and $b_1 b_2 < 1$. In this, we may neglect the terms of the order of b_1^N and $b_1^N b_2^N$. Then

$$Q \cong \frac{1}{(1-b_1)^2(1-b_1b_2)^2} + \frac{(b_1b_2^2)^N b_2^2}{(1-b_2)^2} \qquad (6.63)$$

The average number of bonded nucleotides is

$$2\bar{r} = \frac{\partial \ln Q}{\partial \ln b_2} \cong \frac{\dfrac{2b_1 b_2}{(1 - b_1)^2 (1 - b_1 b_2)^3} + \dfrac{2N(b_1 b_2^2)^N b_2^2}{(1 - b_2)^2} - \dfrac{2(b_1 b_2^2)^N b_2^2}{(b_2 - 1)^3}}{\dfrac{1}{(1 - b_1)^2 (1 - b_1 b_2)^2} + \dfrac{(b_1 b_2^2)^N b_2^2}{(1 - b_2)^2}} \tag{6.64}$$

If $b_1 b_2^2 < 1$, then $2\bar{r} \cong b_1 b_2/(1 - b_1 b_2)$, and if b_1 is small, $\bar{r}$ is also very small. If $b_1 b_2^2 > 1$, then $2\bar{r} \cong 2N - 2/(b_2 - 1)$, that is, $\bar{r}$ is close to N. Consequently, the point at $b_1 b_2^2 = 1$ is the point of reduplication. In open form this condition coincides with the thermodynamic condition [Eq. (6.34)]. At the transition point, $\bar{r}$ is equal, not to $N/2$, as in the calculation on the basis of a cooperative model, but to $0.2N$. This is due to the role of intermediate states during an incomplete separation of the chains when the synthesis of the new chain is less favorable, since the breaking of a given hydrogen bond may lead to the growth of only a single chain. Just as before, the transition is very sharp at large values of N (the dotted line in Fig. 123).

The statistical thermodynamic theory does not contradict experimental results; however, at present, it cannot be regarded as confirmed. The fact is that we still have no data on the dependence of the DNA synthesis on nucleotide concentration, temperature, etc. The role of metallic ions and of the pH of the medium are not considered in this theory. The role of the enzyme is also unclear at present.

Aposhian and Kornberg have obtained very interesting results in their investigation of DNA synthesis.[95] They studied phenomena which occur during the infection of E. coli cells by the T2 bacteriophage. DNA is synthesized in the cell as a result of the reduplication of the phages; this is accompanied by the appearance of new enzymes which are absent from noninfected cells. In extracts from infected cells, the polymeric activity is 10 times stronger than in an extract from noninfected cells. The experiments of Aposhian and Kornberg have shown that infection of the cells with phage leads to a number of events. When phage DNA is introduced into the cell, there is an immediate inactivation of the chromosome, i.e., the inactivation of the bacterial DNA. As a result, the bacterial proteins are no longer synthesized and bacterial DNA is not reduplicated. What is the physical chemistry of this inactivation? At present this is not known. The phage DNA immediately takes over the genotypic and phenotypic functions. Four minutes after the infection, new enzymatic reactions can be observed; these ensure the reduplication of phage DNA. The rate of DNA synthesis, as has been said already, increases tenfold. Simultaneously, the other enzymatic mechanisms of the starting cell, which are responsible for production of energy and various biochemical processes, continue without changes and support the multiplication of the phage.

It seems that two enzymes are active in these cases; one catalyzes the process of the unwinding of the double helix, and the second, the polymerase, catalyzes the synthesis of new chains. It has been possible to isolate the polymerase of T2. The most amazing fact is that when it is used in the synthesis of DNA *in vivo* with any DNA as initiator, it performs 10 to 20 times more actively if the initiator is introduced in the denatured heated form, in the form of a single chain. An effective initiator in the work with this enzyme is also the single-chain DNA of the phage φX174. These results prove that, during synthesis of phage DNA, reduplication is preceded by chain separation. Even earlier, Bollum[96] had shown that preparations of an enzyme extracted from calf thymus, and not of phage polymerase, are more active during the reduplication of a heated rather than native DNA. Thermal denaturation of the initiator DNA, however, does not speed up the matrix synthesis of DNA if the polymerase used has been extracted from noninfected cells of *E. coli.*

The question whether DNA reduplication occurs after the separation of the double helix or simultaneously with it is solved by these experiments in favor of the first alternative, in the case of phage DNA. It is quite probable that both the first and second possibilities may exist depending on the enzyme. We may ask the question: How does synthesis occur in the first case if the matrix is a single-chain DNA rolled up into a coil? First, possibly it is not rolled up so tightly under these conditions if it has charges which repel each other. Secondly (and this is more reasonable), since according to all the data the synthesis of the new chain proceeds on the old one link by link, this process may induce the sequential unwinding of the coil, starting with one end and the formation of the double helix.

On the other hand, Sinsheimer has shown in direct experiments (with labeled atoms and density-gradient sedimentation) that even the multiplication of the single-chain phage φX174 DNA proceeds through the stage of a double helix.[97] When a phage particle enters the cell, the single DNA chain acts as a matrix for the synthesis of the complementary chain. Then reduplication takes place by the unwinding of the formed double helix and the growth of the new chains. Just before the formation of the phage particle, the two chains separate and only one of them is included in the finished particle. It is interesting that the reduplication of viruses as well proceeds through the intermediate double-chain structure.[98]

The statistical thermodynamics theory just described at first glance corresponds to the second possibility, i.e., to the simultaneous synthesis and unwinding of the double helix. In fact, the theory does not contradict the single-chain mechanism; it does not consider the sequence of the processes as a function of time. In this sense, it is not important whether the states of the system which enter into the partition function are simultaneous or separated in time. One should emphasize that the model examined in the second variant of the calculation (Fig. 125) indicates the existence of a large fraction of DNA (up to 33%) in denatured form. In a recent paper, Rolfe[99] has shown by direct experiments that, in the reduplication of the DNA of a number of bacteria (during the exponential growth of the culture), the DNA is present in the state of an incompletely denatured form, which is in agreement with both the models examined.

Recently, reliable data have been obtained supporting the simultaneous unwinding and reduplication, in complete agreement with the scheme used as the basis of the theory. The replication of the chromosome of *B. subtilis* proceeds according to the mechanism of unwinding from a single end.[100] The autoradiographic investigations of the replication of the cyclic DNA in *E. coli* demonstrate directly the simultaneity of the separation of the starting double helix and the formation of the new ones. The ends of the bacterial DNA are linked and the DNA remains as a ring during the replication. The same applies as well to phages, whose DNA also forms loops.[101]

Let us examine now the kinetics of DNA reduplication; let us examine what controls the rate of synthesis. The expounded thermodynamic and statistical theory does not give and cannot give an answer to this question, since it examines only a change of the equilibrium state, independent of time. The sharpness of the reduplication transition, in the sense of its dependence on the monomer concentration and temperature, is the factor which indicates that a transition occurs according to the all-or-none-principle, and it has no relation to the question of the rate of the transition. This may be explained by the following example: The crystallization of a liquid also proceeds according to the all-or-none-principle, at a well-determined temperature, but it can also proceed quite slowly, since the rate of the process is a function of the rate of the formation of crystal nuclei and of their subsequent growth.

Kinetics of the synthesis of DNA in the cell are of particular biological interest. The thermodynamic conditions of the synthesis are probably always realized within the cell, but synthesis occurs only at that stage of development in which definite regulating kinetic conditions are fulfilled. Therefore, the kinetics of the synthesis are more interesting than their thermodynamics. We are talking here about the general kinetic theory of the synthesis of a polymer on a matrix, a theory which, in the final account, must be applicable to the reduplicating synthesis of DNA and to the synthesis of matrix RNA on matrix DNA, as well as to the synthesis of protein on RNA.

Single chains of DNA act as matrices during their reduplication synthesis. Consequently (and this is in agreement with experiment), the separation of the matrix chains must precede the formation of the progeny double helix. Separation of the double helix during its simultaneous reduplication occurs under more favorable conditions than in simple thermal denaturation. The liberated chains on which new ones have grown can no longer return to the starting state, i.e., the process is clearly irreversible. The free energy liberated during reduplication is more than sufficient to give a rapid unwinding of the helices.

Levinthal and Crane[102] have carried out calculations using the model shown in Fig. 126. In this model the double helices rotate similarly to an automobile speedometer cable. When the rotation is simultaneous in the

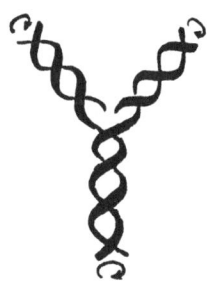

Fig. 126. Levinthal and Crane model.

directions shown by the arrows, the braking caused by the viscous medium is very low. The energy release and the corresponding rotational strain are localized in the Y fork. The rate of unwinding turns out to be very high, greater than during denaturation separation. However, since it is not clear yet whether separations precede reduplication or the two processes proceed simultaneously, it is difficult to speak about the correctness of the Levinthal and Crane model. Their model is not capable of accounting for reduplication according to the mechanism of Fig. 125.

An examination of the reduplication of a cyclic double helix (Fig. 120) requires a solution of special problems in topology and in the mechanics of such a system. This has not been done up to now.

The unwinding of the double helix during the simultaneous reduplication on two (four) or one (two) liberated chain ends may proceed also by a purely diffusional mechanism. In fact, such an unwinding is irreversible and, instead of Kuhn's equation (6.15), it is necessary to write

$$\tau = \tau_1 + \tau_2 + \cdots + \tau_{v/2} \qquad (6.65)$$

where τ is expressed by (6.14) as before. Assuming even Kuhn's value of D (6.13), we obtain

$$\tau = \sum \tau_j \cong \frac{3\pi^3}{2} \frac{\eta_0 q^3}{kT} \int_0^{v/2} j^3 \, dj = \frac{3\pi^3}{128kT} \eta_0 q^3 v^4 \qquad \text{seconds} \quad (6.66)$$

i.e., when $v \sim 1,000$, $\tau \sim 200$ minutes, which is much less than Kuhn's values. In fact, one should not use the values of D (6.13) for an absolutely rigid rod, since long double helices of DNA form porous coils. One should then lower the exponent of v in (6.66) from 4 to about 2.5; this results in a decrease of the time of unwinding to fractions of minutes.[103]

The problem of unwinding cannot be regarded as solved. To take irreversibility into account during chain separation, it is necessary to consider that the unwinding helix is an open system relative to the surrounding medium. During reduplication, the condition $\Delta F = 0$ must be fulfilled for the system as a whole, i.e., the helix plus the solution. The irreversibility of the process may be explained by the fact that the separation of a double helix and the enzymatic synthesis of new chains proceed faster than the reversible process related to the return of the pyrophosphate molecules, which had been liberated into the solution. As a whole, the problem is quite complicated and interesting.

In one way or another, one may assume that even a long helix can unwind over a time of the order of tens of seconds. On the other hand, the magnitudes of the times of reduplication are quite a bit larger, of the order of 10^3 seconds. This value follows both from the experiments by Kornberg on the synthesis of DNA *in vitro* and from the rate of mitosis. It follows from this that the rate of synthesis is limited, not by the rate of unwinding of the double helix, but by the rate of enzymatic polycondensation.

Let us examine the deposition of the nucleotides upon the matrix (a single DNA chain) with the formation of hydrogen bonds, assuming that the irreversible polycondensation proceeds from one end.[104] The processes of nucleotide deposition on the matrix and of its liberation consist of at least two acts: diffusion, i.e., the approach from solution or the departure into solution of the nucleotides, and the binding or removal of the nucleotide, which requires overcoming an activation barrier (it is necessary to break the nucleotide–water hydrogen bonds during the attachment of nucleotides and the nucleotide–nucleotide hydrogen bonds during the breaking away of a nucleotide). The nuclei of the matrix may be in the following states: (a) occupied by polymerized nucleotides (their fraction is x_1); (b) occupied by deposited (i.e., bonded by hydrogen bonds to the matrix) but not polymerized nucleotides (their fraction is x_2); (c) free (their fraction is x_3); and (d) in complexes with nucleotides not attached to them, but located in the immediate vicinity (their fraction is x_4). If the number of nuclei in the matrix is equal to N, then the kinetic energy equations have the form

$$\frac{dx_1}{dt} = \frac{\kappa}{N} \frac{x_2}{1 - x_1}$$

$$\frac{dx_2}{dt} = kx_4 - Kx_2 - \frac{\kappa}{N} \frac{x_2}{1 - x_1} \qquad (6.67)$$

$$\frac{dx_3}{dt} = -k_{\text{dif}} n x_3 + K_{\text{dif}} x_4$$

$$x_1 + x_2 + x_3 + x_4 = 1$$

Here κ, k, and K are rate constants for, respectively, the polycondensation, breaking off, and attachment of nucleotides; k_{dif} and K_{dif} are rate constants for the diffusion of nucleotides within the domain of reaction and outside of it; n is the number of nucleotides per cubic centimeter of solution. In the equations, we have taken into account the fact that the deposition and departure of nucleotides proceed at any point in the matrix which is not occupied by a new chain and that irreversible polycondensation proceeds from one end. An estimate of the coefficients gives $k_{\text{dif}} = \delta^2 u = 10^{-12} \sec^{-1} cm^3$ ($\delta \approx 10^{-7}$ cm, the dimension of the nucleotide, $u \approx 10^2$ cm sec^{-1} is the mean velocity of its diffusion);

$$K_{\text{dif}} = \frac{u}{\delta} = 10^9 \sec^{-1}$$

$$K = v_0 \exp\left(-\frac{E_{HH}}{RT}\right) = 10^2 \text{ sec}^{-1} \qquad v_0 \approx 10^{13} \text{ sec}^{-1}$$

where $E_{HH} = 15$ kcal/mole is the mean activation energy during the breaking of hydrogen bonds in a single pair of nucleotides; $k/K = \exp(\Delta E/RT) \cong 10^2$ ($\Delta E \sim 3$ kcal/mole is the average free-energy difference of the nucleotide–nucleotide and nucleotide–water hydrogen bonds), that is, $k = 10^4 \text{ sec}^{-1}$ and $n = 10^{17} \text{ cm}^{-3}$ (the concentration in the Kornberg experiments).

If enzymatic polycondensation proceeds much faster than the departure of the nucleotides from the matrix, that is, $\kappa/N \gg K$, then it follows from the first two equations of (6.67) that

$$\frac{d(x_1 + x_2)}{dt} = kx_4 \tag{6.68}$$

Taking into account that diffusion processes are faster than the processes on the matrix ($kn \approx 10^5 \text{ sec}^{-1}$, $K_{dif} \approx 10^9 \text{ sec}^{-1}$, while $k = 10^4 \text{ sec}^{-1}$), we obtain

$$\frac{x_3}{x_4} = \frac{K_{dif}}{k_{dif}n} \tag{6.69}$$

from where

$$x_1 + x_2 = 1 - e^{-t/\tau} \tag{6.70}$$

where the time of deposition with subsequent condensation is

$$\tau = \frac{1 + \dfrac{K_{dif}}{k_{dif}n}}{k} \cong 1 \text{ sec} \tag{6.71}$$

This is much too short a time.

Let us examine another case, namely, where the rate is limited by an enzymatic polycondensation, that is, $\kappa/N \ll K$. Then, under stationary conditions

$$\frac{x_2}{x_4} = \frac{k}{K} \qquad \text{and} \qquad \frac{x_3}{x_4} = \frac{K_{dif}}{k_{dif}n} \tag{6.72}$$

and

$$\frac{dx_1}{dt} = \frac{\kappa}{N}\frac{1}{1+\gamma} \tag{6.73}$$

where

$$\gamma = \frac{KK_{dif}}{kk_{dif}n} + \frac{K}{k} \cong 10^2 \tag{6.74}$$

Consequently,

$$\frac{dx_1}{dt} \cong \frac{\kappa}{N\gamma} \cong 10^{-2}\frac{\kappa}{N} \tag{6.75}$$

From an experimental estimate of $dx_1/dt \sim 10^3 \text{ sec}^{-1}$, it follows that $\kappa/N \sim 10^{-1}$, that is, $\kappa \sim 10^{-1}N$.

Thus, this model makes it possible to estimate, from the total rate of DNA synthesis, the rate constant of enzymatic polycondensation. Comparison with experiment shows that the time of the total synthesis, 10^3 seconds, cannot be determined only by the rate at which nucleotides fill the matrix, but that it is a function of the enzymatic process. From (6.73), it follows also that, in this case, the number of polymerized nucleotides increases proportionally with time

$$x_1 = \frac{t}{\tau_N} \tag{6.76}$$

where τ_N is the total time of chain growth

$$\tau_N = \frac{N}{\kappa}(1 + \gamma) \cong \frac{N\gamma}{\kappa} \tag{6.77}$$

Orlov and Fishman have examined the same process without taking into consideration the activation necessary for breaking the bonds and they have obviously obtained a much higher rate of synthesis.[105] In fact, this calculation ignores the participation of the enzyme in the reaction.

Another cooperative model has also been examined.[104] It is assumed that the polycondensation proceeds at any point on the matrix where two or more neighboring nuclei are occupied by deposited nucleotides. Using the general method of the kinetics of cooperative processes,[106] we obtain

$$\frac{dx_1}{dt} = 2\kappa \frac{(1 - x_1)(1 + \gamma x_1)}{(1 + \gamma)^2} \tag{6.78}$$

that is,

$$x_1 = \frac{1 - e^{-t/\tau}}{1 + \gamma e^{-t/\tau}} \tag{6.79}$$

where

$$\tau = \frac{1 + \gamma}{2\kappa} \cong \frac{\gamma}{2\kappa} \tag{6.80}$$

In this case, the number of polymerized nucleotides is a nonlinear function of time, with a considerable induction period. The time τ is $2N$ times less than τ_N (6.77). Experiment shows that DNA synthesis proceeds with a velocity which corresponds to τ_N, and not τ, and without an induction period. Consequently, the mechanism of polycondensation from one end is valid, and not that of overall polycondensation.

As we have stated already, the synthesis of DNA and of polynucleotides (poly AT) without initiator proceeds only after a long latent period, 3 to 5 hours. This is shown in Fig. 127.[21] During the latent period, there is a gradual polycondensation of the monomers which form a matrix; this, then, serves as a basis for rapid synthesis. Calculation of the rate of nonmatrix synthesis with the same constants gives a time for the synthesis of the matrix itself

$$t_N = \frac{N}{\kappa}10^4 \tag{6.81}$$

This is two orders of magnitude greater than the time of synthesis on a matrix [Eq. (6.77)],

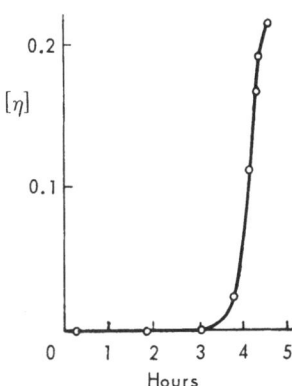

Fig. 127. Change of viscosity of the incubation medium during the synthesis of the AT copolymer.

which is in agreement with experiment. When the enzyme concentration is increased, the rate constant of enzymatic polycondensation κ increases and the length of the latent period t_N must decrease. This is also observed in experiments.[107]

A theory of the kinetics of synthesis of polynucleotides on short oligomeric matrices has been proposed.[108] As has been shown, the lag period in this case is a strong function of the length of the oligomer (AT), decreasing with an increase in n. Such a rate of synthesis is easy to explain if we assume that three processes occur: the filling of the oligomeric matrix, the gliding of the growing chain along the matrix, and a termination of the reaction as a result of the separation of the growing chain from the matrix. The experimental data may be interpreted quantitatively only if we take into account the third process. From a comparison of theory with experiment it follows that very short oligomers can be easily separated from the growing chain; but there must be a critical size of the oligomers, starting with which the probability of separation becomes much smaller. Oligomers with dimensions equal to or greater than the critical dimension form double helices with growing chains.

The statistical and kinetic calculations presented in this section can serve mostly as illustrative material. Still too little is known about the conditions of synthesis of DNA to start developing a real physico-chemical theory. One can hope, however, that these results of the theoretical analysis will be useful for the further development of DNA physics.

Chapter 7

PROTEIN SYNTHESIS

PROBLEMS OF THE GENETIC CODE

The facts presented in the preceding chapter demonstrate that the genetic information is inscribed in the DNA macromolecules and in viral RNA. This is information on the synthesis of protein molecules, i.e., on their primary structure. It is directly evident that the genetic text of the DNA is written linearly by a four-letter alphabet. In other words, information on the sequence of the 20 amino acids in a protein is coded by a sequence of four nucleotides in the DNA. As soon as the composition and genetic role of DNA had been determined, it became possible to formulate very clearly the physical mathematical problems, the solution of which is required for the determination of the nature of protein biosynthesis. The statement of these problems has been the achievement of physicists, and principally of Gamow.[1]

Schrödinger[2] had defined quite clearly the molecular role of the genetic code, but he had no means of proceeding further, since the structure of DNA and its role had not been established at the time when he wrote his book.

What is the genetic code, i.e., what is the correspondence of the four-letter text to the twenty-letter one? This is the basic problem. From the moment of its formulation only 7 years have passed to its almost complete solution. This scientific success in one of the cardinal problems of science may be explained by the joint efforts of a number of investigators—biologists, physicists, and chemists—who tackled with enthusiasm the development of the new field of science.

Let us assume that the code has been established, i.e., it is known which nucleotides and how many nucleotides correspond to the inclusion of a given amino acid within the protein chain. We have then questions relating to the mechanism of transfer of the genetic information, questions of a physical-chemical character. It is known that the matrix for protein synthesis is not DNA, but messenger RNA which is synthesized on DNA. This does not change the problem of the code, since RNA contains the information written in the same four-letter alphabet, and it is not important that one letter has been replaced by another, i.e., RNA contains U where DNA contains T. It is essential, however, to establish the manner in which the synthesis of M-RNA takes place on DNA, in which way the M-RNA is transferred to the ribosomes and forms in them the matrix, and finally, how the protein chain is assembled on this matrix out of the amino acids. These are problems which cannot be solved theoretically. In order to solve them, it is necessary to look inside the cell and find out how it works; it is necessary to carry out experimental investigations.

The problem of the code itself at first glance has a somewhat different character. It is not possible to exclude *a priori* that the code could be found just by examination, that analysis would lead to the deciphering of the manner in which the information included in DNA is transferred to the protein. It is not excluded that considerations of combinatorial calculations may turn out to be the key processes. This is exactly the way Gamow approached this problem.[3,4]

The number of nucleotides is four; the number of amino acids is twenty. It is immediately evident that each amino acid corresponds to a combination of several nucleotides. How many? Two are insufficient, since the number of combinations of two out of four is equal to $4^2 = 16$. Thus, the code relation, i.e., the number of nucleotides which codes a single amino acid, must be at least three. But if each combination of three nucleotides corresponds to one amino acid, then we obtain a number which is too great, $4^3 = 64$.

Notwithstanding the fact that the problem of the code is related to the theory of information, informational analysis may only determine the conditions which must be satisfied by the code. Such an analysis may not yield the code itself. It permits us to answer the question on the minimal number of nucleotides necessary for coding a single amino acid. Information expressed in a binary system per amino acid is

$$i_{aa} = \log_2 20 = 4.322 \text{ bits}$$

and for a single nucleotide

$$i_n = \log_2 4 = 2.000 \text{ bits}$$

Consequently (in view of the additivity of the information), the smallest number of nucleotides per single amino acid is

$$n = \frac{i_{aa}}{i_n} = 2.161$$

that is, $n = 3$, since the number of nucleotides must be a whole number. We have obtained again the same quite trivial result.

Gamow found the necessary number 20 by two methods of combining nucleotides. Assuming that the genetic information is included not in a single chain but in the Watson–Crick double helix, Gamow proposed a rhombic code. Three pairs in sequence within the double helix form combinations which contain one base of the first chain, the second pair as a whole, and one base from the third pair, for example,

$$A \overset{C}{\underset{A}{-}} T$$

The total number of such diamonds is much smaller than the number of random combinations of four out of four; it is equal to $4^4 = 256$. The structure of the DNA itself superimposes limiting conditions; the small diagonal of the diamond always links A with T or G with C. The number 20 is obtained if we consider identical, from the point of view of coding, the right and left forms of asymmetric diamonds, for example,

$$A \overset{A}{\underset{C}{-}} T \equiv T \overset{A}{\underset{C}{-}} A$$

The 20 diamonds are shown in Fig. 128. Of these, 8 have twofold symmetry, while 12 are devoid of it. It is possible to find stereochemical arguments in favor of such an identity.[4] As long as each diamond contains bases from three consecutive pairs, the code is overlapping, i.e., two bases lying on one end of the diamond are common for two neighboring diamonds. This places quite rigid limits on the possible sequences of the 20 letters, i.e., of the amino acids within the protein. The possible combinations of the nearest neighbors in the rhombic code are shown in Table 22 (see also Fig. 128).

Gamow attempted to verify the overlapping rhombic code, by comparing the possible relations of the diamonds with known primary structures of insulin and adrenocorticotropic hormone.[4] This led to insoluble contradictions; the amino acid sequences could not be deciphered by the rhombic code. The concept of a double-chain matrix also is not in agreement with experiment. The point is that the matrix is not DNA but M-RNA, which does not have a regular double-helical structure.

Various types of triangular codes have been proposed as single-stranded codes. It is assumed that the amino acid is coded by a triplet of bases located at the apices of a triangle. The number 20 is obtained if it is assumed that all the triangles are identical, from the point of view of the

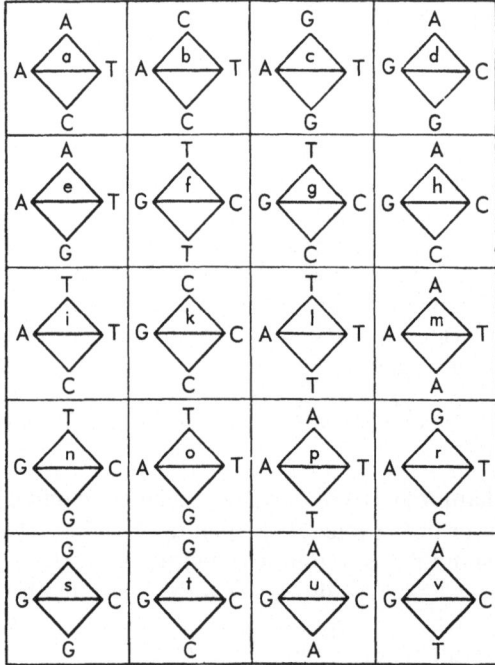

Fig. 128. Twenty rhombic combinations of nucleotides.

code, if they can be obtained from each other by rotations and reflections, such as

$$\begin{array}{cccccc} & G & & A & & T & & G & & A & & T \\ A & T, & T & T, & G & A, & T & A, & G & G, & A & G \end{array}$$

and

$$\begin{array}{ccc} & G & & A & & A \\ A & A, & A & G, & G & A \end{array}$$

A triangular code is single stranded and it can be densely overlapping, it can be loosely overlapping, and it can be not overlapping at all. In the first

Table 22 Combinations of Nearest Neighbors in a Rhombic Code

a, e, i, o	Combine with	$a, d, e, f, g, h, i, l, m, n, o, p, u, v$
d, g, h, n	Combine with	$a, b, c, d, e, g, h, i, k, n, o, r, s, t$
l, m, p	Combine with	a, e, i, l, m, o, p
k, s, t	Combine with	g, h, k, n, s, t
b, c, r	Combine with	d, f, g, h, n, u, v
f, u, v	Combine with	a, b, c, e, i, o, r

case, the sides of the triangle are common for two neighboring triangles; in the second, two of the three apices are common for two neighboring triangles; while in the third, each amino acid corresponds to its own triangle. These three possibilities are shown schematically in Fig. 129. In the first two cases (a) and (b), there are limitations on the amino acid sequence; in the third case (c), (d), there are no such limitations; the question arises, however: What plays the role of the comma, which factor indicates that the nucleotide text must be read as c and not d in Fig. 129?

A number of factors indicate that an overlapping code is absolutely impossible. The question of the comma is discussed in the interesting paper by Crick, Griffith, and Orgel.[5] In the case of the nonoverlapping code, a triangle loses all significance; the only point of significance is the sequence of a triplet with a given arrangement of bases, such as ATT. By assuming that there are no commas, it is possible to obtain the number 20 out of the 64 possible triplets by applying the following considerations. Some of the 64 triplets make sense; others are devoid of it. If the code is singular, i.e., if it may be read only in one way and there are no commas, then sequences of the type of AAA, TTT, GGG, and CCC make no sense since we obtain the same result whether we read the sequence AAAAAA starting from the first or the second letter. Therefore 4 triplets may be rejected. The remaining 60 triplets may be grouped into 20 sets of 3, each set of 3 containing only cyclic transpositions, such as ATC, TCA, and CAT. Only one of these three sequences must be selected, since if the amino acid h corresponds to TCA, a sequence hh corresponds to TCATCA. Since there are no commas, CAT and ATC must be rejected. Since only one triplet may be valid out of a set of cycles, we obtain 20 triplets.

The presence of amino acid sequences gh and hg does not decrease this number, since it is possible to write a sequence which obeys all the stated rules and codes 20 different amino acids. One of the possible solutions has the form

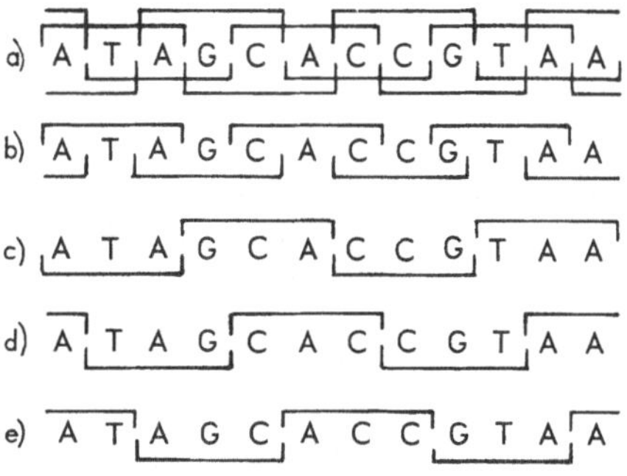

Fig. 129. Triplet codes: (a) completely overlapping; (b) partly overlapping; (c), (d), (e), non-overlapping.

where AT_T^A means ATA and ATT, etc. Any pair of triplets of such a series may be placed next to each other without setting up nonsense triplets. The given solution is one of 288 possible ones.

A code without commas may be valid if, for example, the amino acid becomes attached more strongly to a sense triplet than to a nonsense one, or if the attachment of amino acids goes triplet by triplet from one end of the matrix to the other.

It is evident that the absence of overlaps in the code precludes the possibility of its deciphering on the basis of known amino acid sequences. Several attempts were made to overcome these difficulties. It is possible, for example, to consider that the code is not completely unique, by assuming that only the one middle nucleotide of the triplet is rigidly fixed in the code, while the first and the third nucleotides are less important and may be replaced without changing the coded amino acid. This senior–junior code has been developed by Gamow et al.[4] and Feinman; it has been found to be contrary to experiment. Teller has proposed a code based on the principle that an amino acid is coded by two nucleotides and the preceding amino acid. In the Teller code, the amino acid sequence has certain limitations, and thus it is similar to an overlapping code. Not speaking even of the absence of physical-chemical bases for such an idea, the Teller code once again does not correspond to experiment.

These attempts to deduce the genetic code by reasoning, as well as some others, did not meet with success. This is not surprising; after all, we do not know yet a single primary structure of DNA or RNA, and these clever hypotheses are attempts to read a nucleotide text on the basis of a few known protein texts. Gamow and his followers have approached the problem in the same manner as scientists who decipher ancient texts written in unknown languages. In archeology, there are a number of such brilliant achievements, starting from the deciphering of the Egyptian hieroglyphics by Champollion, to most recently the deciphering of Mayan texts, carried out in Siberia with the help of electronic machines. The most important event which made possible the deciphering of Egyptian hieroglyphics was the discovery of the Rosetta Stone on which the same text had been written in hieroglyphics, in a simplified Egyptian alphabet, and in Greek. Champollion, who knew the ancient Greek and Coptic (akin to ancient Egyptian) languages, read these texts and thus laid the foundation of Egyptology. In our case, there is no Rosetta Stone, we have no knowledge of any kind of the nucleotide language, and, as a result, it is necessary to use other methods to solve this problem.

Should we conclude from this that the combinatorial attempts at deciphering the genetic code were completely useless? Such a conclusion would be completely erroneous. This analytical study not only made possible the formulation of the problem quite clearly, but it also established certain facts. In this sense, the studies of Gamow and other scientists have played as important a role as the studies of the predecessors of Champollion in the realm of Egyptology. First of all, these studies have established that the code must be at least triplet. Second, the problem of the comma has been stated in the nonoverlapping code and possible methods for its solution have been proposed. Third, it has been established that the genetic text is written out in the chain of a nucleic acid and not in a double helix; comparison of overlapping codes with experiment has forced them to be abandoned. This is what makes impossible a purely theoretical solution of the code. What is the experimental refutation of overlaps?

As we have seen, mutations may consist in a replacement of just a single amino acid by another. Wittemann[6] and Tsugita and Fraenkel-Conrat[7] have shown that chemical mutants of TMV caused by treatment with HNO_2 are the results of the replacement of only a single amino acid in the virus protein. It is also known that HNO_2 acts on individual nucleotides of RNA. However, if the codes are overlapping, then the replacement of a single nucleotide of a triplet by another must lead to the replacement of several amino acids in the protein: three in the case of Fig. 129a; two in the replacement of the end nucleotides of the triplet in Fig. 129b. It is true that, if in the last case, only the central nucleotide of the triplet is replaced, only one amino acid will be changed.

The senior–junior code makes it possible to replace one amino acid as a result of the replacement of one nucleotide. For any overlapping codes, however, a certain limitation is placed on the sequence of amino acids in proteins. Data on the primary structure of the investigated proteins and a number of polypeptide fragments do not testify to the existence of such limitations, i.e., to the existence of correlations between neighboring amino acids. This question has been examined by Gamow, Rich, and Ycas,[4] and also by Ycas.[8]

Let us construct a grid in which we plot along the vertical coordinate the first amino acids and along the horizontal one the second amino acids following the first ones. The grid is constructed in such a way that the N terminal bases are given by lines and the C terminal bases by columns. In the grid there are $20 \times 20 = 400$ cells. In each cell, we cite the number which shows how many times we meet a corresponding pair along the known dipeptide sequences (Fig. 130). It has been examined[4] whether this distribution of the numbers corresponds to a Poisson distribution, i.e. to a completely random distribution. This correspondence was found. However, as has been pointed out,[8] it is not possible to use this method, since it does not take into account the fact that different amino acids are found with completely different frequencies.

A more rigorous analysis has shown, however, the same thing, namely, the absence of any correlation among the amino acids.[8] In this sense, the protein text differs radically from the text in any language. Thus, in the Russian language, it is more probable to have a vowel following a consonant than to have a sequence of two vowels or two consonants. The

1st amino acid \ 2nd amino acid	Ala	Arg	Asp	Asp-NH₂	Cys	Glu	Glu-NH₂	Gly	His	Ileu	Leu	Lys	Met	Phen	Pro	Ser	Thr	Try	Tyr	Val
Ala	4			1			1	3		1	3	3	2	1		3	1		1	1
Arg	1	3		2	2			4		1	2			1			1			1
Asp	1	2	4			2				1		1		1					1	
Asp-NH₂	2	2			4		3	2		2	3			1	1				2	1
Cys	4	2		3	1		1	2	1	1		1		1	1				1	
Glu	4	1	1	2	1	1		1	1	1			1			1				
Glu-NH₂	4	1	1	1	1	1			2	1	3	1			2		1			
Gly	3	1		1	1	2	1			1		2	1	1	2	2	1		1	1
His									1	3			1	1	1	1	1			
Ileu	1					2				1		1	1	2						1
Leu	1		1	1	2	2		3			1			1	2		1		1	1
Lys	1	1	2		2	1		2		1	1	1		3	1	1			1	1
Met			2		1												1			
Phen		1		1		4	1	1			1	1			1			1	1	2
Pro	2	1			1	1		1		1	2		1	1					1	3
Ser	1	4	1	3		1	1	2	1	1	2	1	1	1	2				2	4
Thr	2			2			1	1			1		1	3	4					1
Try						1			1											
Tyr		1		1		2	1		1	2	1		1	1	2	1				1
Val	1		1	2	3	3	4	1			1		1		1	1		2		

Fig. 130. Pairs of neighboring amino acids within proteins.

probability of the sequence of two consecutive identical vowels is very small, and that of three is practically equal to zero. On the other hand, in the Estonian language, two identical vowels frequently follow each other. It is as a result of such a correlation that a normal text may be regarded as an example of a Markoff chain. In this sense, i.e., in the sense of the primary structure, a protein chain is not a Markoff chain; the correlation between amino acids in proteins is related only to their conformation and it determines the secondary, and not the primary, structure.

One should emphasize that the number of studied dipeptide sequences is limited so far. This probably explains the presence of empty cells in the grid of Fig. 130. If such cells remain even after the examination of a considerably greater amount of material, this will indicate some correlation between neighboring amino acids.

Augenstine[9] has carried out an informational analysis of amino acid sequences and has come to a conclusion different from those previously cited.[4,8] According to his evaluation, there is some correlation between amino acids which decreases the information contained in the primary structure of the protein to 0.85 to 0.95 of the maximal possible value. But even such a degree of correlation cannot be reconciled with an overlapping nucleotide code. Brenner has shown in a general form that an overlapping code is impossible.[10]

In this manner, it is possible to formulate two general conclusions. First, the code must be nonoverlapping. Second, informational analysis of amino acid sequences within proteins must be performed by starting with a nucleotide code that has been established experimentally.

As a conclusion, let us examine just one more study by Gamow and Ycas.[11] They assumed that an amino acid is determined by three nucleotides, without taking their positions into account. There are 20 such triplets, and they are not overlapping. The number of triplets of the type of AGC which determines the same amino acid is equal to 6, the number of transpositions, the number of triplets AAG is equal to 3, and that of triplets AAA is equal to 1. Having determined the nucleotide composition of a DNA or RNA (in a virus), we can find the frequencies of occurrence of such triplets and represent them as a graph on the axis of which we plot the 20 triplets. It is found that a similar graph of the occurrence of amino acids in the corresponding protein has a similar shape.[4,8,11]

These and other data force us to think that the nucleotide code is triplet and nonoverlapping. This evidently permits the degeneracy of the code, i.e., the correspondence of several triplets (for example, the three triplets, AAG, AGA, and GAA) to a single amino acid and also the correspondence of several amino acids to a given triplet.

MATRIX SYNTHESIS OF PROTEINS

In order to advance further the solution of the problem of the genetic code, it is necessary to study concretely the physical-chemical nature of protein synthesis. In essence, this is the principal task of molecular biology, while the problem of the code is only part of the total problem.

As we know, protein synthesis is coded within the DNA. But in what sense? We have said already that the DNA is localized within the nucleus of the cell, within chromosomes, while the protein is synthesized in the ribosomes and the cytoplasm. This means that the DNA is not the direct matrix for the synthesis of the protein, i.e., the genetic information must be transferred in some way from the DNA to the ribosome.

The ribosomal particles of *E. coli* can be classified according to their sedimentation coefficients. The dimensions of the particles are a function of the concentration of divalent ions. At a Mg^{++} concentration of 10^{-4} M, particles sedimenting with $50S$ and $30S$ predominate (the molecular weights of these particles are 1.85×10^6 and 0.95×10^6, respectively). At $[Mg^{++}] > 10^{-3} M$, these particles coalesce and form larger ones with a constant of $70S$. When $[Mg^{++}] > 10^{-2} M$, the $70S$ particles dimerize and form particles with $100S$. There is practically no correlation between the RNA composition extracted from ribosomes and that of the DNA;[12] therefore, there is no reason to consider that the ribosomal RNA is the matrix of protein synthesis. At the same time, it was natural to seek the matrix substance among the RNA molecules, since there is a great deal of similarity between RNA and DNA; DNA is not contained in cytoplasm, but it is the genetic material.

Volkin and Astrakhan[13] have studied the events which take place within an *E. coli* cell after infection with T2 phage. This phage contains DNA and not RNA; during its multiplication, however, i.e., during the duplication of DNA and synthesis of phage protein, a certain amount of RNA is also synthesized. This RNA has an increased metabolic activity relative to that of ribosomal RNA; ^{32}P is rapidly incorporated. The composition of the newly synthesized RNA is similar to the composition of the phage DNA; this is determined by the ratio of the specific radioactivities; thus, in DNA we have 18C:32C:32T:18G; while in RNA we have 18C:30A:30U:22G.[14] It turns out that, during the process of synthesis of the phage particles, there is some RNA which plays a role intermediate between the basic phage DNA and the protein.

Gros and Hiatt established the existence of such a special RNA in uninfected cells of *E. coli* and studied its properties in detail.[15] It was found that this RNA (which we shall call *M-RNA*) has a sedimentation coefficient of 12S. It is characterized by a high metabolic activity; this is demonstrated by the following interesting experiments.

The bacteria were labeled with a short-lived isotope; they were grown in a medium which contained ^{14}C uracil. The extract of such bacteria was ultracentrifuged in a density gradient and the distribution of optical density at 260 mμ was measured as well as the radioactivity of the fractions. The results are shown in Fig. 131. The radioactivity is concentrated in the 12S fractions; the other fractions are not radioactive. At high Mg^{++} (10^{-2} M), M-RNA is bound to the 70S ribosomes. According to Watson's data, Mg^{++} ions form bridges between ribosomes, M-RNA, and T-RNA by neutralizing the phosphate groups. The kinetics of inclusion of radioactive amino acids show that it is the stable form of the 70S ribosome

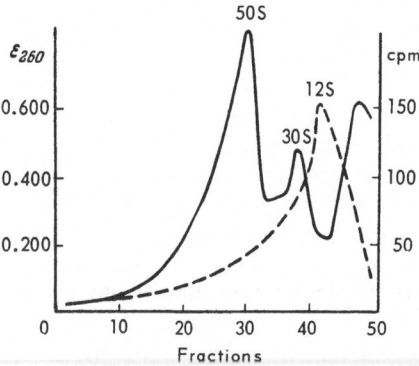

Fig. 131. The distribution of radioactivity and the optical density in a bacterial extract tagged with ^{14}C, after sedimentation in a density gradient. The radioactivity is shown by the dotted line.

which is responsible for protein synthesis. Such ribosomes are called active. Investigations using the same method of labeled atoms show that some M-RNA combines reversibly with 70S ribosomes. The contents of M-RNA in bacteria after 1 hour of growth are approximately 3% of the total RNA; the molecular weight of M-RNA is 2×10^6.

These and similar results, plus indirect evidence, obtained from studies of induced enzyme synthesis, demonstrate the existence of a special M-RNA, which carries the genetic information from DNA to the ribosomes, i.e., from the construction office of the cell factory to the production lines which build the finished products, namely, proteins. Many different names have been proposed for this new RNA; it is called informational RNA, intermediate RNA, and messenger or matrix M-RNA. We shall call it *matrix M-RNA*.

M-RNA is synthesized in nuclei and its composition is similar to that of the DNA. This immediately brings up questions on the structural interactions between DNA and RNA. This problem has been examined by a number of authors, among them Rich.[16] Rich had already spoken of this before the discovery of M-RNA, i.e., in 1960. Now we know that, during the simultaneous renaturation of M-RNA and the corresponding single-chain DNA obtained by heating native DNA, hybrid complexes of M-RNA with DNA are formed. These double helices are composed of one chain of M-RNA and one chain of DNA.[17] This proves that not only the composition of M-RNA is similar to that of DNA, but also the base sequences of the two chains are complementary. The U in RNA combines with A in DNA, which follows from direct experiments with synthetic polyribonucleotides.

DNA–RNA complexes have been isolated directly also from cells.[18,19] Finally, enzymes have been found with the help of which RNA can be synthesized on DNA as on an initiator matrix. The RNA formed in this manner is complementary to the DNA.[20–22]

What complementarity are we talking about? This question is directly related to the important problem of the synthesis of M-RNA on DNA. It is evident that such a synthesis may take place in three ways: M-RNA is synthesized on both mutually complementary DNA chains and copies their text in entirety; M-RNA copies the text of both chains but always in different regions—there is a sort of selective copy; the matrix for the synthesis of M-RNA is only a single chain.

If the first method is true, then there may be different possibilities for the further synthesis of protein. If both chains of M-RNA print proteins on the basis of a single code, then each protein must correspond to another one which is complementary to it in amino acid composition. This is not observed. It is possible to assume then that complementary coding groups, for example, AGU and UCA, are responsible for a single amino acid. This is contrary to experiment. Finally, it is possible that, of two complementary chains of RNA, only one chain is genetically functional. The second chain either does not function at all (then it is not clear why it is needed in the cell) or it serves as a matrix for the synthesis of the functional chain. The last assumption introduces a secondary mechanism for the matrix synthesis in addition to the synthesis on DNA. Is this not excessive?

Such considerations cannot, however, give a convincing answer to our question. Here we need direct experimental data.

Spiegelman et al. have recently carried out such studies.[23] As has been pointed out already, during the duplication of the phage φX174, its single-stranded DNA builds a second complementary chain from the material of the infected cell. Using labeled atoms and density-gradient sedimentation, Hayashi and Spiegelman have established that the M-RNA synthesized by the phage hybridizes only with one of the two DNA chains. This is not the starting chain of the phage, but a newly synthesized complementary chain. This proves that, in the case of the φX174 phage, M-RNA is printed only by one DNA chain: the second chain is necessary for replication. This is the manner in which the process occurs in vivo.

On the other hand, in vitro experiments with DNA preparations and the necessary enzymes have shown that both DNA chains are functional in the synthesis of M-RNA. The difference is due to the fact that, in the developing phage, the DNA double chain has a cyclic structure which is destroyed during the extraction of the DNA; separate fragments of DNA act in vitro. It has been shown that the synthesis of M-RNA occurs in vivo on one of the two DNA chains not only in the case of φX174 phage but also in other systems.[24,25] So far, it is not known what factors determine the functionality of the single-chain DNA and the absence of the functionality of the second chain. Maaløe established that the synthesis of M-RNA has an influence on the reduplication of DNA.[26] A further investigation of the problems of reduplication requires an examination of the entire process of protein synthesis and examination of feedback in this process.

Warner, Rich, and Hall have shown by electron microscopy that several ribosomes combine with each other during protein synthesis by means of the M-RNA molecules.[27,28] The scheme of such a combination is shown in Fig. 132.[29] A system is formed which contains from four to eight and even more ribosomes; it is called a polysome. An electron-microscopic picture of the polysome is shown in Fig. 133.

The existence of polysomes has been demonstrated by several methods. Their necessity is evident also from logical considerations. Let us assume that a rather small protein molecule is being synthesized, containing, for example, 150 amino acids. Since, genetically, the code is a triplet (proof will be presented later), such a protein corresponds to an M-RNA containing 450 nucleotides, i.e., having a length of about 1,500 Å. On the other hand, the diameter of a functional ribosome is no greater than 220 Å. Consequently, several ribosomes may participate in the synthesis of a protein and be connected by a single chain of M-RNA. In fact, such a protein chain is found in hemoglobin, i.e., a protein synthesized in reticulocytes. Here

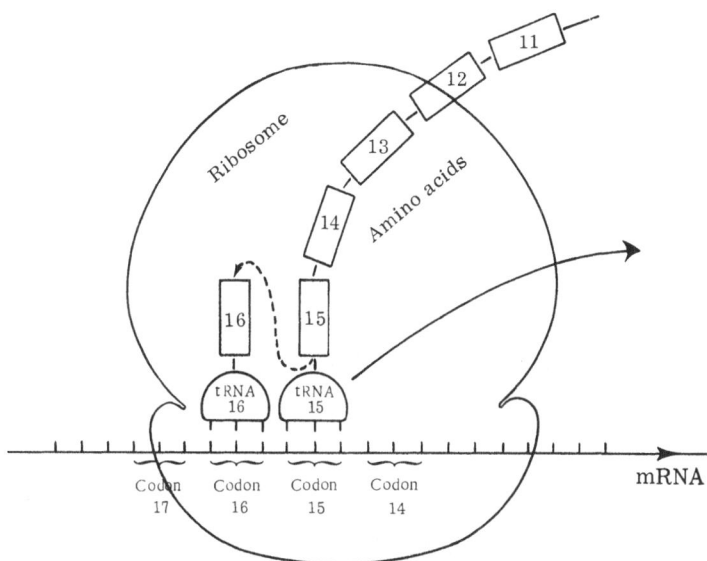

Fig. 132. Scheme of protein synthesis on a ribosome.

polysomes of the length of 1,500 Å have been found; they contain from four to six ribosomes. A single protein chain is synthesized on each ribosome and, in this manner, the polysome works as a conveyor belt.

The protein synthesis on the polysome must occur as a result of the displacement of the ribosome along the M-RNA chain, which is in turn a result of the sequential transcription of the text coded by the nucleotides. On the average, each ribosome and a polysome must contain about half of a protein chain; the ribosome which has proceeded to the end of the M-RNA has synthesized the entire chain, while a ribosome which has just become attached to the M-RNA does not yet possess any protein chain at all. This is confirmed by direct experiments.

How does a protein synthesis take place in a polysome? The protein chain is synthesized from the N terminal. Two molecules of T-RNA are present simultaneously on the ribosome; these carry amino acids. On one of these T-RNA molecules, which is attached to nucleotide triplet number n of M-RNA, a protein chain, n amino acids in length, has grown. The second T-RNA molecule, which carries residue number $n + 1$, is attached to a triplet number $n + 1$. The bond between the first T-RNA molecule and the protein chain is broken and the chain becomes attached by its N terminal to the $(n + 1)$th residue on the second T-RNA molecule. After this, the first T-RNA molecule separates from the M-RNA and migrates into the cytoplasm. This scheme is shown in Fig. 132. It cannot be regarded as fully proven, but it is quite probable.

In cells which synthesize more complicated and large proteins, polysomes have been found with up to 50 ribosomes. It is quite interesting to

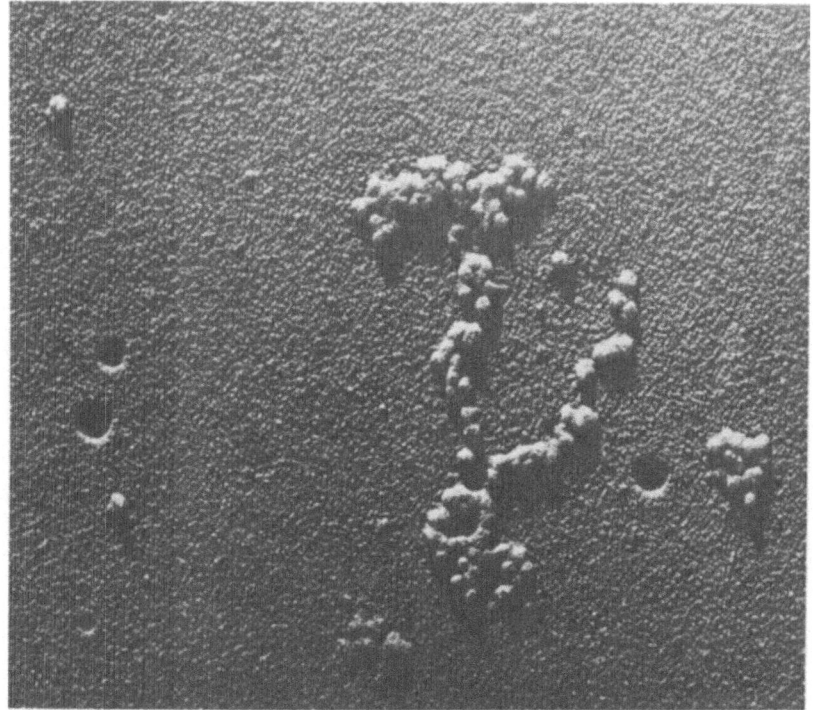

Fig. 133. Electron micrograph of polysome.

look at data on the synthesis of gamma globulins, i.e., of proteins which form antibodies. Polysomes of two types have been found in cells which synthesize such proteins; these polysomes contain 7 to 8 and 16 to 20 ribosomes. These numbers correspond to molecular weights of polypeptide chains of about 25,000 and 50,000 to 60,000. In fact, it is known that gamma globulins have a quaternary structure in which each molecule consists of two chains with molecular weights of 23,000 and two with molecular weights of 55,000.

Displacement of the ribosomes along an M-RNA chain in the polysome is a mechanochemical process, possibly similar to muscle contraction (see Chapter 9). Qualitative speculations have been proposed on the basis of elementary mechanical analogies; the 70S ribosome, consisting of two parts (30S and 50S), has been regarded as a turning block which pulls an M-RNA chain. These speculations have not been supported in any way up to now.

The kinetics of protein synthesis on polysomes have been investigated.[30] The dependence of the amount F_n of incorporated ^{14}C lysine per unit of mass of RNA has been examined as a function of time for polysomes containing various numbers of ribosomes. This study was not sufficiently complete; however, it produced data which enabled theoretical calculations to be carried out. This has been done.[31]

It is evident that, in accordance with the described model, the average velocity v of the motion of a ribosome along an M-RNA chain must be proportional to the average velocity w of amino acid incorporation into a protein on a single ribosome.

$$v = \lambda w \tag{7.1}$$

If v is determined by the number of nucleotides by which the ribosome is displaced and w is determined by the number of amino acids incorporated at the same time, then λ is equal to the number of nucleotides in the coding group, i.e., in the codon. If over a period of time t there are $K_i(t)$ polysomes, containing i ribosomes, then Y amino acids will be incorporated over a unit of time

$$Y = \sum_{i=1}^{n} iK_i(t)w \tag{7.2}$$

The summation is carried out over all types of polysomes.

The ribosomes have different positions within the polysome. Let l be the maximal distance between the first ribosome and the end of the M-RNA chain of the polysome. The number of n-somes, the first ribosome of which is at a distance $l - \xi$ from the end of the chain, is equal to $K_n(t, \xi)$. Then,

$$K_n(t) = \int_0^l K_n(t, \xi)\, d\xi \tag{7.3}$$

The rate of change of $K_n(t)$ is proportional to the number of n-somes in which the first ribosome has reached the end of the chain and is ready to break off.

$$\dot{K}_n(t) = -\kappa K_n(t, l) \tag{7.4}$$

where κ is the probability of the breaking off of the end ribosome from the M-RNA chain. The rate of change of $K_n(t, l)$ is

$$\dot{K}_n(t, l) = -\kappa K_n(t, l) + \int_0^l k(\xi) K_n(t, \xi)\, d\xi \tag{7.5}$$

where $k(\xi)$ is the probability that at time t the ribosome will have reached the end of the chain, after starting at position $l - \xi$ from the end. It follows from (7.4) and (7.5) that

$$\dot{K}_n(t) = \dot{K}_n(t, l) = \int_0^l k(\xi) K_n(t, \xi)\, d\xi \tag{7.6}$$

If there is no agglomeration of ribosomes at the end of the chain, then $\dot{K}_n(t, l) = 0$, that is, we have a stationary state. Assuming a stationary state and an even distribution of ribosomes in the interval $(0, l)$, that is

$$K_n(t, \xi) = \frac{K_n(t)}{l} \tag{7.7}$$

we obtain

$$\dot{K}_n(t) = -k K_n(t) \tag{7.8}$$

where

$$k = \frac{1}{l} \int_0^l k(\xi)\, d\xi \tag{7.9}$$

is the average value of $k(\xi)$ in the interval $(0, l)$. For a polysome with $i < n$, we obtain the equation

$$\dot{K}_i = k'(K_{i+1} - K_i) \tag{7.10}$$

where $k' = k$ if l is equal to the average distance between ribosomes on the polysome. Equation (7.10) describes the formation of an i-some from an $(i + 1)$-some, as a result of the breaking off of one ribosome and the break-up of the i-some, i.e., its transformation into an $(i - 1)$-some by a similar breaking-off.

We now solve the system of Eqs. (7.8) and (7.10) with the initial conditions

$$K_n(0) = K_n^0 \qquad K_i(0) = 0 \qquad (1 \leqslant i \leqslant n - 1)$$

The solution has the form

$$K_i(t) = K_n^0 e^{-kt} \frac{(kt)^{n-i}}{(n-i)!} \qquad (1 \leqslant i \leqslant n) \tag{7.11}$$

The rate of introduction of ^{14}C leucine per unit of mass of RNA is

$$I_n(t) = \dot{F}_n(t) = \frac{1}{nK_n^0 M} \sum_{i=1}^{n} i K_i(t) w x y \tag{7.12}$$

where M is the RNA mass of a single ribosome, x is the fraction of labeled leucine in the entire leucine, y is the ratio of incorporated leucine to the total number of incorporated amino acids. It follows from (7.11) and (7.12) that

$$I_n(t) = \frac{W}{n} e^{-kt} \sum_{i=1}^{n} i \frac{(kt)^{n-i}}{(n-i)!} \tag{7.13}$$

where

$$W = \frac{wxy}{M}$$

Carrying out calculations for the concrete values of $n = 2, 4, 6, 8$, we obtain theoretical curves which are in good agreement with experimental data (Fig. 134). The slopes of the curves at $t = 0$ are identical and the tangent is W, while the limiting value at $t \to \infty$ is proportional to W/k. From these experiments we find $W = 284$ counts/min^2 and $k = 0.3$ min^{-1}.

If we set $L = vk^{-1}$, that is, the distance traveled by the ribosome over time k^{-1}, we obtain

$$\lambda = \frac{Lkxy}{M} \tag{7.14}$$

If the motion of the ribosome is uniform, $L = 0.5l$. The average distance between ribosomes, l, is 90 nucleotides according to sedimentation data and electron microscopy; M is of the order of 2×10^6. Unfortunately, the values of x and y are unknown; λ is found to be equal to three nucleotides if $x = \frac{1}{2}$ and $y = \frac{1}{20}$. In any case, λ agrees with the dimensions of the codon in order of magnitude and, as a whole, the theory does not contradict experiment.

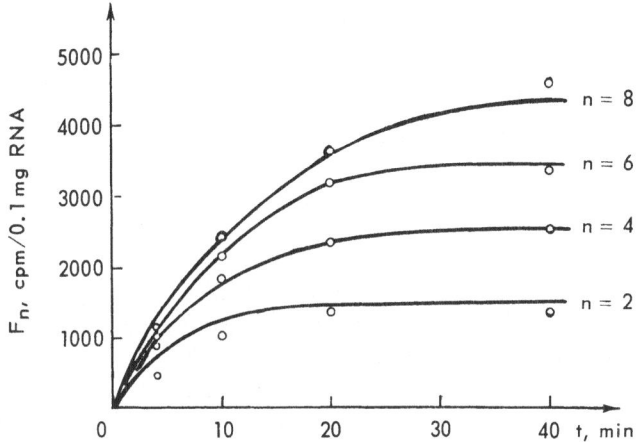

Fig. 134. The dependence on time of the uptake of labeled leucine into a protein: n is the number of ribosomes in the polysome; solid lines are theoretical curves; and circles are experimental points.

Energy is necessary for carrying out such a mechanochemical process of motion of the ribosome along an M-RNA chain. It is possible to assume that the source of energy is a high-energy GTP bond. Its energy is about 8 to 13 cal/mole. Part of the energy, 3.5 kcal/mole, is used up in the synthesis of a peptide bond, and 4.5 to 9.5 kcal/mole is left for the mechano-chemistry if there are no other as yet unknown dissipation processes.

The motion of the ribosome takes place in one direction, which obeys an as yet unknown regulating action carried out probably by an enzyme or enzymes.

Ribosomal RNA has no direct matrix role. It is incorporated into the complex regular structure, which has been discussed recently by Spirin and Kiselev.

The matrix of protein synthesis is one-dimensional; it is the linear M-RNA macromolecule. The biological significance of a one-dimension-ality is evident in this case. Let us assume that we have a three-dimensional matrix. This is somewhat similar to three-dimensional chess, a game that has been proposed by an imaginative mind. It is really impossible to play such chess. It would also be impossible to extract from such a three-dimensional matrix network the biological macromolecules which would be synthesized in it. A two-dimensional matrix, which at one time was proposed by Haurowitz, is also very inconvenient, since it must meet with a great resistance during displacement within the cell and its con-formational mobility could not be sufficient for the formation of functional structures of the type of M-RNA–ribosome complexes. A one-dimensional M-RNA matrix is necessary, evidently, for the formation of the linear structure of the synthesized biopolymer, i.e., the protein.

The problem of protein synthesis, which has been stated with great clarity by Crick,[32] is not exhausted, however, by the problem of the matrix synthesis of a linear polymer.

The protein is synthesized from amino acids. The polycondensation of amino acids is an endergonic process and is thermodynamically unfavorable. Therefore, it must be conjugated with exergonic processes which furnish free energy to the reaction. Just as in other biological reactions, it is dephosphorylation of ATP which releases the necessary chemical energy. In other words, amino acids enter into the protein-synthesis reaction in an activated form. As has been shown in the investigations by Hoagland *et al.*[33] and by Meister,[34] the activation of the amino acid takes place through the formation of what is called *aminoacyladenylate*: amino acid + enzyme + ATP $\overset{Mg^{++}}{\rightleftharpoons}$ enzyme–aminoacyladenylate + PP.

The structure of aminoacyladenylate is shown in Fig. 135. The reaction proceeds with the participation of a specific enzyme. Two phosphate groups of ATP are released in the form of pyrophosphates. The reaction scheme is shown in Fig. 136, where the shaded sections are the surface of the enzyme molecule. There is a belief that the cofactor of this enzyme is vitamin B_{12}. It has been shown that each amino acid requires its own enzyme. Consequently, for the synthesis of a single protein molecule, it is necessary to have 20 enzymes. The question of the source of these 20 enzymes, also of the source of protein molecules, is directly related to the problem of the origin of life. It is evident that they are synthesized also with the participation of DNA and RNA.

The next step in protein synthesis consists in the transport of the activated amino acid to the matrix. It has been shown that the activated amino acid combines with the end of the transport RNA molecule. The bonding occurs by the formation of an extra linkage as a result of the interaction with one of the two adjacent hydroxyl groups of the adenosine ribose which is always found at the end of T-RNA molecules (Fig. 137).[35,36] The interaction takes place with the mediation of the same enzyme. The

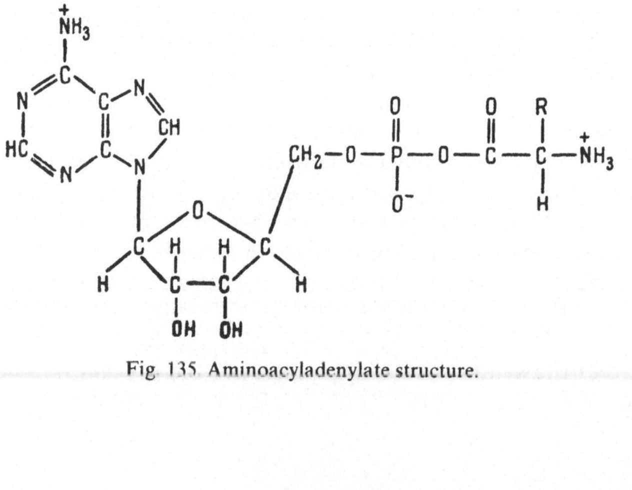

Fig. 135. Aminoacyladenylate structure.

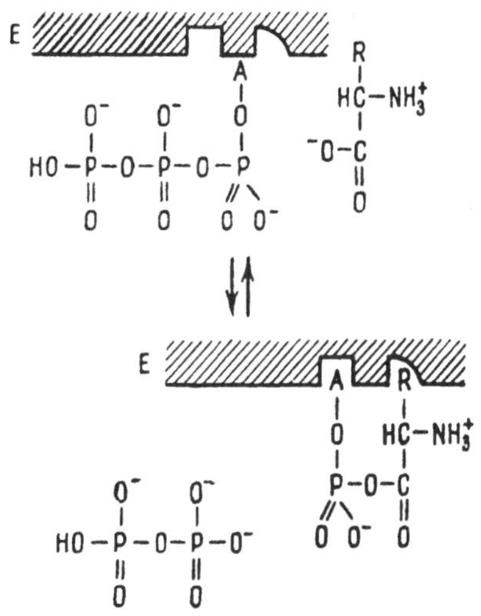

Fig. 136. Scheme of formation of aminoacyladenylate.

number of different types of T-RNA is also not smaller than 20, that is, the
number of amino acids. In other words, each amino acid has its own
enzyme and its T-RNA. All 20 or more types of T-RNA have identical end
groups. One end of the chain has the sequence APCPC, and the amino
acid is attached to that end; the other end is PG. The ends of the T-RNA
molecule are thus nonspecific. The specificity of T-RNA, which is manifested twice, namely, in the addition of a given amino acid and in the
attachment to the proper place in the M-RNA, must be localized, not in
the end of T-RNA, but in its internal nucleotides.

It has been stated already that the primary structure of a T-RNA has
been determined.[37] By the end of 1966, the structures of five such RNAs
had been established: alanine, tyrosine, phenylalanine, and two serines.
The secondary structure of T-RNA seems to have the shape of a clover leaf
(Fig. 106); the anticodon appears to be located in one of the leaflets.

It appears that all three types of RNA, and not only M-RNA, are
synthesized on DNA. It has been shown that DNA has a nucleotide
sequence which is complementary to ribosomal RNAs with 23S and 16S.
However, these cistrons occupy only a small fraction of the chromosome
of E. coli. The 23S cistron corresponds to 0.2% of the entire DNA, that

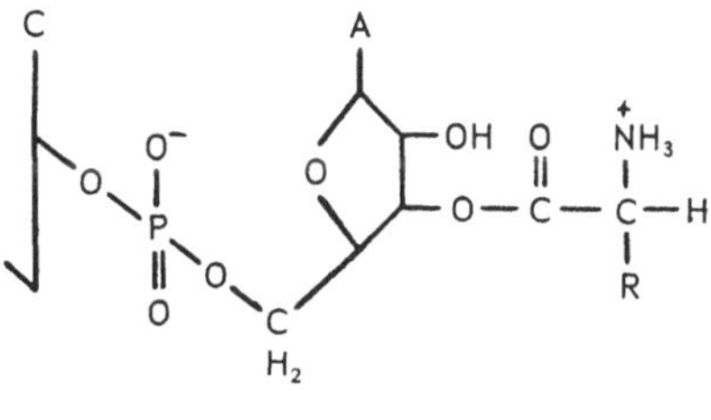

Fig. 137. Scheme of bonding of amino acids to
T-RNA.

of the 16S corresponds to 0.1 %. Transport RNA corresponds to 0.02 % of DNA on the basis of hybridization experiments.[38]

It is essential to emphasize that the double specificity of T-RNA has probably similar bases in both the interaction with the enzyme and the interaction with the M-RNA. Both are determined apparently by the structural complementarity of T-RNA and M-RNA. It has been shown recently by direct experiments that these two functions are performed by different regions of T-RNA. The functions of these two segments can be disturbed to different extents by bromination.[39]

The next stage in protein synthesis consists in the formation of the peptide chain. Aminoacyl T-RNAs are attached in turn to the matrix by their ribonucleic ends. The reaction of protein-chain growth seems to occur as

$$H_2N \cdot CHR \cdot CO \sim T\text{-}RNA + H_2N \cdot CHR' \cdot CO \sim T\text{-}RNA' \longrightarrow$$

$$H_2N \cdot CHR \cdot CONH \cdot CHR' \cdot CO \sim T\text{-}RNA' + T\text{-}RNA$$

The end group, which contains the T-RNA and plays the role of a catalyst, remains reactive all the time, i.e., it is ready for further chain growth. In this way, the reaction is similar to a number of well-studied processes of catalytic polymerization in the chemistry of synthetic polymers.

It is not difficult to write out such a reaction; at present, however, there is still much unknown in this realm. How and why does chain growth start on the matrix, and what determines its end? What are the three-dimensional characteristics of this process?

Let us once again enumerate the principal stages of protein synthesis.

Molecules of M-RNA are synthesized on the DNA of the chromosome; they are transported to the ribosomes where they form a matrix. The amino acids are enzymatically activated by ATP to form aminoacyl-adenylates, which, in turn, are attached to the T-RNA molecules by the same specific enzymes. The amino acid–T-RNA complexes are assembled on the matrix. During this process the protein chain is formed; it leaves the matrix and migrates into the cytoplasm and to its ultimate destination. The fate of the M-RNA and T-RNA molecules which participated in the synthesis is not quite clear. The first seem to disintegrate.

A general simplified scheme of the process is shown in Fig. 138. It is quite complicated, but it is nevertheless efficient in the sense that protein synthesis is carried out by compounds of one type, namely, DNA, M-RNA, ATP, T-RNA, and ribosomal RNA, which plays the role of a support for M-RNA.

Since the starting materials for protein synthesis are nucleic acids, it is possible to say that the optical asymmetry of proteins and amino acids is determined by the asymmetry of DNA. The last exists always as a right-handed helix; this, in turn, is dictated by the asymmetry of the carbohydrate molecule, namely, deoxyribose. The asymmetric complexes of T-RNA

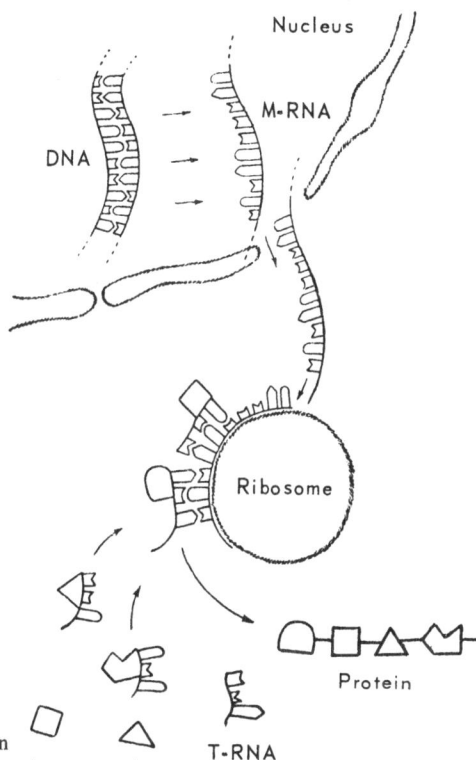

Fig. 138. Simplified scheme of protein
synthesis.

with amino acids are assembled also on an asymmetric matrix; the amino
acids always belong to the same stereochemical types. In this sense, the
asymmetry of amino acids is a function of the asymmetry of carbohydrates.

Spirin has found in the embryos of the groundling and the sea urchin
nucleoprotein complexes of M-RNA with proteins, which contained more
protein than ribosomes (80 % protein and 20 % RNA). These complexes,
which are characterized by different sedimentation coefficients (from 8S
to 58S), have been called *informosomes*. It is assumed that the informosomes
are a reverse transport form of M-RNA in which the M-RNA is inactive
in protein synthesis, i.e., it is repressed by the protein. When the M-RNA
is transformed into the active form, the protein of the informosome must
become separated. A hypothesis has been stated on the important role of
these structures in processes of regulation and in morphogenesis.[40] If these
observations are confirmed on a broader basis, the informosomes will
occupy an important place in molecular biology.

The biological role of the various types of RNA is now becoming
clear. These are auxiliary substances which are necessary for protein synthesis and are directed by DNA. The genetic information passes into the

protein from DNA through RNA; however, once it has entered the protein, it can no longer leave it,[32] i.e., the transmittal of information from one protein to another or from a protein to a nucleic acid does not occur.

One might speculate on the subject of the evolutionary origin of the described mechanism. It is quite possible that at the beginning of life it was much simpler, let us say, similar to the synthesis of the protein of plant viruses, which takes place without the participation of DNA. As has been pointed out already, some scientists regard viruses as primitive living organisms which have become degraded to a parasitic form of life after the creation of more perfect forms of life.

It is also clear that this mechanism of the synthesis does not leave room for any precursors of the protein, such as peptides containing a few amino acids, and it is thus evident why such peptides are not found in cells. This is actually more economical since, otherwise, there would have been needed, not 20 enzymes and 20 types of T-RNA, but a much larger number. Thus, if a protein had been synthesized from dipeptides, their total number could have been $20^2 = 400$; in the case of tripeptides the number would have been $20^3 = 8,000$.

There have been important advances in the understanding of the synthesis of proteins, but we are still far from understanding it completely. Even now, there is no real physical-chemical theory of the process; its thermodynamics and kinetics have not been established. The rate of protein synthesis is not very great; the peptide chain of hemoglobin is synthesized in 1.5 minutes, that is, two amino acids are copolymerized in 1 second.[41] What determines this rate? We shall have complete answers to these questions as soon as science learns how to carry out protein synthesis in a nonliving system, in a system of nonbiological origin.

As we shall see, protein synthesis is regulated by a complex system; the mechanism described here is only a simplified scheme. Very little is known yet about one of the most important factors in protein synthesis, namely, the action of the enzymes which take part in this process. On the other hand, data obtained over the last few years have made it possible to elucidate the nature of another fundamental factor necessary in the synthesis: The genetic code has been deciphered. Now we may return to this problem and approach it, not by a combinatorial or reasoning process, but by one based on the concrete physical-chemical principles of protein synthesis.

THE GENETIC CODE

The problem of the genetic code was solved as a result of direct genetic and biochemical investigations. The very posing of these questions was certainly related to the preceding theoretical investigations.

Crick and coworkers[42] undertook the genetic investigation of the B cistron of the r_{II} region of T4 bacteriophage which grows in E. coli. The genetic map of this region had already been studied by Benzer; this made it possible to establish the principal characteristics of the code. Among the point mutations of r_{II} there are some which are determined by the deletion of nucleotides and some which are determined by the insertion of nucleotides. Such mutations are caused, for example, by acridine dyes.[43] The mutants are unable to give plaques of a given type on the corresponding E. coli cultures. The FCO mutants, studied by Crick et al., are able to revert to the wild type as a result of a second mutation in the same cistron. Such reversions of the wild type turned out to be, in fact, double mutants with a wild phenotype. Each of the two mutations results in the independent loss of the ability to synthesize a certain protein; the combination of the two mutations in a single cistron, however, restores this ability. About 80 independent r mutants of the B cistron were studied, including their double and triple combinations. Many of the double and triple combinations turned out to be related to the wild type. The results obtained can be interpreted directly on the hypothesis of a triplet nonoverlapping code which is read from a particular nucleotide of the cistron or the DNA molecule. If the code is read triplet by triplet, then the deletion or addition of a single nucleotide disturbs the entire subsequent text.

In Fig. 139, the DNA chain is shown as a sequence of the letters ABCABC... The reading of the code, starting from a given nucleotide, means the superposition on this sequence of a frame with holes such as is shown in Fig. 139b–g. If one of the letters is deleted (−) or, to the contrary, added (+), we find a different sequence already appearing through the frames (Fig. 139b and c). If we have a double mutant, in which the letter has been deleted in one spot and added in another spot (− +), the sequence is disturbed only in the region between the two mutations, while the rest

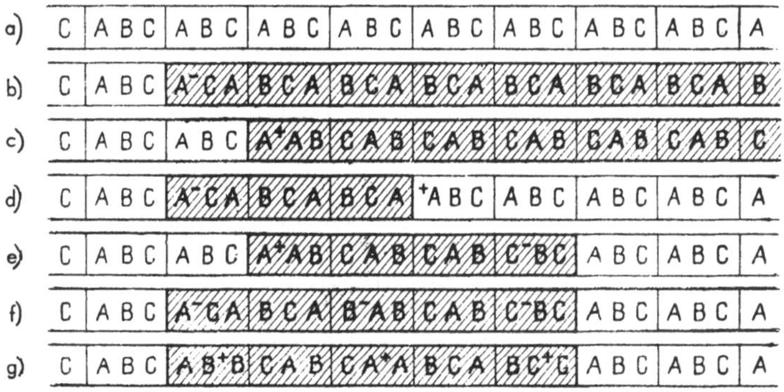

Fig. 1.39. The DNA text: (a) wild type, (b) mutant with a single deletion, (c) mutant with an addition, (d) double mutant of the − + type, (e) triple mutant − − −, (g) triple mutant + + +. The disturbed portion of the text is marked by the crosshatched area.

of the text remains correct (Fig. 139*d* and *e*). If the incorrect region is not too great, then the synthesized protein will maintain its function.

This is in agreement with the data by Levinthal and coworkers,[44] who have studied the mutants of *E. coli*. Revertent mutants were obtained which formed the enzyme (alkali phosphatase), with a normal activity but with a changed amino acid composition. The protein may contain regions in which amino acid substitutions have occurred without an accompanying change in the enzymatic activity. All combinations of the $(+ +)$ or $(- -)$ types must be related to the mutant of the *r* type; on the other hand, a set of $(+ -)$ combinations may be related to the wild type. Furthermore, one should keep in mind that any long sequence of nucleotides may be read correctly only in one way, and incorrectly in two ways depending on whether the frame has been displaced by one space to the right or to the left $(- +$ or $+ -)$. It is evident that the sequence of incorrect triplets between $-$ and $+$ and also between $+$ and $-$ will be different. An analysis of double mutants shows that a $(- +)$ displacement is acceptable everywhere within the examined region (acceptable from the point of view of the appearance of a wild enzyme), while the $(+ -)$ displacement is acceptable only within a more limited region.

On such a hypothesis, Crick has predicted the properties of 28 mutants which have been completely verified by experiments.

Six triplet mutants of the $(+ + +)$ and $(- - -)$ types, selected in such a manner that there should be no displacement over unacceptable regions of a $(+ -)$ type, are found to possess again the wild enzyme (Fig. 139*f* and *g*).

Thus, the hypothesis is in agreement with the known facts. It follows, therefore, that the code is not overlapping and that the nucleotide text is read from a given position. As far as the code relationship is concerned, i.e., the number of nucleotides corresponding to a single amino acid, it may be equal to three or a multiple of three. It is evident that nothing is changed if the frames are placed with openings which contain not three, but six or even nine nucleotides. The most reasonable number, however, seems to be three, i.e., the nucleotide code is a triplet.

This important study established the general character of the code but it did not give its decoding. The decoding has been obtained as a result of biochemical investigations. The establishment of the general character of the code permits the molecular dimensions of a gene to be finally determined; the cistron can now be measured while, prior to that, it had been expressed in probabilities of recombination or in minutes. If the degree of polymerization of the protein under synthesis is 500, which corresponds to a molecular weight of the order of 50,000, the length of the cistron, i.e., of the region of the DNA chain corresponding to the synthesis of this protein, corresponds to 1,500 nucleotides. Consequently, the molecular weight of the cistron is $1,500 \times 330 \times 2 = 10^6$ (330 is the average molecular weight of a nucleotide, 2 is the number of chains in a double helix). A muton then is a single nucleotide, since a change in it changes the entire codon and consequently the amino acid residue.

In 1961, Nirenberg and Matthaei described very interesting experiments on the incorporation of amino acids in proteins in a cell-free biological system.[45-47] They took a ribosomal fraction isolated by centrifugation from broken *E. coli* cells. They added to it the supernatant liquid containing transport RNA and ATP and a system which generates ATP. This nuclei-free mixture represents an incomplete factory capable of synthesizing proteins. Such a factory is deprived of the regulating and planning office, but, in principle, it can manufacture its product if it is furnished with blueprints. In a normal factory, such blueprints are the molecules of M-RNA. Nirenberg and Matthaei introduced into the system ribosomal RNA and synthetic polyribonucleotides and they studied the incorporation of ^{14}C-labeled amino acids. It was found that in the presence of RNA the radioactive label was incorporated much more strongly than in the absence of RNA. The incorporation process was suppressed by ribonuclease, an enzyme which destroys RNA, as well as by puromycin and chloramphenicol. It was established that the RNA stimulated the incorporation of ^{14}C valine, while a synthetic polynucleotide stimulated the incorporation of ^{14}C phenylalanine. In the absence of poly U, 44 radioactive counts per minute were found in 1 mg of protein, while in the presence of poly U, they obtained 39,800 counts!

Thus, in these experiments the biological system was fooled for the first time, i.e., instead of natural matrix RNA, a synthetic poly U matrix was introduced into it; this synthetic matrix coded most probably phenylalanine. If a code is, in fact, a triplet, then phenylalanine corresponds to the triplet UUU within the RNA matrix.

Subsequently, similar studies were carried out by Lengyel, Ochoa, and coworkers, who studied a number of other synthetic polyribonucleotides along with poly U,[48-55,58] and also further studies by Matthaei, Nirenberg, and coworkers.[56,57,59]

It was established that poly U causes the synthesis of acid-insoluble polyphenylalanine; furthermore, it was found that a single mole of phenylalanine required 3.25 moles of U residues within poly U. When poly A was added to the system which contained poly U, phenylalanine was not incorporated; this is explained by the formation of poly-A–poly-U double helices in which the poly U cannot act as a matrix. A number of copolymers were investigated: U with G and A, C with C, G with A and G, and also A with G and C, taken in various proportions. This made it possible to find code triplets for all amino acids, which resulted in the determination of the code triplets for all amino acids. The method of determining the triplet may be illustrated by the example of serine and leucine. The copolymer UC in a 3:1 ratio causes the incorporation of phenylalanine, serine, and leucine. In the absence of polyribonucleotides, phenylalanine is incorporated in an amount of 0.06 mμ of a mole per gram of ribosomal protein; serine, in the amount of 0.04; and leucine, of 0.06. In the presence of poly UC (3:1), phenylalanine is 2.2; serine, 0.9; and leucine, 0.6. Within the assumption of a triplet code, the ratio of the number of triplets in poly UC (3:1) is UUU/UUC (or UCU or CUU) = 3. A similar number is obtained if we compare the fractions of incorporation obtained: Phen/Ser or (Leu) = 2.2:0.75 = 2.9.

Consequently, Ser and Leu are coded by triplets which contain 2U and 1C. On the other hand, Pro is coded by 2C1U. Poly C does not stimulate the incorporation of Pro, while poly UC does stimulate it. Poly UC (5:1) was investigated. The results obtained were Phen/Pro = 7:0.4 = 18 or the UUU/2U1C = 5 and UUU/1U2U = 25. It appears then that Pro corresponds to 2C1U.[37]

The synthetic polyribonucleotides were then subjected to treatment[60] with HNO_2, which causes a deamination of the nitrogen bases, with the result that G is transformed into xanthine and hypoxanthine and C into U. The deaminated polymers have a low activity (the reason is not clear). It has been established that hypoxanthine may replace G in the coding triplet. The coding triplets turned out to be consistent with the base substitution that took place. Thus, poly UA (5:1) does not stimulate the incorporation of Val (triplet 2U1G). After treatment with HNO_2, A is transformed into hypoxanthine, like G, and the deaminated polymer stimulates the incorporation of valine. Analysis of a number of TMV mutants, obtained by treatment with NHO_2, also confirms the code.

The "fooling" of the biological system in these experiments is very important. It becomes possible to combine the ribosomal fraction of one source with a supernatant liquid from another source.[38] This system, containing ribosomes from rat liver and a supernatant liquid from E. coli cells, works almost as well as a system obtained solely from E. coli. This is not true of a system of the opposite composition or of a system obtained solely from liver cells; in these, poly U is found to be less active. The reason for this is not clear.

Preliminary data have been obtained[61] which point to the possibility of forcing the E. coli system to synthesize a protein, the structure of which is dictated by the introduction of a given matrix RNA. The RNA of TMV was introduced into the system; this stimulated the incorporation of a number of labeled amino acids and the synthesis of a protein similar to the TMV protein. These data, however, still require confirmation.

The studies just described, started by Nirenberg and Matthaei and continued by a number of other investigators, have a fundamental significance for molecular biology. The biological mystery of the cell as a biosynthetic system has been destroyed. The possibility of interfering directly in protein synthesis has been found; the technological process of the cell factory, which was proposed earlier, has been demonstrated by direct experiments. The discovery of the genetic code itself is of still greater significance; this is the solution of a problem that was posed in the early

1950's. The finding of the coding triplets means the finding of the structure of the gene, namely, of the cistron.

Even before the discovery of the code, Sueoka studied the correlation of the composition of the total protein of a number of bacteria with the composition of their DNA.[62] As has already been stated, the GC contents of the DNA of bacteria vary within wide limits, namely, from 25 to 75%. Sueoka constructed graphs for 16 amino acids showing the dependence of their contents in the total protein on the GC contents of DNA. The dependence was found to be linear. Furthermore, the amino acids were found to fall into three groups. In the first group, the correlation coefficient which expresses the slope of the straight line is positive, i.e., the amino acid contents of the protein increase with an increase in the GC contents of DNA; in the second group, the coefficient is close to zero; and in the third group, it is negative. Sueoka himself tried to deduce the code (not necessarily triplet) from these data; but he was not successful.[62,63] However, the combination of Sueoka's data with the code triplets found by the method of Nirenberg and Matthaei is quite instructive.[64] Such a comparison is given in Table 23. It contains the composition of the codons corresponding to the amino acids examined by Sueoka. We see that three groups of amino acids correspond to three types of triplets in the sense of their GC contents. This confirms the validity of the codons found.

The DNA of higher organisms contains about 45% G + C. The G + C contents of all the functional codons is 51%, which is close to 45%. This explains the contradiction of the thermodynamic theory of mutations. An excess of G + C would signify the cessation of the incorporation of a number of amino acids into proteins, namely, the cessation of protein synthesis. Consequently, the AT → GC mutations, while thermodynamically advantageous, are biologically nonadvantageous; starting from certain contents of G + C, such mutations would be lethal. The evolution of DNA composition is determined, not by thermodynamics, but by the entire chemistry of protein synthesis and by natural selection. Bacteria with high G + C contents do not contradict what has been said, since their proteins are not as diverse as those of higher organisms.

It is worth noting that the correspondence between the data of Sueoka and the composition of the codons was already established before the discovery of a number of codons.[64] Furthermore, the predictions which were made have been completely verified. For example, one such prediction was the finding in the codons of the first group of a large number of G and C molecules and of the possibility of the substitution of A for U in the codons in the second and third groups (see also Lanni[65]).

All the available data point to the universality of the code. The codons found are in agreement both with data on TMV mutants and with data on the mutants of human hemoglobin.

It has been shown that poly U stimulates the incorporation of phenylalanine in proteins in systems obtained from cells of rats, rabbits, humans,

Table 23   Comparison of Codon Composition with Data on the Correlation of the Amino Acid Composition of Proteins and the G + C Contents of Bacterial DNA

Residue	Correlation coefficient, b	Codon composition	Number of G + C in codons	Mean fraction of G + C in codons
		Group 1, $b > 0$		
Ala	0.164	1A1C1G, 2C1G, 1C2G, 1C1G1U	2, 3, 3, 2	0.83
Arg	0.089	2A1G, 1A2G, 1A1C1G, 2C1G, 1C2G, 1C1G1U	1, 2, 2, 3, 3, 2	0.72
Gly	0.051	1A2G, 1C2G, 3G, 2G1U	2, 3, 3, 2	0.83
Pro	0.024	1A2C, 3C, 2C1G, 2C1U	2, 3, 3, 2	0.83
		Average for group	2.39	0.80
		Group 2, $b \cong 0$		
Val	0.008	1A1G1U, 1C1G1U, 2G1U, 1G2U	1, 2, 2, 1	0.50
Thre	0.008	2A1C, 1A2C, 1A1C1G, 1A1C1U	1, 2, 2, 1	0.50
Leu	−0.006	1A1C1U, 1A2U, 1C1G1U, 2C1U, 1C2U, 1G2U	1, 0, 2, 2, 1, 1	0.39
His	−0.010	1A2C, 1A1C1U	2, 1	0.50
Ser	−0.017	1A1C1G, 1A1C1U, 1A1G1U, 1C1G1U, 2C1U, 1C2U	2, 1, 1, 2, 2, 1	0.50
		Average for group	1.41	0.48
		Group 3, $b < 0$		
Met	−0.024	1A1G1U	1	0.33
Phen	−0.040	1C2U, 3U	1, 0	0.17
Tyr	−0.047	1A1C1U, 1A2U	1, 0	0.17
Gly	−0.052	2A1G, 1A2G	1, 2	0.50
GluN		2A1C, 1A1C1G	1, 2	0.50
Asp	−0.053	1A1C1G, 1A1G1U	2, 1	0.50
AspN		2A1C, 2A1U	1, 0	0.17
Lys	−0.083	3A, 2A1G	0, 1	0.17
Ileu	−0.098	2A1U, 1A1C1U, 1A2U	0, 1, 0	0.11
		Average for group	0.83	0.24

and the green algae *Chlamydomonas*.[66] The code guarantees the transfer of information necessary for the synthesis of any protein and there are no reasons to believe that it could change during the course of the evolutionary process.

The code is degenerate. Almost all the amino acids may be coded by several codons. The degeneracy, evidently, is genetically advantageous in the sense that it decreases the number of nonfunctional noncoding triplets.

The second stage of the investigations which have led to the complete deciphering of the code consisted in the use, not of polynucleotides, but of trinucleotides with a given sequence of nitrogen bases.[67] The formation of ribosome-trinucleotide–T-RNA amino acid (^{14}C) complexes was followed in a cell-free system. Polypeptide synthesis under such conditions, of course, does not take place, but as long as a trinucleotide imitates the codon, the very formation of the complex permits us to read it. To do this, it is necessary to investigate all the T-RNAs, which have been complexed in turn with each amino acid labeled with [14]C. Nirenberg *et al.* investigated all the 64 trinucleotides and have established the entire code dictionary; they determined not only the composition of the codons, but also the sequence of the nucleotides in them. Furthermore, it was found that several triplets are excessive.

An independent verification of the code was carried out in a series of very elegant studies by Khorana.[68,70] First, olygodeoxyribonucleotides containing 8 to 12 links per chain were synthesized. The sequence of links in such an oligomer is known; it is a sequence of repeating triplets [for example $(TTC)_4$]. This DNA-like oligomer was used then as a primer in the template synthesis of a DNA-like polymer, according to Kornberg, in a system containing nucleoside triphosphates of T, C, G, and A, the necessary concentration of ions, and DNA polymerase. A DNA-like double helix was synthesized; it consisted of two complementary chains in each of which the triplet was repeated n times. Then two chains of polyribonucleotides with a given sequence of codons were synthesized on such a DNA-like polymer for the template with the use of RNA polymerase. Since the experiment was carried out *in vitro*, both chains of the double helix served as templates. Finally, each of the two polyribonucleotides which imitate M-RNA was introduced into the cell-free system and the incorporation of amino acids into the polypeptide chain was investigated by Nirenberg's original method. Since, under these conditions, the code may be read from any position, a single synthesis permits the verification of six codons according to the scheme

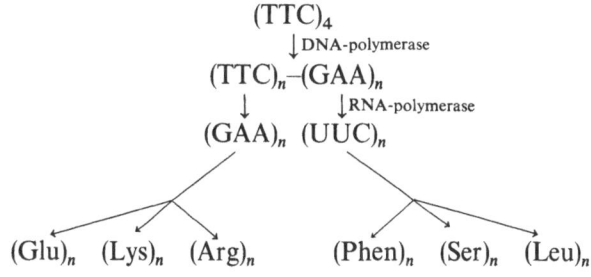

In fact, the $(GAA)_n$ polymer contained the codons GAA, AAG, and AGA, while the polymer $(UUC)_n$ contained the codons UUC, UCU, and CUU. Since the functionality of the homopolymers $(AAA)_n$, $(CCC)_n$, $(GGG)_n$, and $(UUU)_n$ had already been determined in a singular manner, it was necessary to have 10 syntheses of such a type in order to read the remaining 60 codons.

Tables 24 and 25 contain the entire code established by these methods and confirmed by investigations of mutations. The tables are constructed in the sequence ACGU, according to the Latin alphabet. The GUG (Val) and AUG (Met) codons in Table 25 are marked by asterisks. These codons code the indicated amino acid residues in the middle of the chain. On the other hand, these codons serve as initiators of the synthesis of a protein chain; furthermore, on the end of the chain they do not code any of the 20 canonic amino acids, but formylmethionine

$$H_3C-S-CH_2-CH_2-CH\begin{array}{c} NH-CO-H \\ \\ COOH \end{array}$$

In the absence of these codons in the synthetic templates, the chains are synthesized from any starting point and thus different N chain ends are formed. On the other hand, in the presence of GUG or AUG, standard formylmethionine N ends are formed. Khorana showed that poly-UG forms a chain formylMet-$(His-Val)_n$. On the other hand, native proteins do not contain formylmethionine. The question is what happens to it. It is known that formylmethionine is found on the chain ends synthesized

Tabe 24   Codon–Amino Acid Dictionary

1. AAA	Lys	17. CAA	GluN	33. GAA	Glu	49. UAA	–
2. AAC	AspN	18. CAC	His	34. GAC	Asp	50. UAC	Tyr
3. AAG	Lys	19. CAG	GluN	35. GAG	Glu	51. UAG	–
4. AAU	AspN	20. CAU	His	36. GAU	Asp	52. UAU	Tyr
5. ACA	Thre	21. CCA	Pro	37. GCA	Ala	53. UCA	Ser
6. ACC	Thre	22. CCC	Pro	38. GCC	Ala	54. UCC	Ser
7. ACG	Thre	23. CCG	Pro	39. GCG	Ala	55. UCG	Ser
8. ACU	Thre	24. CCU	Pro	40. GCU	Ala	56. UCU	Ser
9. AGA	Arg	25. CGA	Arg	41. GGA	Gly	57. UGA	–?
10. AGC	Ser	26. CGC	Arg	42. GGC	Gly	58. UGC	Cys
11. AGG	Arg	27. CGG	Arg	43. GGG	Gly	59. UGG	Trp
12. AGU	Ser	28. CGU	Arg	44. GGU	Gly	60. UGU	Cys
13. AUA	Ileu	29. CUA	Leu	45. GUA	Val	61. UUA	Leu
14. AUC	Ileu	30. CUC	Leu	46. GUC	Val	62. UUC	Phe
15. AUG	Met*	31. CUG	Leu	47. GUG	Val*	63. UUG	Leu
16. AUU	Ileu	32. CUU	Leu	48. GUU	Val	64. UUU	Phe

Table 25   Amino Acid–Codon Dictionary

1.	Ala	GCA, GCC, GCG, GCU
2.	Arg	AGA, AGG, CGA, CGC, CGG, CGU
3.	Asp	GAC, GAU
4.	AspN	AAC, AAU
5.	Val	GUA, GUC, GUG*, GUU
6.	His	CAC, CAU
7.	Gly	GGA, GGC, GGG, GGU
8.	Glu	GAA, GAG
9.	GluN	CAA, CAG
10.	Ileu	AUA, AUC, AUU
11.	Leu	CUA, CUC, CUG, CUU, UUA, UUG
12.	Lys	AAA, AAG
13.	Met	AUG*
14.	Pro	CCA, CCC, CCG, CCU
15.	Ser	AGC, AGU, UCA, UCC, UCG, UCU
16.	Tyr	UAC, UAU
17.	Thre	ACA, ACC, ACG, ACU
18.	Trp	UGG
19.	Phe	UUC, UUU
20.	Cys	UGC, UGU

in cell-free systems, but not *in vivo*. Thus, in endogenic matrices of *E. coli* in a cell-free system, one finds the N terminals formylMet-Ala, formylMet-Ser, but not formylMet-Met. The native proteins of *E. coli* have on their N chain ends usually Met, Ala, Ser. These results lead to the conclusion that two enzymes operate in the living system. One of these splits formylMet off from the rest of the chain, the other splits the formyl group off from methionine. At present, it is still not understood why formylmethionine plays such a special role.

In Table 24, dashes are placed next to the codons UAA, UAG, and UGA. These codons are nonsense; they are not functional, they do not correspond to any amino acid residues, and the polypeptide chains are broken when they reach them. Khorana established that the synthesis of polypeptides does not proceed on synthetic matrices $(GAUA)_n$ and $(GUAA)_n$. This is understandable, since, in these cases, one quite frequently meets nonfunctional triplets.

Table 25 shows that the degree of degeneracy of the codons (the number of synonyms, i.e., the number of different codons which code the same amino acid residues), varies from one (Met, Trp) to six (Arg, Leu, Ser).

Is there any intrinsic law in the code, are there definite principles of construction of the code alphabet, or is it completely random? Let us note first of all that any transposition within a codon xyz must change the coded amino acid. In other words, the number of different amino acids which correspond to a given codon composition is equal to the number of transpositions within a triplet. What has been said is illustrated in Table 26. This fact alone testifies to the nonrandom character of the code. Pelc[71] and also Rumer[72] have remarked that the replacement of z within the codon xyz with constants x and y, does not change the coded amino acid in the majority of cases. It is possible to call x the prefix, y the root, and z the suffix of the word, of the codon. The 16 xy doublets are broken into two groups of 8 doublets each. In the first group, the coded amino acid is completely independent of z; in the second group (with a single exception for Ileu), the character of the amino acid depends on whether z is a purine or a pyrimidine nucleotide. This is shown in Table 27. Consequently, replacements of the type of

$$xy\mathrm{Pu}_1 \longrightarrow xy\mathrm{Pu}_2$$
$$xy\mathrm{Py}_1 \longrightarrow xy\mathrm{Py}_2$$

do not change the coded amino acid (with the exception of AUA → AUG, Ileu → Met), while the replacement $xy\mathrm{Pu} \rightleftharpoons xy\mathrm{Py}$ changes it in the second group of the table. The first and second groups are noticeably different in the composition of the doublets.[73] In the first group, the prefixes and roots contain predominantly G and C; in the second group, A and U. In the first group, the roots y are predominantly pyrimidines; in the second

Table 26   Genetic Code

Codon composition	Number of transpositions	Amino acid residues	Number of residues
3A	1	Lys	1
2A1C	3	AspN, Thr, GluN	3
2A1G	3	Lys, Glu, Arg	3
2A1U	3	AspN, Ileu	2
1A2C	3	Thr, His, Pro	3
1A2G	3	Gly, Arg, Glu	3
1A2U	3	Ileu, Leu, Tyr	3
1A1C1G	6	Ser, Thr, Ala, Asp, GluN, Arg	6
1A1C1U	6	Ileu, Thr, His, Leu, Tyr, Ser	6
1A1G1U	6	Val, Asp, Ser, Met, –, –	4
3C	1	Pro	1
2C1G	3	Ala, Pro, Arg	3
2C1U	3	Ser, Leu, Pro	3
1C2G	3	Gly, Ala, Arg	3
1C2U	3	Phe, Ser, Leu	3
1C1G1U	6	Cys, Ser, Val, Ala, Leu, Arg	6
3G	1	Gly	1
2G1U	3	Trp, Val, Gly	3
1G2U	3	Leu, Val, Cys	3
3U	1	Phe	1

Table 27   Two Groups of Codons

First group				Second group				
							z = A, G amino acid residue	z = C, U amino acid residue
			z = A, C, G, U					
x	y	N	amino acid residue	x	y	N		
C	C	6	Pro	A	G	5	Arg	Ser
C	G	6	Arg	C	A	5	GluN	His
C	G	6	Ala	G	A	5	Glu	Asp
G	G	6	Gly	U	G	5	Trp(z = G)	Cys
A	C	5	Thr	A	A	4	Lys	AspN
C	U	5	Leu	A	U	4	Ileu(z = A), Met(z = G)	Ileu
G	U	5	Val	U	A	4	......	Tyr
U	C	5	Ser	U	U	4	Leu	Phe

group, they are purines. We have

First group

$$\frac{G + C}{A + U} = 3$$

$$\frac{C + U}{G + A} = 3$$

Second group

$$\frac{G + C}{A + U} = \frac{1}{3}$$

$$\frac{C + U}{G + A} = \frac{1}{3}$$

The intrinsic meaning of the code alphabet becomes clear on the basis of logical deductions which stem from modern concepts on the space structure and function of proteins. Let us repeat the basic principles (Chapter 5).

The biological function of a protein is determined by its space structure.

In turn, this structure is the result of two principal factors, the amino acid sequence, i.e., the primary structure, and the interactions of the residues.

A particularly important role is played by hydrophobic interactions which are determined by the aqueous surroundings of the protein molecule, by the special properties of water. Roughly speaking, hydrophobic interactions force the nonpolar, n, base residues to be located inside the globule, and the polar ones, p, to be located on its surface. Thus, the space structure and also the biological function of the protein is a strong function of the ratio and mutual disposition of the p and n residues.

Let us pass now to genetic deductions. From what has been said, it follows that mutational replacements which change the class of the residue

$p \to n$ and $n \to p$, must be more dangerous for the structure and, consequently, for the biological function of the protein than replacements in which the class remains the same, i.e., $p \to p$, $n \to n$.[74] Thus, proteins which have the same function but differ in their primary structure (for example, the same proteins of different species) must have an analogous sequence of amino acids p and n. This is, in fact, found to be the case with hemoglobin.[74] Mutational replacements in human hemoglobin (Table 3) are found in an overwhelming number of cases to be of types $p \to p$, $n \to n$ (correct ones), and not $p \to n$ and $n \to p$ (incorrect). Margoliash[75] has noticed this in the analysis of the composition of the cytochromes; he noted the constancy of the positions of p and n residues within the cytochromes c of five types of vertebrates (man, horse, pig, rabbit, and chicken). Margoliash asked: Is this an evolutionary constancy, a result of the selective pressure on the structural particularities important for the function of the cytochrome, or even in part a result of genetic variability? Now we can find an answer to this question.

It is evident that there are two possibilities. The first is that the correct and incorrect substitutions are equivalent, but mutations of the second type are lethal and, therefore, not observed. The second possibility is that the very structure of the genetic code is such that incorrect mutations are less probable. This would mean that the coding of the primary structure redetermines directly the coding of the space structure and biological function of the protein, that the evolutionary origin of the code is related to the particular properties and role of water.

To answer the question asked by Margoliash, it is necessary to examine the manner in which the p and n amino acids are coded. Even before the establishment of the complete code alphabet, it was shown that the composition of the codons for the p and n residues are in general quite different.[74,76] The p residues are coded primarily by triplets which contain adenine and cytosine; the n residues, by triplets which contain guanine and uracil. The ratio $(A + C)/(G + U)$ is equal to 1.86 for the p residues and 0.50 for the n residues, while as an average over all the residues, it is equal to 1.00. An examination of the entire code alphabet reveals the meaning of these differences.[77] In Fig. 140, the coding alphabet is shown in the form of a table. We see that the polar and nonpolar residues are located within different regions of the table. Figure 140 shows that the replacement of flection z with constant prefix x and root y never changes an amino acid from class p to class n and vice versa. This permits us to limit our examination, not to Fig. 140, but to a simplified key scheme of the genetic code containing, not 64 triplets, but only 16 doublets (Fig. 141). We see that if the root $y = A$, the amino acid residue is always polar, while if $y = U$, it is nonpolar. In this scheme, the tyrosil residue is regarded as a polar one; it contains the phenolic group C_6H_4OH. It is necessary to note, however, that undissociated tyrosine can also participate in hydrophobic interactions. In all we may have $64 \times 3 \times 3 = 576$ single substitu-

y / x	A	C	G	U	z
A	Lys	Thr	Arg	Ileu	A
	AspN	Thr	Ser	Ileu	C
	Lys	Thr	Arg	Met	G
	AspN	Thr	Ser	Ileu	U
C	GluN	Pro	Arg	Leu	A
	His	Pro	Arg	Leu	C
	GluN	Pro	Arg	Leu	G
	His	Pro	Arg	Leu	U
G	Glu	Ala	Gly	Val	A
	Asp	Ala	Gly	Val	C
	Glu	Ala	Gly	Val	G
	Asp	Ala	Gly	Val	U
U	Ochre	Ser	Nonsense?	Leu	A
	Tyr	Ser	Cys	Phe	C
	Amber	Ser	Try	Leu	G
	Tyr	Ser	Cys	Phe	U

Fig. 140. Full table of the genetic code. Ochre and amber are arbitrary names of nonsense triplets.

tions in 64 codons. Not one of the 192 substitutions for flection z results in the transformation $p \rightleftarrows n$, while 176 substitutions for z correspond to transitions $p \rightarrow p$ and $n \rightarrow n$, 7 to transitions nonsense → sense ($b \rightarrow 0$), 7 to transitions $0 \rightarrow b$, 2 to transitions $b \rightarrow b$; 114 substitutions for the prefix x correspond to transitions $p \rightarrow p$, $n \rightarrow n$, 60 to transitions $p \rightarrow n$ and $n \rightarrow p$, 9 to transitions $b \rightarrow 0$, and 9 to transitions $0 \rightarrow b$. The most dangerous mutational substitutions from the point of view of changes in protein structure are in root y, where 102 of the 192 are related to transitions $p \rightarrow n$ and $n \rightarrow p$, and only 74 to transitions $p \rightarrow p$ and $n \rightarrow n$. Furthermore, 7 substitutions correspond to transitions $b \rightarrow 0$, 7 to transitions

y / x	A	C	G	U
A	Lys AspN	Thr	Arg Ser	Ileu Met
C	GluN His	Pro	Arg	Leu
G	Glu Asp	Ala	Gly	Val
U	– Tyr	Ser	–? Cys	Leu Phe

Fig. 141. Key table of genetic code doublets.

$0 \rightarrow b$, and 2 to transitions $b \rightarrow b$. As a whole, 364 of 549 substitutions ($576 - 27$ transitions $b \rightarrow 0$ and $b \rightarrow b$, since in a protein there are no amino acids which correspond to the b codons), i.e., two-thirds do not result in transformation of the amino acid residues from one class to the other; they are correct transitions $p \rightarrow p$ and $n \rightarrow n$. In this sense, the code has quite a strong stability against disturbances. In fact, if all the mutational substitutions for the amino acid residues in the protein had been genetically equiprobable, the ratio of the number of correct and incorrect substitutions would have been not $2:1$ but $0.9:1.0$ (there are 10 p residues and 10 n residues; a given residue may be replaced by 9 residues of the same class and 10 of the other class).

The correlation described between the codons and the amino acid residues points to the genetic coding of the space structure of the protein and its function. The molecular sense of this correlation should be sought in the correlation between the anticodon of T-RNA and the aminoacyladenylate which corresponds to the specific enzyme. It is possible that such an enzyme has two active centers, one of which interacts with the aminoacyladenylate and the other with the anticodon. At present, we know very little about this.

On the other hand, the code characterizes some idiosyncrasies of the interactions of the codon with the anticodon. As has been stated already, the greatest danger of substitution is in the root of the codon, i.e., the second nucleotide y. In Table 27, we have listed the numbers N which express the number of hydrogen bonds formed by the prefix x and the root y of the codon with the corresponding nucleotides of the anticodon. It is characteristic that, in the first group of codons, these numbers are equal to 6 and 5 (with an average of 5.5), while in the second group, they are 5 and 4 (with an average of 4.5). One may suppose that, when $N = 6$ for the prefix and root xy, the interactions $z-z'$ of the codon with the anticodon are not important since the bond $xy-x'y'$ is sufficiently strong and ensures the necessary complementarity. It is just because of this that the corresponding codons are indifferent to the flections.

We see that the genetic code has a *water* origin; its structure, to a great extent, is determined by the specific structure of water which gives rise to hydrophobic interactions within the protein.

In this way, the genetic code is deciphered and the important rules found within it are established. However, to understand the molecular sense of these rules, it is necessary to carry out further investigations. Independent confirmations of the code and an elucidation of the nature of its action are found in studies on point mutations.

POINT MUTATIONS

As has been stated already, it is necessary to differentiate between chromosomal and point or gene mutations. The first are expressed in changes of the supramolecular structure, namely, the chromosomes, which are visible under the microscope; the second can be reduced to a change of the molecular structure of the genes. Here we are interested only in point mutations. Point mutations may be spontaneous, i.e., they may arise randomly as a result of the thermal motion within the nucleotides, or they may be induced. Induced mutations are brought about by different chemical substances (chemical mutagenesis) and by shortwave radiation (radiational mutagenesis).

It is evident that changes in the sequence of nucleotides within DNA as a result of a mutation must find an expression in corresponding (within the code alphabet) changes of the structure of the synthesized proteins. At present, a complete chemical investigation of a number of mutated proteins has been carried out, and, in particular, of the protein of the tobacco mosaic virus (Wittmann, and Fraenkel-Conrat),[70] alkaline phosphatase of *E. coli*,[70] tryptophan synthesis of *E. coli*,[70] etc. In a number of cases, it was possible to establish directly a correlation between changes of the nucleotides and the corresponding changes of the amino acid residues. It is reasonable to differentiate between three types of point mutations, namely: (*a*) mutations which change the sense of the codon (*missense mutations*) and also result in the replacement of a single amino acid in the protein by another one; (*b*) mutations which destroy the sense of the codon (*nonsense mutations*), i.e., transitions, indicated above by $0 \rightarrow b$, from sense codons to three nonsense codons, UAA, UAG, and UGA (such mutations result in an interruption of the synthesized protein chain); (*c*) deletions and insertions, i.e., the deletion of nucleotides or their insertion into the chain (mutations of this type, *frame-shift mutations*, result in a displacement of the reading of all the codons and a considerable disturbance of protein synthesis).

A high mutational activity is possessed by some analogs of the nitrogen bases of the nucleic acids. Such, for example, are 5-bromouracil (BU)

Keto form   Enol form

and 2-aminopurine (AP)

$$
\begin{array}{c}
\text{H} \\
|\\
\text{C}
\end{array}
$$

These compounds are incorporated within the DNA and replace the usual nitrogen bases. Thus, in the synthesis of DNA *in vitro* according to Kornberg, BU is incorporated in the chain instead of T, 5-bromocytosine (BC) instead of C, Hx instead of G. The mutagenesis is explained probably by errors in the DNA reduplications which contain an anomalous base instead of a normal one.[80]

It appears that BU can imitate T and form a pair with A if it is not present in the usual keto form (Fig. 142*a*). On the other hand, in its rarer enol form, BU imitates C and forms a pair with G (Fig. 142*b*). These two cases differ from each other. The first may be regarded as an error in incorporation which occurs during the reduplication of the starting DNA; A combines, not with T, but with BU. In the second case, we are dealing with an error in reduplication; the DNA chain, which already contains BU, forms in this position, not a BU–A pair (which would have corrected the error of incorporation), but a BU–G pair as a result of the tautomeric transformation of BU:[81]

$$
\begin{array}{llll}
\text{A—BU} \longrightarrow \text{A} & \text{BU} \longrightarrow \text{A—T} & \text{BU—A} \longrightarrow \text{A—T} \\
\text{BU} & \text{A} \longrightarrow \text{A—T} & \text{BU—A} & \text{A—T} & \text{etc.}
\end{array}
$$

The probability of tautomeric transformation to the enolic form increases, thanks to the presence of the Br atom which attracts electrons from the pyrimidine ring, as a result of which the H transfers more easily to O than it does in U, which does not contain Br.

In view of the above phenomena, it is necessary to emphasize the importance of experimental and theoretical investigations of tautomeric transitions in purines and pyrimidines. Here we need even orientational quantum-mechanical calculations (it is impossible to expect exact calculations in the present-day status of quantum-mechanical chemistry) of the relative contents of keto and enol forms. Such calculations may shed light on the problems of mutagenesis, including spontaneous mutagenesis. The point is that even the normal DNA bases may exist in tautomeric forms and form non-Watson–Crick pairs. At present, we have no other principles for examining the possibility of pairs, except for an evaluation of the hydrogen bonds formed. It is evident that it is necessary to take

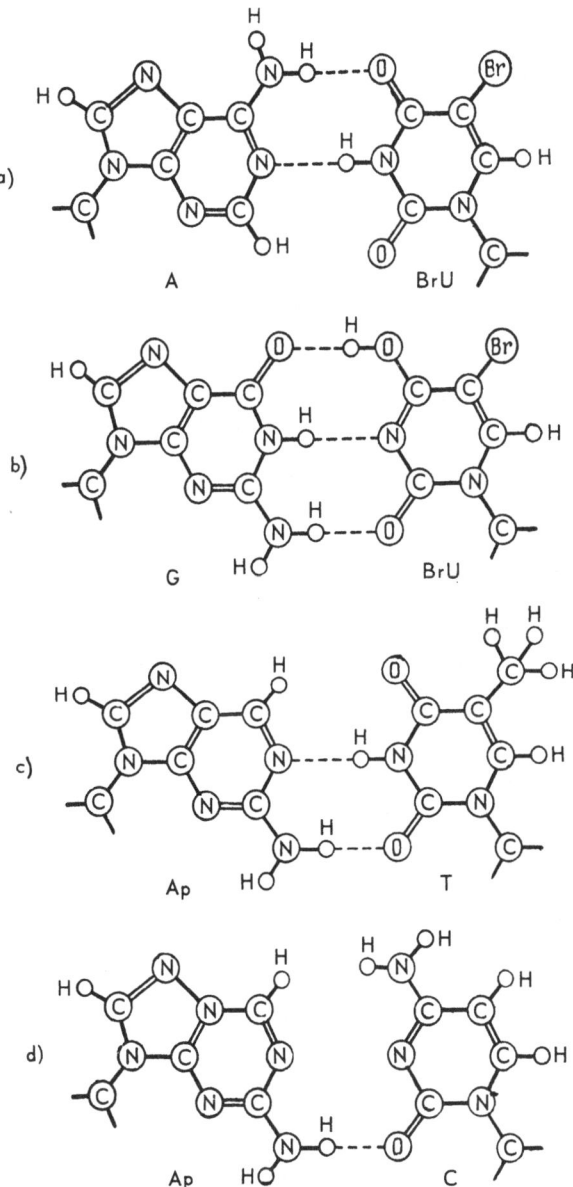

Fig. 142. Non-Watson–Crick pairs: (a) keto BU–A; (b) enol
BU–G; (c) AP–T; (d) AP–C.

tautomerism into account in order to construct a molecular theory of mutagenesis.

In the normal form, AP can form pairs with T involving two hydrogen bonds, and with C by a single hydrogen bond (Figs. 142c and d); therefore, it is capable of imitating both A and G.[82]

In view of what has been said, direct structural investigations of nucleotide pairs are of great interest. X-ray investigations were carried out on monocrystals built from methyladenine–methylguanine, ethyladenine–methyluracil, and methylguanine–methylcytosine pairs and also from a number of pairs which include analogs of nitrogen bases endowed with a high mutagenic activity.[81–85] In these investigations, the nitrogen bases were methylated in those positions in which the ribose of deoxyribose was attached to the nucleotides in order to exclude additional hydrogen bonds. It was found that, in the presence of one tautomeric form of the base, it is possible to have different dispositions of hydrogen bonds. In Fig. 143, we see the Watson–Crick A–T pair, the structure of the MA–MT complex found by Hoogsteen,[81] and the structure of the nucleoside complex of adenosine with 5-bromouridine found by Haschmeyer.[83] In all three cases, the pyrimidine exists in a diketo form, but the hydrogen bonds are different. A strong mutagen, bromouracil (methylated) has been investigated in pairs with ethyladenine[84] and methyladenine.[85] It is noteworthy that these two cases are quite different from each other. In the MBU–EA pair, the principal structure contains the hydrogen bonds O_2—N_6 and N_3—N_7, while in the MBU–MA pair, it contains the hydrogen bonds O_4—N_6 and N_3—N_7 (Fig. 144). These results show that it is not only tautomerism but also the ability to form different hydrogen bonds which can be of importance in mutagenesis.

It is noteworthy that it is not possible to obtain monocrystals of noncomplementary bases. Infrared spectra and nuclear-magnetic-resonance spectra of solutions have shown that the formation of hydrogen bonds occurs only in the pairs of A–U (or A–BU), A–T, and G–C, and not in the pairs of G–U, A–G, and A–C. It is possible that the selectivity

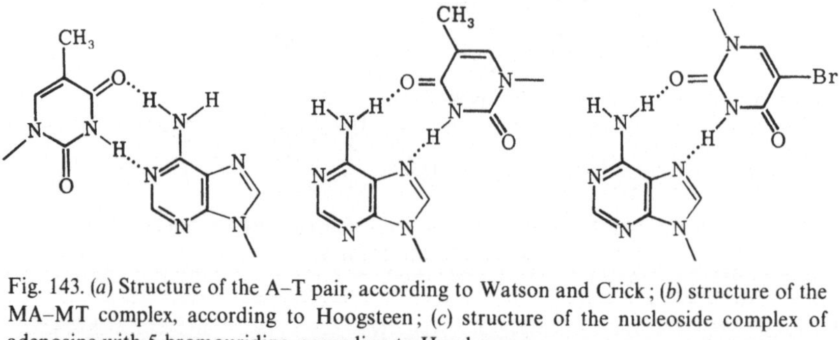

Fig. 143. (a) Structure of the A–T pair, according to Watson and Crick; (b) structure of the MA–MT complex, according to Hoogsteen; (c) structure of the nucleoside complex of adenosine with 5-bromouridine, according to Haschmeyer.

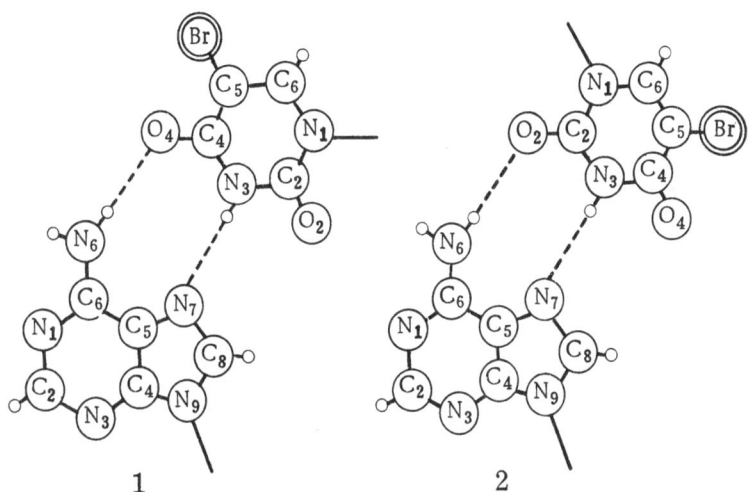

Fig. 144. (a) Structure of the MBU–MA complex; (b) structure of the MBU-EA complex.

during the formation of pairs is related to the interactions of a dipolar and hydrophobic nature. Another group of mutagenic agents is the acridine type of dyes, for example, proflavin

The interaction of compounds of this type with DNA has already been discussed.

The mutagenic action of nitrous acid HNO_2 can be reduced to the deamination of nitrogen bases. In DNA, it is G which is deaminated most frequently, and A least frequently; in the case of RNA, U does not deaminate.[88] The action of HNO_2 on TMV has been studied in detail. The TMV-RNA was subjected to deamination *in vitro*; then the plants were infected with a chemically changed RNA. New mutant forms of the virus were obtained in a number of cases.[83,89-91] In other cases, the RNA had lost its infectivity. To form a mutant TMV, it was sufficient to deaminate only one nucleotide out of 6,500. By varying the experimental conditions, it was possible to obtain any necessary number of deaminated bases in a single molecule of virus RNA.

In the experiments with the nitrite mutants of TMV, a long-standing wish of biologists was finally realized: it was possible to carry out a direct and directed action on genetic characteristics. It is not important that we are talking here about the genetics of such primitive organisms (if they are really organisms) as viruses. The basic problem was solved and new possibilities are open now to science-fiction writers. Now there is no longer any doubt that chemical genetics will result quite shortly in a complete upheaval in medicine and agriculture.

In fact, we know exactly the changes which occur in the nucleotides during nitrite mutagenesis. As has been pointed out already, deamination leads to the following transitions: A → Hx (hypoxanthine) → G, C → U, G → X (xanthine). Also, Hx forms a pair not with T but with C, U with A and not with G; X forms a pair both with C and G, but it forms only a single hydrogen bond.

Wittmann studied 100 nitrite mutants of TMV and established a correlation between the changes obtained and the triplet code.[6,90]

Diethylsulfate $(C_2H_5)_2SO_4$ is a strong mutagen. It causes the ethylation of G with its subsequent removal from DNA. Hydroxylamine NH_2OH interacts with C to form compounds which imitate T. In these cases, as well, we are talking about action on single bases. Freese studied in detail the effect of chemical mutagenesis on the genetics of T4 phage. A number of recent data on chemical mutagenesis have been summarized in a review article by Freese.[91]

Recently new strong mutagens which result in the mutations of the first two types have been found. Among these it seems necessary to point out NN'-nitrosonitroguanidine

$$O_2N-NH-\underset{\underset{NH}{\|}}{C}-NH-NO$$

Mutations brought about by this reagent revert easily. In other words, the mutagenic activity of this substance is reversible. New combinations of old mutagens, namely, of acridine compounds with nitrous yperite, for example,

$$HN-(CH_2)_3-NH-CH_2-CH_2-Cl$$

are also quite effective mutagens. The so-called "suppressor" mutations are quite important for verifying the codons *in vivo*. Let us assume that a mutation of the first or second type, namely, missense or nonsense, has occurred. There are suppressor mutations which correct the error which has arisen during this. Yanofsky has shown that suppressor mutations take place in a region of the DNA chain which codes T-RNA and that these are reduced to a change in the anticodon of T-RNA. Thus, if we have the mutation GGU (Gly) → UGU (His), the *E. coli* strain *Su*78 corrects this error. This strain synthesizes a T-RNA which has an anticodon ACA to UGU, but binds Gly instead of His. Khorana isolated such a T-RNA from *Su*78 and used it for the synthesis of a polypeptide on a template of polyUG according to the method described above. The cell-free system contained ribosomes and enzymes of the normal nonsup-pressed strain of *E. coli*. Under these conditions, instead of (Val–His)$_n$, (Val–Gly)$_n$ was actually obtained. A number of suppressor strains of *E. coli* which correct nonsense codons have been found.

Other substances, such as antibiotics, also have a strong influence on the translation of the code; these cannot be called mutagens in the exact sense of this word. It has been shown recently that it is possible to distort the code and to introduce errors into its translation by introducing streptomycin[76–78] into a cell-free system which synthesizes proteins. Antibiotics, including streptomycin, kill bacteria by influencing in some manner their protein synthesis. As has been stated already, resistance or susceptibility to an antibiotic may be the result of a mutation. It was found that streptomycin acts on the ribosomes by entering the 30S units.

In the presence of streptomycin, poly U stimulates to a much weaker extent the incorporation of phenylalanine; the synthesis of the polymer is 50 to 85% blocked. To the contrary, the incorporation of other amino acids, which are not stimulated normally by poly U, continues also in the presence of streptomycin.

Isoleucine is incorporated; serine and leucine are incorporated to a lesser extent. Similar effects have also been observed during the introduction of other synthetic polynucleotides and under the action of two anti-biotics, polymycin and neomycins B and C. Thus, the blocking of protein synthesis by antibiotics is related to errors in translation of the code, i.e., to modifications of the ribosomal region which bind T-RNA. A modifica-tion of this region, caused by the binding of streptomycin, permits the error T-RNA to complement the template well, so that the error amino acid enters the peptide chain.[77] A further modification caused by the mutation changes the structure of this region in such a way that the correct T-RNA pairs, independently of the presence of streptomycin. Here we are talking about the formation of mutants which are resistant to strepto-mycin. Finally, still another modification is possible; in it, the ribosomal region requires the streptomycin molecule in order to function correctly. This explains the existence of mutants which depend on antibiotics.

The coding by means of poly U is also a function of the concentration of Mg^{++} ions.[79] At concentrations which are considerably different from physiological, the code is read incorrectly.

It has been shown recently that streptomycin can result also in the correction of mutational errors of the first and second types (pheno-typic and not genotypic). Under the effect of streptomycin, the T-RNA molecules start recognizing doublets instead of triplets. This is particularly interesting in relation to what has been said about the secondary role of the flection z in the codon. Such changes caused by streptomycin may remove nonsense mutations, since nonsense triplets cease to interfere with protein synthesis.

Direct investigations of different types of T-RNA (in particular of three valine T-RNAs) have shown that these distinguish the prefix and the root, xy, quite well; the reaction with z, with given xy, differs only quantitatively. It has been established independently that there are such T-RNAs the anticodons of which are better adsorbed on the codons of M-RNA if the z is a purine nucleotide, and also such T-RNAs the anticodons of which interact with z pyrimidine nucleotides.[70]

The investigations of point mutations have resulted in strong confirmations of the correctness and universality of the genetic code. At the same time, it was found that the scheme of protein synthesis that has been described above requires serious additions. The work of the biosynthetic template system is a function of the state of the surrounding medium and it obeys complicated regulatory rules. If there were no such rules, then it would be absolutely impossible to understand such phenomena as the differentiation of cells and morphogenesis, which is the basis for the ordering in time and space of protein synthesis.

PROBLEMS OF REGULATION

As we have seen, proteins are synthesized according to a program coded on DNA with the participation of various types of RNA. The scheme is, in general, rather simple, notwithstanding a number of yet unsolved problems. But is it possible really to assert that the entire problem of protein synthesis and, consequently, the problem of the behavior of the organism on the molecular level can be reduced to the transfer of information by DNA → RNA → protein? No, in fact, the situation is much more complicated. Protein is synthesized, not continuously, but only when it is necessary to the cell, during definite stages of its development. DNA reduplication also occurs not continuously, but only during the interphase of mitosis. During cell differentiation of the multicellular organism, its proteins become specialized and its entire biochemistry becomes specialized. Consequently, protein synthesis is regulated in some way within the cells. The same also evidently applies to the reduplication of DNA.

The simplest method of regulating the action of any machine is feedback. The work of the regulated system depends on the results of this work. Living systems are very well regulated. One of the most important present problems of molecular biology and molecular biophysics is the investigation of the nature of this regulation and the discovery of the molecular contours of the feedback. The first stages in the solution of this problem were laid with the investigation of the induced synthesis of enzymes.

Three types of influence of mutagenic substances on the synthesis of proteins are known. These are, first, the synthesis of antibodies under the action of foreign biopolymers, namely, antigens; this is the basis of immunity. The second is the induction of enzyme synthesis. The third is the repression of such a synthesis by certain substances. Immunity, antibodies, and antigens will be discussed in the next chapter. Here we shall talk only of the closely related phenomena of induction and repression of enzyme synthesis.

The nature of enzymatic action is also described in the following chapter. Looking ahead, let us point out that a given enzyme catalyzes a certain chemical reaction. The substance which enters into the reaction is a substrate; the substances obtained as a result of the reaction are the products of the enzymatic reaction. The phenomenon of induction is that cells, which do not synthesize a certain enzyme, start synthesizing it under the influence of a specific chemical compound which almost always is a substrate of the given enzyme or a compound analogous to such a substrate. The phenomenon of repression consists in the stopping of the synthesis of the enzyme by specific chemical compounds, which are the products of the reaction catalyzed by the given enzyme or analogs of the product.

Induced enzyme synthesis has been well studied in a number of bacteria and, in particular, in *E. coli*. These cells can produce the enzyme β-galactosidase (β-Gal) which catalyzes the hydrolysis of β-D-galactosides, i.e., the reaction

E. coli bacteria of the wild type, which grow in a medium of inorganic salts with a nongalactoside carbon source (with succinic acid), practically do not synthesize β-Gal. The addition of a suitable galactoside to the growing culture increases the rate of synthesis of this enzyme one thousand-fold. The high rate of synthesis of β-Gal persists as long as an inducer is available within the nutritional medium, namely, a galactoside or a related compound. When the inducer is removed, i.e., when the growing cells are transferred to a new medium which does not contain the inducer, the rate of synthesis falls to the original value. This situation is illustrated in Fig. 145, which shows the dependence of the differential intensity of synthesis P, shown as the ratio of the increase of the amount of enzyme Δz to the total increase of cell mass ΔM,

$$P = \frac{\Delta z}{\Delta M}$$

on the total amount of protein within the bacteria.[96,97] Stimulation of the synthesis of β-Gal takes place very rapidly; P increases one thousandfold in less than 2 minutes after the introduction of the inducer.

The enzymes obtained during the induced synthesis are called *induced enzymes*. This term is not exact since we are not talking about some special enzyme but about a special synthesis.

Along with wild induced cells, there are also *E. coli* strains which synthesize β-Gal in the absence of the inducer. In this case, we speak of the *constitutive enzyme* or, more exactly, the *constitutive mutant*.

The first question which arises during the investigation of induced synthesis is the following: Does the inducer act on some protein precursors of β-Gal and activate them in some form, or is the protein synthesized anew, *de novo*, to use the traditional biological term?

The answer to this question was obtained in experiments with the aid of labeled atoms. This is not the first time we have met the possibilities that have been made available to biology by artificial radioactivity. In these experiments, the *E. coli* bacteria were grown in a medium which contained ^{35}S in the form of a sulfate. Labeled radioactive proteins were synthesized within the cells. Then, these cells were transferred to a medium which contained unlabeled sulfate and the inducer, namely, methyl-β-D-thiogalactoside (the sulphur in this compound replaces oxygen in the pyranose ring of the galactoside). The cells grew and synthesized β-Gal. It was found that there was no radioactive sulphur in β-Gal. Consequently, the induced β-Gal did not have any protein precursors. Simultaneously,

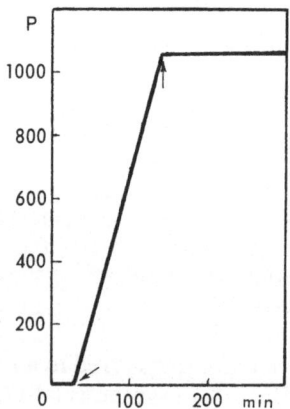

Fig. 145. Induced synthesis of an enzyme.

this result shows that other proteins synthesized in the first stage of the experiment, in the absence of the inducer and in the presence of ^{35}S, do not become decomposed again into amino acids which could be incorporated into β-Gal during the second stage of the experiment. Protein synthesis, including that of β-Gal, is essentially an irreversible process. Similar results are obtained also with the use of ^{14}C.

Thus, the induced synthesis takes place *de novo* from amino acids. If we consider that protein synthesis is always determined genetically and proceeds with the constant participation of DNA and RNA, we must come to the conclusion that the inducer acts on the genetic system of the cell, on the nucleic acids. Jacob and Monod carried out the genetic analysis of the induced synthesis, starting with a very elegant hypothesis which has been substantiated by facts and has now taken on the aspects of a scientific theory.[98,99]

The hypothesis is that the induced synthesis is suppressed by a specific compound present in cytoplasm. This compound may be called a "repressor", i.e., a suppressor of the synthesis. The repressor is synthesized under the control of a special regulator gene, which is different from the usual structural genes, responsible for the primary structure of the given enzymes. Furthermore, the cell contains an operator gene which governs the transfer of the information simultaneously from several structural genes to several proteins. The operator can interact with the repressor. This interaction suppresses the work of the operator and the synthesis of protein ceases. The action of the inducer (metabolite) consists in chemical influence on the repressor as a result of which the repressor blocks the operator.

This hypothesis explains the phenomena of both the repression and the induction of enzyme synthesis. The repressing metabolite (the product and analog of the product of the enzyme reaction), in this case, must activate the interaction of the repressor with the operator. This model is presented schematically in Fig. 146. The model appears, at first glance, to be extremely abstract. It permits the formulation, however, of a number of conclusions that are susceptible to experimental verification.

If the regulator and operator genes actually do exist, they must be subject to mutations. Let us first examine such mutants of *E. coli* as verify the existence of the regulator gene. In the wild type of *E. coli*, it is possible to cause an induction; let us call it R^+. There is also a mutant R^- which produces β-Gal constitutively; it is evident that it has lost the regulating system, i.e., it lacks a repressor. If this is correct, then diploidal heterozygotes R^+/R^- must be induced and not constitutional; the repressor, as a substance which enters the cytoplasm, must act also on the operator gene of the R^- chromosome. In this sense, the R^+ allele is dominant. This is found to be true.

Another verification is the existence of mutants R^t, the properties of which depend on temperature. At low temperature, repression is active

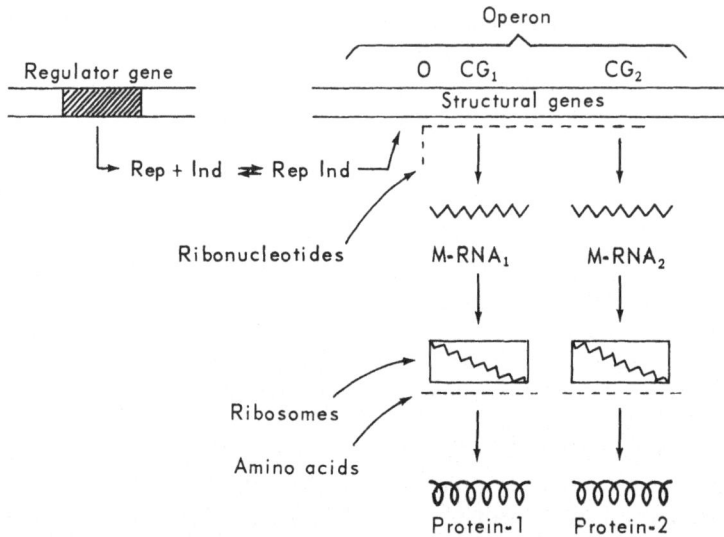

Fig. 146. Schematic representation of the action of the operon system: regulator gene, operator gene, and structural genes.

and the system performs as R^+, that is, it is inductive. At high temperature, the repressor is inactivated and the system is similar to the constitutive mutant R^-. The fact that the introduced metabolite, the inducer, reacts with the repressor in the cytoplasm is proved by the existence of a super-repressed allele R^s. The R^s mutants are unable to synthesize β-Gal even in the presence of the inducer. The heterozygote R^s/R^+ is similar to R^s, that is, in this case, the dominant fact is not the presence of some function, but its loss. The properties of R^s can be explained by the fact that, in this case, the regulator gene controls the production of a repressor with a reduced chemical analogy to the inducer.

The interaction of the repressor with the inducer may be written out in the form of Rep + Ind $\rightleftharpoons$ Rep–Ind, i.e., it corresponds to the reversible association with the complex Rep–Ind. In the induced system, the repressor Rep is active and represents the synthesis of proteins, while the Rep–Ind complex is inactive in this case. On the other hand, in the repressed system, the repressor and inducer change places, i.e., the substance produced under the control of the regulator gene is inactive, but the Rep–Ind complex is active and blocks protein synthesis.

Actually a reverted mutant of *E. coli*, R^r, has been found recently. This mutant is partially constitutive, but the addition of galactosides, which are inducers for the wild type R^+, results in a decrease of production of β-Gal in R^r, that is, it causes repression. Thus, a mutation of the regulator gene may change an inducing system into a repressing one.

These facts are summed up in Table 28. In the last column of this table, we have shown the properties of lysogenic systems formed during the

Table 28[99] Mutations of the Regulator Gene of β-Galacto-sidase Synthesis in *E. coli* and of the Regulator Gene of λ Phage

Allele	Product	Mutant properties	Moderate λ phage
R^+	Wild type of repressor.	Induced by specific inducers (β-galactosidase).	Capable of lysogeny: The lysogenic system is induced for phage production by irradiation with ultraviolet light and chemical interactions.
R^-	Inactive repressor.	Constitutive; induced R^+/R^- heterozygote.	Incapable of lysogeny: A lysogenic system R^+/R^- is obtained on mixed infection with R^+.
R^s	Superrepressor, not suppressed by the inducer.	Incapable of producing the enzyme even in the presence of inducer. R^s/R^t and R^s/R^- also do not produce enzyme.	Capable of lysogeny: The lysogenic system is not induced. The R^s/R^+ and R^s/R^- systems are similar to R^s.
R^t	Repressor system sensitive to temperature.	Induced at low temperature; constitutive at high temperature.	Capable of lysogeny at low temperature, but not at high temperature: Lysogenic systems, grown at low temperature, produce phage on heating.
R^r	Reverted system.	Partly constitutive; repressed by β-galactosides.	

infection of *E. coli* cells with λ phage. The phage DNA becomes incorporated in the genetic system of the host cell, which multiplies during lysogeny until it falls under the action of the inducer. At this point, the phage starts being synthesized within the cell and the cell perishes. The table shows that the transformation of the prophage into the phage is subject to the same laws as the synthesis of β-Gal, i.e., it is controlled by a regulator genome. This is not surprising, since multiplication of the phage is also related to protein synthesis.

The operator gene can also mutate. The O^0 mutant is able to synthesize neither β-Gal nor a second enzyme which guarantees the penetrability of the cell by galactosides, i.e., β-galactosidepermease. The O^c mutant is constitutive; the O^+ mutant is a normal induced type. The O^+/O^c heterozygote synthesizes both proteins constitutively; the same is true of O^0/O^c.

Consequently, the action of the operator gene occurs, not through the product which passes into the cytoplasm, but directly in the chromosome. The operator gene controls the function of the structural genes which are connected to it. This entire system is called an *operon*. It has been shown in experiments on the crossing of bacteria that the regulator gene is spatially and structurally independent of the operon; they are located relatively far from each other.

Similar phenomena have been found not only in *E. coli*, but also in other types of bacteria, for example, in *Salmonella*. In this case, eight enzymes are necessary for histidine synthesis. The structural genes which control their synthesis are part of a single operon.

Experiments have verified a number of predictions by Jacob and Monod's theory. It was found, for example, that the separation of structural genomes from the operator gene as a result of chromosomal crossovers, may release these genes from the control of the starting operator and link them to another operator, either to one next to the original one (in the case of deletion) or to one separated from it (in the case of translocation). From what has been said, it follows that the operator gene determines the place from which the reading of the genetic text will start on the DNA, i.e., the place in which the action of the enzyme RNA polymerase may start.

What is the nature of the repressor synthesized under the control of the regulator gene? Up to now, this substance has not been isolated, but a number of properties of induced systems have led Jacob and Monod to believe that the repressor is a protein. The repressor, like the enzyme, interacts with the low-molecular-weight metabolite, namely, the inducer.

In this manner, the function of the genetic system of the chromosome is regulated by the cytoplasm, i.e., through the repressor.

Investigations of the regulation of biosynthesis have undergone broad development in recent years. It has been found that operons have a definite polarity, which guarantees regulation of the reading of the genetic code. The first cistron which is attached to the operator gene organizes the synthesis of the protein 10 times more effectively than the next cistron. The number of synthesized enzymes decreases monotonically with an increase of the distance from the operator. An interruption (as a result of nonsense mutations) of the synthesis of the first protein influences the synthesis of subsequent proteins and, in this case, mutations also have a definite polarity.

The T-RNA molecules also participate in regulation. There are incomplete types of T-RNA, which are synthesized in limited amounts. If there is a codon which requires for its actions such a T-RNA, then, in its absence, the synthesis of the protein on the condon stops. Consequently, such a codon, not being nonsense, can modulate biosynthesis.

More complicated systems of regulation are also known. To synthesize a certain metabolite, it is necessary to have a number of chemical reactions, each of which is catalyzed by a given enzyme. The end product of the entire reaction chain locks the action of the enzyme, which catalyzes the first reaction of the chain, notwithstanding the fact that this product has nothing in common with a substrate or product of the first reaction. The nature of this most interesting phenomenon, called *allostery*, is described in the following chapter. It is clear that there is feedback in the true sense of the word.

Protein synthesis is regulated with the help of the cytoplasm. Unlike cybernetic machines, in which regulation is carried out by electromagnetic interactions, the cell regulation is carried out chemically. Here is seems that an extremely important role is played by the structural complementarities of the interacting molecules.

Up to now, we have talked of the regulation of protein synthesis. The synthesis of M-RNA is also regulated on the molecular level. The reduplication of DNA and the reduplication of chromosomes is also subject to regulation. After all, these processes also occur in a concerted manner with the development of the cell.

Jacob and Monod[99] and Maaløe[100] have proposed a hypothesis according to which DNA reduplication must be induced by some substance of a proteinaceous nature. The genetic element (the chromosome or episome) is reduplicated as a whole after such an induction. They have called such an element a *replicon*. It is important that, unlike the operon, a replicon starts to operate under the action of an activator–initiator, and not under the action of an inducer which influences a repressor. The scheme of the replicon is shown in Fig. 147. The cyclic chromosome (the DNA of a bacterium or a phage) contains a structural gene responsible for the synthesis of the protein activator which acts on the replicator gene. It is probable that this is accompanied by the breaking of the ring and the chromosome is replicated then as a whole.

The hypothesis on the replicon leads to a number of genetic conclusions on the nature of mutants of the structural genes and of the replicator gene. These conclusions have been verified experimentally.[101] A number of data, including the results of direct experiments by electron microscopy, have shown that the initiator is bound to the cell membrane during the

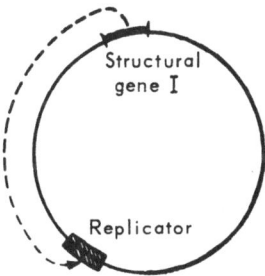

Fig. 147. Scheme of the replicon.

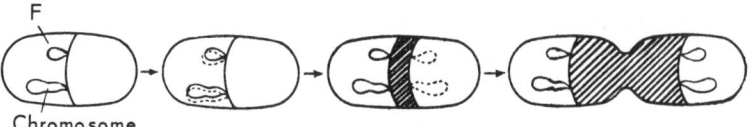

Chromosome

Fig. 148. Scheme of the reduplication of replicons, and cell division.

division of bacterial cells. The division of the cells occurs according to the scheme: signal for chromosome reduplication, growth of the cell surface, signal for chromosome reduplication. Such a scheme is shown clearly in Fig. 148.

The conjugation of bacteria also occurs as a result of a direct reduplication of the genetic material (Fig. 149).

The system which governs the synthesis of proteins within the cell is also responsible for the development of the cell and for the differentiation of the cells in multicellular organisms. The different genes do not act simultaneously. Khesin[102] has investigated the laws of synthesis and regulation of specific M-RNAs in *E. coli B* cells, infected with T2 phage. It was found that the first M-RNA molecules synthesized are those which determine the synthesis of enzymes formed at the beginning of the infection, and then those M-RNA molecules are synthesized which are needed for the synthesis of the *viral antigens*, as they are called. The early and late M-RNAs are synthesized on different sections of the phage DNA. The regulation of M-RNA synthesis, the suppression of some genes, and the activation of others are coordinated with the reduplication of phage DNA and depend on the formation of proteins. Initially those DNA genes which are responsible for the synthesis of late M-RNA are blocked.

According to the Jacob and Monod's theory, the consecutive reading of the genes must be regulated by the action of repressors. It has been found, however, by Khesin *et al.*[103] that, even in purified DNA, the RNA polymerase reads only some genes, even though there is no repressor in the system. It follows from this that the specificity of RNA synthesis is a function first of all of the interactions of RNA polymerase with definite

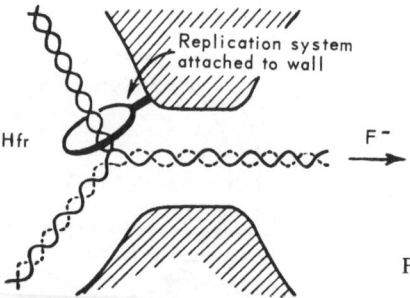

Replication system attached to wall

Hfr

F⁻

Fig. 149. Scheme of DNA reduplication during the conjugation of bacteria.

points on DNA. In all probability, these points correspond to operator genes.

In order to understand these details of extremely important regulator processes within the cells, it is essential to study the interactions of DNA with proteins. Recently, data have been obtained which indicate that the histones (the proteins of the chromosomes of higher organisms) affect the rate of RNA synthesis. This means that the regulation of the synthesis may be carried out through the histones as intermediates; the turning-on and turning-off of the synthesis takes place at different times in different loci of the chromosomes.[104,105]

According to the analysis by Allfrey and Mirsky,[104] only 15 to 20% of the DNA of calf-thymus lymphocytes are genetically active, i.e., synthesize M-RNA. The addition of thymus histones to the suspension of isolated nuclei of the thymus inhibits RNA synthesis; on the other hand, selective removal of histones increases this synthesis. The addition of histones to isolated chromosomes of the lampbrush type results in the contraction of the loops and a decrease of RNA synthesis.

In this manner, regulatory processes are evidently responsible for the differentiation of the cells; they consist in the blocking or deblocking of parts of the genetic information. Histones, which are protein components of the chromosome, play an important role in this process.

Solution of the problem of regulation will make it possible to pass over to a cellular and organism level of structure, to questions of cell differentiation.[106] In the human body, there are approximately 10^{15} cells which belong to 10^2 different types. Each somatic cell contains the entire set of genes analogous to those which existed in the initial zygote. However, in any specialized cell, only a small fraction of the entire stored genetic information is active; the other genes are evidently blocked. Recently, direct proofs of this important position have been obtained.

The cells of the salivary glands of two-winged insects (among them, the drosophila) contain gigantic chromosomes, enormously magnified copies of the chromosomes of other cells. The nonuniform distribution of the genetic material is seen quite clearly in such chromosomes under a regular microscope (see Fig. 24). Here the genes can be seen directly. We see puffs on the chromosomes. It turned out that these are the places in which, during a given stage of development, strong synthesis of RNA takes place; the puffs are active genes[107] in contrast to the other parts of the chromosome. It has also been possible to establish that structurally different parts of the chromosomes of the so-called lampbrush type, which are found in the developing egg cells of amphibia,[108] namely, the loops and granules, correspond to genes of different activity.[104,106,109] It is also interesting to give an account of experiments carried out with carrots.[110,111] Small cylindrical samples of 1 to 2 mg have been cut out from thin sections of the nongrowing region of the carrot root. They were placed in a nutritional medium, for example, coconut milk, to which were added growth stimulators, such as adenine, and to which air was supplied. The carrot tissue started to grow rapidly; after 20 days, its weight had increased eightyfold! However, the growth took place in a nondifferentiated manner; round colorless cells multiplied. They divided. Then an individual cell was again placed in a nutritional medium and developed independently. Under these conditions, in a number of cases, the cell behaved like a zygote; the result was the development of an entire organism with roots and leaves!

It was shown in this manner that, under certain conditions, specialized cells of plant tissue can release a vast amount of genetic information. It is evident that genes become unblocked. Probably we are dealing with similar phenomena in the production of potatoes from eyes, of begonias from leaves, etc. It is quite evident that these experiments promise important practical results.

The model of the operon may be put as the basis of the model theory of cell differentiation. The process of differentiation means that cells, which have acted identically in the past, start acting under different conditions after differentiation, as a result of the repression of part of the genetic information. In other words, the system is of a trigger type. Jacob and Monod have proposed a trigger scheme built from two antagonistic operons (Fig. 150).

The product P_1 is synthesized from the substrate S_1 with the help of enzyme E_1; E_1 is a co-repressor for the synthesis of enzyme E_2; P_1 activates the inactive form of the repressor R_2 by combining with it and forms an active repressor Re_2, which blocks the operator of operon 2 and which suppresses in this manner the synthesis of E_2. In time, P_2 represses the synthesis of E_1.

This scheme has been subjected to experimental investigation.[112,113] However, it was possible for Chernavskii et al.[114] to obtain valuable

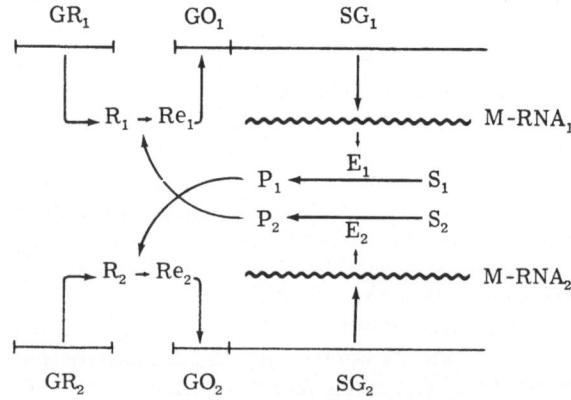

Fig. 150. Model of two interacting operons.

results only after simplification of the calculations. The simplification consisted in the application of the principle of minimization; the kinetics of a process are determined by the slowest step, which consists in the synthesis of the M-RNA and the enzyme. Analysis shows that the system shown in Fig. 150 can exist in two stationary states, in one of which the concentration of E_1 is much higher than that of E_2, while in the other one the inverse relation is true. Maximal changes in the concentration of one or several specific substances, i.e., the co-repressor, the enzyme, or M-RNA, shift the system from one stationary state to the other, i.e., the system is actually a trigger one.

Later, a unidimensional model of the cell system was examined, a set of cells in a narrow and long tube, with nutritional medium entering the system from one of the ends of this tube. Equations for the diffusion of the products and the substrates must be added to the initial kinetic equations describing the two-operon trigger system. The solution (obtained on an electronic calculator) shows that various layer distributions of the concentrations of E_1 and E_2 may arise as a function of the ratio of the diffusion coefficients, i.e., differentiation may take place.

A general conclusion from this quite interesting investigation is that specific structures may be obtained without a prior programming of the time and place of the introduction of the signal into each cell. It is sufficient that a given ratio of the parameters of the trigger system and the diffusion coefficients be included in the code. In other words, it is sufficient to code only the ability of the tissue to form a structure, and not the time and nature of action of the specific inducers.

One should emphasize that any molecular triggered system is actually a cooperative system; it requires the interaction of the elements of structure (for example, of two operons).

The theory of differentiation is, at present, in its initial state. This is an extremely interesting and important realm of investigation for both theoreticians and experimenters. We are talking here about one of the basic problems of modern biology.

The problem of differentiation approaches also the problem of cancer, that is, of malignant new growths. Cancer cells are characterized by the very absence of differentiation. If it became possible in some way to block part of the genetic system of these cells, they would lose their malignancy.

Investigations of regulation and differentiation are evidently more biological than the investigations of isolated biopolymers, and even of the template synthesis of protein. Regulation takes place in an entire living cell, and up to now, we have not found the means to carry it out in simplified model systems. As has been pointed out by Khesin, investigations of somatic cells *in vitro* are quite important and full of possibilities for genetics. What has been said shows that molecular biophysics penetrates to the very essence of the regulatory processes; entire biological systems become amenable to physicochemical investigations.

Chapter 8

BIOLOGICAL SPECIFICITY AND STRUCTURE OF MOLECULES

IMMUNITY

This and the following chapters treat the action of proteins, namely, ways in which they perform their specialized tasks. These tasks are varied and have been listed already. We shall dwell on some of the most important functions of proteins without expecting to treat this problem exhaustively. Let us start from the immunological, or protective, function.

It has been known for a long time that humans and animals who have been sick once with an infectious disease are no longer susceptible to a new attack. The great success of bacteriology is related, on one hand, to the discovery of chemical agents for fighting microorganisms (salvarsan, sulfamides, antibiotics) and, on the other hand, to the discovery of sera and vaccines which cause immunity against various diseases. Millions of human lives have been saved as a result of the work of Pasteur, Koch, Mechnikov, Khavkin, and others, whose work was truly heroic. Here science approached the border between life and death. It is impossible. therefore, to read calmly either novels about doctors and bacteriologists or their specialized scientific papers. We shall not tell here, however, about the enormously enticing history of the scientific fight for human health.

When foreign proteins or some other polymeric compounds enter an organism, specific proteins begin to be produced in the organism as a reaction: these proteins impede the action of the pathogenic compounds. The foreign substances which we are discussing are called *antigens* (this term has no relation to the gene), while the reaction proteins are called *antibodies*. Antigens are introduced into an organism by viruses, bacteria, and other unicellular organisms. Antibodies are produced by blood-forming organs, namely, by lymphatic nodes, the spleen, and bone marrow and they penetrate into the blood and into the lymph. They belong to the class of globular proteins, to the group called *gamma globulins* (it is well known that gamma globulin is frequently injected into an infant who has been in contact with another one infected with measles; this prevents or alleviates the sickness). The molecular weights of antibodies are high. Thus, in man, in rabbits, and in a number of other species, the antibodies have molecular weights of the order of 160,000; in the horse, pig, etc., the antipneumococcus antibodies have molecular weights of the order of 900,000.

The interaction of antibodies with antigens may manifest itself in a number of phenomena :[1] (a) Antibodies to toxins may neutralize the toxic antigens; antibodies to viruses neutralize infectious antigens. (b) Antibodies to soluble proteins and other soluble antigens may form with them insoluble precipitates (the precipitin reaction). (c) Antibodies to micro-organisms and foreign erythrocytes may cause their adhesion to each other and their precipitation, namely, agglutination. (d) Antibodies to foreign cells may cause their decomposition, namely, lysis; for these reactions to take place, it is necessary also to have the participation of other components of blood plasma which, as a whole, are called comple-ments. (e) Antibodies cause the swelling of membranes and of the capsules of a number of microorganisms. (f) The interaction of antibodies with foreign cells renders these more susceptible to phagocytosis, namely, to their attack by leucocytes. (g) Complement fixation, i.e., complement adsorption on an antigen–antibody complex during a reaction may serve as a criterion for the existence of such a reaction: for example, the Was-sermann reaction is based on this; this is the reaction used for the detec-tion of syphilis. (h) The antigen–antibody reaction may induce the release of histamines and other toxic substances from the host tissues; this results in symptoms of illness, namely, anaphylaxis and allergy. Symptoms of allergenic sicknesses caused by foreign biopolymers are well known : hay fever, child diathesis, asthma, etc. Individuals suffering from allergy must renounce strawberries, crabs, eggs, and other foods which cause in the individual serious pathological phenomena.

If an antigen is introduced into the organism of the rabbit, the produc-tion of antibodies specific for this antigen occurs after several days. What is called a crisis during a sickness means, in part, the formation of a large amount of antibodies in the blood. The antibodies rapidly remove the antigens, but their contents also decrease after a certain amount of time. If, however, a new dose of the antigen is introduced into the same rabbit, then, after approximately 3 days, antibodies are formed again, the concentration reaches a high level, decreases slowly, but remains appreciable for a long time. This is a secondary reaction with antigens, namely, the formation of strong immunity. Vaccination against smallpox must be repeated several times during a life, but yellow fever, for example, gives immunity practically for the entire life. It is quite fortunate however that one very rarely becomes sick for a second time with an infectious disease, such as measles, scarlet fever, or poliomyelitis. In this manner, the immunity of the organism indicates the existence of a certain memory about the antigens; this manifests itself in the continuous synthesis of the antibodies.

The organism of the newly born rabbit is incapable of producing antibodies; furthermore, if a large amount of antigen is introduced into it, it may survive, but, even as an adult, this rabbit will not produce antibodies to the initial antigen in its blood-forming organs. This interesting phenomenon is called *immunological tolerance*. X-ray irradiation interferes with the production of antibodies. If large amounts of antigen are introduced after irradiation, the inability to produce antibodies persists even after all other effects of irradiation have been eliminated. To the contrary, if the antigen is introduced before irradiation, the irradiation does not interfere with the formation of antibodies. If a rabbit receives a first portion of antigens several days before irradiation, the secondary reaction is not blocked.

A very important example for man of the interaction of antigens with antibodies is related to Landsteiner's discovery of blood groups in 1900. Serious pathological consequences may stem from blood transfusion without taking blood groups into account. This is due to the agglutination of erythrocytes by the antibodies present in the blood plasma. There are four principal blood groups, called A, O, B, and AB; these differ from each other by the presence of antigens (agglutinogens) to A and B in the erythrocytes and antibodies (agglutinins) a and b in the plasma: a agglutinates A, b agglutinates B. Therefore, it is impossible to mix group A blood with group B blood. The principal relationships for blood groups are shown in Table 29.[2]

The possible ways of blood transfusion are shown in Fig. 151.[1] Reactions of the type $A + a \rightarrow Aa$ are used in court medicine for determining blood groups. These reactions are so sensitive that with their use it has been possible to determine the blood groups of long-dead mummified humans from ancient Egypt, Mexico, and Peru. The blood of these ancient ancestors of modern man turns out to be the same as that of modern man. Human races differ in the frequency of different blood groups, but are not characterized by any given group.

Table 29

Blood group	Approx. % of population U.S.A.	Agglutinogens in erythrocytes	Agglutinins in plasma	Groups to which blood may be transfused	Groups from which blood may be transfused
O	45	....	a and b	O, A, B, AB	O
A	42	A	b	A, AB	O, A
B	10	B	a	B, AB	O, B
AB	3	A and B	....	AB	O, A, B, AB

Blood groups are hereditary, genetically determined characteristics, which follow Mendel's laws.[3] This is quite understandable; we are dealing here with specific erythrocyte and plasma proteins, the structure of which is determined genetically. It was further established later that there are two types of A antigens, namely A_1 and A_2, which are important in court medicine and anthropology, but not in blood transfusion. The same is true of the antigens M, N, and S.

In 1940, Landsteiner and Wiener discovered another very important blood agglutinogen, namely, the rhesus factor Rh, which was first discovered in the rhesus monkey (*Macacus rhesus*). Its presence may lead to the death of the infant in the mother's womb because of antigen–antibody interactions. Of the humans of the white race, 85% have blood containing an Rh antigen which is Rh positive. If the woman has Rh negative blood and the father of the child is Rh positive, then the infant may inherit from the father the Rh antigen. If the blood of the fetus has penetrated the maternal blood stream, antibodies to the Rh antigen are formed in the latter. During the second pregnancy. the interaction of the antibodies with the antigen leads to the agglutination of the erythrocytes of the fetus and it perishes from what is called fetal erythroblastosis, or shortly after birth.

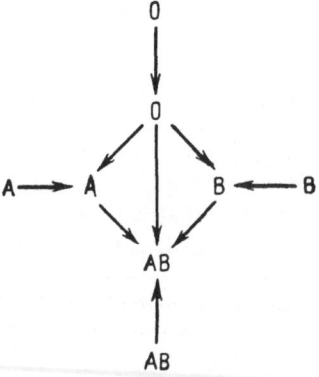

Fig. 151. Relation of blood groups.

In the second case, the infant may sometimes be saved by total blood transfusion. The best way of avoiding this misfortune is to determine the blood groups before marriage. The probability of a dangerous combination of $\male$Rh($+$) $\times$ $\female$Rh($-$) is fortunately small: it is 3% of all marriages and the probability of tragic consequences is no greater than 0.25%. For further details of the biology and medicine of immunity, the reader is referred to the monograph literature.[4]

The phenomena just described bring up two groups of questions. The first one is: How are antibodies made and what is the nature of the memory? The second one is: What is the nature of the antigen–antibody interaction? The first group of questions is biological in the true sense of the word, but in this case a number of important positions have been established by molecular biology as well. The second group relates to physical-chemical problems. Let us start with it.

Antibodies are produced only *in vivo*, but their interaction with antigens may be investigated *in vitro*. As a result of this, immunochemistry has become a field of exact physical-chemical studies. If we introduce into a test tube the blood serum of an immunized organism and we add to it the corresponding antigen, then the precipitation or agglutination reactions may be observed directly. It is easy to follow the formation of a precipitate, of a turbid zone, in an agar gel during the diffusion of solutions of antigen and antibody toward each other. In other words, antibodies may be titrated. This opens wide possibilities for the quantitative investigation of immunologic reactions.

A very important property of antigens is their great specificity. Immunity toward one disease does not prevent infection by any other one. As a result, it is necessary to get immunizations against a large number of diseases prior to travel in tropical countries. Antibody specificity is high, but not absolute. Thus, a man who has been infected with the typhus caused by *Rickettsia mooseri* carried by fleas becomes immune to the serious typhus carried by lice and caused by *Rickettsia prowozekii*.[1]

What causes the specificity of antibodies? How is it determined by their molecular structure? The answers to these questions have been given in the studies by Landsteiner.[5] In 1919 he discovered a method of producing artificial, synthetic antibodies. His method is based on the ability of diazotized aromatic amino acids to combine with proteins in weakly alkaline solutions.

The Landsteiner reaction is

$$\text{R}-\underset{\bigcirc}{\bigcirc}-\text{N}=\text{N}-\underset{\bigcirc}{\bigcirc}-\text{OH} \quad -\text{CH}_2-\overset{\overset{\displaystyle\text{NH}}{|}}{\underset{\underset{\displaystyle\text{CO}}{|}}{\text{CH}}}$$

The *R*-phenylazo compound reacts with the amino acid residues of the protein (mostly with tyrosines) and introduces any radical R into the protein. An R-phenylazoprotein is obtained; this may be used as a synthetic antigen. Proteins modified in this manner cause the production of antibodies and lead to specific immunological reactions (these reactions are also called *serological*, from the Latin word *serum*). Thus, long before the existence of modern biology, studies were carried out in which the biological activity of an organism was modified by artificial chemical means.

A comparison of the immunological properties of the radicals R attached to proteins and of the proteins themselves has shown that the determining factor in antigenic specificity is the radical, while the protein is secondary. This was demonstrated by cross-reaction experiments. An antiserum, obtained against R—⟨ ⟩—N=N—Protein 1, reacts with any other antigens which contain R but are otherwise serologically quite diverse proteins; it does not react with the antigen R′—⟨ ⟩—N =N—Protein 1, which contains another radical, R′. Consequently R is the determining group. R-phenylazoprotein gives a highly insoluble precipitate with the corresponding antibody. If small molecules containing the same determining group, the same radical such as

$$\text{R}-\underset{\bigcirc}{\bigcirc}-\text{N}=\text{N}-\underset{\bigcirc}{\bigcirc}-\text{R}''$$

are added to the system R antigen–R antibody, the precipitation reaction is retarded; as the concentration of the low-molecular-weight compound is further increased, the precipitation ceases completely. Small molecules of a given type do not cause the formation of antibodies in experimental animals and, consequently, they are not antigens. However, they interact with previously formed antibodies to form soluble complexes. This is what is called the *hapten reaction*; these small molecules are called *haptens*. The haptens probably compete with the antigens and, consequently, they interact with the same reactive regions of the antibodies. These reactions have the nature of usual chemical reactions and obey the law of mass action.

Thus, an antigen contains a determining group and a macro-molecular carrier; a hapten contains only the determining group. With the gradual destruction of the protein component of an antigen, its action is at first maintained; then, this is replaced by haptenic action.

An investigation of the immunochemical reaction shows that natural antigens contain a number of determining groups and in this sense are polyvalent. The groups may be different; therefore, during immunization they may cause the formation of a number of different antibodies. Using synthetic incomplete antigens it is possible to separate selectively different antibodies from each other.

In immunochemical reactions, a determining role is played by the space structure and the dimensions of the determining groups. A synthetic antigen of the type

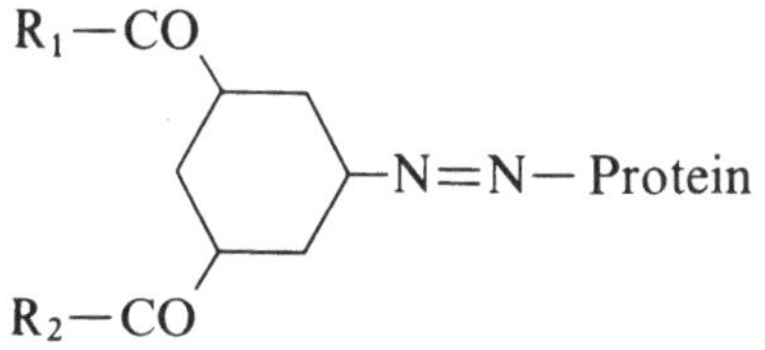

$$R_1-CO$$
$$\langle\rangle-N=N-\text{Protein}$$
$$R_2-CO$$

(I)

where R_1 is $-NH-CH_2-COOH$, that is, Gly, and R_2 is

$$-NH-CH-COOH$$
$$\quad\quad\quad|$$
$$\quad CH_2-CH(CH_3)_2$$

that is, Leu, results in the formation of two types of antibodies, one to the antigen

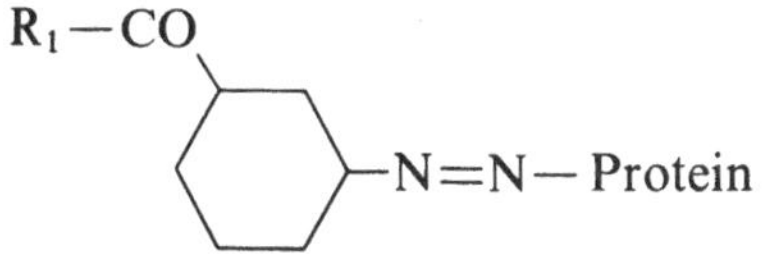

$$R_1-CO$$
$$\langle\rangle-N=N-\text{Protein}$$

(II)

and the other to the antigen

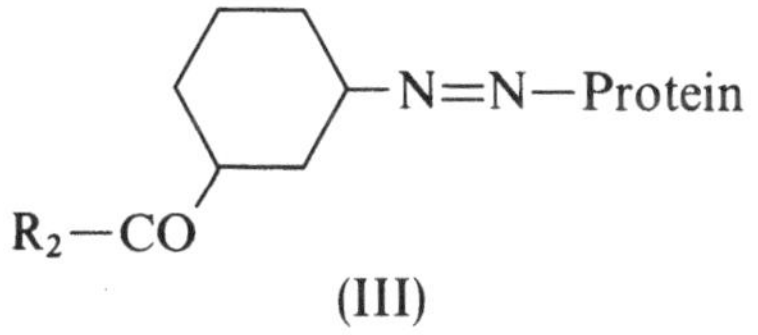

$$\langle\rangle-N=N-\text{Protein}$$
$$R_2-CO$$

(III)

but not to I as a whole. This shows that the dimensions of the determining groups and of the active regions of the antibodies which are adequate for them are relatively small.[6]

The presence of special reacting groups in antibodies is demonstrated directly by the elegant experiments of Pressman and Sternberger.[7] The rabbit was immunized against synthetic antigens which contained residues of p-azobenzoic acid and p-azophenylasonic acid as the determining groups.

$$HOOC-\!\!\left\langle\bigcirc\right\rangle\!\!-N{=}N-Protein$$

$$H_2O_3As-\!\!\left\langle\bigcirc\right\rangle\!\!-N{=}N-Protein$$

The antibodies obtained were then iodinated; this resulted in the complete suppression of their immunological activity. However, if the iodination took place in the presence of haptens, it did not disturb the serological activity. This means that the introduction of iodine into the active region

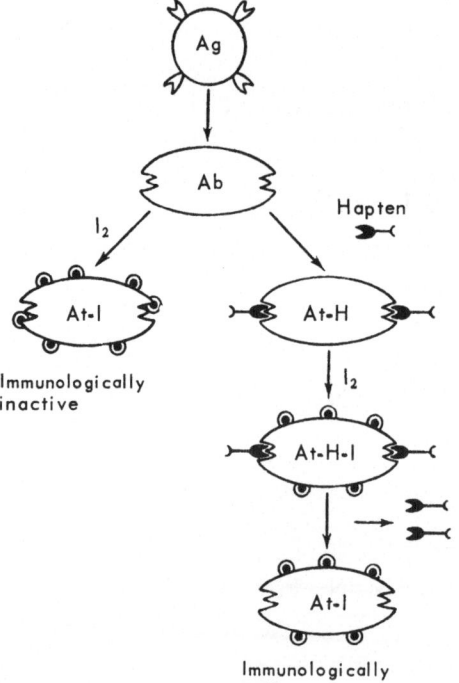

Immunologically
inactive

Immunologically
active

Fig. 152. Scheme illustrating the action of haptens.

of the antibody destroys its action, as the hapten protects the active region from the iodine. This is illustrated in Fig. 152. It is evident that there is a structural relation between the active region of the antibody, on one hand, and the hapten or determinant group, on the other hand; they fit each other as a lock and key. The action of haptens also becomes understandable; by replacing the determinant group of the antigen, the hapten plays the role of a plug and the antigen key can no longer enter the antibody lock.

In this case the situation is the same as would happen if the lock had been jammed with wax; in order to open the lock it is first necessary to remove the haptenic group.

The degree of unit specificity is quite high. It may be illustrated in Table 30. Here 0 indicates the absence of precipitate; $^+$, $^{+++}$, and $^{++++}$ are increasing amounts of precipitation.

If the determinant group (let us refer to it as a hapten) of a chicken serum is devoid of a carboxylic group, there is no interaction. It is observed only if the carboxylic group is in the meta position. A chloride or methyl group has practically no effect on the interaction. Replacement of the COO^- group by SO_3^- sharply depresses the interaction.

The antigen–antibody interaction responds also to the stereochemical structure of the hapten and its conformations.[5,6,8,9] Stereoisomeric tartaric acids linked to protein by Landsteiner's method,

$$HOOC-CH(OH)-CH(OH)-CO-NH-\hspace{-0.5em}\bigcirc\hspace{-0.5em}-N=N-Protein$$

give different specific antibodies (Table 31).

The structure of the three tartaric acids is shown in Fig. 153.

The molecular theory of antigen–antibody interaction, which explains immunological specificity and gives it a quantitative treatment, has been developed by Pauling and his coworkers.[10–14]

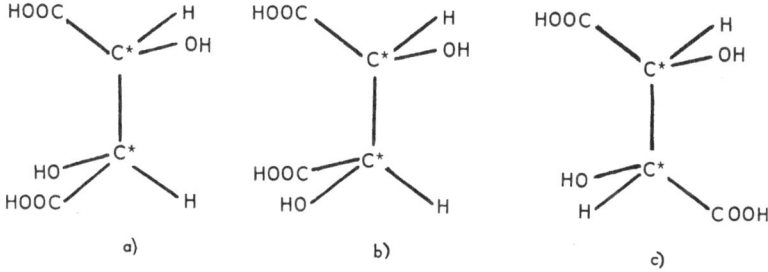

Fig. 153. Stereoisomeric tartaric acids, (a) dextro, (b) levo, (c) meso-Tartaric acid.

Table 30[8] Effect of Structural Changes in the Determinant Group on Antigen–Antibody Interactions

Determinant group coupled to horse serum antigen used in preparation of antibody	Intensity of the precipitation reaction upon introduction into the antigens of chicken serum with various determinant groups					
	$-N{=}N-$⬡	$-N{=}N-$⬡$-COO^-$	$-N{=}N-$⬡(Cl)$-COO^-$	$-N{=}N-$⬡(H_3C)$-COO^-$	$-N{=}N-$⬡$-COO^-$	$-N{=}N-$⬡$-SO_3^-$
$-N{=}N-$⬡$-COO^-$	0	+++	++++	+++	0	+
$-N{=}N-$⬡$-SO_3^-$	0	0	0	0	0	++++

Table 31   Serological Specificity of Three Isomeric
Tartaric Acids

	Antigen with					
	D(−)-Tartaric acid		L(+)-Tartaric acid		meso-Tartaric acid	
	Dilution of antigen					
Immune sera against	1 : 2,000	1 : 10,000	1 : 2,000	1 : 10,000	1 : 2,000	1 : 10,000
D-Tartaric acid	+ + +	+ +	(±)	0	+	(±)
L-Tartaric acid	0	0	+ + +	+ +	+	(±)
meso-Tartaric acid	(±)	(±)	0	0	+ + +	+ + +

ANTIBODIES AND ANTIGENS

According to the ideas of Pauling, complementarity, i.e., the structural correspondence between the antigen and antibody, expressed by the words *lock* and *key*, means that, with such correspondence, various weak forces effectively bind the antigen with the antibody.[8,15] These weak forces, which decrease rapidly with distance, are Van der Waals forces, hydrogen bonds, electrostatic interactions, etc.

The structural complementarity is determined, thus, by the Van der Waals surface of the hapten group. The cross sections of such surfaces for ortho-, meta- and paraazobenzarsonate are shown in Fig. 154.[15] The surface is drawn according to the Van der Waals radii of the atoms and groups, i.e., by using those distances to which they can approach each other. The surface of the antibody cavity corresponds to the surface of the hapten.

The hapten retardation constant K_0', also introduced by Pauling, may serve as a measure of the structural correspondence. This is a relative hapten–antibody interaction constant, determined from the retardation

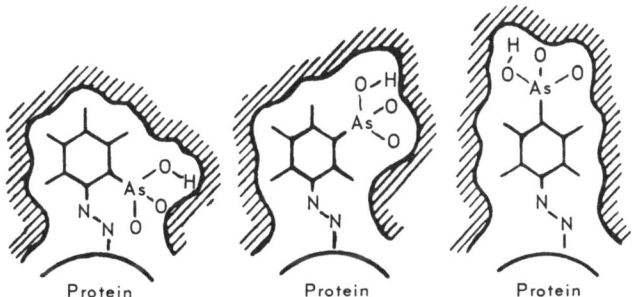

Fig. 154. Van der Waals surfaces of ortho-, meta-, and
paraazobenzarsonates.

of the precipitation reaction of the antigen by a low-molecular-weight hapten.

The values of K'_0 for the interactions of the anti-o-azobenzarsonate, anti-m-azobenzarsonate, and anti-p-azobenzarsonate with different haptens are shown in Fig. 155. The K'_0 varies over several orders of magnitude; furthermore, an antibody of the ortho type interacts strongly with ortho-haptens but weakly with meta- and parahaptens, etc.[14,15]

The antigen–antibody reaction is similar to usual chemical reactions; it follows the law of mass action; therefore, by studying the true and not relative equilibrium constant K_0, it is possible to determine the free energy of the interaction. This can be done directly by a measurement of the heat of the reaction and by other methods. If oppositely charged groups participate in the reaction, the interaction constant must be a function of the pH of the medium; this has been found to be true in a number of cases. The dependence of the equilibrium constant on pH is

$$\log\left(\frac{1}{K} - \frac{1}{K_0}\right) = \log\frac{K_H}{K_0} - \mathrm{pH} \tag{8.1}$$

where K is the measured equilibrium constant, K_0 is its value at neutral pH, K_H is the dissociation constant of the acid group of the protein.[16] The dependence of $\log(1/K - 1/K_0)$ on pH is linear in this case. In the case of the ovalbumin–antiovalbumin interaction, $K_H = 10^5$; this is in agreement with the participation in the reaction of an ionizable carboxyl group, $COOH \rightarrow COO^- + H^+$.

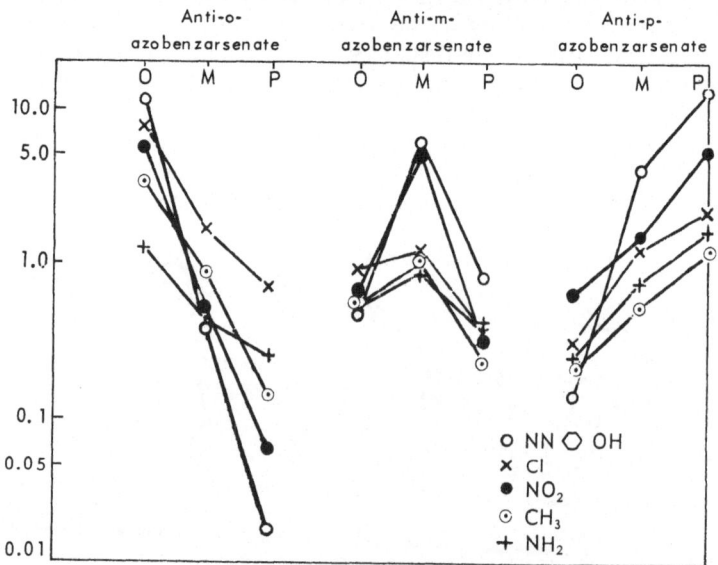

Fig. 155. Values of the hapten retardation constant K'_0.

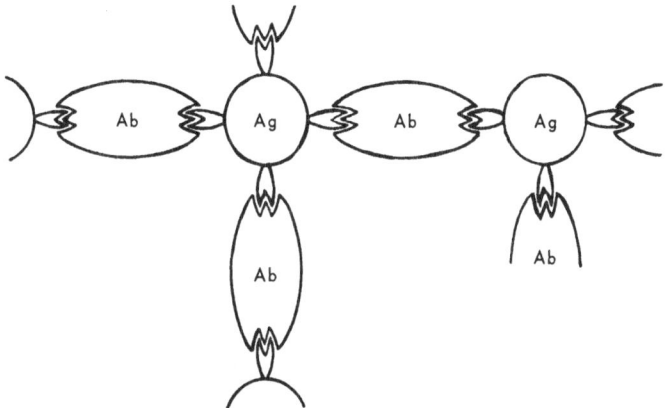

Fig. 156. Scheme illustrating the precipitation reaction of antigen with antibody.

The majority of antibodies are bivalent, while protein antigens are polyvalent. This explains precipitation and agglutination (Fig. 156); the interacting molecules form large insoluble aggregates which precipitate.

In the presence of free antigen, the equilibrium constant may be expressed as[17]

$$\frac{[Ab-(Ag)_2]}{[Ag][Ab-Ag]} = K \tag{8.2}$$

This corresponds to the reaction $Ab-Ag + Ag \rightarrow Ab-(Ag)_2$. Here Ab is the antibody and Ag is the antigen.

We are interested, however, in the energy of the single bond Ab–Ag. The compound $Ab-(Ag)_2$ contains two identical bonds, and 1-mole of Ag contains v moles of hapten groups, where v is the valency of Ag. The constant K' which corresponds to the equilibrium constant is

$$K' = \frac{Ag-Ab \text{ bonds}}{Free\ Ag \times Free\ Ab}$$

and it is equal to

$$K' = \frac{2[Ab-(Ag)_2]}{v[Ag][Ab-Ag]} \tag{8.3}$$

i.e.,

$$K' = \frac{2}{v}K \tag{8.4}$$

The decrease of free energy during the formation of a single bond is

$$\Delta F = -RT \ln K' = -RT \ln K + RT \ln \frac{v}{2} \tag{8.5}$$

Values of ΔF, ΔH, and ΔS are given in Table 32.[1]

Table 32 Thermodynamic Properties of Serological Reactions

	Reaction	Method	ΔF, kcal/mole	ΔH, kcal/mole	ΔS, cal/mole^{-1} deg^{-1}
Hemocyanin............	A	direct measurement	-10	-40	-100
Human plasma albumin	C	direct measurement	$-7.5; -8.0$	-3.66	$13.1; 14.8$
Bovine plasma albumin .	D	$+III+IV$ III	-5.5 ± 0.2	0 ± 2	20 ± 8
Ovalbumin............	D	III	-5.6 ± 0.2	0 ± 2	$20 + 8$
Carboxypeptidase	B	II	-9	-7	7
Bovine R globulin	B	II	~ -8.5	-2	21
Bovine R albumin	D	I	-5.2 ± 0.2	0 ± 2	18 ± 8
Bovine S albumin	B	II	-8.5	-2.8	21
Hapten T	F	IV	-7.4 ± 0.2	-0.8 ± 2.6	22 ± 9
ε-Dinitrophenyllysine ...	E	I	-6.8	-1.6	17
D-J	E	I	-7.24	-7.1	0.3
Lac		I	-7.25	-9.7	-8.8
Dye B		I	-6.54	-3.93	8.75

The notations in the second column refer to the reactions

A: $nAb + Ag \rightleftharpoons (Ab)_n Ag$

B: $Ab + (Ab)_{n-1} Ag \rightleftharpoons (Ab)_n Ag$

C: $Ab + 2Ag \rightleftharpoons Ab(Ag)_2$

D: $Ab-Ag + Ag \rightleftharpoons Ab(Ag)_2$

E: $Ab + 2H \rightleftharpoons Ab(H)_2$

F: Group H + Active group of Ab $\rightleftharpoons$ H – Ab bond

Here H is hapten.

The notations in the first column refer to the compounds

The methods of determination (third column) are: I, equilibrium dialysis; II, direct analysis of the precipitates and supernatant liquid; III, electrophoresis and sedimentation in the ultracentrifuge; IV, light scattering.

Table 32 shows that the changes of the free energy and enthalpy are small during the Ab–Ag interaction. Their order of magnitude corresponds to the formation of one or two hydrogen bonds or to a number of weak Van der Waals interactions. At first glance it is surprising that the entropy increases during the interaction (with two exceptions). It would seem that the combination of Ab with Ag should decrease, and not increase, the entropy. This may be explained by the liberation of water molecules which have been linked with the surface of the antigen and antibody.[1,17] Further details on the thermodynamics of serological reactions are given in the literature.[1,8]

These reactions, then, are determined by the structural correspondence and the geometry of the surfaces of the antigen and antibody molecules. The surface is controlled by the secondary and tertiary structures of the protein (or another biopolymer). It is quite characteristic that the protein loses its antigenic properties upon denaturation.

The immunoglobulins, antibodies, belong to the group of gamma globulins; they consist of two light chains with molecular weights of the order of 20,000 and two heavy chains with molecular weights of the order of 50,000. These chains are linked with each other by disulfide bonds. The active center, which interacts with the haptenic group, is in the heavy chain; its dimensions must probably be commensurate with those of the haptens. Karush[18] has estimated that the active center is constructed of

10 to 20 amino acids residues, i.e., approximately 1% of all the residues in the molecule. The shape of the active center is different for various antibodies and it is unknown.

The model which illustrates the interaction of a divalent antibody with antigen is shown on Fig. 157.[19] The interactions between the light and heavy chains at the end of the molecule permit the antibody to open up in correspondence with the shape of a given antigen.

Pauling believes that the structural complementarity between the antigen and antibody is dynamic, and not static. The globulin is attached by some bonds to the antigen molecule and the tertiary structure of the antibody is formed by the antigen in such manner that its surface forms a cavity that corresponds to the haptenic group of the antigen. This model is quite attractive and, as we shall see later, it agrees well with modern concepts of the enzymatic activity of proteins. Within the realms of this model, however, the number of antibody molecules has a definite relationship to the number of the antigen molecules and it remains a mystery how a permanent antibody character remains in the plasma even in the absence of antigens.

We have now approached the second basic problem of immunity, namely, the question of the nature of the synthesis of antibodies within the organism.

A number of hypotheses have been proposed in this realm. There is the dilemma: Is the antibody synthesis genetically predetermined and only enhanced in the presence of antigens, or do the antigens cause anew the formation of the antibodies? We have already run into a similar dilemma during the discussion of the inductive synthesis of enzymes. The second possibility can still be divided into two parts: Does the very synthesis of the antibody occur anew, or does the antigen only rebuild the tertiary structure of a ready-made globulin, as has been proposed by Pauling?

Haurowitz, starting with the precept that the matrix of protein synthesis is a protein,[20] assumed that the antigen becomes included in the matrix and, being fixed in it, causes the further production of antibody. This point of view was supported by experiments with labeled proteins containing ^{35}S, ^{14}C, and ^{131}I in iodine compounds attached to the protein. It was found that the labeled atoms remained for a long time within the cell.[21] It is noteworthy that the required immunity is not

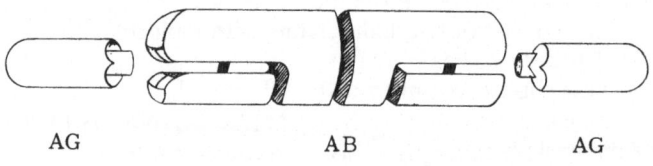

AG   AB   AG

Fig. 157. Model of antigen–antibody interaction.

transferred to the progeny by heredity. From these facts, Haurowitz concluded that the matrix action of the antigen is limited to the cytoplasm of special somatic cells. Haurowitz considered that the antigens differ from normal plasma globulins only in the secondary structure (it is known that some of the end-group amino acids are the same in these two cases). As a result, he accepted the Pauling hypothesis on the antigen as a factor which would ensure a definite folding of the globule; he placed this factor, however, in the matrix. This removes, of course, the objection to the Pauling hypothesis.

In this way, according to the theory of Haurowitz, the antigen is a mutagen of the tertiary structure of the protein; it is responsible for the somatic mutations of the cell in which the globulins are synthesized.[21,22]

The objections to the theory of Haurowitz are, first of all, that the matrix for protein synthesis is not a protein, but M-RNA, and that a protein cannot be synthesized by means of direct copying. If the antigenic action is, in fact, of a matrix nature, it may not be examined in such a simple manner. As a matter of fact, the antigen and antibody complement each other; they do not copy each other.

In recent years it has been established that antibodies differ in their primary structures, namely, in their amino acid sequences. Recently the primary structure of a number of Bence-Jones proteins has been established.[23] These proteins are microglobulins; they are fragments of immunoglobulins and are found in the urine of myeloma patients. It turned out that the primary structure of Bence-Jones proteins varied; approximately 30 amino acids out of 250 may be replaced.

Haurowitz modified his theory by taking into account differences in the primary structure of antibodies.[24] He proposed that the antigens modify not only the tertiary but also the primary structure of the globulin. They do this by combining with ribosomes or with M-RNA. The antigen forms the structure of the antibody by not permitting the incorporation into the synthesized polypeptide chain of inadequate amino acid residues brought by the corresponding T-RNA.

The ideas of Haurowitz on the matrix function of antigens are contradicted by theories of the genetic preexistence of the information on specific antibodies.

Burnett[25] took as a starting point the fact that antibodies continued to be formed over a quite long time after immunization, up to 50 years in the case of yellow fever. It is difficult to imagine that the antigen could remain trapped for such a long time within the matrix system which synthesizes the protein. The role of antigens from Burnett's point of view can be reduced to an enhancement of enzyme synthesis, necessary for the synthesis of antibodies. These are enzymes of the induced type; once the antigens have induced their synthesis, the enzymes are produced in an unlimited way, not only in the original cells but also in the progeny. As has been shown by experiments, the introduction of an antigen induces a strong activation of mitosis in lymph nodes, a rapid multiplication of cells; furthermore, it is the cells which produce antibodies to the given antigen that are affected.

In this manner, Burnett pointed out for the first time the similarity between the synthesis of antibodies and the induced synthesis of enzymes.

Answering Burnett, Haurowitz was perplexed by the question of how the cells by themselves could synthesize antibodies to exotic synthetic antigens, such as the ones containing phenylarsonic groups. Haurowitz regarded as completely irreconcilable the theory of Burnett's and the theory of the matrix action of antigens.[21]

Still earlier, Jerne[26] regarded an antigen as some agent of natural selection. Lymphatic cells normally synthesize a large number of globulins. antibodies to any conceivable antigens. Introduction of an antigen stimulates the synthesis of the necessary antibodies; it acts as some sort of catalyst. There is a strengthening effect; the newly introduced antigen needs a higher concentration of specific molecules and causes their still faster production. There is no physical-chemical consideration of any kind in the work by Jerne. The physicist Leo Szillard became quite interested by the problems of antibody synthesis. He developed a mathematical theory based on a number of assumptions. The principal idea of Szillard is that, just as with Burnett, the synthesis of antibodies is similar to the induced synthesis of enzymes.[27] According to Szillard, an antigen is an inducer, i.e., it removes the repression of the synthesis of a specific gamma globulin from the cell and, simultaneously, it induces the synthesis of other proteins important for the development of the cell. As a result, antibodies are synthesized in growing cells. This constitutes the primary reaction with the antigen. The memory of the primary introduction of the antigen is maintained since the number of the lymphatic cells able to synthesize the given antibody has been increased. This, in turn, explains the secondary reaction.

In recent years, data have been obtained on differences in the primary structures of antibodies produced by the same organisms against different antigens. This is direct confirmation of Burnett's ideas.

Hereditary immunity in plants and man has been studied by Efraimson.[28] The immunity of plants to parasites, for example, to rust fungi, is of great importance for agriculture. A new type of wheat, resistant to rust, is the result of the selection of the necessary genetic traits, the selection of wheat mutants. In turn, the development of such a type serves as a means for the development of new aggressive mutants of the rust and further selective work takes place in the fight between the mutants of the host and the parasite. It is important that the immunity is determined by the genome of the plant host, the aggression by the genome of the fungus parasite; the laws of genetic division are maintained during hybridization. In this important phenomenon we are quite evidently talking about the synthesis within the wheat cells of proteins which work as antibodies against the proteins of the rust; in the aggressive mutants, other new proteins are developed.

We have seen already that the genes of sickle-cell anemia cause the immunity of man to the plasmodia of the malarial mosquito. In this case, it has been shown exactly that the appearance of the immunity is related to the substitution of a single amino acid in the hemoglobin, i.e., to a point mutation in a single codon.

The interactions between the erythrocytes of different blood groups also belong to hereditary immunity.

One might surmise that there must be great similarity between hereditary and acquired immunity. Antibodies which differ in their primary structure may be regarded as mutant globulins. The number of such mutants must be very large. The organism of an animal contains 10^{11} to 10^{12} lymphoid cells, characterized by a high frequency of somatic mutation. The active group of an antibody consists of 10 to 20 to 30 amino acids. If each amino acid can be replaced, as a result of mutations, by just four others, then the number of different mutants would be 4^{10} to 4^{30}, that is, it is very large. Consequently, it is possible to have a set of mutants which contains antibody molecules to all possibly imaginable antigens, even to the most exotic ones, to such that the cell never encounters. This is no more surprising than the adaptation of microorganisms, namely, the selection of mutants which are stable toward poisons that are not present under native conditions.[29]

The introduction of an antigen into the organism causes an increase in the multiplication of those lymphatic cells which produce the given antibodies. This point of view, presented by Efraimson[28] is close to the hypothesis by Jerne;[26] it is based, however, on a deeper genetic foundation.

It is possible to formulate general theories of antibody formation. First of all, such a theory cannot be constructed independent of the general theory of protein synthesis. Second, it must be related to the theory of cell differentiation and regulation of cell division. There are convincing experimental data which indicate that antibody synthesis takes place *de novo* and not through some transformation of presynthesized gamma globulins.[30,31]

It has been shown by direct experiments that when an antigen is introduced into an animal organism (a tetanus antitoxin was introduced into a mouse), leucocytes, called *neutrophils*, migrate to the region of this introduction. These cells absorb the antigen and then perish. The destroyed neutrophils and pieces of antigen are absorbed by lymphocytes and monocytes which approached at this time. The lymphocytes and monocytes swell and become macrophages, which grow and multiply by division. During this process, there is synthesis of DNA and of those enzymes which break down the antigen. This is the picture of an inflammation caused by the antigen.

However, part of the antigen remains intact and is kept for a long time inside some of the macrophages. Together with these, the antigens are transported into the lymphatic tissues, a phenomenon which has been proved by the method of autoradiography.

The number of neutrophils decreases and the number of macrophages increases during the secondary introduction of the antigen: furthermore. many of these cells contain the labeled antigen that had been introduced previously. The macrophages start dividing earlier, and large numbers of plasmatic cells are formed; these actively synthesize antibodies.[32,33]

The mechanism of sensitization and immunity proposed by the Spiers (R. and E.) is that an antigen that has penetrated into a macrophage is maintained by forming a complex with RNA. This complex is stable, but is destroyed during the repeated introduction of antigen. A new macrophage absorbs the old one; its enzymes act mostly on the antigens which were complexed with the RNA and on the free antigens of the old cell. Plasmacytes, which begin synthesizing antibody, are then formed.

This is a matrix model in which the matrix is RNA. However, it has been shown in experiments carried out with cultures of individual plasma-cytes which produce antibodies, that their production starts under the direct control of genes, i.e., of DNA.[34,35] It has become apparent that the antigen does not penetrate at all into the plasmacytes which synthesize the antibodies. In these studies as well, the data were obtained by auto-radiography; these data may be explained only on the basis of the clone-selection theory of Burnett. It would appear that the cells which form the antibodies do not need to obtain direct information from the antigen; it is only necessary to have the presence of the antigen which acts as an inducer. Large numbers of cell clones are formed in the organisms (clones are progeny of a single cell); furthermore, a given clone interacts only with a single antigen. It is only then that antibodies are synthesized. It is possible that induction takes place during the surface contact of the cells with the antigen.

This, then, confirms Szillard's ideas on the similarity of the synthesis of antibodies and the induced synthesis of enzymes.

This question is evidently still quite far from a complete solution. We have seen that detailed investigations of the induced synthesis of enzymes have led to a considerable complication of the concepts of protein synthesis and to the discovery of the regulator and operon genes. The induced synthesis of enzymes could be explained in general terms within the frame-work of molecular biology, but we are still far from deciphering various segments of DNA which are responsible for these processes. The synthesis of antibodies is a more complicated and specialized phenomenon than the induced synthesis of enzymes. To understand it, it is necessary to have a molecular interpretation of the processes of differentiation. However, there are grounds for believing that the present-day concepts are probably quite correct.

Investigations of immunochemical processes are quite important for the future development of molecular biology and biophysics. On one hand, antibody synthesis is an easy and direct example of protein synthesis which can be studied by quantitative methods. On the other hand, antigen–antibody interactions clearly demonstrate the importance of structural factors, which play an important role in biology.

For further details, the reader is referred to the monograph literature.[36]

ENZYME REACTIONS

The most important biological function of proteins is enzymatic action. Enzymatic reactions also show the structural complementarities of molecules, which, to some extent, are similar to those of the antigen–antibody interactions that we have just described; they also may be thought of as lock and key.

Enzymes are catalysts. To understand their action, it is first necessary to discuss the mechanism of the action of normal catalysts. A variety of solids, from pure metals to quite complicated compounds, can act as catalysts by accelerating chemical reactions which take place in the gas or liquid phase. This is heterogeneous catalysis; in this case, the catalyst forms a different phase and reaction takes place on its surface. Homogeneous catalysis exists where the catalysts and the reagents are found in a single phase, for example, in a solution. As a result, homogeneous catalysis takes place, not on the interface between phases, but from interaction between molecules in solution, usually by means of intermediate compounds.

The process of heterogeneous catalysis is the adsorption of reacting molecules on the surface of the catalyst, the reaction between them, and the desorption, i.e., the separation from the surface of the product obtained. Adsorption brings reacting molecules into a state of close approach: it modifies the structure of their electronic shells and, in this manner, lowers the activation barrier which separates the reagents from the products. The energy curves for the reaction $S \rightarrow P$ (S are reagents, P are products) in the absence of catalysts are compared with those in a catalytic process in Fig. 158. In the second case, the activation barrier is lowered and the rate of the reaction is considerably increased. Let us recall that the rate of a reaction depends on the height of the barrier according to the exponential relation

$$k = Ae^{-E_a/RT} \tag{8.6}$$

As a result, a lowering of the height of the barrier by, let us say, 6000 cal/mole means an increase of the rate by e^{10}, that is, by about 30,000-fold.

It is evident that a change in the state of the molecule during adsorption is necessary for catalytic action. Balandin has shown that a very

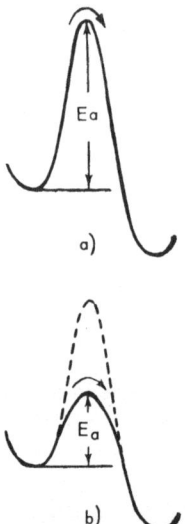

Fig. 158. Energy profile for a reaction (a) in the absence of catalysts, and (b) in the presence of catalysts.

important role is played in this by the structural complementarity between the surface of the catalyst and the adsorbed molecule. A metallic catalyst is crystalline. The atoms on its surface are distributed in a definite manner, at fixed distances from each other. If the symmetry of this distribution and the interatomic distances correspond to the geometry of the molecules to be adsorbed, the latter can be adsorbed effectively by the metal surface and the state necessary for the reaction is attained as a result of the interaction of its electrons with the metal atoms. Thus, the reaction of benzene hydrogenation

$$C_6H_6 + 3H_2 \longrightarrow C_6H_{12}$$

is catalyzed by platinum, nickel, and some other metals, but is not catalyzed by iron, silver, etc. This is due to the fact that the benzene molecule is an exact hexagon with distances between neighboring carbon atoms equal to 1.4 Å. The atoms on a nickel surface are also distributed in the shape of a hexagon and the distances between them are similar to those in benzene (Fig. 159).[37] The same is true of other metals which catalyze the hydrogenation of benzene. On the other hand, metals which do not catalyze this reaction either do not have a hexagonal surface distribution of their atoms, or they have such a distribution but their atoms have unsuitable dimensions.

When they are in a state of structural complementarity with a metal, the chemical bonds in benzene undergo the following transformation

$$\cdots HC\cdots CH\cdots \qquad -HC-CH- $$
$$Ni \; Ni \qquad\qquad Ni \; Ni$$

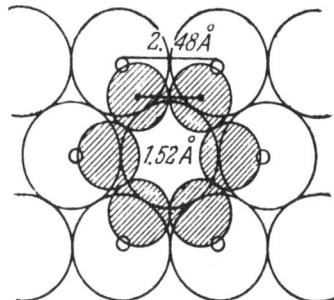

Fig. 159. Model of the adsorption of benzene on nickel.

This indicates the breaking of π-electron bonds in benzene and the formation of bonds with the metal atom through the π-electrons. In this state, benzene reacts easily with hydrogen

$$
\begin{array}{c}
\text{H—H} \\
\text{—HC—CH—} \quad \longrightarrow \quad \text{—H}_2\text{C—CH}_2\text{—} \\
\text{Ni\ \ Ni} \qquad\qquad\quad \text{Ni\ \ Ni}
\end{array}
$$

The theory of the structural complementarity of low-molecular-weight reagents with solid catalysts has been developed in detail by Balandin. This is called the *multiplet theory*.[38] It develops the mechanism of heterogeneous catalytic reactions and points out scientific ways of selecting catalysts.

Once the reaction has taken place on the surface, the reaction products leave the surface, they become desorbed, and the catalyst can act again. In this manner, six nickel atoms are capable of transforming a large number of benzene molecules.

Enzymes are globular proteins. They carry out their catalytic function either in solution or by participation in organized solid systems. Thus, for example, the enzyme of oxidative phosphorylation acts within mitochondria. In both cases, however, the action of an enzyme is more like that of a heterogeneous catalyst, since the reaction takes place on the surface of a macromolecule, which is determined by the tertiary structure of the protein. There are also some similarities to homogeneous catalysis, since it is possible to establish the formation of intermediary compounds of the enzyme with reacting molecules.

At present, a number of enzymes and enzymatic reactions have been studied. A large amount of material has been summarized in the monograph literature.[39,40] Here we cite only the principal results of a general nature, concentrating on structural complementarities as important factors which determine enzymatic activity.

First of all, let us talk about the basic principles. A substance which enters into the reaction under the influence of an enzyme is called a substrate. In many cases, in addition to an enzyme and a substrate, it is necessary to have still another compound, namely, a cofactor, in order to carry out the reaction. Cofactors are coenzymes; they are complicated organic compounds which usually participate in a catalytic reaction as carriers of given chemical groups. Quite an important source of a number of coenzymes is found among vitamins. It is well known that the absence of vitamins from food leads to serious consequences, to scurvy, pellagra, and other sicknesses. Animal or human organisms cannot synthesize vitamins; they must obtain them in a ready state. Fortunately, vitamins are required in very small quantities and normal food contains a sufficient amount. The B group of vitamins are essential as sources of coenzymes which participate in the fundamental processes of oxidative phosphorylation, in which the battery is charged, or ATP is synthesized. Coenzymes I and III contain heterocyclic rings with nitrogen; they are introduced by vitamins B_1 (thiamine) and B_2 (riboflavin).

Other cofactors are enzyme activators. These are usually simple compounds such as inorganic ions.

In a number of cases, enzymes also contain what are called *prosthetic groups* of nonprotein nature. Such a group, for example, is a heme which participates in the oxidation–reduction processes in the cytochromes.

Prosthetic groups often contain metal atoms, such as iron in heme, etc.

Thus, when we are studying enzymes, we are dealing with quite complicated chemistry. However, this is still chemistry, and not biology. Enzymatic reactions may be studied by chemical and physical-chemical methods; they occur not only *in vivo*, but also *in vitro*. This is quite understandable; enzymes are individual proteins which may be obtained in pure form and even may be crystallized.

The structures and functions of enzymes may be investigated directly by physical methods which determine the structure of the protein and the substrate or coenzyme, inhibitor, etc., and follow changes during the catalytic process. On the other hand, since the action of enzymes is centered in their influence on the rate of reaction, an investigation of the kinetics of enzymatic reactions is a very important source of information on enzymes.

Let us discuss the principal relations of stationary kinetics, which are valid in the simplest case.[39,41,42]

When an enzyme E interacts with its substrate or substrates S, a complex ES is formed; in this complex the reactivity of the substrate is increased, since the substrate is transformed into the reaction product and the enzyme returns to its initial state

$$E + S \; \rightleftharpoons \; ES \; \longrightarrow \; E + P$$

Let k_1, k_{-1}, and k_2 be the rate constants for the direct reaction of complex formation, the reverse reaction of its decomposition, and the reaction of the transformation of the substrate into the product, respectively. Usually we are dealing with a large excess of substrate; its concentration is considerably greater than the amount of the enzyme, $S \gg E$. Under stationary conditions of the process, the concentrations of the free enzyme molecules, indicated by F_1, and of the enzyme molecules tied up in the complex with the substrate, F_2, are constant.

Using the scheme of Fig. 160, we may write out the kinetic equations

$$\dot{F}_1 = -k_1 S F_1 + (k_{-1} + k_2) F_2 = 0$$
$$\dot{F}_2 = k_1 S F_1 - (k_{-1} + k_2) F_2 = 0 \tag{8.7}$$

$$E = F_1 + F_2 \tag{8.8}$$

The expression for a stationary rate process has the form

$$v = -\dot{S} = \dot{P} = k_2 F_2 \tag{8.9}$$

We find F_2 from (8.7) and (8.8). We obtain

$$v = \frac{k_1 k_2 E S}{k_{-1} + k_1 + k_1 S} = \frac{k_2 E S}{K_M + S} \tag{8.10}$$

where

$$K_M = \frac{k_{-1} + k_2}{k_1} \tag{8.11}$$

Equation (8.10) is called the *Michaelis–Menten equation*.

As the substrate concentration S is increased, the rate of the reaction, v, increases; it approaches a maximal rate asymptotically

$$v_{\max} = \lim_{S \to \infty} v = k_2 E \tag{8.12}$$

If two substrates participate in the enzymatic process studied, the concentration of one of them, S', enters into the rate expression as a comultiplier, i.e., in this case, instead of (8.12), we have

$$v_{\max} = k_2 S' E \tag{8.12a}$$

An investigation of the dependence of v on E and S permits us to find the reaction rate constants. We have

$$v = \frac{v_{\max} S}{K_M + S} = \frac{v_{\max} S}{K_S + (v_{\max}/k_1 E) + S} \tag{8.13}$$

where

$$K_S = \frac{k_{-1}}{k_1} = \frac{1}{K} \tag{8.14}$$

and K is the equilibrium constant of the reaction $E + S \rightleftharpoons ES$.

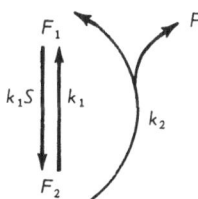

Fig. 160. Scheme of the stationary enzymatic reaction.

The curve of the dependence of v on S, according to Eq. (8.13), is shown in Fig. 161. It follows from this equation that

$$\frac{v_{max}}{v} - 1 = \frac{K_S}{S} = \frac{v_{max}}{k_1 ES} \qquad (8.15)$$

When the concentration of the substrate, S, and enzyme, E, are constant, but the concentration of substrate, S', is variable, the dependence of $(v_{max}/v) - 1$ on v_{max} follows a straight line which extrapolates to a point on the ordinate at K_S/S and on the abscissa at $-k_1 E/K_s$. The tangent of this straight line is $1/k_1 ES$ (Fig. 162). Since $v_{max} = k_2 S'E$, it is possible to find $k_1 E, k_{-1} E$, and $k_2 E$ from the straight lines obtained at various values of S, or if the enzyme concentration E is known, it is possible to determine all three rate constants, k_1, k_{-1}, and k_2.

Thus, during the dehydrogenation of succinic acid to fumaric acid by adding the enzyme dehydrase, we have

$$\begin{array}{ccc} \text{CH}_2\text{COOH} & \xrightarrow{\ 2\text{H}\ } & \text{HOOC}\cdot\text{CH} \\ | & & || \\ \text{CH}_2\text{COOH} & & \text{HC}\cdot\text{COOH} \end{array}$$

$$k_1 = 1.15 \times 10^{-2}\ \text{sec}^{-1} \qquad k_{-1} = 3.6 \times 10^{-7}\ \text{mole}^{-1}\ \text{sec}^{-1}$$

$$k_2 = 5.15 \times 10^{-6}\ \text{mole}^{-1}\ \text{sec}^{-1}$$

Usually one determines, not all three constants, but only two constants v_{max} (that is, k_2) and K_M. This is done by means of the Lineweaver–Burke equation (8.13) which is reduced to the form

$$\frac{1}{v} = \frac{1}{v_{max}} + \frac{K_M}{v_{max}}\frac{1}{S} \qquad (8.13a)$$

The straight line which describes the dependence of v^{-1} on S^{-1} extrapolates on the ordinate to v_{max}^{-1}, while the tangent of the straight line is equal to arctan $K_M v_{max}^{-1}$.

The basic idea of Michaelis (1913) on the formation of an enzyme–substrate complex has been confirmed indirectly by the agreement between the observed rate of the reaction and the kinetic equations; in a number of cases it has been confirmed directly. Thus, the complexes of catalase and peroxidase (enzymes which contain a prosthetic group of the heme type) with substrates have been observed spectroscopically. Free catalase absorbs light in the region of 410 mμ to a much greater extent than its complex with hydrogen peroxide H_2O_2 and with methyl hydroperoxide, CH_3OOH.[43]

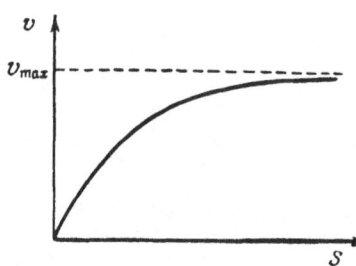

Fig. 161. Dependence of the steady-state rate of the enzymatic reaction on substrate concentration according to Michaelis–Menten.

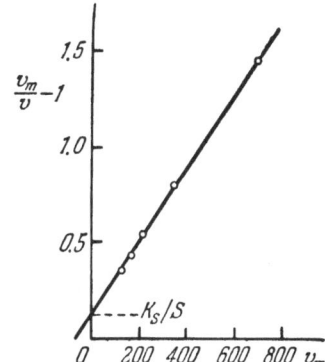

Fig. 162. Determination of the kinetic constants of an enzymatic reaction.

Let us remark that everyone is familiar with catalase; one observes it every time that one disinfects a cut with hydrogen peroxide. When it comes in contact with blood, oxygen is strongly evolved and H_2O_2 decomposes. This is due to the fact that blood contains catalase. The constants for the reaction are $k_1 = 5 \times 10^6$ mole^{-1} sec^{-1}, $k_{-1} = 0.02$ sec^{-1}, $k_2 = 1.8 \times 10^7$ mole^{-1} sec^{-1}. The reaction in this case has to form

$$\text{Catalase} + H_2O_2 \rightleftharpoons \text{Catalase} - H_2O_2$$

$$\text{Catalase} - H_2O_2 + H_2O_2 \longrightarrow \text{Catalase} + 2H_2O + O_2$$

The second reaction is the more rapid one in this case.

In many cases, one has to deal with much more complicated relations. But in one way or another a kinetic analysis makes it possible to determine the reaction rates for a number of steps of an enzymatic reaction. In its turn, an investigation of the dependence of the rate constant on temperature, pH and other factors yields values of free energies, enthalpies, and entropies which determine the energetic profile of the enzymatic reaction, similar to that shown in Fig. 158. In Fig. 163a, we have shown the profile of the reaction free energy for the transformation of fumaric acid into L-hydroxycinnamic acid, which is catalyzed by fumarase.

$$\begin{array}{l} \text{HOOC} \cdot \text{CH} \\ \parallel \\ \text{HC} \cdot \text{COOH} \end{array} + H_2O \longrightarrow \begin{array}{l} H_2C \cdot \text{COOH} \\ | \\ \text{HO} \cdot \text{HC} \cdot \text{COOH} \end{array}$$

In Fig. 163b, we have shown the enthalpy profile.[44] The reaction is more complicated than those examined earlier; it is necessary to take into account, as well, the enzyme–product complex EP, that is,

$$E + S \rightleftharpoons ES^* \rightleftharpoons ES \rightleftharpoons EX^* \rightleftharpoons EP \rightleftharpoons$$

$$EP^* \longrightarrow E + P$$

The stars indicate the activated states at the apices of activation barriers.

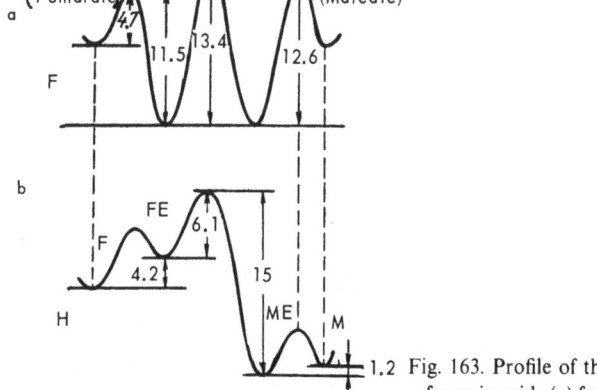

1.2 Fig. 163. Profile of the enzymatic reaction of fumaric acid: (a) free energy, (b) enthalpy.

The highest barrier corresponds to the transition

$$ES \; \rightleftharpoons \; EX^* \; \rightleftharpoons \; EP$$

This stage in the transformation of the substrate attached to the enzyme into a product attached to the enzyme is the rate-limiting step, since its rate is minimal.

The great difference between the curves of Fig. 163a and b points to the considerable role of the entropic factor, since

$$T \, \Delta S = \Delta H - \Delta F$$

In a number of enzymatic reactions, the free-energy profile is due mostly to entropy. In this way, enzymatic reactions are considerably different from usual catalytic reactions.[45]

What are the forces which determine the interaction of the enzyme with a substrate? In general, they are the same as those found in the antigen–antibody interactions. First, these involve the formation of chemical bonds, such as salt bonds (such bonds are rather weak, since they are in competition with water molecules); secondly, there are hydrogen bonds; thirdly, there are ion–dipole interactions; and fourthly, there are hydrophobic and Van der Waals bonds.

Finally, in a number of cases the interactions occur through electron exchange. The surface of the enzyme molecule, in contrast to the surface of a metal, is quite complicated and nonhomogeneous. The adsorption of the substrate may occur on one set of groups on the protein, while catalysis may be caused by other groups. The scheme of action of fumarase is shown in Fig. 164.[46] Adsorption occurs as the result of the electrostatic interaction of the COO^- ion of fumarate with the positive charges on the surface of the protein, fumarase. The catalytic groups are R' and R (the latter is probably a histidine), located in another region of the surface. It is evident

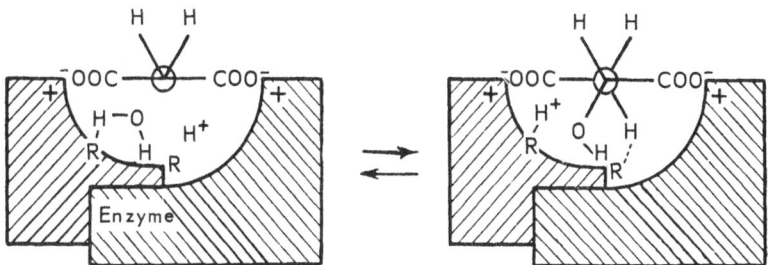

Fig. 164. Scheme of the action of fumarase.

that the structural geometric complementarity between substrate and enzyme is extremely important. It requires that the adsorbing and catalytic groups be brought close to the substrate molecule simultaneously.

Enzymatic action may be suppressed by a number of substances, called *inhibitors*. It is necessary to differentiate between competing and noncompeting inhibitors. The competing inhibitor interacts with the surface of the enzyme in a similar manner as the substrate; it occupies the place of the latter. It is a case similar to that found with a hapten in immunochemistry. Such an inhibitor is similar in structure to the substrate, but it is not capable of reacting. The noncompeting inhibitor interacts with the already formed *ES* substrate and prevents its transformation into the product and the enzyme. Cases of a mixed action are also known.

The relations of the steady-state kinetics are easily derived here as well. It is necessary to solve systems of two linear equations with two unknowns. During competing retardation, we have (Fig. 165a)

$$\dot{F}_1 = -(k_1 S + kJ)F_1 + (k_{-1} + k_2)F_2 + k_2 F_2 + k'F_3 = 0$$

$$\dot{F}_2 = k_1 S F_1 \qquad\qquad -(k_{-1} + k_2)F_2 \qquad\qquad = 0 \qquad (8.16)$$

$$\dot{F}_3 = kJF_1 \qquad\qquad\qquad\qquad -k'F_3 = 0$$

$$E = F_1 + F_2 + F_3 \qquad (8.17)$$

Here J is the inhibitor concentration, and F_3 is the concentration of the complex EJ. By solving (8.16) and (8.17), we find

$$v = k_2 F_2 = \frac{k_2 SE}{K_M + K_M K_J J + S} \qquad (8.18)$$

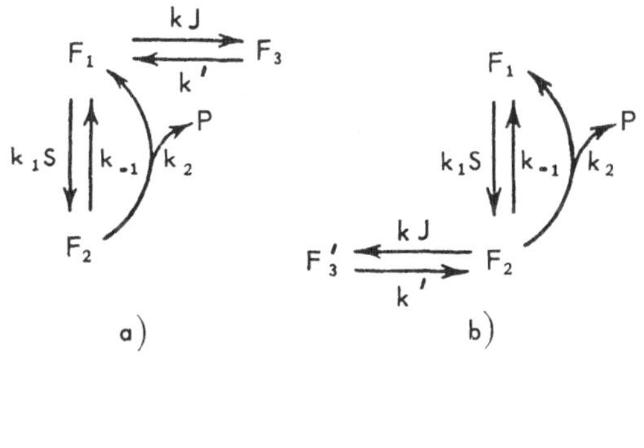

Fig. 165. Scheme of the stationary state of enzyme reaction in the presence of an inhibitor: (a) competing inhibitor; (b) noncompeting inhibitor.

a)   b)

where $K_J = k/k'$. During noncompeting retardation (Fig. 165b), we have

$$\dot{F}_1 = -k_1 S F_1 + (k_{-1} + k_2)F_2 = 0$$

$$\dot{F}_2 = k_1 S F_1 - (k_{-1} + k_2 + kJ)F_2 + k'F'_3 = 0 \qquad (8.19)$$

$$\dot{F}'_3 = kJF_2 - k'F'_3 = 0$$

$$E = F_1 + F_2 + F'_3 \qquad (8.20)$$

Here F'_3 is the concentration of the complex ESJ. We find

$$v = k_2 F_2 = \frac{k_2 SE}{K_M + K_J SJ + S} \qquad (8.21)$$

In the first case v_{max} is independent of J; in the second case it is dependent.

We see that the schemes of the processes shown in Fig. 165 are more complicated than the simple Michaelis scheme (Fig. 160). Therefore, it is necessary to use a large number of constants and to seek a rate equation for the reaction by solving a more complicated system of linear equations. It is found that, in a number of cases, the number of states of the enzyme, F_i, and the number of equations may be quite large. The solution of such problems is greatly simplified by using a new mathematical algorithm based on the theory of graphs.[47] Graphs are systems of points connected by directed lines, which indicate flow from one point to another. In fact the schemes of Fig. 165a and b are graphs. The theory of graphs has been developed in topology: earlier it had found quite effective applications in calculations of electric networks, since simple rules make it possible to write out equations for the current without calculations just from an examination of the graphs. This method is found to be quite effective and applicable for complicated enzymatic reactions, and first of all under conditions of steady-state kinetics. However, it can be applied also to pre-stationary-state conditions, in which the substrate concentration is almost constant, while the enzyme concentration in its different states is not.[48]

It is quite evident that it is impossible to determine the individual values of kinetic constants from kinetic experiments when there is a vast number of constants. A solution of the entire set of equations, however, is still quite useful, since a knowledge of the dependence of the rates on the concentration of substrate, inhibitor, activator, etc., renders quite far-reaching deductions possible on the nature of the process, and on the nature of the action of the enzyme.

NATURE OF ENZYMATIC ACTION

A number of factors indicate that enzymatic action is determined by the structure of the protein macromolecule as a whole. It would seem that the active center of the enzyme, at which the substrate is adsorbed and undergoes its chemical transformation, contains amino acid residues which are far removed from each other along the chain. One of the most thoroughly studied enzymes is ribonuclease. This enzyme breaks down RNA; it contains 124 amino acid residues, the sequence of which is known. Part of the molecule, which is locked by disulfide bridges, forms a nucleus to which are attached two tails, the N terminal, consisting of 25 residues, and the C terminal, consisting of 14 residues. In Fig. 166 is shown a scheme of the space structure of ribonuclease.[50] If 20 residues are broken

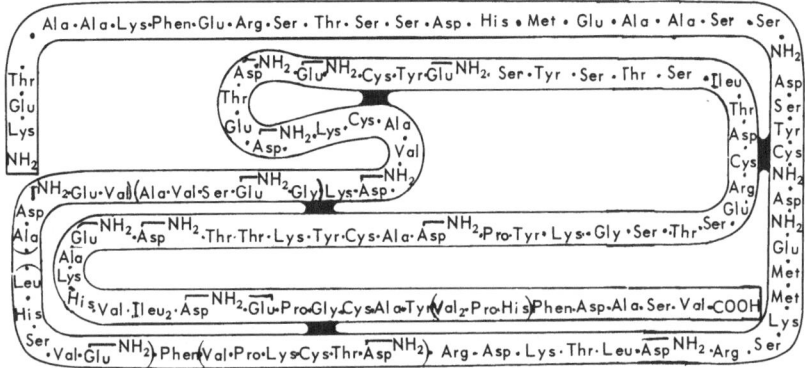

Fig. 166. Structure of ribonuclease. The black solid regions indicate S—S bonds.

off the N tail of the protein, its activity is maintained if the cut-off tail is bonded to the remaining chain by nonchemical bonds. The same can be done with the C tail. The biological activity of the protein vanishes completely when its tertiary structure is destroyed, for example, when the four S—S bonds are broken by reduction to SH groups. The activity of the protein is maintained (although it is reduced) when one or even two S—S bonds are broken. If the denaturation is carried out reversibly and the S—S bonds are restored, the activity is restored. In this case, activity requires the presence of both tails and a nucleus, but it is evident that not all amino acids participate in the activity.[49,50,51]

The denaturation of the enzyme and, in particular, the helix–coil transition destroys its catalytic activity. This extremely important biological function of the protein is directly related to the highest levels of its structure.

Let us cite another interesting example. The digestive proteolytic enzymes, pepsin, trypsin, and chymotrypsin, are synthesized as inactive precursors, namely, pepsinogen, trypsinogen, and chymotrypsinogen. It has been shown that the activation of trypsinogen, namely, its transformation into trypsin, occurs as a result of the breaking of a single peptide bond Lys-Ileu. A hexapeptide, Val–(Asp)$_4$–Lys is broken off during this process. The splitting-off of such a chain results in a change of the tertiary structure, as shown in Fig. 167. The splitting-off of the hexapeptide destroys those nonchemical interactions (hydrogen bonds or electrostatic forces) which had held the long chain end in the trypsinogen; once it is freed, the latter coils up and forms the active center of trypsin. The structural changes which take place during the transition of chymotrypsinogen to chymotrypsin have also been studied.[52,53]

Up until recently, one could find, in the literature, data according to which the splitting-off of a large part of the protein chain in an enzyme (papain, trypsin) did not reduce its activity. Later on, it was found that

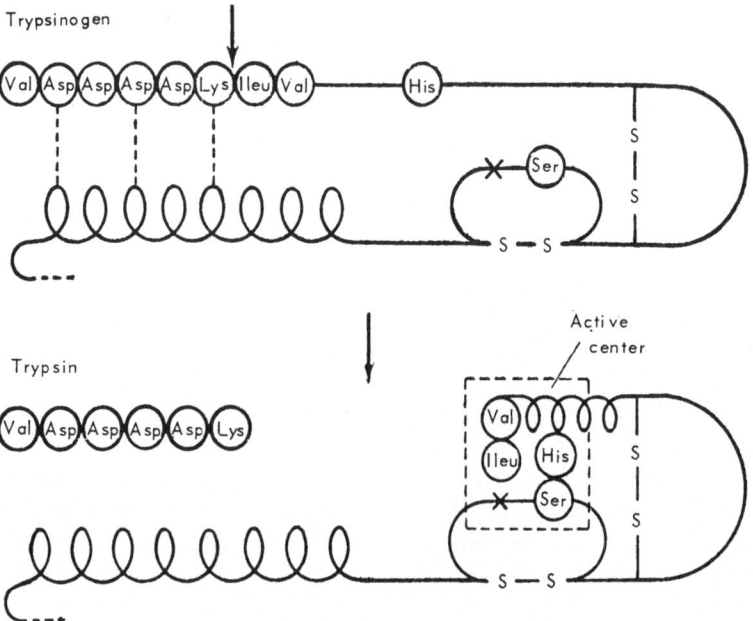

Fig. 167. Trypsinogen → Trypsin transformation.

these results had been in error. All the available experiments indicate that the entire protein chain is necessary for enzymatic activity.

The enzymatic activity requires the space structure of the protein molecule. The renaturation of ribonuclease and other similar facts show that the space structures, i.e., the tertiary and quaternary structures, are uniquely determined by the primary structure. In other words, these structures are dictated genetically.

To form an active center, the protein chain must fall into a strictly determined tertiary structure. This is illustrated in Fig. 168. The possibility of forming such a structure is found in the primary structure. What is the manner in which this folding occurs?

The specificity of an enzyme to the substrate is quite high, but not absolute. A number of enzymes catalyze reactions of the type

$$B - X + Y \longrightarrow B - Y + X$$

If B is a methyl group, such a reaction may be regarded as passing through the intermediates

$$Y + H-\overset{\displaystyle H}{\underset{\displaystyle H}{C}}-X \longrightarrow Y\cdots\overset{\displaystyle H}{\underset{\displaystyle H}{C}}\cdots X \longrightarrow Y-\overset{\displaystyle H}{\underset{\displaystyle H}{C}}-H + X$$

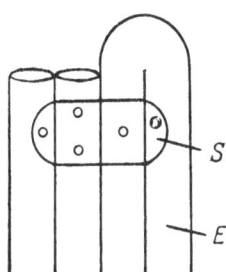

Fig. 168. Scheme of the formation of the active center.

If B is not a methyl group, but an asymmetric group and BX is an optically active compound, then the reaction is accompanied by inversion, the transformation of a right antipode into the left one; this is actually observed.

In explaining the specificity of enzymes which catalyze transfer reactions, one meets with difficulties. Phosphorylases and kinases catalyze the transport of hydroxyl groups in a number of compounds. However, water H—OH, is not reactive in these cases. It would appear that the molecule of H_2O should be easily adsorbed and activated by the reactive center. since it is smaller than the usual substrates. Phosphotransacetylase acts on acetate, propionate, and butyrate, but does not act on formate. Some compounds are adsorbed by the active center, but do not react.

These and other facts were explained by Koshland, who proposed an interesting qualitative theory of enzyme specificity.[54-56] The theory stems from the following: The substrate interacts with the active groups of the enzyme and changes the geometric structure of the protein; enzymatic action requires a special orientation of the catalytic groups, which is brought about by those changes in the protein geometry which are caused by the substrate. The Koshland model is shown in Fig. 169. In Fig. 169a are shown the complementarities of the normal substrate and the enzyme, the catalytic group C of which is located next to a mobile bond, because S has come into close contact with the flexible molecule, E. In Fig. 169b the violation of this correspondence is shown when the normal substrate is replaced by a larger compound: in Fig. 169c the same is shown when the substitution is for a smaller group. In both cases, the position of the functional group C is changed and it is no longer present in spacial complementarity with the mobile bond. Finally in Fig. 169d, the influence of an inhibitor is shown.

Thus, it is not only the enzyme which changes the state of the substrate, but also the substrate which changes the state of the enzyme. In the first process, the electronic cloud of the substrate is changed, while in a simultaneous second process the conformational state of the enzyme macromolecule is changed.

The cooperation of several catalytic groups can be found also in some phenomena of homogeneous acid-base catalysis. Thus, 2-hydroxy-

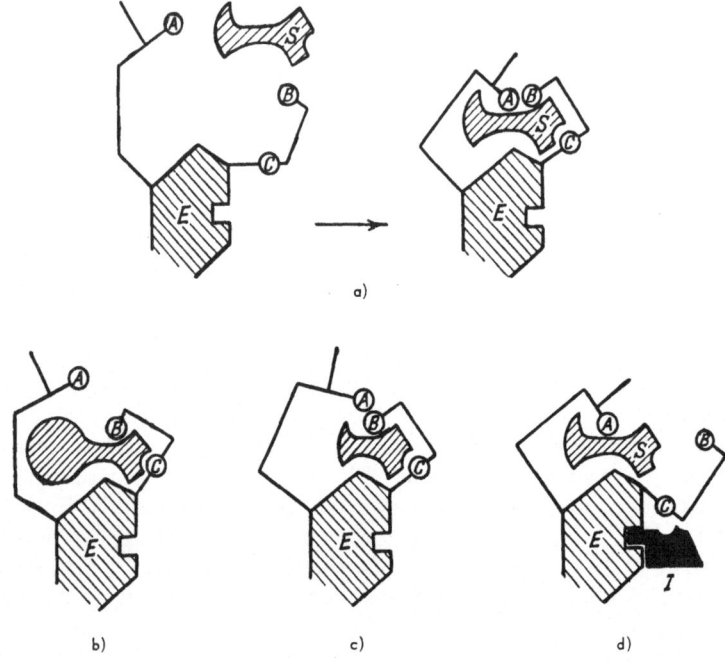

Fig. 169. The Koshland model.

pyridine catalyzes the mutarotation of glucose by breaking its six-membered ring and undergoing a tautomeric transformation

The same reaction with a mixture of pyridine and phenol occurs $1/7,000$ as fast.[57] In order to form an intermediate compound, it is necessary that the functional groups be close together in space; this is true in the case of hydroxypyridine.

We see that the Koshland theory examines the complementarity of the enzyme with the substrate, not as a static relationship of an immobile lock and key, but as a dynamic adaptation of one to the other.

Even earlier, Karush had come to the same general conclusions in studies of the interaction of serum albumin with various organic compounds: The protein possesses configurational adaptability, i.e., its structure may adapt itself to the structure of the substrate.[58] This explains the ability of serum albumin to combine with the most diverse materials. The idea of Pauling on the dynamic structural complementarity of antigens with antibodies turned out to be quite valid for another group of phenomena, namely, in the theory of enzyme action.

It is evident that the starting assumption of these studies consists in an examination of the macromolecules of the protein as flexible systems. We know in fact that, in contrast to synthetic polypeptides, globular proteins are only partly ordered. Some parts of the chain exist in the state of a rigid helix bonded by hydrogen bonds, while other segments are melted and possess a statistical set of conformations. If this point of view is correct, then it is the macromolecular flexibility which is responsible for the extremely important biological function of proteins. i.e.. their ability to undergo rotational isomerization. We are talking here about quite general properties of macromolecules. One may assume that enzymatic action and the resilience of rubber have similar physical natures.

In fact, there are a number of direct experimental proofs of conformational changes of enzymes during their interaction with substrates. These proofs were developed in particular for two reactions catalyzed by chymotrypsin (Ct), namely, the reaction with diisopropylfluorophosphate, which gives diisopropylphosphoryl-α-chymotrypsin and HF, and the reaction of the hydrolysis of n-nitrophenylacetate.[59] The intermediate complex of the second reaction can also be isolated. The evidence for the conformational transformations of chymotrypsin in these complexes consists in the following:[60]

1. Kinetic investigations show that monoacetyl Ct may exist in two forms, of which only one is the intermediary complex ES of the hydrolysis reaction.
2. The second form can aggregate, as shown by light scattering.
3. Spectroscopic investigation of ES complexes in the two reactions demonstrates a change in the position of the tryptophan.
4. It has been shown that there is a change in the reactivity of some amino acid residues of the enzymes during the formation of the complex.
5. The titration curve of the enzyme changes.
6. The parameters of optical activity change.
7. The thermodynamic parameters which characterize the conformational transitions of the protein change.

Indirect, but no less convincing proofs of the conformational lability of the enzyme. as a factor which determines its catalytic properties. is given by the above-mentioned phenomenon of allostery.

Similar facts have been established by Koshland in his investigation of phosphoglucomutase.[55] It has been shown polarimetrically[61] that the change of the activity of chymotrypsin during the change of pH is paralleled by a change of the degree of helicity of the protein.

Braunshtein, Torchinskii, and Koreneva[62] have studied the anomalous dispersion of optical rotation in the absorption band of aspartate-glutamate-transaminase; they were able to interpret the change of the anomalous dispersion of optical activity during the interaction of the enzyme with the substrate and with inhibitors as being due to conformational transitions of the enzyme.

Koshland has made some quantitative calculations; he estimated the increase of the rates of reaction during the approach and the orientation of the reactive groups of the substrate. Such an approach and orientation are attained as a result of induced structural complementarity between enzyme and substrate.[63]

A number of indirect confirmations of the Koshland hypothesis have been obtained in studies with phosphoglucomutase.[64]

The conformational transitions during the interactions of the enzyme with a substrate and the coenzyme were found recently during an investigation of dehydrogenase.[65,66] It was shown by spectropolarimetry that the contents of α-helical regions changed during such interactions.

It has been known for a long time that the maximal catalytic activity of an enzyme is attained at some optimal pH. An increase or decrease of the pH relative to the optimum is accompanied by a decrease of enzymatic activity. This has a bell-shaped pH dependence. A simple explanation of this effect was given by Michaelis and Davidson who believed that the interactions of the enzyme with a substrate in the active center are determined by the ionization states of ionizable groups. If there are several such groups and they are characterized by different degrees of dissociation. then there must be an optimal value of the pH.[39] Kirkwood and Schumaker assumed that the interaction of the enzyme with a substrate has an ion–dipole contribution. Since there are many ionizable groups in the molecule, it is possible to have fluctuations of charge and the charge distribution on the macromolecule may vary. This model also results in a bell-shaped dependence of the enzymatic activity on pH.[67]

Spectropolarimetric investigations of dehydrogenase have shown that the change of the degree of α-helicity with pH is parallel to the change of the enzymatic activity. If the α-helices are constructed of ionogenic groups, then, for a given sequence of anionic and cationic groups, the dependence of the amount of α-helix on pH can, in fact, be bell-shaped: but this is not a general rule, since the shape of such a curve must be a function of the relations between such groups and of their distribution.[68] This correlation proved directly the relation between enzymatic activity and the conformation of the macromolecule. These results do not vitiate the considerations by Michaelis and Kirkwood, they extend them.

Phillips investigated the interaction of the enzyme with a substrate by a direct X-ray diffraction method. The substrate of lysozyme, a carbohydrate, was introduced into the cleft present in the three-dimensional structure of this enzyme. This is not accompanied by any considerable changes in the conformation of lysozyme, but the substrate molecule undergoes a rotational isomerization. It is evident that the structural complementarity between the enzyme and substrate can take place only as a result of the change of conformation of the enzyme or of the substrate or of both.

As a whole, it is possible to consider that conformational transitions are, in fact, very important for enzymatic activity and that their existence has been demonstrated. On the other hand, one should not consider the Koshland model proved in its exact sense. Most probably it is valid in some cases and invalid in others, since the interactions between enzymes and substrates and other ligands may be quite diverse.

To what extent does the dynamic model of the enzyme lend itself to a theoretical physical analysis? The first steps in this direction were made by Linderstrøm-Lang and Schellman.[52] The free-energy change during the conformational melting of the secondary structure of a protein is small, since the enthalpy increase is compensated for by an entropy increase. It follows, however, that the structure of a protein must fluctuate constantly: some regions of the chain will be melting, while others will be forming ordered structures without any considerable change in the total free energy. During interaction with a substrate, such conformations are selected which are most favorable in the sense of complementarity between enzyme and substrate. This changes the conformational free energy of the protein and the degree of fluctuations. These changes are cooperative: namely, the residues in the protein chain are linked to each other and the conformational transformation in a given position is transmitted along the chain.

If the substrate is macromolecular, the action of the enzyme induces similar changes in it. Proteolytic enzymes split proteins. One can imagine that the enzymatic attack is directed principally at the disordered rather than the helical regions, since the latter are regions of the macromolecule which are protected by hydrogen bonds. As soon as any peptide bond of the hydrolyzed protein is broken, however, the cooperative interaction weakens the entire structure linked together by hydrogen bonds and increases in it the degree of fluctuations. This permits the secondary attack, etc. The splitting of a protein by proteolytic enzymes is a cooperative process.[70] It follows from the dynamic concepts of enzyme catalysis that the thermodynamics of such reactions must be particular. The observed changes of the free energy, enthalpy, and entropy must include changes in the thermodynamic parameters of the protein chain of the enzyme, related to changes in its configuration.

One can imagine the following picture. When a substrate approaches some region in the chain, the free energy decreases. The emitted energy is transmitted cooperatively along the chain and causes its partial melting, i.e., it increases its entropy. The fluctuations increase, the number of chain conformations increases, and conformations arise which are capable of maximal interaction with a substrate. i.e.. which have the best structural complementarity with it. The enzyme–substrate complex is formed. In such a complex. the state of the enzyme is unusual. Of all the possible conformations. those are selected which are specific for this interaction and. as a result. the entropy of the protein chain decreases. There is an entropy resilience: the protein tends to return to the state of maximal entropy. The resulting conformation of the enzyme is favorable to inter- action of the substrate with the active groups on the surface of the protein molecule and to the change in the structure of the electronic shell of the substrate. As a result, the substrate molecule turns out to be. so to say. stretched and the activation barrier of the reaction decreases.

Quite evidently, in addition to the decrease of the barrier as a result of interactions with the active groups of the enzyme. there is also a decrease due to the formation of the resilient entropic force. This contribution. however, is insignificant: the modulus of entropic resilience is always very small. The entropic resilient force cannot break or significantly deform the chemical bond. Furthermore, the entropy change of the protein, due to its conformational change, influences considerably the entropic profile of the reaction.

As soon as the reaction has taken place, desorption occurs: the enzyme ejects the product of the reaction. The thermal energy, which is excessive over the starting state of the enzyme, is dissipated into the sur- rounding medium and the entire process may start anew.

This is a qualitative model which stems from the general principles of the modern physics of macromolecules. The properties of the protein catalyst which differentiate it from usual heterogeneous catalysts are relat- ed to the cooperative conformational properties of the polymeric chain.

The considerations on the fluctuations of the conformation do not at all contradict the hypothesis of Koshland. The attainment of structural complementarity between enzyme and substrate as a result of conforma- tional changes may be regarded just as a selection of the necessary con- formation from a set of structures between which the protein may fluctuate.

Analysis of the thermodynamic properties of enzymes does not favor the accumulation of excess energy within the active center of the enzyme. as was proposed. The role of the conformational changes is structural. rather than energetic.

A quantitative theory which explains the activity of enzymes, starting from conformational fluctuations, has not been developed yet. We meet considerable difficulties here. First of all. the basic question has not been answered: To what extent is the mechanism of enzyme action universal? Is there a single physics for all enzymatic reactions or do the different enzymatic processes have different physical natures? Notwithstanding the abundance of experimental material in this realm. it is still insufficient to answer this question. As a result, the dynamic concept of enzyme action must be examined today as a hypothesis. Its value lies in enabling us to make a number of deductions which are subject to experimental verification. One of the interesting consequences of the hypothesis relates to the behavior of an enzyme in a field of external force. If an enzyme molecule is stretched, the set of its conformations must change. This means that the enzymatic activity must change as well. As will be shown later, such processes may, in fact, be observed. This concept is important in the examination of mechanochemical problems.[71]

The difficulties of formulating a model theory of enzyme action is due to the difficulty of combining thermodynamic, kinetic, and structural factors. It is necessary to investigate the interaction of a heterogeneous, aperiodic macromolecule with a substrate. This occurs in an open system. The purpose of a theory is the quantitative explanation of the lowering of the activation barrier for the reaction substrate → product, i.e., the problem is basically kinetic. Consequently, it is insufficient to apply, in this case, methods of statistical physics which are applicable to systems that exist in thermodynamic equilibrium.

The concept on fluctuations in a protein macromolecule was applied in a somewhat different manner by Kirkwood to explain enzymatic action.[72] Kirkwood examined fluctuations of charges in the macromolecule. At a given pH, the charge distribution in a polyelectrolyte chain may have different atoms. If the distributions differ little in free energy, the charges fluctuate. Introduction of the substrate selects the suitable charge distribution, which results in a specific electrostatic interaction between the enzyme and the substrate. On the basis of these concepts, it was possible to carry out calculations of the effect of pH on constants which determine the activity of the enzyme; these were found to be in acceptable agreement with experimental results. The Kirkwood theory, however, may not be regarded as proved. It has been applied only to particular cases and it does not take into account fluctuations of the conformation of the various structures. The quantitative calculations are questionable, since it is necessary to use quantities that have not been measured directly in experiments, for example, the local dielectric constant.

As a whole, the specific catalytic ability of enzymes is determined by a number of factors, which have been listed by Braunshtein:[73]

1. The high affinity between the enzyme and substrate, i.e., the high probability of forming the complex ES, which is equal under normal conditions to a sharp increase in the concentration of reagents (the effect of juxtaposition).

2. The rigorous mutual orientation of the reagents, cofactors, and the active center of the enzyme (the orientation effect), in contrast to the normal homogeneous reactions, where the probability of the strict orientation of three or more interacting molecules is very small.

3. The action in the region of contact of the ES complex of nucleophylic and electrophylic groups of the active center (the effect of the synchronous intramolecular oxidation–reduction catalysis).

4. As a result of these three effects, the sharp increase in the absolute rate of the reaction, due to the replacement of the low-probability higher-order reactions, which require the simultaneous collision of three or more molecules, by the highly effective first-order reactions (the effect of intra-molecular polyfunctional catalysis).

5. The activation of the substrate by the redistribution of the electron density, under the action of the electroactive groups of the enzyme (the polarization effect).

6. Finally, that factor which has been discussed most extensively, the change of the conformation of the protein during interaction with the substrate (the induced-contact effect).

One can assume that the last effect determines the possibility of all the others.

The artificial modeling of an enzyme by a synthetic polymeric catalyst would be an extremely important event for science and technology. If this could be done, it would be possible to obtain catalysts which had an extremely high activity and specificity and which worked under normal conditions, without the use of high temperatures and pressures. This would lead to a complete revolution in chemical technology.

What are the bases on which we can hope to develop enzyme models? In order to do this, it is necessary to investigate the chemical properties of polymers, and first of all, the reactions in their chains, which up to now have been the subject of very little work of a quantitative kinetic and ther-modynamic nature. The point is that the functional group of a macro-molecule has fixed neighboring groups and, as a result, its chemical reac-tivity must differ from the reactivities of similar groups in a small molecule. Its catalytic activity must also differ.

The chemistry of polymers makes it possible to approach the prob-lem of the nature of enzymatic action from another side. The *theory of specific polymerization*, which leads to the formation of isotactic polymers, takes place on the surface of solid catalysts. This surface is asymmetric, which makes possible the formation of the given space configuration of the polymeric chain.[74] The asymmetry of the active center of the enzyme results in the stereospecific course of the catalyzed reaction. It has been shown with the use of labeled atoms that the reactions of molecules of the type CAABB occur in an asymmetric manner on the surface of an enzyme. This has been shown, for example, for the transformation of aminomalonic acid (containing a labeled carbon in one of the two carboxyl groups) into glycine,

$$
\begin{array}{ccc}
\text{C*OOH} & & \text{C*OOH} \\
| & & | \\
\text{H}_2\text{N}-\text{C}-\text{H} & \longrightarrow & \text{H}_2\text{N}-\text{C}-\text{H} \\
| & & | \\
\text{COOH} & & \text{H}
\end{array}
$$

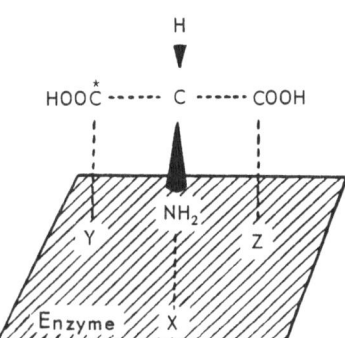

Fig. 170. Asymmetric enzyme–substrate complex.

The reaction occurs in only one group, which is indistinguishable from the other one both chemically and geometrically. This is explained by the asymmetry of the enzyme surface (Fig. 170). Since groups Y and Z of the enzyme are different, the reactivity of the two carboxyls is different.[56] Thus, a study of the stereospecific catalysis during polymerization gives valuable information on the possible mechanism of the stereospecificity of enzymes.

From what has been said, it follows that enzymes distinguish stereoisomers very well, and the optical antipode of the given substrate is no longer a substrate. Since enzymatic reactions are at the very basis of life, it is quite natural that, in the process of biochemical evolution, a selection of stereoisomers of all substances necessary for metabolism has taken place.

ALLOSTERY AND OXYGENATION OF HEMOGLOBIN

We have already mentioned allostery several times. This seems to be an extremely important phenomenon in biochemistry. It demonstrates very clearly the cooperative properties of the enzyme macromolecules.

The simple relationsips of steady-state kinetics of the Michaelis–Menten type [Eq. (8.10), (8.18), and (8.21)] are not always followed. Instead of the curve of Fig. 161, which is similar to a Langmuir isotherm for the adsorption of gases on a solid surface, in a number of cases we find S-shaped curves (Fig. 171) and even curves with maxima. An attempt has been made to explain these curve shapes by the interaction of the substrate with two or more molecules of the enzyme. This explanation turned out to be incorrect; the reaction occurs under a high excess of substrate (it is only under such conditions that one may apply the Michaelis theory—and no deviations from it), and the probability of triple collisions of the necessary type is very small. The attempt to regard as bimolecular the reaction of the interaction of a substrate with two regions (or a greater

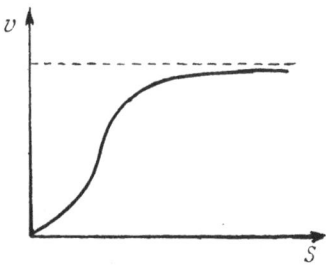

Fig. 171. The curve of v as a function of S with non-Michaelis kinetics.

number of regions) of the active center of the enzyme[75] does not save this hypothesis. In fact, the reaction of S with E is unimolecular and rigorous calculations once again result in equations of the type of (8.10).[76]

The true causes of the anomalous kinetics result from the cooperative interaction of several active centers on the enzyme. If the enzyme molecule binds two or more substrate molecules at two or more active centers and if the binding of S and the reactivity of the complex ES in the given active center depend on the binding of S by other active centers, then one obtains $v(S)$ curves with inflection points and maxima. We are dealing, then, with the interaction between active centers which may take place by conformational changes of the macromolecule. These considerations are confirmed by rigorous mathematical analysis.[76] From the shape of the kinetic curves, it is possible to judge the detailed processes which occur in the protein during the adsorption of the substrate.

The phenomenon of allostery is directly related to this. The synthesis of a metabolite, for example, an amino acid, within the cell occurs stepwise with the participation of a number of enzymes. As has been shown by Novick and Szillard,[77] quite frequently the end product of the entire reaction chain, namely, the metabolite, inhibits the activity of the first enzyme, which catalyzes the first reaction of the chain.[78,79] This is the *allosteric effect*. In Fig. 172 we have shown the (incomplete) scheme of the synthesis of CTP within the *E. coli* cell. The end product CTP inhibits the action of the first enzyme, namely, aspartate-transcarbamylase (ATC). It is quite important, and quite surprising, that CTP is structurally unrelated to the substrates of ATC and, consequently, it is not a usual competitive inhibitor. On the other hand, the inhibition is of a competitive nature. The scheme of the kinetic processes of allosteric inhibition is shown in Fig. 173. Fig. 175 shows the model proposed by Jacob and Monod: the allosteric inhibitor changes the conformation of the enzyme, making impossible its interaction with substrate S.

The lack of correspondence between the allosteric inhibitor and the substrate causes us to believe that the allosteric inhibitor is attached to another active center, this attachment bringing about the general change of the space structure of the protein. In the changed state, the protein does not interact any longer with the substrate.

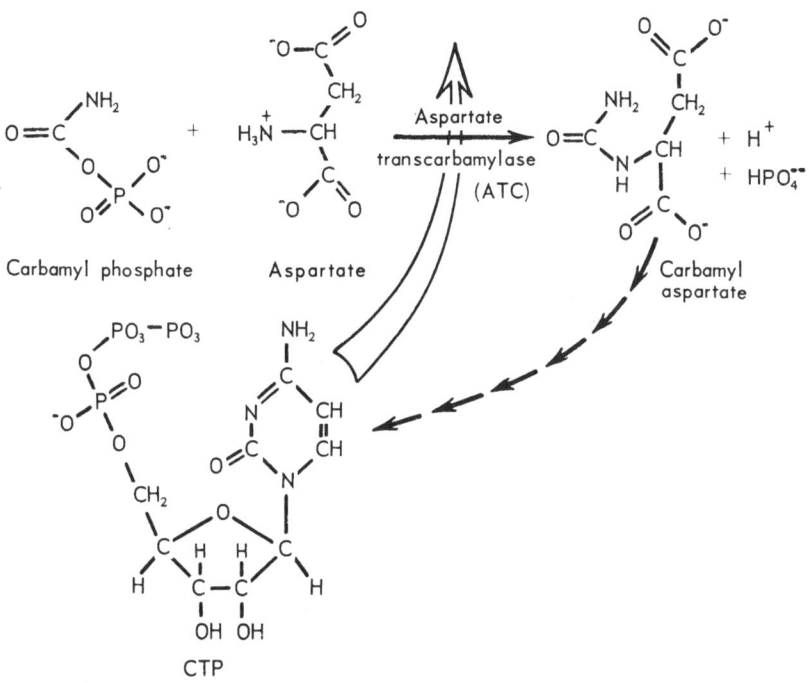

Fig. 172. Feedback in the reaction of the synthesis of CTP.

Experiments show that the kinetics of the reaction catalyzed by allosteric enzymes are cooperative; we find inflection points on the $v(S)$ curves (Fig. 174). These facts, plus the results of a direct investigation of such enzymes by physical methods, have shown that they have a quaternary structure, i.e., they are composed of several subunits. Monod, Wyman, and Changeux have examined the model of allosteric enzymes as a symmetric aggregate of interacting subunits.[80]

The cooperative properties of such a system are shown by the following model. Let us assume that the system is a dimer which may exist in two different states, A and B. In each of these states, the dimer may bind 0, 1, or 2 molecules of the ligand. Thus, we obtain six states: $A_{00}, A_{10}, A_{11}, B_{00}, B_{10}, B_{11}$ (there are two active centers in the dimer; zero indicates a free center, 1 indicates a center occupied by the molecule S).

Let us write out the conditions of equilibrium:

$$B_{00} = LA_{00}$$

$$A_{10} = 2A_{00}\frac{S}{K_A} \qquad A_{11} = \frac{1}{2}A_{10}\frac{S}{K_A} = A_{00}\frac{S^2}{K_A^2} \qquad (8.22)$$

$$B_{10} = 2B_{00}\frac{S}{K_B} \qquad B_{11} = \frac{1}{2}B_{10}\frac{S}{K_B} = B_{00}\frac{S^2}{K_B^2}$$

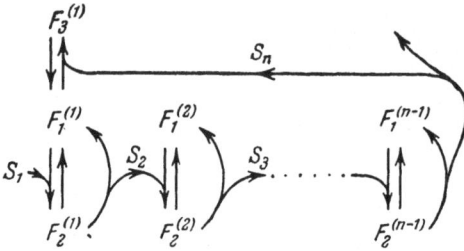

Fig. 173. Scheme of a stationary process with allosteric retardation: $F^{(1)}, \ldots, F^{n-1}$ are enzymes of the corresponding state of the reaction, $S_1, \ldots, S_n$ are metabolites, and S_n is the allosteric inhibitor.

The total amount of enzyme is

$$E = A_{00} + A_{10} + A_{11} + B_{00} + B_{10} + B_{11} \tag{8.23}$$

Here K_A and K_B are the dissociation constants of S in states A and B, and L is the equilibrium constant of states A and B in the absence of S. Now it is easy to calculate the function of the saturation of the protein by substrate S

$$\bar{Y} = \frac{A_{10} + 2A_{11} + B_{10} + 2B_{11}}{2(A_{00} + A_{10} + A_{11} + B_{00} + B_{10} + B_{11})}$$

$$= \frac{K_A[(1 + Lc)/(1 + Lc^2)]S + S^2}{K_A^2[(1 + L)/(1 + Lc^2)] + 2K_A[(1 + Lc)/(1 + Lc^2)]S + S^2} \tag{8.24}$$

where $c = K_A/K_B$. The $\bar{Y}(S)$ curve has an S-shaped form, namely, an inflection point. The same may be true of the rate of the stationary-state reaction

$$v = k'E\bar{Y} \tag{8.25}$$

where k' is some rate constant.

The inflection point, i.e., the cooperativity, vanishes when $c = 1$ or when $L \to 0$, since, in this case,

$$\bar{Y} = \frac{K_AS + S^2}{K_A^2 + 2K_AS + S^2} \equiv \frac{S}{K_A + S} \tag{8.26}$$

or when $L \to \infty$

$$\bar{Y} = \frac{(K_A/c)S + S^2}{(K_A^2/c^2) + 2(K_A/c)S + S^2} = \frac{S}{(K_A/c) + S} \tag{8.27}$$

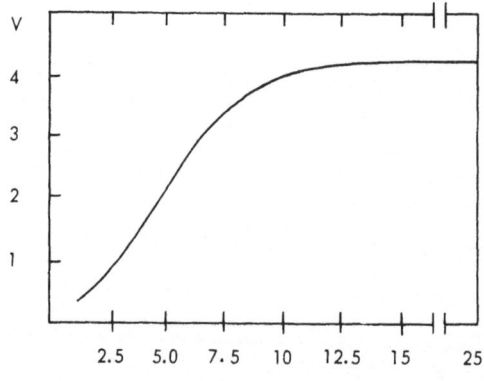

Concentration of aspartate transaminase, moles

Fig. 174. Dependence of $v(S)$ on concentration of aspartic transaminase.

Let us assume that each of the two subunits has two active centers, one for the substrate S and the other for the allosteric inhibitor J. Let us assume, for the sake of simplicity, that $c \ll 1$ and that J has complementarity only with the B state of the dimer. Then

$$\bar{Y} \cong x \frac{1 + x}{L' + (1 + x)^2} \tag{8.28}$$

where $x = S/K_A$, $L' = L(1 + J/K_J)^2$, and K_J is the constant of interaction with the inhibitor. We see that the allosteric inhibitor influences the interaction with the substrate.

The examination of an alternative model leads to the same results; this does not take into consideration two states A and B, but it assumes a direct influence of one bound molecule of S or J on the binding of the second molecule.[81]

Hemoglobin is an interesting analog of an enzyme consisting of interacting subunits. Hemoglobin is, of course, not an enzyme; its role is not catalytic; it acts as a transporter of oxygen. The binding of oxygen molecules to hemoglobin, however, is similar to the formation of the enzyme–substrate complex. The heme groups take the role of prosthetic groups.

The hemoglobin molecule consists of four subunits, each of which contains a heme. In this manner, the hemoglobin molecule is capable of binding four oxygen molecules. Each subunit taken separately is similar to a molecule of myoglobin, as has been shown by X-ray diffraction. The interaction of hemoglobin with oxygen occurs in a cooperative manner. The isotherm of the reaction

$$Hb + 4O_2 \rightleftharpoons HbO_8$$

is not similar to the Langmuir isotherm, but it has a sharp inflection at a definite partial pressure of oxygen; the p curve is S shaped and it is similar to that shown in Fig. 171 with the one difference that, on the ordinate, we plot the percentage of oxygenation and on the abscissa we plot p. The empirical equation which describes this isotherm has the form

$$Y = \frac{Ap^n}{1 + Ap^n} \tag{8.29}$$

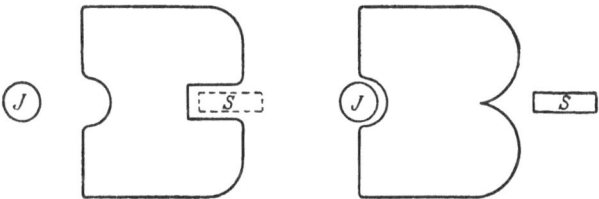

Fig. 175. p. 398: Schematic representation of the allosteric phenomenon.

where $n = 2.8$; this is the Hill equation. Furthermore, the isotherm for myoglobin is similar to that of a single subunit of hemoglobin; it is a Langmuir isotherm:

$$Y' = \frac{Bp}{1 + Bp} \tag{8.30}$$

During the oxygenation of hemoglobin (not oxidation, since entire molecules of O_2 are bonded by coordination bonds), we observe the Bohr effect, namely, protons come off from the protein chain. In the case of myoglobin, there is no Bohr effect. Pauling has shown that the curve of Eq. (8.29) is, in fact, explained by the cooperative interaction of the subunits of hemoglobin during its oxygenation.[82]

Let us give the calculations in a modified form. The four subunits of hemoglobin are located in an approximately tetrahedral array. The intermediary stages of oxygenation contain the set of molecules shown in Table 33. Here, in the second column, we have shown the number of ways of binding oxygen; F_1 is the free energy of interaction of a subunit containing oxygen with a free subunit, and F_2 is the free energy of interaction of two subunits containing oxygen. Both F_1 and F_2 include the free-energy change of the oxygen. By setting

$$a = e^{F_1/kT} \qquad b = e^{F_2/kT}$$

we obtain an expression for the partition function of the system

$$Q = 1 + 4a^3 + 6a^4b + 4a^3b^3 + b^6 \tag{8.31}$$

Let us assume that $b = fa^2$, then

$$Q = 1 + 4a^3 + 6fa^2 + 4f^3a^9 + f^6a^{12} \tag{8.32}$$

The partition function for the bound oxygen molecules is

$$Q' = 4a^3 + 2(6a^4b) + 3(4a^3b^3) + 4b^6 \tag{8.33}$$

and if $b = fa^2$

$$Q' = 4a^3(1 + 3fa^3 + 3f^3a^6 + f^6a^9) \tag{8.34}$$

Table 33   Hb + 4O$_2$ Equilibrium

Type of molecule	Statistical weight	Relative free energy
Hb	1	0
HbO$_2$	4	$-3F_1$
HbO$_4$	6	$-4F_1-F_2$
HbO$_6$	4	$-3F_1-3F_2$
HbO$_8$	1	$-6F_2$

The average degree of oxygenation is

$$Y = \frac{1}{4}\frac{Q'}{Q} \tag{8.35}$$

As the free energies of Hb, HbO_2, etc. contain respectively $kT \ln p$, $2kT \ln p$, etc., we can write $a^3 = cp$ and we get

$$Y = cp \frac{1 + 3fcp + 3f^3c^2p^2 + f^6c^3p^3}{1 + 4cp + 6fc^2p^2 + 4f^3c^3p^3 + f^6c^4p^4} \tag{8.36}$$

If $f = 1$, there is no cooperativity, and $Y = cp/(1 + cp)$. Choosing suitable values of c and $f > 1$, we can obtain results similar to those obtained with the Hill equation (8.29).

In this simple calculation, we do not take into consideration the differences between the structures of the hemoglobin subunits. Hemoglobin contains two protein chains of type α and two chains of type β, which differ somewhat in their primary structures.

When Pauling carried out similar calculations in 1935, nothing was known yet about the conformations of macromolecules. Therefore, it was unclear how the hemes could interact, since they were far removed from each other. At present, we have every reason to believe that these interactions are carried out by conformational changes of the protein chains, brought about by changes in the electronic states of these hemes during their oxygenation. This is also supported by the Bohr effect. The electronic states of the heme have been described, for example, in Ref. 83. For further details of the oxygenation of hemoglobin see Refs. 84 to 86.

Muirhead and Perutz[87] have demonstrated by X-ray crystallography a change in the quaternary structure of hemoglobin during direct oxygenation. They distinguish between the α and β subunits of hemoglobin. During oxygenation, the distance between the iron atoms of hemes diminishes by 7 Å (between β_1 and β_2, from 40.3 to 33.4 Å; between β_1 and α_2, from 37.4 to 30.4 Å). As has been remarked by Engelhardt, the molecule of the hemoglobin breathes by changing its dimensions. The X-ray diffraction method has not permitted to establish as yet whether the tertiary and secondary structures of the protein change during the oxygenation of hemoglobin. It has been shown by spectropolarimetry[88] that the degree of α-helicity in Hb and HbO_8 is identical. Notwithstanding the considerable similarities of the subunits of hemoglobin to those of myoglobin, there are significant differences in the corresponding curves of the anomalous dispersion of magnetic rotation. Such curves in the α band of Hb and HbO_8 are shown in Fig. 176a; in Fig. 176b the same is shown for Mb and MbO_2. This method is quite sensitive to oxygenation and permits its kinetics to be followed. The curves for Hb and Mb are different; namely, in the first the two minima have sharply different depths, while in the second one, they are almost identical. It is evident that this difference is related to the cooperative interaction of the hemes within hemoglobin.[89]

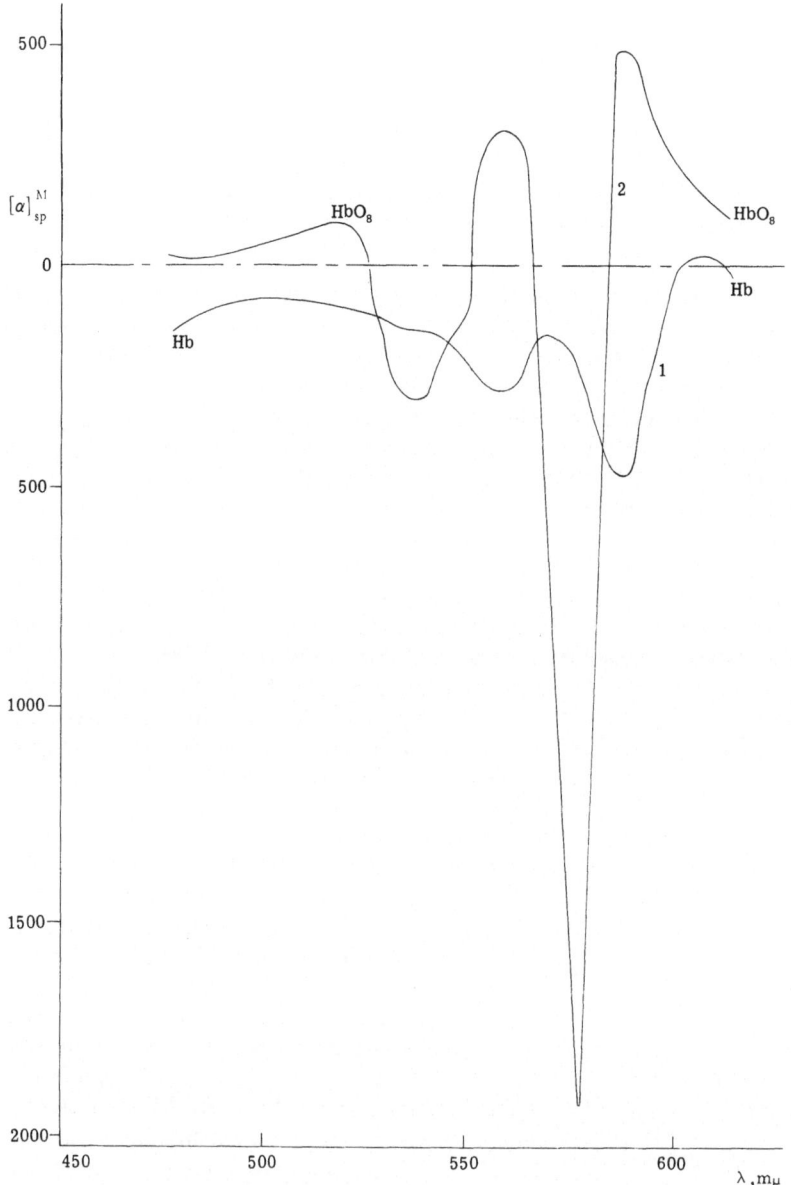

Fig. 176a. Anomalous dispersion of magnetic rotation in the α absorption band. Curve 1, Hb; curve 2, HbO$_8$.

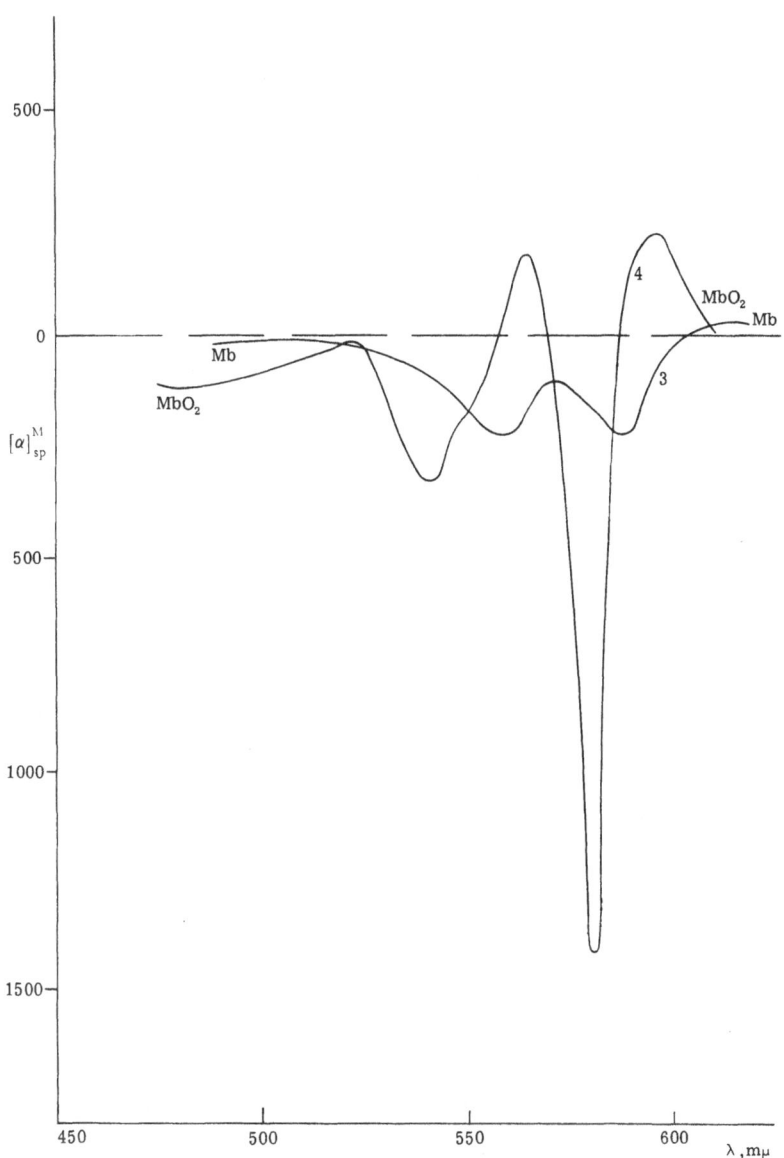

Fig. 176b. Anomalous dispersion of magnetic rotation in the α absorption band. Curve 3, Mb; curve 4, MbO₂.

Thus, the properties of hemoglobin deserve detailed attention. Many problems in this realm have not been solved yet. Hemoglobin is an important analog of allosteric enzymes; its investigation will certainly shed light on their nature also.

Atkinson assumes that not only do allosteric enzymes carry out a feedback in metabolic chains, but that their very synthesis by the operon is repressed by the allosteric effect.[90] In other words, there are interactions between the allosteric and genetic regulations which lead to homeostasis in a much more efficient way than allostery by itself. This idea has not yet been subjected to theoretical analysis. In the same study, Atkinson notes that, in many cases, it is the adenylates (AMP, ADP, and ATP) which are the allosteric effectors and, thus, there is a link between allostery and the most important biochemical processes in the cell, such as the Krebs cycle.

So far we have examined three phenomena of structural complementarity which have an important role in biology. These are, first, the complementarity of the antiparallel chains of DNA, the complementarity of DNA with template RNA, and the complementarity of template RNA with the transport RNA which bears the amino acid; second, the complementarity of the antigen with the antibody; and, finally, the complementarity between the enzyme and the substrate. In all cases, the interactions occur in the presence of definite conformations of the macromolecules, the formation of which requires the primary structure and the higher levels of structure determined by it. The spacial closeness of atomic groups is required for interactions; with this condition, relatively weak forces of a nonchemical nature, such as Van der Waals forces, electrostatic forces, and hydrogen and hydrophobic bonds, result in a significant gain of free energy. One can say that the principle of structural complementarity is universal in molecular biophysics.

The regulatory processes in cells are determined by the cooperative conformational transitions of proteins through variations in the structural complementarities. The lability of a protein molecule permits it to recognize different metabolites and to change its behavior accordingly. This is demonstrated quite clearly by the phenomenon of allostery. While in cybernetic machines information is transmitted by electrical or magnetic processes, in organisms it is transmitted by structural complementarity. Here we are talking about informational, and not energetic, properties; when the conformation of a macromolecule changes, its energy does not change significantly, while the entropy changes considerably. Thus, the inhibition of an enzyme during allostery means the transformation of negentropy into information. In fact, the transcription of the DNA molecules by M-RNA and the very duplication of DNA has a similar character. The transmittal of structural information leads to the corresponding energetic processes; structural information is some sort of signal. Later on, we shall develop the general theory of biological regulatory processes, starting, on the one hand, from concepts of molecular physics and physical chemistry and, on the other hand, from concepts of the theory of information.

In the last section of this chapter, we shall examine some other proposed or established cases of structural complementarity in molecular biophysics.

NUCLEOPROTEINS AND SYNAPSIS OF CHROMOSOMES

The high-molecular-weight nucleic acids, DNA and ribosomal RNA, exist in the cell, not as pure substances, but as complexes with proteins in nucleoproteins. The complex of DNA with a protein is a deoxynucleoprotein (DNP), while the complex of RNA with a protein is a ribonucleoprotein (RNP).

The principal component of chromosomes is DNP. DNA is complexed with basic proteins, namely, with protamines (in fish sperm) or with histones (in other cases). There is also a small number of nonhistone acid proteins. The basic properties of histones and protamines are due to the amino acids arginine and lysine:

$$
\begin{array}{ll}
\overset{|}{C}O & \overset{|}{C}O \\
\overset{|}{C}H(CH_2)_3C\overset{\nearrow \overset{+}{N}H_2}{\underset{\searrow NH_2}{}} & \overset{|}{C}H(CH_2)_4\overset{+}{N}H_3 \\
\overset{|}{N}H & \overset{|}{N}H \\
\quad\quad Arg & \quad\quad Lys
\end{array}
$$

Protoamines are relatively simple proteins with molecular weights of the order of 4,000 to 10,000. Two-thirds of the amino acids of a protamine are arginines. The other amino acids are not basic in nature and are present in pairs in the protamine chain. The DNA is linked with a protamine by electrostatic bonds. The positively charged amino groups of protamines interact with the negatively charged phosphate groups of DNA. This is proved by the fact that the complex containing DNA is totally dissociated at a concentration of NaCl higher than 1 M: the place of the amino group is taken by Na^+, that of the phosphate group by Cl^-.[91]

The X-ray investigation of deoxynucleoprotamine has shown that relatively short chains of the protamine wind themselves around the DNA double helix, locating themselves in the shallow groove on its surface (Fig. 177). In this, the arginine residues are located close to the phosphate

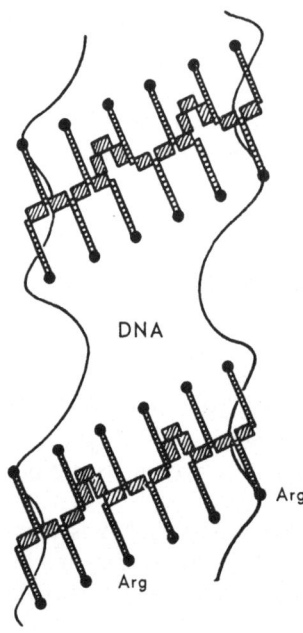

DNA

Arg

Arg

Fig. 177. Structure of deoxynucleoprotamine.

groups, while the other amino acids are found in pairs in the folds of the polypeptide chain, forming loops, if geometry requires this. In DNA, the distance between the nucleotides is 3.4 Å; this is the same as the period of a polypeptide chain in the β form. Consequently, protamines are present within DNA in the β form. The amino acids in the loops probably interact with the nitrogen bases of DNA.[92] In this manner, we are dealing again with structural complementarity, in this case of DNA with protamine.

A similar interaction occurs also in nucleohistones. Histones are proteins with molecular weights somewhat higher than 10,000; they contain approximately 30% basic amino acids. It appears that histones bind together different DNA molecules. The nucleohistone isolated from calf thymus has a molecular weight of 18.5×10^6; it contains DNA with a molecular weight of 18.0×10^6. Zubay and Doty regard such nucleohistones as the fundamental intrachromosomal structure.[93]

The role of the protamines and histones has not been explained as yet. The function of a chromosome consists in the storage of the genetic information, transfer of it by synthesis of M-RNA, and its reproduction by reduplication. It would seem that all this could be carried out by the DNA molecule as such. In fact, DNA stores the genetic information and reduplicates even *in vitro*. The histones and protamines must certainly participate in the reduplication of chromosomes. It has been stated already that histones have an effect on the synthesis of M-RNA and regulate the activity of genes.

The complexes of chromosomal DNA form helical threads with a diameter of 20 Å. These threads, in turn, are built up into microfibrils, namely, fibers with a diameter of 200 Å. In turn, the elementary components of the chromosomes, the chromonemes, which can be seen in a microscope, have a thickness of 0.5 to 1 μ and consist of a large number of microfibrils. In this manner, a chromosome contains an entire hierarchy of structures. As has been already stated, it has been shown by the dichroism of dyes adsorbed onto chromosomes that the double helices of DNA are parallel to the axes both of DNP and of the chromosome.

Divalent ions Ca^{++} and Mg^{++} play an important role in the structure of the chromosomes. Substances which form complexes with these ions (called *chelates*) break up the chromosomes of the sea urchin into pieces 4,000 Å in length and 200 Å in thickness, namely, microfibrils or their segments.

The chromosome has similarities with particles of bacteriophage in that it contains DNA (probably in its nucleus) and protein (probably as an external shell). To the contrary, the ribosomes, which are made up of RNA, are similar to particles of plant viruses. The latter may be regarded as molecules of RNA. A particle of tobacco mosaic virus is a cylinder 3000 Å in length and 170 Å in diameter: its molecular weight is 4×10^7. It contains 6 % RNA with a molecular weight of 2.5×10^6 in the form of a single helical molecule with a diameter of 80 Å. The RNA is surrounded by protein consisting of 2,300 subunits with a molecular weight of 17.000. In Fig. 178 is shown the scheme of the protein envelope of the tobacco mosaic virus: in Fig. 179 we have shown its molecular model.

Plant and animal viruses form symmetrical geometric structures. such as cylinders. or spheres. Their general geometric properties have been analyzed by Crick and Watson.[94] The regular structure is obtained as a result of the packing of identical units, protein molecules, around a nucleic acid. It is characteristic that the spherical viruses form cubic crystals. The basic structural requirement is that there be a protein envelope which apparently is for the protection of the hereditary substance. namely, the viral RNA, from external interactions. One may assume that the same role is played by the histones and protamines in chromosomes.

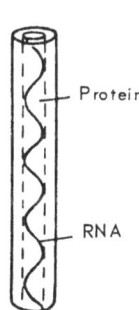

Fig. 178. Schematic representation of the structure of tobacco mosaic virus.

Protein

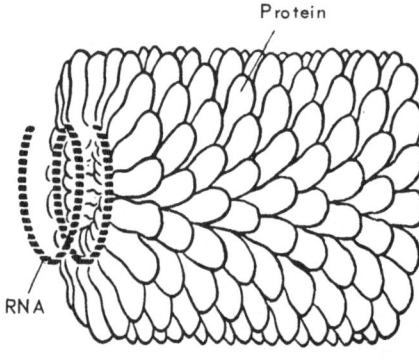

RNA

Fig. 179. Protein on the tobacco mosaic
virus.

The membrane must be quite large and it seems convenient to construct it from a large number of identical small protein molecules, rather than from a few large chains. It is possible to show theoretically that small molecules must become organized into spheres and cylinders. The number of small protein molecules in a cylindrical envelope is unlimited. while in a spherical envelope it is a multiple of 12. The point is that, in fact, we are dealing, not with a sphere, but with a highly symmetrical multiangular structure, namely, an octahedron, a dodecahedron, an icosahedron. It is characteristic that the heads of bacteriophages as well, which in general have a more complicated structure than viruses, are multihedral in structure. The structures of different viruses and phages are shown on a single scale in Fig. 180.[95]

There are data which indicate that structural complementarity is important for the adsorption of phage onto the surface of the bacterial cell.

Spirin and Kiselev have elucidated a number of factors on the structure of ribosomes.[96] It has been possible to open up ribosomes into long flexible chains of RNA. For this. the 70S ribosomes were treated with 0.5 M NH_4Cl and then were placed in a solution of low ionic strengths in the absence of Mg^{++} ions. This unwinding is reversible; Mg^{++} ions restore the structure of the ribosome.

Electron microscopic investigations of folded and unfolded ribosomes have shown that the structural basis of each ribosomal subunit (50S and 30S) is a macromolecule of RNA with an uninterrupted polynucleotide chain, which forms a large number of short double-helical regions as a result of interactions between chain sections. The RNA is bonded to a histonelike structural protein which is localized in the grooves of the helical regions of the RNA and between these grooves. Such an RNA contains 75% RNA and 25% protein. It has a diameter of about 30 Å and a length of several hundred angstroms, and it is folded into a compact structure. These proribosomes, in turn, are surrounded by protein linked by means of Mg^{++} ions. The final ribosomal particles are formed from the proribosomes.

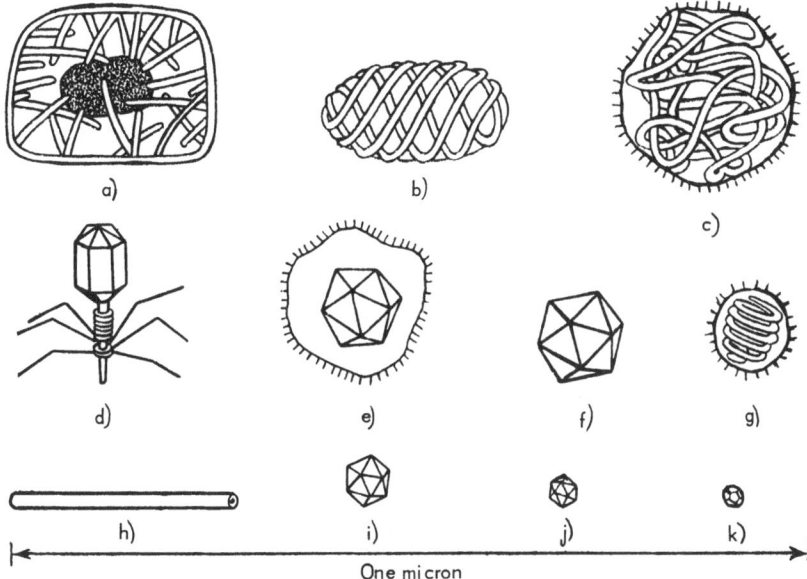

Fig. 180. Schematic representation of various viruses and phages. (a) *Vaccinia* virus; (b) pustulous dermatitis virus; (c) mumps; (d) T-even phage; (e) herpes; (f) *Tipula iridescent*; (g) influenza; (h) tobacco mosaic virus; (i) adenovirus; (j) polyoma; (k) poliomyelitis.

The simple molecular structure of the ribosome is determined by such interactions of RNA with protein, in which structural complementarity probably plays an important role.

Thus, the most important elementary supramolecular structures of life, the chromosomes, the ribosomes, the bacteriophages, the viruses, consist of nucleoproteins. It is quite evident that a study of these complexes is one of the most timely problems of molecular biophysics. Up to now, only a few studies have been carried out in this realm, immeasurably fewer than in the domain of the nucleic acid and protein molecules.

With a general knowledge on the nature of DNA duplication, we do not have a clear picture of the replication of the chromosome. The main difficulty consists in the transition from molecular to supramolecular levels of organization. It is possible to assume that the principle of structural complementarity will have an important role also in the elucidation of the mechanism of chromosome replication. But up to now all of this is only in the realm of assumptions.

Let us now examine one special problem related to the properties of chromosomes, one which attracted the attention of physicists long ago. This is the synapsis or conjugation of chromosomes. During meiosis (cellular division which results in the formation of haploidal sex cells), a pairwise blending of similar chromosomes takes place in the prophase.

This process is absolutely exact and specific. The chromosomes combine gene to gene by homologous regions. During synapsis, the chromosome behaves as a single unit.

Here, there is a physical problem (or more exactly two problems). What are the forces which cause the chromosomes, located at large distances, to approach each other and then to combine by identical regions? It seemed interesting to examine this process with simple physical models. The first attempt at such an examination was made by Jordan, a theoretical physicist working in quantum mechanics.[97,98] His idea was that quantum-mechanical resonance forces act between similar regions of two homologous chromosomes. We have already talked about such a homology. If one of two identical molecular structures is present in an excited state, the energy of excitation may be transferred to a second structure; this results in a general lowering of energy. The force of the resonance interaction is inversely proportional to the fourth power of the distance between the chromosomes. The identity of the interacting structures is a necessary condition for resonance. Thus, long before the formulation of molecular biology, Jordan proposed a theory which certainly belongs to the realm of molecular biophysics.

This theory is, however, incorrect. The principal objection is that the resonance forces may have considerable importance in interactions only when a significant number of molecules are excited. The excitation energies of the electronic states of molecular structures which form the chromosomes are quite large; they are much larger than the energy of thermal motion, kT. As a result, the fraction of electronically excited molecules is actually very small. The excitation of the vibrational states of the same structures requires a much lower energy, of the order of kT; however, the resonance interaction of vibrational levels is also very small, for another reason. The point is that the vibrational motions of atoms are related only to small changes in the distribution of the electric charges, and, in the final account, it is these changes which are manifested in a resonance interaction.[99,100]

Jehle turned away from the resonance interaction and attempted to explain the attraction of homologous regions of the chromosome by intermolecular interactions, namely, by dispersion forces caused by polarization. In fact, it can be shown that the dispersion interaction of identical particles is greater than that of different ones. The interactions depend on the similarity of the polarizabilities of the particles, i.e., on the frequencies in their electronic spectra (also in the vibrational spectra, but their role is insignificant). The dispersion interaction forces are inversely proportional to the seventh power of the distance between the chromosomes.[101,102]

The studies of Zyryanov are related to those of Jehle.[103] Using the method of collective interactions, Zyryanov examined, in the final account, the same dispersion forces. The physical pattern of the interaction consists in the following: The chromosomes are examined as electrically neutral systems, but since the electrons and the atoms of the nucleus are not immobile, they create electromagnetic fields. The motions of the electrons and nuclei can be represented as vibrations with such frequencies as are determined by the polarizability. The detailed mathematical analysis shows that interactions arise between the electromagnetic fields of the two systems, and that these interactions result in their attraction to each other. The magnitude of the forces of attraction increases as their sets of frequencies approach each other; these forces are maximal when the frequencies coincide. Here also the interaction is of a resonance character, but it is independent of the existence of excited states, since both the electrons and the nuclei are never at rest and even in a non-excited state they have a nonzero energy. The Zyryanov theory is a particular case of the general theory of interaction of solid bodies developed by Lifshitz.[104] Zyryanov has also examined the influence of the surrounding medium on the chromosomes which interact with each other. The medium is an aqueous solution of electrolytes. It is assumed that a contact potential difference and a polarized ionic cloud are formed between the chromosome and the electrolyte. These ionic clouds, which surround the chromosomes, naturally cause their mutual repulsion; this competes with the electromagnetic attraction. If the attraction still predominates, then it may be overcome by changing the concentration of ions.

The studies by Jehle and Zyryanov are not convincing. They do not contain any attempt at a correlation between the theoretical calculations and the concrete structure of chromosomes, namely, the electrical properties of nucleoproteins. The DNP consists of DNA which contains essentially similar pyrimidine and purine bases and a protein which contains similar amino acids (67% arginine in the protamines). Thus, one should not expect any particular specificity in the spectra of individual regions of the chromosomes. It is sufficient to remember that all the Watson–Crick pairs of nitrogen bases are characterized by quite similar bands with an absorption maximum at 260 mμ. If we take cistrons as a whole, then their polarizabilities may not differ drastically. Consequently, a dispersion resonance mechanism does not explain the preferred attraction between homologous regions. It is quite probable that the energy difference between the attractions of different regions of the chromosomes is less than the thermal energy kT and, consequently, it cannot guarantee a specific interaction; it will be violated by thermal motion. Furthermore. it does not explain the interaction between the chromosomes at large distances; these forces decrease rapidly with distance. Here, we encounter a situation which is quite frequent in theoretical physics. The theory predicts a certain effect and all the theoretical deductions are correct. But, when quantitative calculations are carried out, when quantities found experimentally are introduced into the corresponding equations, it turns out that the phenomenon does not occur since the magnitudes of the energies, forces, etc., are negligibly small. The purpose of a physical theory is not to find a qualitative explanation which may be wrong, but rather a quantitative interpretation of the observed phenomena.

Thus, at present, there is no satisfactory theory of the synapsis of chromosomes. It is clear that, to formulate such a theory, it is necessary to have much more detailed information on the structure of the chromosomes and on the state of the cell during meiosis than is available at present. It is possible, however, to make some general remarks.

First of all, this problem should be subdivided. The mutual attraction between the chromosomes at large distances must not be specified at all. It is necessary that forces which decrease slowly with distance act within the chromosomes. Such forces may be either electrostatic interactions between charges and dipoles, or hydrodynamic forces which arise from the flow of liquid between two masses that flow in it. Let us examine the second effect. If a liquid flows in a tube, then the velocity of the flow of the liquid increases as the tube becomes narrower. This is accompanied by a decrease in the pressure of the liquid. In fact, any element of the liquid which moves in such a tube with a gradually decreasing cross section must increase its rate. For this, the element must be under the influence of a force in the direction of its motion and, consequently, the pressure on the forward end of the element must be less than that on the rear end.[105] If liquid flows between two chromosomes, then the pressure between them is less than outside and they are pushed toward each other. It is well known that such phenomena result in the collision of ships.

But is there motion of the liquid, of the protoplasm, within cells? This was discovered as early as 1774 by Corti. Protoplasm does undergo motion; it is characterized by a great diversity and its study presents an extremely interesting problem.[106] It has not been investigated in relation to the synapsis of chromosomes. As a result, we may only assume that a flow of liquid arises at a certain stage of meiosis.

Electrostatic attraction between the chromosomes, which as a whole are neutral or similarly charged, could result from a change in the concentration of the ions between them; this effect has been noted by Zyryanov. If the concentration of ions in the mutually repulsive ionic atmospheres of the chromosomes decreases during the stage of synapsis, the chromosomes may attract each other.

These interactions are not specific. But if the chromosomes have approached each other sufficiently, then there is no need to seek exceptions from the principle of structural complementarity. The difficulty is that the chromosomes are identical and do not complement each other. However, if we take into account the action of ions, and in particular that of polyvalent and hydrated ones, then it is possible to conceive of a specific attraction of regions with similar surface profiles (Fig. 181). A confirmation of this concept is the effect of the removal of divalent ions on the integrity of the chromosome; removal of Ca^{++} and Mg^{++} ions results in

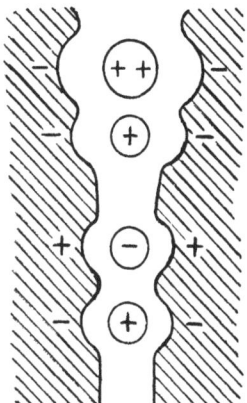

Fig. 181. Hypothetical model of the synapsis of chromosomes.

the breaking of the chromosomes into microfibrils. The presence of ions is also necessary for the formation of polynucleotide double helices.

All these considerations are only qualitative; they can, however, serve the purpose of working hypotheses for the formulation of a future theory of the synapsis of chromosomes.

Chapter 9

MECHANO-
CHEMICAL
PROCESSES

THE NATURE OF MECHANOCHEMICAL PROCESSES

We have examined two functions of the protein, a catalytic function (of enzymes) and a protective function (immunological). Both are related to the attainment of structural complementarity of the protein with another molecule. The possibility of forming structural complementarity, in turn, is the result of the multitude of conformations which are present in protein macromolecules. This chapter treats still another very important function of proteins, namely, mechanochemical work.

A very important property of living organisms and of their functional components is their ability to move, to carry out mechanical work under conditions of constant temperature and pressure. Each cell, each organism contains some engine whose work is expressed in cell division and motion, and in the motion of organs and of the organism as a whole. Some forces separate the parent and progeny chromosomes and carry them to different ends of the cell, cause independent motion of spermatozoids or infusoria, cause muscle contraction when the writer moves the pen along the paper or the boxer knocks out his opponent. Is the nature of such an engine, of such forces, identical in all cases? There is no full answer to this question, but most likely the answer will be positive. In one way or another, all these motions occur without changes in temperature and pressure and the working substance involved is a protein. It is quite clear that the source of the mechanical work in organisms and cells is chemical energy drawn by them from the environment during the processes of nutrition and respiration. In this sense, living systems may be called mechanochemical systems.

Transformations of chemical energy into mechanical work occur also in a steam engine or in an internal-combustion engine. But, in these cases the transformation is not direct. The combustion of the fuel is a chemical reaction of oxidation, similar to respiration. However, in these cases, it is not the chemical reaction which is used as such, but the emitted heat. It is unimportant what molecules enter into the reaction and what molecules come out of it; it is only necessary that the fuel have a sufficient efficiency. In the second stage of the process, the heat is transformed into work. This is possible only with heat transfer; i.e., for the engine to function, a refrigerator is necessary along with the heater. The working material of a heat engine, for example, steam, does not undergo chemical changes. The coefficient of useful action of a heat engine is always low; it is determined by the temperature difference between the heater and the cooler.

On the other hand, in a mechanochemical system, the chemical energy is directly transformed into mechanical work, bypassing the heat stage. Consequently, there is no necessity for a change in temperature or pressure. How does such a process occur?

It is possible to point to several ways in which chemical energy may be transformed into mechanical work in systems in which the working substances are polymers. Let us examine these ways.

1. A polymeric gel may change its volume as a result of swelling, i.e., of the interaction with some low-molecular-weight solvent. The change in volume may be used to create work. Processes of such a type may be important in the motion of protoplasm.

2. As has already been mentioned, when the charge on the polyelectrolyte changes, the length of its chain also changes. One may carry out the following demonstration experiment: A polyelectrolyte fiber prepared from a polyanion, for example, from polyacrylic acid, may be taken and stretched in a vessel in such a way that its end is held by a weight over a pulley. The fiber is immersed into a liquid. Let us now add acid to the solvent, which lowers the pH. The charges on the polymeric chains of the fiber will become neutralized and, as a result of the decrease of the electrostatic repulsion, the distance between the links of the chain will decrease and, consequently, the fiber will shrink. It will lift the weight and carry out mechanical work. It is evident that here the electrochemical action had resulted in a direct change of the chemical energy into mechanical work.

A number of attempts have been made to explain muscle action by the polyelectrolyte nature of muscle proteins. This will be discussed in detail below.

3. If the polymer chain is lengthened by polymerization or poly-condensation, i.e., by introducing new links into it, and if some weights are attached to the ends of the chain, then the chain growth will result in the separation of these weights, i.e., in mechanical work. There are data indicating that such a process and its reversal are possible during mitosis.

4. The set of conformations of polymer chains may change also without changing the charges, as a result of another chemical or physical-chemical interaction with other molecules. This may also result in a change of the chain length and, consequently, this may produce work.

Using principles of thermodynamics, we come to the absolutely necessary conclusions that reverse processes are possible, namely, a change in the chemical state of the engine under the action of an external force, or a transformation of mechanical work into chemical energy. Mechanical work may be transformed into heat in just the same manner: by reversing a heat engine, we make it into a refrigerator. By stretching a poly-electrolyte fiber with a weight, we change its degree of ionization. Among such reversible processes, a particularly interesting one is enzymatic mechanochemistry, which, so far, has not been carried out experimentally but which may play an important role in muscle action.

5. The stretching of a polymeric catalyst by an external force must result in a change in the catalytic activity and, consequently, in a change in the chemical state of the system.

Are such similar processes possible in low-molecular-weight substances? Certainly yes, but in biological systems they probably do not play a role. If the chemical reaction of a low-molecular-weight substance is accompanied by a change in the volume of the system, then the chemical energy may be transformed into mechanical work. In an internal-combustion engine, the oxidation of hydrocarbons takes place. The change in the volume of the gas, however, is determined, not by the direct change in the specific volume, but by thermal expansion, namely, the reaction proceeds with a large evolution of heat.

A low-molecular-weight substance may change its volume both with a change in charge and with a polarization of the molecule; this is the phenomenon of electrostriction. Such phenomena, however, are not large.

It is noteworthy that the mechanochemistry of small molecules may be determined principally by an isotropic change in the volume, while the mechanochemistry of polymeric chains is related to directed anisotropic motions.

A physical analysis of mechanochemical processes must start, just as other problems of molecular physics, with a statement of general pheno-menological rules, which are independent of the concrete nature of the working substance undergoing the chemical reactions. The thermo-dynamics of mechanochemical processes have been developed by Katchal-sky and coworkers.[1]

In the general case of chemical and heat transformations, the change of the internal energy of the system is expressed by

$$dE = T\,dS - dA + \sum_i \mu_i\,dn_i \tag{9.1}$$

where dS is the increase in the entropy of the system at temperature T, dA is the mechanical work carried out by the system, dn_i is the number of moles of the ith component introduced into the system at a chemical potential of μ_i.

If the system consists of uniform fibers of length L and volume V, stretched by force f, then

$$dA = p\,dV - f\,dL \tag{9.2}$$

and

$$dE = T\,dS - p\,dV + f\,dL + \sum_i \mu_i\,dn_i \tag{9.3}$$

This is a general expression.

In the usual mechanical process, for example, in the stretching of a string, $dS = 0$ (the purely mechanical process is always adiabatic), $dV = 0$, $dn_i = 0$. Consequently,

$$dE = f\,dL \tag{9.4}$$

In the usual heat engine

$$dE = T\,dS - p\,dV \tag{9.5}$$

and in an isothermal process $dE = 0$; consequently

$$dA = p\,dV = T\,dS \tag{9.6}$$

i.e., all the heat is transformed into work.

In the isothermal stretching of rubber

$$dE = 0 \qquad dV = 0 \qquad dn_i = 0$$
$$dA = f\,dL = -T\,dS \tag{9.7}$$

Finally, in the isothermic mechanochemical process,

$$dE = 0 \qquad dS = 0 \qquad dV = 0$$
$$dA = f\,dL = -\sum_i \mu_i\,dn_i \tag{9.8}$$

Let us construct a working cycle of the mechanochemical engine. For the sake of simplicity, let us assume that the reaction takes place with only one chemical component and, consequently, the system has two degrees of freedom and it is described by two independent variables μ and f; n and L are functions of these. Let us construct a cycle within the variables f and L (Fig. 182). It will be formed of isopotentials, i.e., curves corresponding to a constant value of μ, and isofores, i.e., curves corresponding to a constant value of n. All the points of an isopotential curve may be obtained from a single experiment in which the dependence of the deformation L on the force f is measured in a large reservoir with a constant chemical potential μ. It is evident that the isopotentials are similar to isotherms, which are found by measuring the mechanical changes of a heat engine and which occur during its contact with a large reservoir at a constant temperature, i.e., a constant heat potential. On the other hand, the

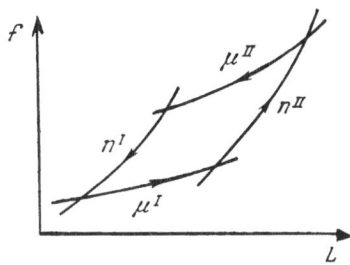

Fig. 182. Mechanochemical cycle built of isofores
and isopotentials.

isofores, which express the behavior of an isolated fiber with a constant amount of reacting material, are similar to adiabatic curves, i.e., the curves which show the change of the state of the heat engine at constant entropy, i.e., in the absence of heat exchange. The mechanochemical cycle, in this manner, is similar to the Carnot cycle; μ plays the role of T and n plays the role of S.

The simplest mechanochemical cycle is an engine which works between two reservoirs containing the same reacting component at chemical potentials $\mu^I \neq \mu^{II}$. The work is expressed in the following manner (see Fig. 181):

$$A = - \oint f\, dL = \int_I \mu\, dn + \int_{II} \mu\, dn$$

$$= \mu^I \int_I dn + \mu^{II} \int_{II} dn = (\mu^I - \mu^{II})\Delta n \qquad (9.9)$$

Δ_n is the amount of the reactive material removed from reservoir II and transported into reservoir I. Let us recall then that, in a heat engine in a Carnot cycle,

$$A = (T_1 - T_2)\,\Delta S \qquad (9.10)$$

The coefficient of useful action of a heat engine is

$$\eta = \frac{T_1 - T_2}{T_1} \qquad (9.11)$$

It would appear that, by analogy, one could write for a mechanochemical engine

$$\eta = \frac{\mu^I - \mu^{II}}{\mu^I} \qquad (9.12)$$

This expression, however, is devoid of sense since a chemical potential does not have an absolute zero. Therefore, the ratio of the work obtained in a cycle to the work introduced at the high potential $\mu^I \Delta n$, is a function of the selected standard value of μ. It is still possible, however, to express the efficiency of a mechanochemical engine in the form of the ratio of the work obtained in a real cycle to the work which could have been produced in a reversible cyclic process, namely,

$$\eta' = - \frac{\oint f\, dL}{\sum_i \oint \mu_i\, dn_i} \qquad (9.13)$$

Let us cite, finally, some characteristic thermodynamic relations which have analogies in heat processes:

$$\left(\frac{\partial f}{\partial \mu}\right)_n = -\left(\frac{\partial n}{\partial L}\right)_\mu \tag{9.14}$$

$$\left(\frac{\partial f}{\partial L}\right)_n - \left(\frac{\partial f}{\partial L}\right)_\mu = \left(\frac{\partial f}{\partial \mu}\right)^2 \left(\frac{\partial \mu}{\partial n}\right)_L > 0 \tag{9.15}$$

This is thermodynamics. What, then, is the molecular nature of mechanochemical processes?

Mechanochemical phenomena were observed for the first time by Engelhardt and Lyubimova in myosin gels, a protein extracted from muscle fiber.[2] When ATP is added, the myosin fibers shrink in a reversible manner. The source of the mechanical work, in this case, is chemical energy stored in the high-energy bond of ATP, since, when it interacts with myosin, ATP becomes dephosphorylated. Later, mechanochemical processes were observed in systems already mentioned, namely, in fibers and films made from synthetic polyelectrolytes.[3-5]

Pasynskii and Blokhina observed for the first time the reverse process, the transformation of mechanical work into chemical work, i.e., the displacement of the isoelectric point of a protein (keratin) during its deformation.[6] Vorob'ev investigated the change of pH of a medium during the mechanical stretching of a nucleohistone fiber immersed in it.[7]

In order for a polymeric system to perform mechanical work, it must be able to undergo reversible deformations and its individual molecules or molecular systems, as a whole, must be able to change shape under the action of an external force and the chemical environment. The model systems investigated and, first of all, the simplest one, polyelectrolytes, possess these capabilities. The cyclic deformation of a fiber of nucleohistone during periodic changes of the pH of the medium is shown in Fig. 183. The scheme illustrating the direct and reverse processes of transformation of chemical energy into mechanical work for a polyanionic fiber is shown in Fig. 184.[7] The fact that a change in the length of a polyelectrolyte fiber during a change in pH is related to a change in the conformation of the chain appears quite evident. The detailed mechanism of such processes, however, is not at all trivial.

The dissociation and recombination in a polyelectrolyte chain is a cooperative phenomenon. The dissociation constants of the ionizable groups in a polymer differ from the same constant in a monomer because of the interaction of neighboring charged groups, i.e., their electrostatic repulsion. In turn, the repulsion is a function of the conformation of the chain.[8,9] These phenomena are reflected in the potentiometric titration curves of polyelectrolytes.

There are two types of effects of free charges on the properties of macromolecules in solution.[10] First, a long-range interaction, i.e., the

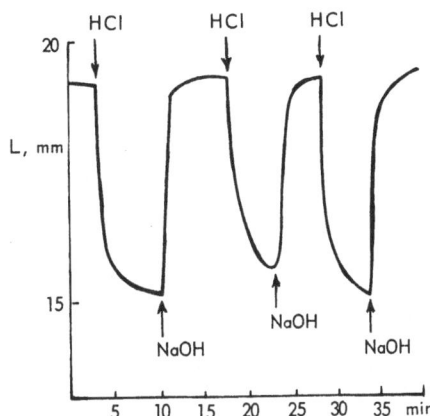

Fig. 183. Cyclic deformation of a nucleo-histone fiber under periodic changes of pH.

interaction between charges which are far removed from each other along the chain. Second, a short-range interaction, i.e., the interaction between charges which are near each other. Theory shows that when the force f is not very small and is sufficient to orient the chain as a whole, the first effect does not play a role and the length of the chain is expressed by the equation

$$\bar{x} = \frac{f}{3kT}\overline{h_0^2} \qquad (9.16)$$

Here $\overline{h_0^2}$ is the mean-square end-to-end distance of the chain, with only short-range interactions taken into consideration. This quantity is evidently a function of the degree of ionization α, that is, of the pH of the medium. If as α increases, the quantity $\overline{h_0^2}$ increases, that is, $\overline{dh_0^2}/d\alpha > 0$, then the polyelectrolyte chains become stretched during ionization in a field of constant force; they contract during the decrease of the degree of ionization if $\overline{dh_0^2}/d\alpha < 0$, that is, all the processes are reversed. The second case is found with synthetic polypeptides; $\overline{h_0^2}$ decreases with an increase in α during helix–coil transitions, since an increase in the degree

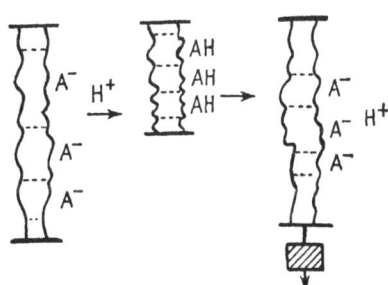

Fig. 184. Schematic representation of the mechanochemical cycle for a polyanionic system.

of ionization destroys the stretched helical structure. On the other hand, in the case of synthetic atactic polyions, $\overline{h_0^2}$ increases with an increase of α. The actual dependence of $\overline{h_0^2}$ on α may be expressed in the general case in the form[10]

$$\overline{h_0^2} \cong (\overline{h_0^2})_{\alpha = 0}\, e^{-\alpha^2 \Delta E/kT} \qquad (9.17)$$

where ΔE is the difference between the interaction energies of free charges in the coiled and elongated conformations of the chain. It is evident that ΔE is a measure of the cooperativity of the system; if $\Delta E = 0$, the chain is not cooperative and changes in the state of ionization do not have an effect on the probabilities of different conformations; consequently, the length of the chain is independent of α when f is a constant. The sign of the derivative $d\overline{h_0^2}/d\alpha$ is determined by the sign of ΔE. The same theory, based on the cooperative interaction of charges, enables us to determine the manner in which the pH of the medium must change during the stretching of a chain.

It should be noted that the described cooperative mechanism cannot be considered unique; there exist other polyelectrolytic mechanochemical processes determined by a change in the degree of binding of ions.

In this manner, changes in the conformation of a chain, brought about by changes of the chemical environment under the action of a constant force, result in mechanical work. Conversely, the acting force must change the conformation of the chain. Flory[12] and Birshtein[13] have investigated the influence of an external force on helix–coil transitions in polypeptide chains. Theoretical investigations of a one-dimensional model, based on the application of the method described above,[13] have shown that the external force stabilizes the helical conformation of the chain if the force is not very large. On the other hand, a large force stabilizes the conformation of a highly stretched coil. The displacement of the transition temperature may be as great as 20 to 30°; the sharpness of the transition is practically independent of the force. In charged chains, a rather small external force results in an increase of the degree of dissociation which brings about a transition at a fixed temperature. When the external force is large, the degree of dissociation which brings about the transition may be smaller than in the absence of the force. The stretched coil may result in an increase, as well as a decrease, of the dimensions of the chain in the direction of the force, depending on the magnitude of the applied force during the helix–coil transition.

A number of deductions have been made from a statistical examination of mechanochemical processes; these are subject to further experimental verification.

Kachalsky has been able to demonstrate a cyclically acting mechanochemical system with the use of collagen fibers. In these polyelectrolytic fibers, the helix–coil transition occurs under the influence of LiBr. The fiber contracts and develops a considerable force. When the fiber is transferred to a water bath, it releases the salt and stretches. If it were possible to find a fiber which would undergo a similar transition under the influence of NaCl, such a system could be used for the desalting of sea water. In this case, the work would use up only the entropy of mixing, while, in the usual distillation of the water, a very large amount of heat energy is used for evaporation.

MUSCLE ACTION

What are the mechanochemical processes which we encounter in cells and organisms? They are quite manifold. On the cellular level, these include motions of the protoplasm, motions of flagella and cilia and, finally, of mitosis and meiosis. On the level of organisms, these include the muscle action of animals and the motions of plants; for example, a sunflower rotates its flower toward the sun.

Let us start our examination of biological mechanochemical processes with muscle action. This problem has interested scientists for a long time. Let us acquaint ourselves with the basic facts.

There are muscles of different types and behavior. Skeletal muscles of vertebrates are attached to bones and are controlled by the nervous system. The muscles of the intestinal wall do not work randomly, the same is true of the heart muscle. We shall discuss mostly the skeletal muscles. The classical object of experimental investigations has been the striated muscle called the *sartorius*, of the frog. The muscle consists of sheaves of fibers, each of which is a very long cell containing many nuclei and possessing a cell wall, called a *sarcolemma*. Capillary blood vessels are located between the fibers. A schematic representation of a single muscle fiber is given in Fig. 185.[14] The fiber is split into what are called *myofibrils* which have a quite characteristic structure that has been thoroughly investigated by electron microscopy. These are separated by the so-called "Z" membranes into individual cylindrical regions with lengths of approximately 2.5 μ. These regions are called *sarcomeres*. The medium inside a myofibril, namely, the sarcoplasm, contains mitochondria and the endoplasmic reticulum, as is normal for cytoplasm. The structure of a myofibril is shown in Fig. 186. An electron micrograph of the myofibril is shown in Fig. 187.

It has been shown experimentally that all the fundamental contractive properties of a muscle are properties of individual myofibrils. In this case, we easily reach the cellular level. Since this is so, it becomes normal to seek the molecular mechanism of muscle action. First of all, we must familiarize ourselves with the molecular composition and the structure of a myofibril.

The principal components of a muscle are evidently proteins. They may be classified into four groups. The soluble proteins of a muscle, the

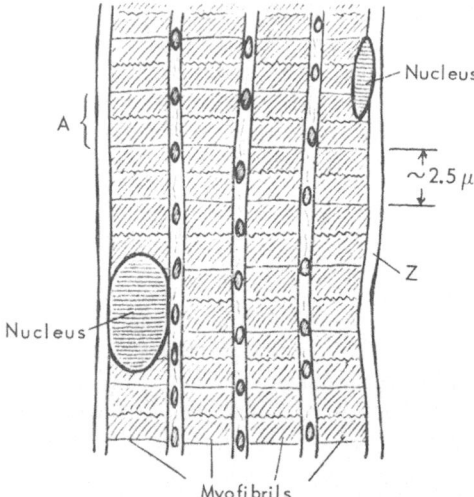

Fig. 185. Schematic representation
of the structure of a muscle fibril.

proteins of the sarcoplasm, participate in metabolism, but not in muscle contraction. The proteins of the granules which are found in the sarcoplasm, i.e., the mitochondria, the nuclei, and the ribosomes, participate in muscle action but are not the working materials of the muscle; they are not the contractive proteins. The proteins of the stroma enter into the sarcolemma and the Z membranes; these are proteins similar to collagen, but they also are not contractive. The contractive muscle proteins are in the myofibrils. These are the ones which are of greatest interest.[15] First of all, these encompass myosin, a protein with a molecular weight of 400,000 to 500,000, which forms dimers with a molecular weight of 850,000. Its molecules are highly asymmetric; they have a length of the order of 1600 Å and a width of 37 to 47 Å. Another very important protein is actin, which exists in two forms, G actin and F actin. Molecules of G actin are globules with molecular weights of the order of 57,000 (in the absence of ions, and 70,000 in 0.6 M KI). The G actin can be polymerized into F actin and it forms fibrillar protein of quite a high molecular weight (1.5 to 3×10^6). The transition G actin $\rightarrow$ F actin occurs in the presence of salts. When solutions of actin and myosin are mixed, a protein complex is formed, actomyosin, with a molecular weight of the order of 2×10^7;

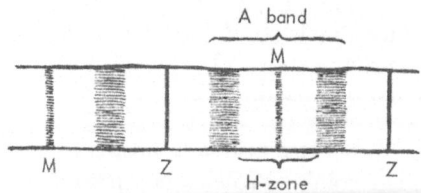

Fig. 186. Structure of a myofibril.

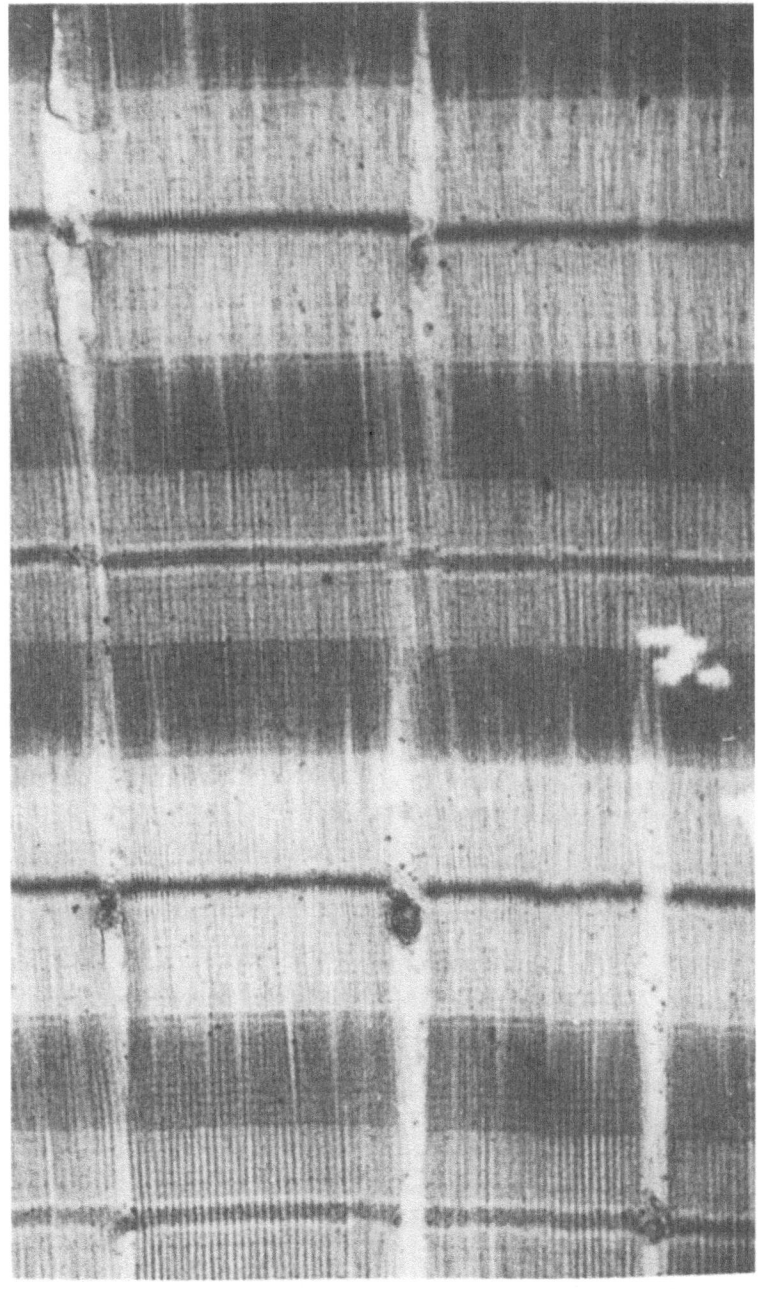

Fig. 187. Electron micrograph of the myofibril.

1 g of actin interacts with 4 to 5 g of myosin. This means that F actin with a molecular weight of 3×10^6 combines with 25 to 30 molecules of myosin.

In addition to these two proteins which serve as working substances in the muscle machine, the myofibril contains in rather small numbers tropomyosin (with a molecular weight of about 60,000) and paramyosin (with a molecular weight of close to 137,000). The role of these proteins in muscle action is not clear.

A very interesting property of myosin is its enzymatic activity, which is usually not found in fibrillar proteins. This was discovered by Engelhardt and Lyubimova, who showed that myosin is ATP'ase; it splits ATP and so extracts energy from it.[2,16]

The finding by these Russian biochemists forms the basis of all the modern considerations on the nature of muscle contraction; it is of great importance to biophysics.

Actin is not an ATP'ase; actomyosin, on the other hand, possesses this enzymatic activity. If an extracted myofibril is immersed in a salt solution containing Mg^{++} ions, it contracts if ATP is added; this is a very striking demonstration experiment. It is clear that any theory of muscle activity must take into account the role of ATP; furthermore, it is necessary to establish whether this role is determining.

Myosin and actin are located within the myofibril in a very specific ordered manner. Huxley and Hanson have carried out a detailed electron-microscopic investigation of the myofibril[17] (see Figs. 187 and 189). The myofibril region between the two Z membranes is divided into several zones (Fig. 185). The central zone A is anisotropic; it is birefringent. Two isotropic zones I adhere to it symmetrically on both sides. In the striated muscle, myosin accounts for 30 to 40% and actin for 10 to 15% of the entire protein. According to Huxley and Hanson, the protofibrils of the myosin are located within the A zone; these have a thickness of about 100 Å and a length of 10,000 Å; the thinner (diameter, 40 Å) actin protofibrils are located between the myosin protofibrils; they completely occupy the A zone and the I zone. During contraction of the muscle, the actin fibers slide between the myosin fibrils; the A zone does not change its dimensions, while the I zone decreases up to total disappearance (Fig. 189). The

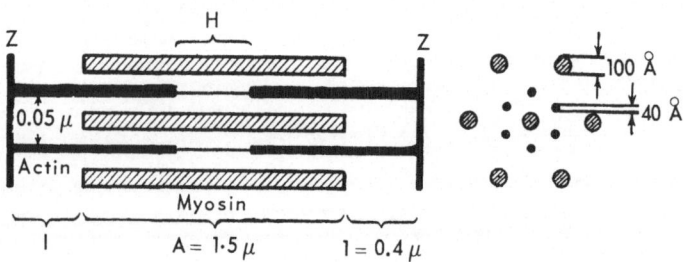

Fig. 188. Molecular structure of the myofibril.

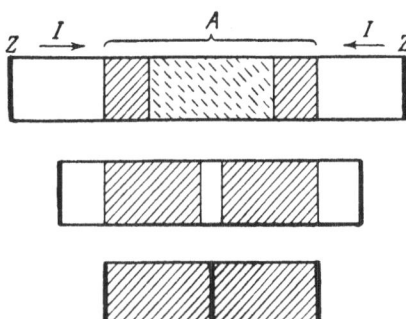

Fig. 189. Schematic representation of the
shrinking of the myofibril.

birefringence of the muscle fibril decreases during the process; it becomes
more isotropic. It should be pointed out that such an interpretation of the
magnificent electron-microscope pictures may not be regarded as unique
and final.[18,19]

Biochemical and physiological experiments carried out over many
years have led to the knowledge of the basic chemical phenomena which
occur during muscle contraction.[20–22] It should be pointed out from the
very beginning that the process is extremely complicated and that there
are serious difficulties which are encountered in all attempts to explain
these chemical phenomena by a single picture.

The muscle is at rest. This state corresponds to a definite length of the
muscle fiber. Then, under the action of a nervous impulse (the nature of this
action is still not quite clear[23]), the muscle contracts. The word contraction is used in the case of muscles in a broad sense. A muscle which lifts a
weight becomes shorter; in this case we are speaking of isotonic contraction, expressed as a shortening, a decrease in the length. If when the
muscle holds the weight, it becomes neither contracted nor elongated,
we must talk of an isometric contraction, expressed in the development of
tension in the muscle. Having performed its work, the muscle must return
to its initial state of rest. This return is known as the relaxation process.

We may consider that it has been established that the following
conditions must be fulfilled for a muscle fibril to contract in a liquid
medium:

1. The pH must be physiological, close to neutral (between 6.5 and 8).
2. The ionic strength must also correspond to physiological conditions,
   i.e., it must be above 0.15.
3. Mg^{++} ions must be present.
4. ATP must be present in a concentration not greater than $5 \times 10^3\ M$.
5. ATP must be split with sufficient rapidity.

These conditions have been established as a result of the investigation
of the properties of isolated muscle fibers, obtained by treating the muscles

with glycerine. Such fibers are good models for the action of the muscle as a whole.

It has been shown that the starting biochemical process consists in the enzymatic splitting of ATP by actomyosin. Other nucleoside triphosphates also bring about the contraction of isolated fibers *in vivo*, but they are much less affected. The required rate of splitting is not less than 0.02 μmole/mg of protein per minute. The role of ATP is twofold. On one hand, it is split, being a source of energy and work ; on the other hand, it serves as a plasticizer to soften the muscle and ease both the contraction and relaxation. The Mg^{++} ions are necessary for muscle contraction : these activate the enzyme actomyosin and, consequently, they play the role of a cofactor. Thus, ATP is split into ADP and phosphate. Furthermore, ADP, as well, serves as a good plasticizer in the presence of Mg^{++} ions.[24]

How does the split ATP become restored, how does this battery of chemical energy become recharged? This obviously occurs as a result of metabolism ; in the long run, it is the result of oxidative phosphorylation. In addition to ATP, the muscle contains other phosphogens, namely, substances which contain high-energy phosphate bonds, in particular, creatine phosphate (CP)

$$HPO_4 \cdot HN \cdot CNH \cdot N(CH_3) \cdot CH_2COOH$$

By a very sensitive spectroscopic method Chance has found that ADP appears in a contracting muscle in a very small amount, not greater than 2 to 3% of the expected value.[25] This, however, does not contradict the splitting of ATP, since the ADP is immediately regenerated into ATP by CP.[23]

$$ATP \longrightarrow ADP + P$$

$$ADP + CP \longrightarrow ATP + C$$

The relaxation of the muscle fiber is controlled by the action of a special factor, the nature of which is still unclear. This factor is contained in the granules present in the cytoplasm and is probably some sort of enzyme. The relaxing factor is also activated by Mg^{++} ions. This factor is active in the unexcited muscle, and ATP'ase activity is absent. Consequently, it inhibits in some way the splitting of ATP. The relaxation factor, in turn, is reactivated by Ca^{++} ions, which, apparently, interact with it during contraction. During the rest period of the muscle, the Ca^{++} ions are probably bound to myosin.

A definite concentration of ATP is essential. If the ATP concentration is very small, the relaxing factor does not act as an inhibitor of the splitting, even in the absence of Mg^{++}, and an irreversible muscle contraction occurs, namely, *rigor mortis*. On the other hand, when there is an excess of ATP, it is not split ; in this case, the ATP, as a substance, inhibits the enzyme ATP'ase (actomyosin).

The main fact is that muscle contraction and ATP'ase activity develop in parallel fashion. There is a correlation between the rate of contraction and the development of the strength of the muscle, i.e., in the final account, the splitting of ATP.[26] The question of the dependence of splitting on the interaction of actin with myosin, which results in the formation of acto-myosin, is still unresolved. A number of hypotheses have been proposed.

This is a very concisely expressed biochemical picture of muscle action. It is supported, to some extent, by direct electron-microscopic observations. One may ask: How does the work of a muscle manifest itself, what energetic rules direct it? It is evident that the molecular bio-physics of the muscle must be constructed on the basis of all the factors, namely, biochemical, structural, and thermodynamic.

The thermomechanical properties of the muscle have been studied by Hill.[27] Using an extremely sensitive electrocalorimetric method, Hill studied the heat evolution of the muscle during its contraction at definite rates and under different loads. During muscle contraction, energy is released, namely, the mechanical work A and the evolved heat Q. The total change of enthalpy of the muscle in an isobaric process is, according to the first law of thermo-dynamics,

$$\Delta H = Q - A \tag{9.18}$$

Hill found a simple relation between the rate of contraction, v, and the load p. This is a hyperbolic relation

$$(p + a)v = b(p_0 - p) \tag{9.19}$$

This relation is shown in Fig. 190. Equation (9.19) is the characteristic equation of Hill. The load p_0 is the load in the muscle subjected to tetanic contraction. If very frequent exciting impulses, for example, an electric current, are fed to the muscle, it passes into a tetanic state.

It can be seen from Fig. 190 that p_0 corresponds to $v = 0$ if the maximal tension with-stood by the muscle at $p > p_0$ is $v < 0$, i.e., the muscle no longer contracts but expands. To the contrary, in the absence of the load, when $p = 0$, the rate of contraction is maximal and it is

$$v_m = \frac{bp_0}{a} \tag{9.20}$$

where a and b are constants.

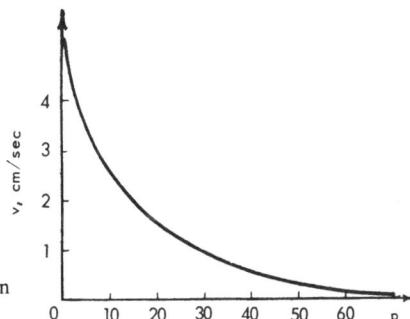

Fig. 190. Dependence of the rate of contraction
of the muscle on the load.

The Hill equation may be rewritten in the form

$$p = \frac{bp_0 - av}{v + b} \tag{9.21}$$

The strength of the muscle, i.e., the mechanical work carried out by it during 1 second, is

$$\dot{A} \equiv \frac{dA}{dt} = pv = \frac{v(bp_0 - av)}{v + b} \tag{9.22}$$

Thus, the rate of contraction controls the rate of evolution of mechanical energy. The higher a load, the higher the generated power.

Simultaneously, heat is evolved. The dependence of the amount of evolved heat on time is shown in Fig. 191. These results, obtained by Hill, are for a muscle 29.5 mm in length, which started to contract under a constant load 1.2 second after stimulation. Curve A corresponds to the isometric contraction (the shortening is equal to zero); curves B, C, and D correspond to contractions of 1.9, 3.6, and 5.2 mm. During contraction, the rate of heat evolution increases at first relative to the rate of an isometric interaction; however, since the curves B, C, and D later proceed parallel to A, the rate of heat production in the contracted state is the same as in the corresponding isometric state. Displacement of curves B, C, and D relative to A, that is, the addition of heat related to contraction, is proportional to the degree of contraction.

The heat emitted by the same muscle, when it is contracted by a constant length at different loads, is shown in Fig. 192. The muscles were released 1.2 seconds after stimulation. The end of contraction is shown by the arrows. Curve E corresponds to isometric contraction. Since curves J, H, G, and F, which are for different loads, are similarly displaced from E, the total heat excess relative to isometric contraction is independent of the load p and, consequently, of the rate of contraction v. It is linearly dependent on the degree of contraction ΔL, but is independent of the performed work.

$$-Q = q_0 + a\,\Delta L \tag{9.23}$$

where q_0 is the amount of heat emitted during isometric contraction, and a is the same constant as found in the Hill equation (9.19).

Consequently, the rate of heat evolution is proportional to the rate of contraction

$$-\dot{Q} = \dot{q}_0 + av \tag{9.24}$$

since

$$v = \frac{d\,\Delta L}{dt}$$

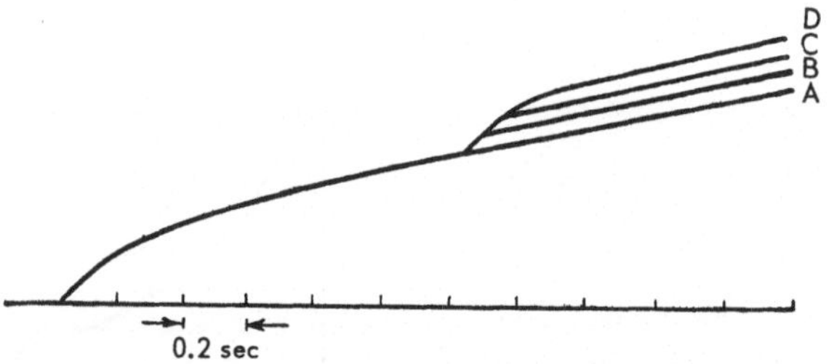

Fig. 191. Heat evolution during muscle contraction.

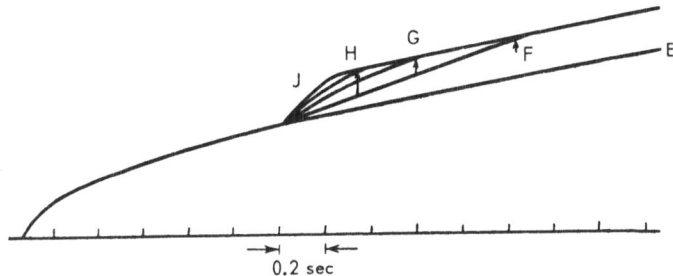

Fig. 192. Heat evolution during muscle contraction.

Thus, muscle activity is quite particular and is completely different from, let us say, the contraction of a spring. A spring obeys the law of conservation of energy,

$$\Delta H = Q - A$$

The more work it produces, the less heat is evolved and *vice versa*, since ΔH is constant. In muscle, the liberated energy is equal to the sum of the heat of activation, q_0 (corresponding to isometric contraction), the heat of contraction, $a\,\Delta L$, and the performed work $p\,\Delta L$:

$$\Delta H = -q_0 - a\,\Delta L - p\,\Delta L \tag{9.25}$$

Since the heat of contraction is independent of the work, the quantity ΔH or $\Delta H + q_0$ is not constant. A muscle works as an electric motor, which uses an increasing amount of current as the load in the circuit increases. A muscle is an open system; it draws its energy from a chemical source.

Thus, the amount of energy emitted per unit of time is

$$-\Delta\dot{H} \equiv -\frac{d\,(\Delta H)}{dt} = \dot{q}_0 + v(p + a) \tag{9.26}$$

The excess energy per second relative to the isometric state is

$$-\Delta\dot{H}' = -\frac{d\,(\Delta H)}{dt} - \dot{q}_0 = v(p + a) \tag{9.27}$$

and from (9.19)

$$-\Delta\dot{H}' = v\,(p + a) = b\,(p_0 - p) \tag{9.28}$$

This is the second Hill equation, which is evidently independent of the characteristic equation. We must add to Eqs. (9.19) and (9.28) the empirical relation between the constants

$$q_0 = ab \tag{9.29}$$

$$p_0 = 4a \tag{9.30}$$

From (9.28), (9.29), and (9.30), it follows that

$$-\Delta\dot{H} = ab + v\,(p + a) = b\,(a + p_0 - p) \tag{9.31}$$

and from (9.21) and (9.30)

$$p = \frac{a\,(4b - v)}{b + v} \tag{9.32}$$

A muscle is a mechanochemical system. The source of the emitted energy ΔH is a chemical reaction which proceeds to a greater extent as the contraction increases and as the amount of produced work increases. Relative to the external work, a muscle is an open system which uses chemical reagents as a source of energy. We may express ΔH in the form[28,29]

$$\Delta H = \xi h \tag{9.33}$$

where ξ is the extent of completion of the reaction, and h is the specific amount of energy emitted at $\xi = 1$. From (9.31), (9.32), and (9.33), it follows that

$$-\dot{\xi}h = \frac{\dot{q}_0 b + [\dot{q}_0 + b(a + p_0)] v}{b + v} \tag{9.34}$$

This equation relates the rate of the reaction, $\dot{\xi}$, to the mechanical rate v. When $v = 0$, in the isometric simplification, we have

$$\dot{\xi}_0 = -\frac{\dot{q}_0}{h} \tag{9.35}$$

At a maximal value of the rate

$$\dot{\xi}_m = \dot{\xi}_0 \left(1 + \frac{bp_0}{q_0} \right) \tag{9.36}$$

and from (9.29) and (9.30),

$$\dot{\xi}_m = 5\dot{\xi}_0 \tag{9.37}$$

i.e., the maximal rate of the reaction during contraction in the absence of a load is five times greater than an isometric contraction.

Designating the ratio of the rates by

$$v = \frac{v}{v_m} = \frac{va}{bp_0} \tag{9.38}$$

we obtain

$$\dot{\xi} = \dot{\xi}_0 \frac{1 + 24v}{1 + 4v} \tag{9.39}$$

The dependence of the rate of the chemical reaction on the rate of contraction is hyperbolic.

Let us indicate finally the dependence of the rate of contraction on the very magnitude of the contraction ΔL. This was found by Ulbrecht and Ulbrecht[26] and turned out to be exponential rather than hyperbolic. It should be noted that Hill worked with a striated skeletal muscle, frog sartorius: the Ulbrechts worked with smooth muscles, the so-called "yellow adductor" of the mollusk *Anodonta*.

Recently, Hill has improved his method and has shown that the constants a in Eq. (9.19) and (9.23) are similar only in order of magnitude. This, of course, changes the conclusions, but does not change the general relationship between the parameters of muscle contraction.[29a]

Pennycuick has shown that the Hill equations may be applied principally to an instantaneous state of the muscle immediately after the arrival of the nerve impulse. After a passage of time following the stimulus, the rate is drastically changed.[29b]

These studies[29a,29b] show that the thermomechanical properties of the muscle cannot be considered finally established; careful detailed experimental studies are necessary in this realm.

The problem of molecular biophysics lies in the search for a process which occurs in muscle, and which would be consistent, on one hand, with the results of biochemical and structural experiments and, on the other hand, with the particular thermomechanical properties of the muscle. This problem turns out to be quite difficult; many studies have been devoted to it, a number of theories have been proposed, but none of these may be regarded as exhaustive, while some are incorrect. As has been stated by Szent-Györgyi, at present the muscle occupies the position of a sacred elephant which has 99 names; the true name is the hundredth one, which is unknown to everybody.[30]

MOLECULAR THEORIES OF MUSCLE ACTION

The theories of muscle action may be subdivided into several groups as functions of their starting assumptions:

1. The principal assumption (an idea which is natural once one is familiar with the properties of polyelectrolytes) is that a muscle fiber at rest consists of charged polyelectrolyte molecules. Contraction causes a neutralization of charges on the fiber, which simultaneously coils up.

2. The principal assumption (again a natural one, once one becomes familiar with conformational transitions of a protein) is that the muscle-fiber proteins have a β form at rest, while a transition into the α form takes place during contraction; the α form would result in a shorter length of the chain.

3. The starting assumption is based on a morphological electron-microscopic picture, rather than on known properties of other macromolecules. This has already been discussed. There are no changes in the conformations of the chains. The contraction of the muscle results from the sliding of chains on each other, namely, the actin chains move in between the myosin chains.

4. The contraction of the muscle is determined by changes in the hydration of the protein, i.e., changes in the structure of water bonded to the protein. This process is carried out by ATP, whose energy is transferred and migrates along the water structure.

5. Muscle contraction is determined by the cooperative transition of the type of the melting of a crystalline polymer.

6. Muscle contraction is determined by a cooperative transition of the type of polymerization.

Theories of types 1 and 3 do not include concepts on cooperative processes; theories of types 2, 5, and 6 discuss sharp cooperative transitions; a theory of type 4, to the contrary, must be quantum mechanical, since it discusses the resonance transmission of energy along a long-range molecular system. We see that the names of the sacred elephant are very numerous. Let us examine these theories in turn.

1. The polyelectrolyte model of the contractive element of the muscle has been examined in a number of studies. Kirkwood and Riseman[31] assumed that such a model can be a flexible protein chain, the charge of which changes during the phosphorylation of serine and threonine residues by ATP. Simultaneously, the chain becomes charged negatively and expands; this explains relaxation. During the splitting of ATP, the actomyosin thread becomes dephosphorylated; it loses its negative charges and contracts. Morales and Botts[32,29] have proposed that the working element of the muscle is the positively charged chain, the negative charges of which are complexed with Mg^{++} ions. Contraction is brought about by the adsorption of ATP, the phosphate groups of which neutralize the positive charges. This model explains the role of Mg^{++} ions; it stems, however, from an assumption that relaxation requires, not the enzymatic splitting of ATP, which is not explained, but an adsorption of the ATP ions by the polycation. Contraction in this model is not energetic, but entropic; namely, the resilient force in the muscle becomes similar to the resilient force in rubber.

This and other studies which stem from polyelectrolyte hypotheses[33,34] are not in accord with experimental facts. The electron-microscopic picture does not testify in favor of the winding-up of the threads during contraction. It is very difficult to reconcile quantitatively electrostatic effects in a medium consisting of a 0.1 M salt solution with a high muscular force. The skeletal muscle develops a tension of the order of several kilograms per square centimeter; it is sufficient to recall the achievements of athletes. The thermomechanical properties of a muscle contradict concepts based on an entropic force. One should note that none of the studies which consider a polyelectrolyte model gives an interpretation of the Hill equations.

2. The theory of the $\alpha \rightleftharpoons \beta$ transition in a muscle has been developed by Astbury[35] on the basis of the studies by Pauling and Corey. The behavior of muscle proteins is examined by analogy with fibrous proteins, namely, keratin, epidemin, fibrinogen, and fibrin. The transition of myosin into the α form, according to Astbury's ideas, occurs during the polymerization of globular G actin into the fibrillar F actin; it is in a complex with the last that myosin contracts. Neither the biochemical processes nor the thermodynamic properties of the muscle are examined in these studies. Nevertheless, the idea itself presents a definite interest and may be used in a modified form during the future construction of a theory of muscle action. Using to some extent Astbury's ideas, Frenkel has proposed

examination of a muscle fiber by analogy with rubber.[36] The contraction of the fiber results from an increase of the modulus of elasticity. As is well known, the modulus of elasticity of rubber increases with an increase of its vulcanization, i.e., with an increase of crossbonds between the chains. Frenkel has proposed that the crossbonds in the muscle are made by ATP molecules. This idea is attractive, but the hypothesis is not consistent with the biochemical or with the structural or with the thermodynamic properties of the muscle. One should remark that most of these properties were still unknown when Frenkel made this hypothesis.

3. Using the electron-microscopic picture of H. Huxley and Hanson, A. Huxley proposed a sliding molecular model of the muscle fiber.[37,17,28] The model by A. Huxley is shown in Fig. 193. The sliding of the actin within the myosin fibrils is determined by the interaction of some active groups of actin and myosin which are shown schematically. The springs indicate Brownian thermal motion. The reaction equation is written as

$$A + M \longrightarrow \text{(rate constant } f)$$

$$AM + ATP \longrightarrow A-ATP + M \qquad \text{(constant } g)$$

$$A-ATP \longrightarrow A + ADP + P$$

Here A is actin, M is myosin, and AM is actomyosin. The constant f defines the bonding of active groups A and M; the constant g, their separation. The activity which splits ATP is ascribed to actin.

The first reaction starts as a result of excitation and is regulated by springs, i.e., by the relative distances and thermal motion. It is necessary to assume that the model is asymmetric; g is far to the left of the central point, then it falls and then increases again slowly from the left to the right. The constant f is equal to zero to the left of the center, then it grows linearly up to a certain distance $+h$, after which it falls again to zero. On the right side, f is greater than g at a distance of $0 - h$.

Weber[21] has proposed a biochemical interpretation of the sliding model. This is shown schematically in Fig. 194. The acid group of actin is

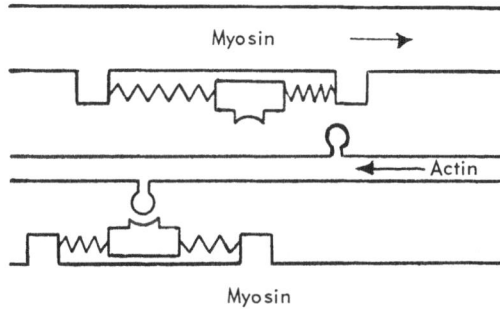

Fig. 193. The Huxley model.

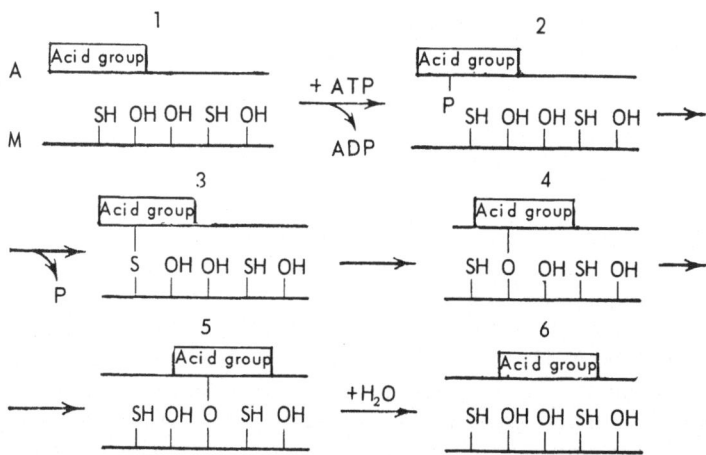

Fig. 194. The Weber model.

phosphorylated during the splitting of ATP and becomes capable of reacting with the sulfhydryl and hydroxyl groups of myosin. The relative displacement of the two proteins becomes favorable. The greater the load on the muscle, the more frequently is the actin thread with its active group displaced by the load into the initial position. This increases the number of stages of the displacement which are necessary for contraction of the fibril. Under isometric conditions there is no contraction, since, under a heavy load, the actin fibers are pulled away into their initial positions after each displacement. The starting assumption consists in the linear distribution of the functional groups of the protein. The attractive biochemistry of Weber is not consistent, however, with the observed experimental relationships within a muscle.

The model of A. Huxley brings about a number of objections. There is no way of verifying the asymmetry of the interactions. The equation contains a large number of parameters, the values of which are adjusted to agree with experiment, but are not obtained from independent molecular physical considerations. The merit of the model, which takes into consideration the construction of the muscles from two types of protein fibers, consists in its explaining directly the dependence of the rate of the driving chemical reaction on the rate of the contraction. The principal idea is that the rate of the biochemical process which furnishes energy to the muscle is determined by the rate of the interaction of the thick and thin filaments. Then, the rate of this interaction must be a function of the rate of contraction.

Such a picture is consistent not only with sliding, but also with coiling of the filaments of one of the proteins[28,38] (Fig. 195). In both cases, a feedback relationship is created between the chemical and mechanical processes.

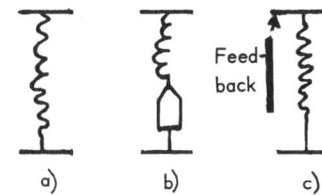

Fig. 195. Three models of contraction: (a) simple contraction, (b) the contracting element connected with the passive resilient element, and (c) coiling with a mechanochemical feedback.

The interesting phenomenon of the rapid decrease of the stress after removal of the load, which is much faster than the subsequent contraction, may be explained with equal success by a two-thread model with sliding and with coiling, but it cannot be explained by a model with a single coiling thread. In this case, it must be assumed that the sliding model consists of a consistently contracting element and a passive resilient element (Fig. 195). If the contracting element is contracted with a velocity which depends on an instant load, but cannot contract faster than v_m, the approach of muscle ends rapidly destroys the load in the resilient element. The load disappears, if the resilient element relaxes completely and it will reappear only with a further contraction of the contractive element. In the two-thread model with coiling at $v < v_m$, the interaction of the threads maintains a constant load (Fig. 195b). However, since a finite time is necessary for the interaction between the threads, at $v > v_m$ the mechanochemical feedback results in a decrease of the load.

Podolsky[38] proposed a special case of the two-thread model shown in Fig. 196. The thin thread in this case is a positively charged polyelectrolyte. During excitation, ATP anions become deposited on it and the thread starts to coil up. Regions of ATP activity on the thick threads remove the ATP molecules from the thin thread, with the result that the number of bonded molecules of ATP and, consequently, the load developed by the thin thread must be a function of the ratio of the rates of bonding of ATP and its decomposition. This special model leads to consequences similar to those found by Hill. It is evident that the Podolsky model is a coiling two-thread model which combines Hill's idea with a polyelectrolyte mechanism. There is no correct molecular basis for this quite curious model and the same objections may be brought against it as against the two polyelectrolyte models.

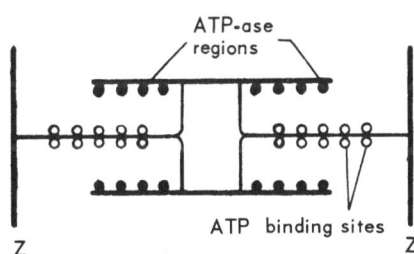

Fig. 196. The Podolsky model.

Eidus[39] has given a new molecular interpretation of the sliding-filament model. It is assumed that muscle contraction is determined by surface-tension forces which arise in the capillary system formed by the myosin and actin threads in contact with each other, immersed in the cytoplasm. The actin threads slide between the myosin ones, since the surface energy σ_{AM} on the actin–myosin interface is less than the sum of the energies on the interfaces of actin–cytoplasm, σ_{AC}, and myosin–cytoplasm, σ_{MC}. An estimate of the energy difference is $\sigma = \sigma_{AC} + \sigma_{MC} - \sigma_{AM} = 30$ to $60 \, erg/cm^2$.

The evolution of the heat of contraction, which is proportional to the contraction, is explained by a hidden heat of formation of a unit of surface, $-T(d\sigma/dT)$. It is possible to obtain reasonable quantitative agreement with Hill's data on the proportionality between the maximal load and the heat of contraction. The heat of activation is explained by evolution of heat during the polymerization of actin, i.e., during the $G \to F$ transition. In the Eidus model (shown in Fig. 197) the chemical energy is used only in the energy of activation, while contraction occurs only at the expense of surface tension.

This model, however, contradicts the results of structural studies which show that myosin interacts with actin only through some cross-bridges. This theory lacks an independent measurement of σ: this quantity is determined, in the final account, from the known value of the load p:

$$\sigma = \frac{p}{2\pi r N K} \tag{9.40}$$

where r is the radius, N is the number of threads per square centimeter of cross-section area, and K is the coefficient which takes into account the

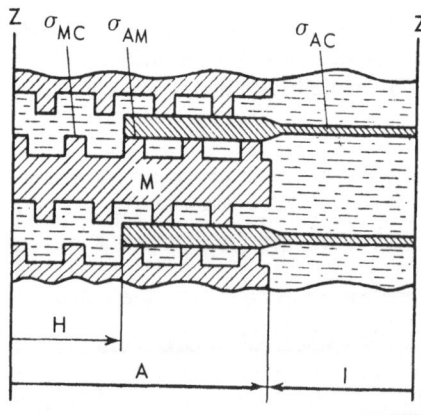

Fig. 197. The Eidus model.

discrete nature of their walls. Eidus assumes that

$$p = 1.5 \, \text{kg/cm}^2 \qquad K = 0.02 - 0.05 \qquad r = 1.5 \times 150 \, \text{Å}$$

$$N = \frac{1.57 \times 10^{11}}{2.2}$$

The transformation of actin and the ATP'ase activity of myosin are examined only qualitatively on the basis of certain hypotheses.

4. The hypothesis proposed by Szent-Györgyi[30] has a completely different character; it stems from a number of ideas related to the transfer of energy. The myosin particles are maintained in a stretched state by a specifically enhanced water structure; contraction occurs as a result of the destruction of this water structure, and its restitution means relaxation. The energy needed to destroy the water structure is supplied by ATP; furthermore, it is necessary to assume for transformation to occur sufficiently rapidly that the energy evolved by ATP passes from one molecule of water to another, that it migrates along a quasi-crystalline network of water molecules. This, evidently, is more a concept than a theory. It is supported by no calculations; there is no attempt to correlate it with the thermodynamic properties of the muscle. This idea is quite speculative; there are no proofs whatever of the existence of a special water structure in the muscle and no proof of energy migration within it. It should be pointed out that the high rate and the sharp character (all or none) of a number of biological processes have led many investigators, and among them Szent-Györgyi, to thoughts on quantum-mechanical resonance interactions and on the migration of energy. We shall discuss this question in the last chapter. At this point, let us state that in all the phenomena described in this book, these peculiarities of biological processes are due to macromolecular cooperativity and not to the migration of energy. It is probably true that this also applies to muscles.

5. In 1946, Engelhardt proposed a very interesting physical hypothesis on the nature of muscle contraction.[40] It is assumed that the protein exists in an ordered crystalline state in the relaxed muscle. The contraction of the muscle is due to the melting of the protein crystal. It follows from this that the active phase of muscle action is the weakening, the relaxation, and not the contraction. The role of the active muscle substance consists, not in resilient tension, but in resistance to compression; the role of ATP must be sought in its action either on the process of crystallization or on the process of melting. Later on, a similar idea was proposed by Pryor.[41]

These ideas are very attractive in the light of investigations on the conformational properties of biopolymers described in this book. The crystallization and melting of a protein are cooperative conformational transformations. Engelhardt's idea, expressed in a general form, is somewhat similar to Astbury's proposals.

A further development of this idea was carried out in the studies by Flory.[12,42,43] These should be discussed in greater detail. Flory noticed the similarity between the behavior of a muscle and that of fibrous proteins, in particular, of collagen. Such proteins have an ordered oriented nearly crystalline structure. During the melting of the structure, the linear dimensions of the sample become reduced in the direction of the original orientation of the chains. It is evident that a phase transition of such a type could be used essentially for the transformation of heat energy into mechanical energy in the heat engine with a working material, namely, with a crystallizable polymer. If this polymer had functional groups which could change easily with a change in chemical environment, then melting, or crystallization, could be brought about, not by heating, but by a change of the chemical potentials and we should be dealing, not with a heat, but with a mechanochemical machine. As long as the functioning of such a system depends on a sharp phase transition, the system must be much more sensitive to chemical processes than a polyelectrolyte system, in which the chemical reactions result in only small isotropic forces.

In fact, it may be shown that, under definite conditions, chemical reactions may alter considerably the melting temperature of a crystalline polymer, i.e., they may bring about its melting at a constant, let us say, room, temperature.[42]

A very interesting situation would exist if a polymer, or a portion of it, possessed enzymatic activity. Then the fiber itself could influence the nature and the concentration of the reagents in the surrounding medium. This, in turn, could influence the chemical potential of the links of the polymer and could displace the crystal–amorphous equilibrium.[43]

The presence of an external force must be reflected both in transitions of the type of crystallization and melting and in the chemical properties of the polymer. We have seen that the external force has an effect on the helix–coil transition, i.e., on the intramolecular melting. On the other hand, concepts of the nature of enzymatic activity, presented in Chapter 8, make us expect that such activity would change under forced changes of the chain conformation.

The idea is quite clear. Contraction of a stimulated muscle may be interpreted as the displacement of the equilibrium between crystalline and amorphous regions. It is proposed that the ATP forms a complex with the protein in its melted rather than crystalline state; in the latter state, the protein is less reactive.

These proposals have received some confirmation in the investigation of the behavior of glycerinated muscle fibers in ATP solutions. The dephosporylation of ATP, in fact, occurs in such a system, but it causes muscle contraction only under certain conditions. Both the temperature and the composition of the solution were varied. The ratio of glycerine to water was varied. A sufficiently sharp contraction occurred only in those cases when the glycerine concentration was decreased to 50%. In solution containing 35% ethylene glycol in water, contraction occurred at 6°C. These data testify to the similarity between the contraction of muscle fibers and phase transitions.

The most significant part of the study by Flory is his analysis of the phase transitions, i.e., of the cooperative nature of the contractions. No matter whether the contraction is related to the melting of crystals or to some other process, for example, to polymerization or depolymerization, its cooperative nature explains the speed and sharpness of the reaction of the muscle to changes of conditions, to the chemistry, and to the load. The search for special mechanisms of the type of energy migration becomes unnecessary.

Flory, however, has made no attempt to relate the proposed mechanism to the biochemical, structural, and thermodynamic data. Such an attempt is described below. Before we pass to it, it seems necessary to consider another possible type of cooperative contraction of the muscle.

6. Oosawa, Asabura, and Ooi have investigated further one of the muscle proteins, namely actin.[45] As has been pointed out already, actin can be polymerized. This is the $G \to F$ transition which occurs in solution in the presence of salts, in particular, of Mg^{++}. For polymerization to take place, it seems necessary for ATP to become attached to molecules of G actin, namely, two molecules of ATP become attached to one of G actin. Dephosphorylation takes place during polymerization (Fig. 198). It has been mentioned already that actin has no ATP'ase activity. The point is that the actin itself changes; it becomes polymerized during the splitting of ATP and, therefore, it is not an enzyme in the true sense of the word. It can be seen, however, that when salts are repeatedly added and removed, the molecule of G actin may catalyze the splitting of ATP many times.

A dynamic equilibrium becomes established between G and F actins and it becomes displaced during the change of salt concentration. The rate of splitting of ATP is maximal, for example, at equal concentrations of G and F actins. The $G \to F$ transition is cooperative; at the point of equilibrium, there are monomers of G and very long (up to several microns) threads of F actin. The polymerization starts at a critical concentration of

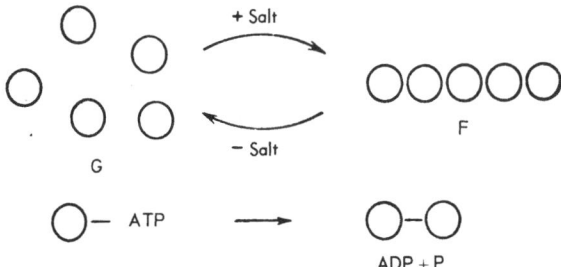

Fig. 198. Scheme of the $G \to F$ transformation of actin.

F actin. The rate of the splitting of ATP is a strong function of the initial concentration of *G* actin and grows proportionately to $[G]^3$ or $[G]$.[4]

This means that the splitting of ATP is not an independent reaction on each molecule of *G* actin, but requires the cooperation of more than two such molecules. The polymerization is not linear; a helical polymer seems to be formed, each link of which is attached to the neighboring one along the chain and simultaneously to the third preceding one and the third subsequent one (Fig. 199). It is evident that a linear polymer of actin may also be formed; it would have a different chain length. The helical polymer passes into the linear one when links between nonneighboring units are broken. This transition may occur as a result of a change in the conditions of the medium and must be accompanied by a lengthening of the chain (Fig. 200). It has been possible, indeed, to detect in solution, in the absence of salt, relatively short polymers of actin with properties different from *F* actin, namely, with a different sign of electrical birefringence.

The Japanese scientists have proposed that such threads in the muscle fiber form linear polymers of actin which go over into helical *F* polymers during muscle contraction as a result of interactions with myosin (Fig. 201). The energy source, as before, is ATP. This qualitative picture is consistent with the result of the electron-microscopic investigations by Huxley and Hanson. A quantitative theory of the cooperative helical aggregation of actin has been described,[46] which gives the thermodynamics and kinetics of the process.

Asakura, Taniguchi, and Oosawa[47] have investigated in detail the mechanochemical behavior of actin. They have subjected its solutions to the influence of ultrasound; this brought about the reversible transforma-

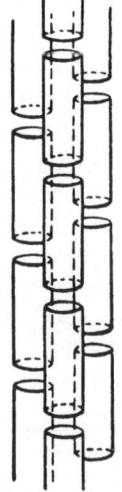

Fig. 199. Helical *F* actin.

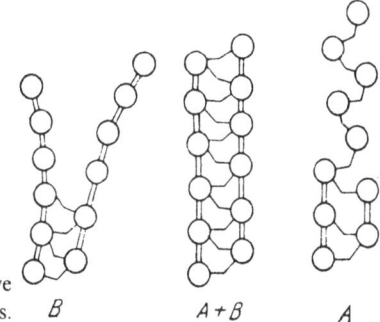

Fig. 200. Bonds of the type of A and B: $A + B$ give a helical polymer; B and A give linear polymers.

tion of F actin into a structurally changed form of the polymeric protein, called f actin. The f chains are longer than the F chains; F actin is helical, while f actin may be a linear polymer. As a result, the mechanical action stimulates the $f \rightarrow F$ transition. In the presence of ATP there is a reversible contracting $f \rightarrow F$ transition which is accompanied by the splitting of ATP.

It is possible that these results could be combined with a sliding model, with the assumption first that only one of the two forms of linear actin may interact with myosin. It is possible also that the cooperative-phase transition, discussed by Flory, is not the melting of a crystalline protein but a cooperative transition from one type of the polymer to another.

An attempt has been made[48] to relate Flory's and the Japanese authors' concepts with the biochemistry and thermodynamics of muscle contraction. Relations similar to those of Hill could be obtained by using concepts of changes of the enzymatic activity of a protein during its contraction and keeping the assumption of the stationary nature of the corresponding chemical process. However, this work is of a purely qualitative character and the assumptions made in it require further verification.

The basic assumption on the change of the enzymatic activity of a protein under the forced change of its conformation has been confirmed experimentally. Vorob'ev and Kukhareva have subjected myosin macro-

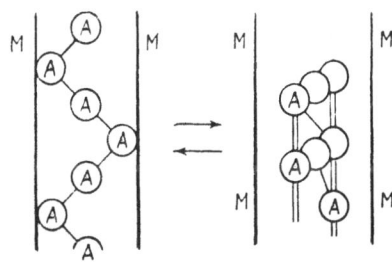

Fig. 201. The Oosawa, Asabura, and Ooi model.

molecules to deformation in a hydrodynamic field, namely, in a dynamometer. This was accompanied by a change in the ATP'ase activity of myosin.[48a]

The most convincing theoretical model of muscle activity has been proposed by Davies.[49] It deserves particular attention, since it is in qualitative agreement with the majority of the known facts. Let us examine it in detail.

This model corresponds to the sliding of actin A relative to myosin M. The A–M interaction takes place in definite regions on the surface of M, which are the loci of ATP which is bonded to M (H-meromyosin) by three positive charges. These probably are the positive charges of Mg^{++} ions. In the absence of A, the conformation of the M—ATP complex is such that the ATP is removed from the part of M which possesses ATP'ase activity. This separation is determined by the electrostatic repulsion of negatively charged groups (Figs. 202 and 203a). When the muscle is activated, Ca^{++} ions penetrate into it; these form chelate bonds between ATP linked, to M and ADP, linked to F actin ($F—A$). The Ca^{++} ions neutralize the negative charges of the extended group M—ATP (Fig. 203b). The Coulombic repulsion vanishes and the functional group of M—ATP changes its conformation, possible coiling up into an α-helix (Fig. 203c). This is accompanied by a shortening of the A—M bridge, and tension develops.

In the new conformation of A—M, the ATP'ase activity of M acts and the ATP of the M—ATP complex is split to ADP and phosphate. The A—M bonds break (Fig. 203d). The tension displaces A relative to M. M—ADP is rephosphorylated by the cytoplasmic ATP and creatine phosphate (Fig. 203e), and the conformation of M—ATP returns to the initial state as a result of the electrostatic repulsion (Fig. 203f). This is followed by a new interaction with A, and the process repeats itself many times. The active state of the muscle stops as a result of the intake of Ca^{++}

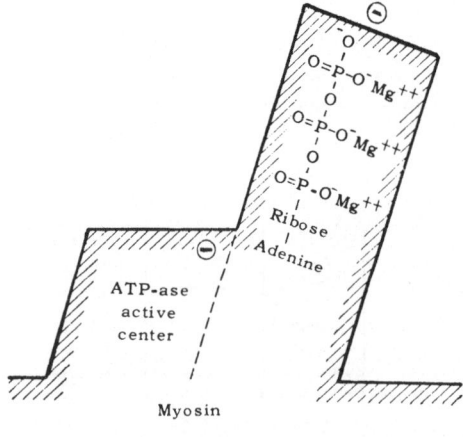

Fig. 202. Functional region of a myosin molecule, which has bound ATP (according to Davies).

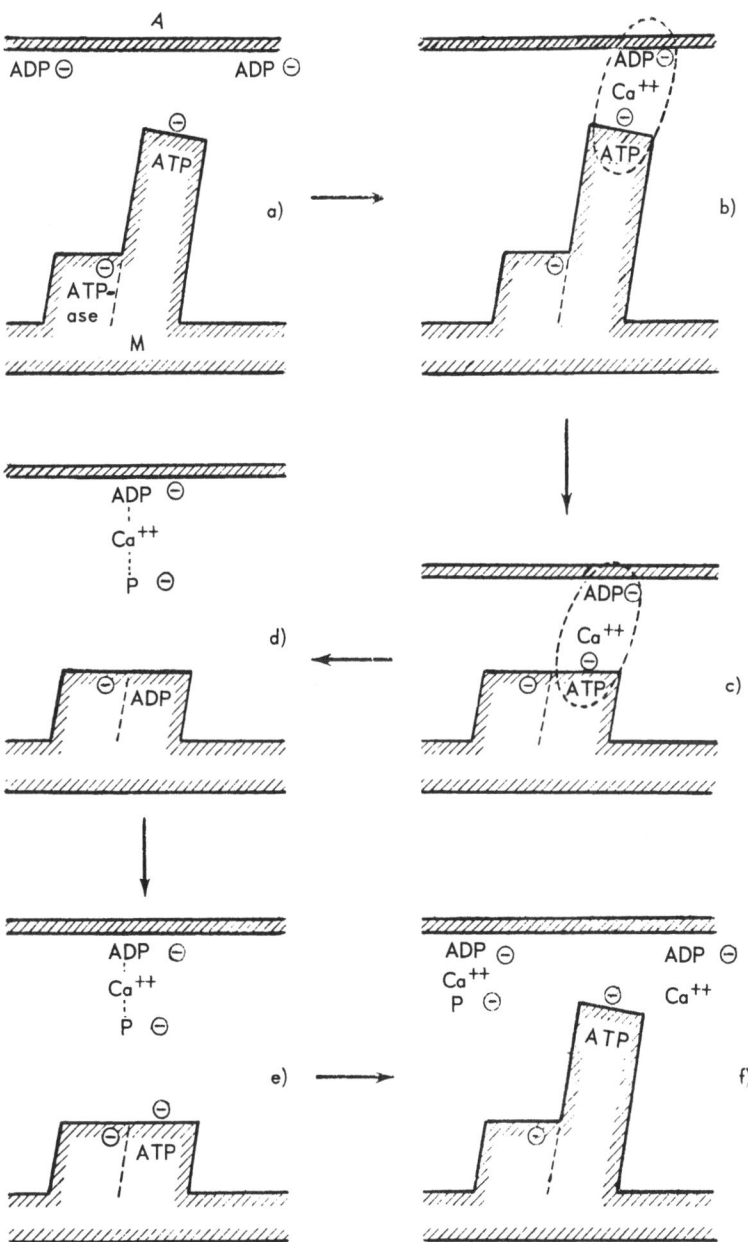

Fig. 203. Molecular model of the contraction of the muscle fiber (according to Davies).

ions back into the sacroplasmic reticulum. This is a process of active transport which requires the expenditure of ATP energy and, consequently, the participation of a specific enzyme, namely, a relaxing factor.

According to this model the ATP'ase activity of myosin manifests itself only at the points of A—M interaction. The Davies model combines structural data quite cleverly with concepts of the conformational dependence of the enzymatic activity and of the cooperative conformational transformations of myosin. The sliding is combined with conformational coiling. Davies reaches a number of conclusions which are in agreement with experiment.

The amount of Ca^{++} introduced into a muscle, necessary to bring about a maximal contraction, must be at least equal to the number of A—M crosslinks. A quantitative estimate gives 0.09 mole/g of Ca^{++} Experimental data are close to this figure (0.125 mole/g).

The latent period corresponds to the time of the passage of Ca^{++} from the cytoplasmic reticulum to the active centers. The experimental duration of the latent period is 10 msec; it is consistent with the diffusion constant of Ca^{++} ions in water (approximately 1 μ).

The heat of activation of the muscle corresponds to the heat of dissociation of Ca^{++} from the cytoplasmic reticulum. If we assume that Ca^{++} is bonded by phosphate ions, then this quantity is about 5 kcal/mole, which gives 0.9 mcal/g of muscle. The experimental value is 1 to 1.5 mcal/g.

The maintenance of tension does not require by itself the decomposition of a large amount of ATP. The ATP decomposes only when the A—M bridges become transposed. The number of such bridges formed per unit of time per unit of length is greater when the rate of contraction is slow with a heavy load than when the rate of contraction is high with a small load. The amount of work obtained depends on the splitting of ATP. Quantitative estimates are in agreement with experiment. The heat of contraction is the heat of formation of a large number of intermolecular bonds; it must not be a function of the length and of the load.

A similar model has been proposed independently.[50]

Let us carry out some calculations.

With a finite rate of formation of the A—M bridge, its formation depends on the rate u of the relative motions of A and M.

At large u, the probability of formation of an A—M bridge is smaller. Let us assume that the displacement of the active region of A relative to the active region of M is equal to x. The probability that the interaction of a given pair of regions has already occurred is n. The probability that such an interaction will occur over the time dt is[28]

$$dn = (1 - n)k(x)\,dt \qquad (9.41)$$

Since the path traveled during the time dt is $u\,dt$, we have

$$dn = (1 - n)k(x)\frac{dx}{u} \qquad (9.42)$$

Let us integrate (9.42). The $k(x)$ is a quite sharp function of x, that is, the bridge can be formed only when the distance between the active regions of A and M is small. If we set

$$\lambda = \int_{-\infty}^{+\infty} k(x)\, dx \tag{9.43}$$

we have

$$n = 1 - e^{-\lambda/u} \tag{9.44}$$

When $u = 0$, $n = 1$; as u increases, n decreases from 1 to 0.

The rate u is related to the rate of muscle contraction (measured in muscle length per second) by

$$u = \frac{sv}{2} \tag{9.45}$$

where s is the length of a sarcomere. If the distance between the active regions is l, then a number of contacts A—M per unit of time will be u/l. The rate of interaction per unit of length of muscle is Mun/l, where M is the number of active regions per cubic centimeter of muscle. Consequently, the rate of the chemical reaction is

$$\dot{\xi} = \dot{\xi}_0 + \frac{Mun}{l} = \dot{\xi}_0 + \frac{Msv}{2l}(1 - e^{-(2\lambda/sv)}) \tag{9.46}$$

When an adequate set of parameters is used, Eq. (9.46) gives the dependence of ξ/ξ_0 on v; this is in agreement with Hill's data.[27]

According to Davies' model, the value of λ must be a function of the concentration of Ca^{++} ions. Assuming that the processes shown in Fig. 203 satisfy the condition of a stationary state, we obtain the following scheme (Fig. 204). Here M_1 is the starting state of M—ATP, corresponding to Fig. 203a, M_2 is the complex M—ATP—Ca—ADP—A (Fig. 203b), M_3 is the conformationally changed complex M—ATP (Fig. 203c), M_4 is the complex M—ADP (Fig. 203d), and M_5 is the rephosphorylated complex (Fig. 203e). Solving the equation of the steady-state kinetics, we find

$$k = \frac{M_2}{\sum\limits_{i=1}^{5} M_i} = \frac{[Ca^{++}]}{\dfrac{k_{-1} + \kappa}{k_1} + [Ca^{++}]\left(1 + \dfrac{k_4\kappa}{k_2} + \dfrac{\kappa}{k_3[S]} + \dfrac{\kappa}{k_4}\right)} \tag{9.47}$$

where $[S]$ is the concentration of ATP in the solution. The dependence of k on x is determined by the dependence of k_1 and k_{-1} on x; k_1 is different from zero only within a narrow interval of values of x. The general nature of the dependence of muscle contraction on time follows from (9.44).

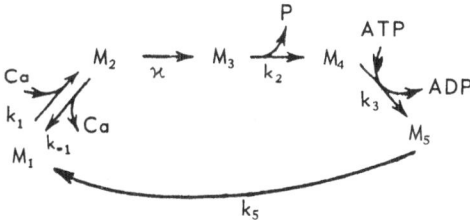

Fig. 204. Scheme of the transformation of myosin under steady-state conditions of the process, according to the Davies model.

The length of the contracting muscle is

$$L = L_0 - \Delta L \tag{9.48}$$

where the contraction ΔL is proportional to the number of A—M bridges, formed during time t,

$$\Delta L = L_0 - L = L_1 n(t) \tag{9.49}$$

where L_1 is a proportionality constant. Introducing Eq. (9.44), we obtain

$$\Delta L = L_1(1 - e^{-\lambda/\mu}) = L_1\left(1 - e^{-\int_0^t k\,dt}\right) = L_1(1 - e^{-k't}) \tag{9.50}$$

where

$$k' = \frac{1}{t}\int_0^t k(x)\,dt$$

The rate of contraction is

$$v = \frac{d\Delta L}{dt} = L_1 k' e^{-k't} = k'(L_1 - \Delta L) \tag{9.51}$$

Equations (9.50) and (9.51) are in qualitative agreement with experiment.

While the Davies model is extremely useful, it is still too early to consider that the theory of muscle action has been fully developed. This is a molecular model, but not a molecular theory. It is necessary to carry out detailed calculations which take into account the results of the structural, biochemical and thermomechanical investigations. First attempts to develop such a theory were made quite recently.[66,67] The aim of the theory is a quantitative determination of the parameters of the Hill equation based on the molecular model. This has been accomplished recently.[67]

The results obtained recently by Frank and coworkers[51] deserve a great deal of attention. It has been shown that, in addition to sliding, the protein filaments can undergo active contraction. Frank believes that it is quite unlikely that muscle contraction may occur only as a result of thin bridges between actin and myosin. He proposes that the sliding of the actin fibers between the myosin ones is a complicated stepwise process during which there is a direct shrinkage of the thick threads which pull the thin threads into the spaces between them. This results in contacts and the formation of complexes. Then, the thick filaments are weakened, etc. To investigate the process of contraction as a function of time, Frank and coworkers developed a clever method of diffraction of visible light, which permits changes of muscle structure to be recorded during a single contraction at intervals of $^1/_{1,000}$ second. In fact, with the help of this method, it has been possible to detect a periodicity of the contraction process. Up to now there is no model picture which would take these facts into account.

While the name of the sacred elephant still remains unknown, ways toward the unraveling of this mystery have been indicated. It appears that cooperative processes of changes at the highest levels of protein structure play an important role in muscle contraction.

The development of the molecular theory faces very serious difficulties. These are due to the heterogeneous nature of muscle fibers, to the presence in them of supramolecular structures and, consequently, of a high complexity of biochemical reactions. The success of a theory must be expressed by its ability to indicate new experiments. Finally, an understanding of muscle action must result not only in the solution of an extremely important biological problem but also in the creation of special mechanochemical model systems which are essential to technology.

MOTION OF CELLS AND MITOSIS

Muscle action is a mechanochemical process which can be observed directly and which is familiar to everybody. Should one look into a microscope, however, one would see a vast number of motions of other cells as a whole and of motions inside the cells. A spermatozoid translates as a result of the motion of its long tail. A paramecium and a number of other unicellular organisms swim as a result of the motion of flagella and cilia; the same mechanism is also used by macroscopic organisms both for motion and for maintaining a flow of liquid which contains substances necessary for its life. Ctenophora regulate the orientation of their bodies by a concerted motion of cilia; a flow of liquid is created in the gills of bivalves and mollusks and in the trachea of man. Intercellular motion occurs in the contraction of mitochondria, in mitosis and meiosis, and in the flow of protoplasm which supplies necessary materials to the cell organoids. All these motions are accompanied by mechanical work as a result of chemical energy, usually stored in ATP.

We shall not discuss the motion of protoplasm, which has been studied in detail by Kamiya.[52] Let us limit ourselves by indicating that the energy for this motion is obtained probably as a result of the nonoxidative enzymatic decomposition of glucose. Hypotheses have been developed which relate the motion of protoplasm to the formation of a pressure difference as a result of the compression of a gel-like substance, or with a displacing force which forces the molecules of the endoplasm to slide relative to each other. In all the hypotheses, the change of the mechanical properties of the gel during its motion plays an important role, i.e., thixotropy of the protoplasm takes place. At present, there is no molecular theory of these phenomena and they require further investigation.

The motion of cilia is quite unique. It is shown schematically in Fig. 205. A cilium moves asymmetrically first in one direction then in the other. A possible mechanism is described in terms of the special structure of

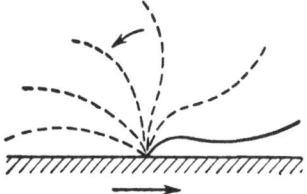

Fig. 205. Scheme of the motion of flagella.

the cilia, which consists of nine pairs of filaments located along a circumference and of two filaments in the center of the cylindrical body of the cilium. It has been proposed that five of the nine pairs of filaments on one side of the cylinder contract rapidly and simultaneously. The reverse motion is related to the slow contraction of the other four pairs. The central filaments serve for the transmittal of the necessary information.[53,54] The motion of flagella is even more wavelike (Fig. 206). Both the flagella and the cilia move in an ordered fashion, both from the point of view of processes which occur inside individual cylindrical bodies and of the coordination of the motion of the totality of the flagella and cilia.

Silvester and Holwill have proposed a molecular model of the action of flagella, based on conformational concepts.[55]

Engelhardt *et al.* have isolated from spermatozoids a protein similar to actomyosin, which they have called *spermozine.*[56] This is a contractual protein. It is important that the flagella contain an enzyme which has the property of an ATP'ase and can release the energy from ATP. Nelson has shown that the ATP'ase activity is possessed only by the nine external pairs of filaments, but not by the two central ones. In this case, as well, contraction occurs at the expense of ATP and it is possible to consider that, once the problem of muscle activity has been resolved, those of the nature of the motion of flagella and cilia will follow. It is evident that muscle activity is easier to study. However, there are definite similarities between muscle activity and the motion of flagella and cilia.

Cell division is a process of fundamental importance for the life of an individual and of a species, whether this is unicellular or multicellular. Mitosis lies at the basis of life, since it guarantees the conservation and multiplication of cells. It is evident that a cell cannot grow indefinitely

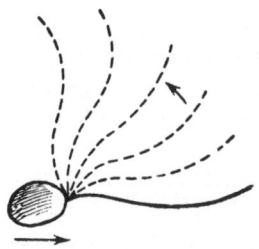

Fig. 206. Scheme of the motion of cilia.

without division. Such a growth would necessarily lead to a violation of the homeostatic state, to an overheating and death of the cell. The absence of growth and division would indicate the absence of life, an extremely important property of which is reproduction.

It is clear that problems related to the nature of mitosis and meiosis are directly related to the most important problems of biology, namely, to the general problem of life, of aging, and of death, to the regulation and the internal linkage of processes in a complicated ordered supramolecular cellular structure, and finally to the phenomena of differentiation. Here we encounter the behavior of the cell as a whole, the study of which falls outside of the domain of modern molecular biophysics. It is possible, however, to separate from the quite complicated and insufficiently studied process of mitosis some events which may be examined by molecular theory.

Cell division is accompanied by mechanical work ; this is the work of the displacement of the chromosomes and the cytoplasmic material. This is again a mechanochemical process; it is undoubtable that the source of the work is chemical reactions which occur under isothermal and isobaric conditions. This is the process which interests us at this point.

The most detailed investigation of mitosis has been carried out by Mazia.[57,57a] Mazia regards the entire process of mitosis in terms of the scheme shown in Fig. 207.[57] The interphase and prophase see a preparation to mitosis, the doubling of the centrioles, the doubling of the amount of DNA within the cell, which results in the formation of the progeny, or sister, chromosomes. Substances are synthesized which serve for the construction of the mitotic apparatus; these are strands which are seen under the microscope and which link the centrioles with each other and the chromosomes with the centrioles. The construction of the apparatus is completed during the metaphase. Simultaneously, the mass of the cell increases. We still know very little about the processes of the construction of chromosomes from DNA, histones, and possibly other substances. Furthermore, the method by which these macromolecules are arranged in the chromosome is as yet unknown. There are experimental data which indicate that the storage of chemical energy necessary for the subsequent mechanical work takes place, not during mitosis, but during the period of preparation for it. From the point of view of cell economics, the cost of mitosis is paid out in energy before the initiation of mitosis. The energy is probably stored in the high-energy phosphate bonds, if not in ATP, then in other nucleoside triphosphates ; one should not consider, however, that the mechanochemical process results directly from the release of this energy, since the level of ATP first increases and then falls sharply before division.

It is necessary to emphasize the large morphological similarity between the centrioles and the bases of the flagella and cilia, namely, of mechanochemical constructions. Furthermore, it has been shown that the

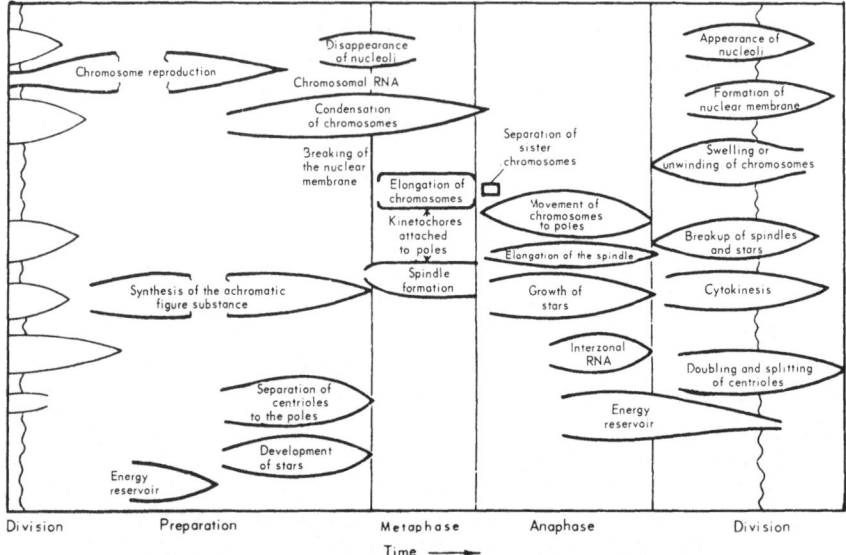

Fig. 207. Sequence of events during mitosis (according to Mazia).

flagella of spermatozoa grow directly from centrioles and the kinetophores of the chromosomes.

The principal working substance of the mechanochemical process in a splitting cell is found in the mitotic apparatus. It consists principally of protein; as much as 11 % of the total protein of the cell enters into the mitotic apparatus. Mazia succeeded in isolating the mitotic apparatus from the cell and he investigated it. The apparatus has a double structure, namely, a system of links which connect the poles with each other and which are known as the *central spindle*, and a system of bonds of the chromosomes with the pole, which are formed at the end of the prophase. The protein of the mitotic apparatus is immunologically different from actomyosin, but its composition has definite similarities with actin. Differing from the proteins of muscle or of cilia, this protein has an unstable gel-like structure. Its mechanochemical activity seems to be related to the formation and decomposition of a supramolecular structure; the mitotic protein polymerizes from protein molecules of a smaller size, and then it depolymerizes. It is possible that this process is in some way similar to the $G \rightarrow F$ transitions of actin. The protein contains thiol groups; it is characteristic that the number of soluble SH groups in the cell decreases before the metaphase, and then increases again. Mazia proposed a hypothesis according to which the formation of the mitotic protein occurs as a result of the transformation of intramolecular disulfide bonds into intermolecular ones. In the first stage of the process, the amount of soluble SH groups decreases, i.e., the amount of glutathione decreases—namely, of the tripeptide Glu–Cys–Gly

$$\begin{array}{c} \qquad\qquad\qquad\qquad\quad CH_2-SH \\ \qquad\qquad\qquad\qquad\qquad\quad | \\ ^-OOC-CH_2-CH_2-CH(\overset{+}{N}H_3)-CONH-CH-CONH-CH_2-COOH \end{array}$$

This is accompanied by a breaking of the intramolecular disulfide bonds. In the second stage, intermolecular S—S bonds are formed, i.e., the protein is polymerized, and the glutathione returns into the solution (Fig. 208). The polymerized mitotic apparatus, namely, the spindle, has optical anisotropy and is, therefore, well visible in a polarizing microscope. The birefringence of the spindle is a function of temperature. At very low temperatures (4 to 6°C) it vanishes. Inoué has proposed that the birefringence is proportional to the amount of oriented material and, consequently, the temperature has an effect on the equilibrium constant K between the oriented and unoriented materials.[53]

If we assume that the total amount of mitotic protein A_0 is constant during the metaphase, we have the equilibrium

$$A_0 - B \; \rightleftharpoons \; B$$

where B is the oriented protein. The equilibrium constant is

$$K = \frac{B}{A_0 - B} = e^{-\Delta F/kT} \tag{9.52}$$

where ΔF is the corresponding free-energy difference.

In fact, Eq. (9.50) has been confirmed by studies of the dependence of birefringence of the protein on temperature; $\ln\,[B/(A_0 - B)]$ is a linear function of $1/T$:

$$\ln \frac{B}{A_0 - B} = -\frac{\Delta F}{kT} = -\frac{\Delta H}{kT} + \frac{\Delta S}{k} \tag{9.53}$$

The values of the thermodynamic quantities obtained in this way are $\Delta F = -1.8\,\text{kcal/mole}$, $\Delta H = 28.4\,\text{kcal/mole}$, $\Delta S = 101\,\text{cal mole}^{-1}\,\text{deg}^{-1}$. The free-energy change is small, which points to the high lability of the mitotic protein in the spindle; the high energy of the reaction, ΔH, is

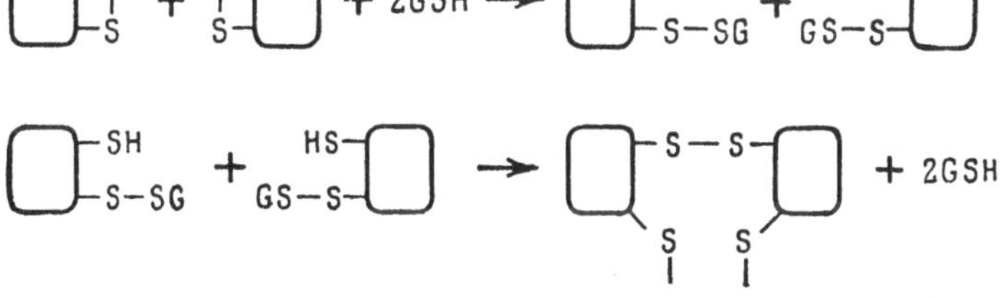

Fig. 208. Scheme of polymerization in the mitotic apparatus (according to Mazia).

explained by the high positive entropy. This is possibly related to the melting of the structured water, linked to the nonoriented protein molecules.
Getting free of the water, the molecules become more easily oriented.

In this manner, processes of polymerization and orientation occur in
the mitotic protein. A theoretical examination of these seems to require an
application of Flory's ideas on the spontaneous ordering of sufficiently
rigid macromolecules. The available experimental material, however, does
not permit the development of such a theory.

If the picture described is correct, then, at still-higher temperatures,
the orientation of the protein molecules should again be destroyed and the
birefringence should fall. It is possible, however, that this could not be
observed, since an increase in the temperature could result in the destruction of the protein.

How are these processes related to the motion of the chromosomes?
A number of theories or, more correctly, hypotheses have been proposed
to explain this motion. It has been proposed that chromosomes are
autonomous and that they are able to move independently. Electrical and
magnetic theories, stemming from the attraction of the chromosomes to
the poles and the repulsion of sister chromosomes from each other, have
been proposed. There is no real evidence for these. Magnetic poles with an
intensity as high as 43,000 gauss have no effect on mitosis. This is not
surprising; as a rule, we do not encounter in biological phenomena any
specific magnetic properties, with the exception of heme and similar
groups which contain iron. Forces of diffusion have been mentioned;
these would bring about the motion of chromosomes; but the high organization and ordering of this motion may not be explained on such a
basis. The same is true of the flow of protoplasm. Already in 1940, Bernal
expressed an idea on the tactoid nature of the spindle, i.e., on the fact that
it is formed from dissolved protein molecules in a manner similar to that
of a liquid crystalline aggregate.[59] This picture already starts to approach
what has been discussed. The mitotic apparatus, however, is formed not
only as a result of the mutual orientation of macromolecules, but also by
polymerization.

According to Mazia, the motion of the chromosomes to the poles is
not related to the contraction of the mitotic fibers which link the kinetophores of the chromosomes with the centrioles. The fibers do not become
thickened during this contraction. As a result, it is reasonable to think that
the contraction is determined by depolymerization. The fiber contracts as
a result of the precipitation of part of the material. On the other hand,
during the separation of the poles, the fiber becomes elongated as a result
of polymerization and pushes the centrioles apart. The kinetophores of
the chromosomes are linked directly with the fiber. In this manner, the
motion during mitosis may be reduced to a pushing-away and pulling-
together of chromosomes and centrioles, not by specific forces, but by
proteinaceous fibers. The mechanochemical character of the process is

demonstrated, in particular, by the fact that addition of ATP brings about an elongation of the mitotic apparatus. The rate of elongation during mitosis is $1.5\,\mu$/minute; it is not great.

This is specific mechanochemistry, different from that which ensures muscle action, and probably also the motion of flagella and cilia, even though the morphological similarity of their basic bodies to the centrioles is quite noteworthy. Let us point out in passing that the rods and cones of the eye retina, i.e., the light-sensitive elements, are also morphologically similar to cilia; the structure "9 + 2" has quite a wide distribution.[60]

The theoretical examination of a model of the process of polymerization and depolymerization as a source of mechanical motion presents a great deal of interest. We are still quite far, however, from an understanding of mitosis as a whole. It is reasonable to expect that future investigations will lead to the ideal situation, namely, a complete understanding of these various phenomena and their interactions.

Let us point out again the mechanochemical processes in mitochondria. It is quite probable that here we encounter a mechanochemistry which is similar to that found in muscle, i.e., one which is determined by conformational changes of enzymes.[61]

Neifakh has extracted from mitochondria a contractual protein similar to actomyosin.[62] In mitochondria, the process of oxidative phosphorylation is related to a complicated mechanochemistry.

Green believes that it is possible to formulate general relations valid for any cell membranes.[3] For example, all membranes contain a contractive protein which possesses ATP'ase activity and plays an important role in the active transport across membranes. Opit and Charnock have proposed a molecular model for the action of the membranes of the nerve fiber, namely, changes of its permeability toward Na^+ and K^+ ions during the passage of a nerve impulse.[64] The basis of this model is a conformational transition of the ATP'ase present in the membrane of the axon.

Another example of a contractive protein system is found in phages. The tail of a phage particle is apparently a contractive organ, the action of which results in the introduction of the phage DNA into the bacterial cell.[65]

Chapter 10

PROBLEMS OF
MOLECULAR
BIOPHYSICS

MACROMOLECULES AND BIOLOGY

Let us summarize. This book is essentially devoted to the physics and biological functionality of the macromolecules of proteins and nucleic acids. Long ago, Schrödinger had already given a well-founded answer to a question which he asked himself; Why do organisms contain a large number of atoms? Let us now go somewhat further and ask another question; Why are the principal biologically functional systems macro-molecular in nature?[1]

It is quite evident from what we have discussed that life, as it exists on this planet, requires macromolecules. A living organism is a self-regulating, self-reproducing, open system. In order to carry out such properties, it is necessary that molecules carry out a number of biological functions.

Genotypic and phenotypic information must be recorded. The simplest way of doing this is by the formation of a linear text, as has been invented by man. But man is a part of nature and his ways of thinking must, in the final account, be determined by general biological laws. Therefore, it should not be surprising that the storage of information in the cell occurs in the same manner, namely, in the primary structure of biological macromolecules, by long linear structures.

This information is transformed and recorded in biological processes. The self-reproduction requires a template synthesis which ensures the transmission of the information to the formed molecules. We have already presented the arguments in favor of a linear macromolecular template.

The storage of information, while it is not of an absolute character (there are mutations), is ensured by the structure of the macromolecular chain which is maintained as a whole, even if individual links change.

The conformational lability of macromolecules determines the formation of higher levels of structure, starting from the secondary. This is a property necessary for biologically functional molecules, since biological specificity is related to structural complementarity. The fact that macromolecules may form completely or incompletely ordered supramolecular structures has an important role in the organization of intracellular processes. We have seen that the specificity and effectiveness of biochemical reactions is determined by the particularities of the structures of the enzyme macromolecules and, in particular, by their conformational properties. It is clear that, in a low-molecular-weight system, the possibilities for conformational transformations of a chemically unchanged substance are very limited relative to a macromolecular system. One may remark that the conformational lability gives additional possibilities for the participation in irreversible processes which take place in an open system and which are related to degrees of free internal rotation.

The same conformational lability of macromolecules ensures, as has been pointed out, the special possibilities of transferring information in regulatory processes, and, in particular, in the phenomenon of allostery. Substances which play the roles of regulators, namely, repressors of protein synthesis and initiators of reduplication of nucleic acids, are macromolecular. The allosteric method of regulation is found only in living systems; it is ensured by the structural complementarities and their changes. One may believe that allostery is one of the key phenomena of life; as a result, its further investigation is of extreme importance.

The ability of semirigid macromolecules to become ordered of their own accord in the absence of forces of specific interaction is important for the formation of intracellular structures. The ability of an organism to be an aperiodic crystal on all levels of structure, starting from the molecular and ending with the organism as a whole, is already found in the structure of the macromolecule. We do not know aperiodic highly ordered systems constructed from small molecules, and it is difficult to conceive of such systems. It is important that the totality of small molecules may be ordered only in the solid phase and under the condition of highly limited chemical heterogeneity. To the contrary, the totality of macromolecules is always ordered, already as a result of the linear structure of each polymer chain. A definite ordering is in the essence of macromolecules and their aggregates, in solution, in a melt, and in the state of an amorphous solid body. Furthermore, macromolecules possess the necessary conformational lability.

Life is impossible without mechanical motion. We have already talked about the special role of macromolecules as the working substances of mechanochemical processes. The macromolecular mechanochemistry may be quite diverse; this is evidenced, for example, by the large number of hypotheses on the nature of muscle action. In addition to their ability to undergo conformational transformations and to form supramolecular structures, we should add also the ability of macromolecules to become polymerized, as is found for example, in the $G \rightarrow F$ transitions of actin and in the mechanism of mitosis.

Life has arisen and exists in salt solutions. Living cells require definite concentrations of small ions and definite pH's. Nature uses changes in the concentrations of ions and in their displacements to carry out a number of biological processes. These processes, which have developed at a very early stage in evolution, are determined by the polyelectrolyte character of biological polymers. The interactions of living systems with the environment, and first of all with sea water, must have been carried out, to a great extent, by the adsorption and transport of ions. One may believe that this is one of the reasons for the formation of living systems out of polyelectrolytic macromolecules, namely, proteins and nucleic acids.

The role of water remains determining in modern organisms, as well. It is impossible to account for the three-dimensional structure of proteins, which are responsible for their biological activity, without taking into account the aqueous medium. In this case, hydrophobic interactions play an important role.

As has been shown in Chapter 7, nature has created a genetic code which is aimed at the stability of a protein structure; there is a strong relation between the particularities of water structure and of the structure of genes. It is evident that the large water content in organisms and its fundamental significance reflect the fact that life has arisen in an aqueous medium.

Let us return now to the aperiodic crystal. Any changes in the states of biological systems are changes in the state of ordering. Examples of such changes on the molecular level are the helix–coil transitions and, consequently, any changes in the secondary and tertiary structures of proteins which occur during the interactions between enzyme and substrate, enzyme and product, enzyme and allosteric inhibitor. Changes of ordering occur during mechanochemical processes and during the template biosyntheses of proteins and nucleic acids, which were already compared with crystallization long ago.[2,3]

Processes of ordering and disordering in a nonliving substance are well known to physics. These are cooperative processes of phase transitions which depend to a great extent on interactions between the elements of the examined system, namely, of atoms and molecules. A cooperative interaction is certainly manifested not only in phase transitions, but also in many other phenomena. Thus, for example, the enthalpy changes by a very sharp jump during the softening of an annealed polymeric glass; this process mimics the melting of a crystal.[4] On the other hand, there is no phase transition in the thermodynamic sense of the word. The structure of the glass is similar to the structure of a liquid, and its difference from a liquid is due, not to thermodynamic factors, but to kinetic factors; namely, it is the result of the very long time necessary for the transformation of the molecular structure at temperatures below the glass temperature. Glass exists in a nonequilibrium state. It is evident, however, that cooperativity manifests itself with extreme sharpness in phase transitions.

If biological processes are so closely related with increases of order and disorder or, to use the language of the theory of information, with storage, loss, and transfer of information, then it becomes evident that, at the very foundation of molecular biophysics, one will find a theory of cooperative processes.[5] Cooperativity, which we encounter in living systems, is an expression of the cooperative properties of macromolecules, of special Markoff chains.

The biological meaning of cooperativity in living systems is determined, on one hand, by the thermodynamic aspect of the necessary high ordering of these systems and by the fact that any motion within them (in a general and not mechanical meaning of this word) requires some changes in ordering. On the other hand, a biological system can live only if it reacts very exactly and in a concerted manner to changes in the external and internal media. Such reactions are carried out in a cooperative manner, occurring as some sort of autocatalytic processes. Consequently, the cooperativity of biological macromolecules is important also in their kinetic relations.

An understanding of this fact removes a number of questions, related, for example, to muscle action and to nerve conduction. Why does the muscle fiber work as a whole? Why does the nerve cell react to a stimulus according to the all-or-none principle? To explain these and other similar phenomena, one frequently invokes concepts related to quantum mechanics, to the migration of energy, etc.[6] It seems, however, that the true explanation of such processes must be based on an examination of the kinetic and thermodynamic cooperative properties of biopolymers. Later we shall talk about the relations between molecular biophysics and quantum mechanics. The autocatalytic nature of cooperative processes indicates the presence of some feedback. In fact, as long as the elements of the given ensemble, let us say, the links of a polymer chain, interact and this interaction leads to a change in the state of the system, it, in its turn, changes the interactions.

In this manner, life on earth is macromolecular, and molecular biophysics are macromolecular biophysics. Life is generated by specific interactions of specialized macromolecules of polyelectrolytes in an aqueous medium.

Let us speculate somewhat. Let us attempt to conceive of a living system, i.e., an open, self-regulating and self-replicating system, constructed not from macromolecules, but from other materials.

It is evident, first of all, that an organized low-molecular-weight system must be crystalline and not liquid, and certainly not gaseous. An unordered system cannot ensure ordered interactions with the environment, as well as the conservation and transformation of information, etc. Changes in the states of the elements of a low-molecular-weight crystalline "living being" could occur, in principle, as a result of partial melting and crystallization or a transformation of the crystal lattice, i.e., phase transitions. Another possibility would be in changes of the magnetization and electrization of crystals, i.e., in changes in the order. Such a system would not be closed; it could certainly exchange energy and information, i.e., entropy, with the surrounding medium; an exchange of matter, however, would be greatly hindered, since chemical reactions in a solid state require high energies of activation, even just because it is difficult for molecules to diffuse into it. In this sense, such a system would be quite inert.

With the presence of conducting or semiconducting properties, however, a crystalline organism could react quite rapidly to changes in the environment, and it could carry out complicated functions similar to the functions of real macromolecular organisms. It could accept any information, store it, transform it, and perform complicated motions. It could acquire conditioned reflexes and perform any types of work. It could reduplicate itself and, furthermore, it could evolve by constructing systems similar to itself and more perfect systems from materials present in the surrounding medium.

We have come at this point to speculations on the cybernetic work of the future. The creation of a nonmacromolecular system which would act as a model for a living organism is definitely possible.

But could such a system be created without the participation of man, could a crystalline low-molecular-weight organism, even primitive, arise by itself, as a result of natural evolution? This appears improbable, just because of the high inertia of such a system in processes of exchange of matter and energy. Macromolecules may be formed as a result of polymerization processes from small molecules which were originally present on earth, in its atmosphere and in its hydrosphere, and this could also happen on other planets. Then, they could form supramolecular structures and living organisms. It is impossible to justify scientifically the possibility of an independent evolutionary transformation of low-molecular-weight metallic and semiconductor crystals into living systems. The macromolecularity of organisms is absolutely related to their evolutionary origin, and to the fact that the living, at some point, arises from the nonliving. This may be the most important consideration in giving an answer to the question why organisms are macromolecular.

Consequently, the cybernetic nonmacromolecular machine, which simulates life, could have been and can be created on earth only by man. Then it could perfect itself without limits. The construction of cybernetic machines reflects a particularity of the modern stage of the biological and social evolution of *homo sapiens*; it reflects the unlimited creative and learning power of the human mind. It would appear that the biological evolution of man has ended; as a result of social life, man is no longer subject to the law of biological natural selection. The brain of a man is so perfect that it may understand the nature of living phenomena and it may organize their artificial reproduction. A further development of cybernetics is related to the social evolution of man and, in this sense, not only does it not diminish his role, but it promises to make him the very creator of life in a direct sense of the word.

MOLECULAR BIOPHYSICS AND QUANTUM MECHANICS

Molecular biophysics is a new extremely interesting realm of physics. It studies molecular phenomena in living systems. It is natural to enquire about the specific quantum-mechanical properties of such systems.

It is possible to present arguments in favor of such a concept. First, all the modern developments of the physics of matter, namely, atoms, molecules, crystals, are absolutely linked to quantum mechanics; without quantum mechanics it would be impossible to explain very important phenomena, namely, the structures and spectra of atoms and molecules, the chemical bonds and the intermolecular dispersion forces, the properties of semiconductors and superconductivity, ferromagnetism and the electrical conductivity of metals. We do not speak here about the properties of the elementary particles and the structure of the atomic nucleus. The microsystems obey, not classical, but quantum laws. Since many of the most important biological phenomena are determined by the structures and properties of molecules, one thinks naturally that these also cannot be explained without the help of quantum mechanics.

Second, a number of definitely quantum-biological processes are known. We know that mutations may arise as a result of the action of quanta of short-wavelength radiation. Photochemical processes lie at the very basis of photosynthesis and vision; these start from the absorption of a quantum, according to the Einstein law of photochemistry, which has played an extremely important role in establishing the quantum nature of radiation.

Third, there is a purely psychological factor due to the enormous achievements of quantum physics. It seems particularly interesting and enticing to apply quantum-mechanical concepts to a new realm, namely, to biology.

One should, however, reject all psychological motives and one should approach soberly the problem of the relation of quantum mechanics to molecular biophysics, or even to biology as a whole.

Longuet-Higgins developed some ideas in this realm which cannot be rejected.[7] Special quantum effects are observed only if quanta of energy, which are exchanged between different parts of the system, are large relative to the heat energy kT. With a few exceptions, however, processes which occur in living substance are dark reactions, i.e., they proceed without the participation of radiation, without large quantum energies. In fact, this is quite true of all phenomena examined in the preceding chapters, of the denaturation and biosynthesis of biopolymers, of enzymatic and immunological activity, of mechanochemical processes. We have already cited the considerations which refute attempts to interpret the synapsis of chromosomes or muscle contractions as specific quantum phenomena. All the principal biochemical processes are dark reactions; this is also true of the processes of active and passive transport, which participate in physiological function.

Quantum processes found in biology are photosynthesis, the first act of visual perception, and bioluminescence. These are phenomena which consist in the absorption and the emission of visible light, i.e., of large quanta; consequently, these are electronic phenomena.

Certainly, quantum mechanics determine, in the final account, all the properties of biological molecules, just as of any other molecules, large or small. The very existence of a molecule, the existence of a chemical bond, cannot be understood without quantum mechanics. The same is true of the dispersion of intermolecular forces. In this sense, all of chemistry is absolutely linked with quantum mechanics. And this means that biology is also linked. The question is different: Are special quantum-mechanical effects manifested in the properties of biological systems, of biological molecules, in addition to those already found at the basis of chemistry? In other words, are the fundamental molecular-biological processes electronic processes?

It is necessary to answer in the negative in relation to the dark reactions. We encounter special electronic properties of molecules when, for example, the molecules contain electrons which are not localized in given chemical bonds but are commonly shared. This occurs, for example, in the conjugation of chemical bonds. The simplest molecule with conjugated bonds is 1,4-butadiene

$$H_2C=CH-CH=CH_2$$

The number of conjugated bonds is large in the porphyrin ring, in carotin, which is the pigment of carrots, Pro-vitamin A. Its structural formula is

Vitamin A is formed within the organism of an animal, which has obtained carotin in its food; this occurs by the breaking of a long carotin molecule into two equal halves and the combination of the ends so formed with two hydrogen atoms and one oxygen atom.

Molecules with conjugated bonds are, to some extent, similar to conductors of an electric current, since the electrons may migrate within them along the entire conjugated chain. In the theoretical examination of such molecules, quantum mechanics uses the metallic model, which directly takes into account the delocalization of electrons.[8] If proteins and nucleic acids, hydrocarbons and lipids had been chains of conjugated bonds, then, because of their great lengths, it could have been possible to expect the manifestation of specific electronic quantum-mechanical properties; such molecules would have been conductors, and their interactions with each other and with other molecules could have been quite special. In biopolymers, however, there is no conjugation of any sort; they are typical dielectrics.

At times one hears of the special semiconductor properties of proteins. This is a misunderstanding; proteins are not semiconductors in their ground nonexcited state. In order to excite a protein electronically, quantum energies much greater than kT are required. In this sense, biopolymers do not differ from the majority of synthetic polymers known to science. The semiconductor properties of polymers are encountered in a few cases of specially synthesized chains with conjugated bonds and with conjugated benzene rings. Certainly these cases are of great theoretical and practical interest.

A rough criterion of the absence of conjugation in a long chain is the absence of light absorption in the visible region, which is one of the special properties of conjugated systems, namely, their color, which is easily explained in terms of the metallic model. Carotin has an intense orange-red color. Proteins, nucleic acids, carbohydrates, and lipids are colorless; they absorb light in a rather far ultraviolet region of the spectrum.

Hemoglobin, however, is colored. Its color, though, is related, not to the protein moiety of the molecule, but to the heme group, which contains the porphyrin ring. The biological function of hemoglobin consists in the transport of oxygen. Here, there are no special quantum effects which are considerably different from those known in the usual chemistry of nonliving systems; the transition of hemoglobin into oxyhemoglobin and its reduction are explained, first of all, by the chemical properties of the heme, and not of the polymeric molecule.

As we have seen, the physics of macromolecules and, consequently, the physics of biological macromolecules are based on the classical quantum-mechanical fact of their existence.

Specific properties of biological macromolecules are determined, first of all, by their conformational cooperativity and not by electronic phenomena. A necessary apparatus of the theory in this case is classical statistical physics and kinetics, and not quantum mechanics. It is just to such nonquantum problems that this book is devoted.

We have seen that the processes of the template synthesis of proteins and nucleic acids are classical if we do not consider the truly chemical phenomena of enzymatic catalysis.

It would seem possible to expect the manifestation of special quantum properties in organized supramolecular structures. In fact, such structures are similar to crystalline solid bodies. To understand certain properties of low-molecular-weight crystals, quantum mechanics is necessary. Quantum conductor and semiconductor properties, however, which may be interpreted by the zone theory of conductivity, are not applicable to molecular crystals, i.e., to bodies, the crystal-lattice positions of which are filled with molecules with saturated chemical bonds. Such crystals are usually dielectric (the semiconductor properties of crystals of anthracene are due to impurities); their structure may be well explained by the principle of a tight packing of molecules, of the characteristics of Van der Waals surfaces.[9] We encounter the quantum-mechanical properties of such systems only during their electronic excitation. Ordered structures, built of macromolecules, are classical also in this sense.

On the other hand, an ordered structure of the chloroplast of a plant cell is constructed in such a way as to permit its electrons to interact with light, and, in this case, the theory must be quantum. But here we are no longer speaking of a dark reaction.

Furthermore, there are a number of biochemical problems of extreme importance, the solution of which requires quantum-mechanical methods. Today, we may already speak of quantum biochemistry.[10,11,12] We speak here about the quantum-mechanical investigations of the properties of biologically functional molecules and of their parts, even if these are not linked into a general chain of conjugation. Quantum chemistry has given much valuable information on the chemistry of nonliving substances, both organic and inorganic, and there are no reasons to doubt its importance in the realm of biochemistry. Let us list some problems to which quantum mechanics have already been applied with some success.[11]

Even though there are no conjugations in the macromolecules of biopolymers as a whole, in a number of cases the functional groups are conjugated systems. In order to understand the chemical properties of such groups, quantum mechanics are necessary. This applies to the purines and pyrimidines of DNA and RNA, to phenylalanine, tryptophan, and tyrosine in proteins. The presence of conjugation in the nitrogen bases of nucleic acids is manifested, for example, in their stability toward ionizing and ultraviolet radiation. All the chemical properties of these groups have a reasonable quantum-mechanical interpretation.

We have spoken already about the significance of the tautomerism of the nitrogen bases in the problem of mutagenesis. It is evident that the theoretical analysis of this phenomenon also must be quantum mechanical.

The chemical phenomena of enzymatic catalysis are also subject to such an analysis; these include the interactions between the substrate and the protein, the enzyme and the transformation of the enzyme–substrate complex. It is quite characteristic that the coenzymes and the prosthetic groups, namely porphyrins, carotinoids, etc., are conjugated systems.

Oxidation–reduction processes which occur with the participation of DPN, cytochromes, etc., are interesting quantum-mechanical problems. It is very important to understand in these cases the electron-donor and electron-acceptor properties of the participating compounds. While the biochemist and the biophysicist may take as a fact the existence of the high-energy bonds in ATP, the investigation of the special electronic structure of this bond is more than just a satisfaction of scientific curiosity.[11,13,14] It is impossible, however, to agree with the opinion of the Pullmans that the principal manifestations of life are deeply related to the existence of highly conjugated systems. All that has been said above testifies to the contrary.

It is also difficult to pay serious attention to attempts at explanations of the carcinogenic properties of benzopyrine and similar compounds by calculations of the electron densities and orders of the bonds in such molecules. Whatever the nature of carcinogenesis, whether it is in the formation of somatic mutations or in modifications of as-yet-not-understood regulatory processes, or in a combination of both, it cannot be explained by the electronic properties of the carcinogenic materials apart from general biological causes which determine the growth of malignant cells.

In all the problems that we have discussed, we do not trespass the limits of the usual applications of quantum mechanics in chemistry. It is correct to speak about quantum biochemistry, but one should not over-estimate the significance of quantum mechanics for biology as a whole.

Quite true, the physical properties of biopolymers related to the motions of the electrons require quantum mechanics and quantum chemistry for their interpretation. This applies absolutely to the optical properties of DNA and proteins, to their hypochromous effect, to changes in the optical activity during helix–coil transitions. In this sense, bio-polymers are specific, but their quantum chemistry does not differ in principle from that of any other group of compounds.

It is impossible to assert that the further development of molecular biophysics and biology, as a whole, will not require the development of quite new physical theories and concepts, and particularly of quantum-mechanical ones. But all that we know today in this realm does not testify in favor of such a possibility and its discussion is meaningless at present.

Let us examine now some biophysical problems already related, not to dark reactions, but to processes of electronic excitation. The migration of energy by resonance interactions or by other mechanisms has been invoked frequently for the explanation of biophysical and biochemical processes.[10,12,15] As has been stated already, there are no reasons for talking about the existence of sufficiently close, shared electronic levels in biopolymers. There are also no foundations for the migration of electronic excitations along a structured water; hydrogen bonds do not create any conjugation.

At the same time, migration of energy seems to occur in photochemical biological processes. Chlorophyl molecules, which are regularly packed in chloroplasts, form some sort of crystalline structure and the entire aggregate may be collectively excited by the absorption of each photon. Such excitations spread quite rapidly in the normal molecular crystals, such as in the crystals of aromatic compounds. An absorbed quantum results in the formation of an exciton, a quantum particle, whose migration signifies the migration of the exciting energy along the crystal. Thus, for example, at some place in the crystal, the photon removes an electron from an atom or from a molecule; an exciton is formed, i.e., the combination of an electron and a cavity which is the position from which an electron is missing. As a result of resonance interaction, the electron is displaced along the crystal up to a point at which there is an electron acceptor, i.e., a trap for this electron, or a trap for a cavity, namely, an electron donor. The propagation of the exciton leads to the rapid migration of energy, on the one hand, and, on the other hand, to special optical phenomena, namely, to changes in the luminescence and absorption spectra.[10,12,15-17] The concept of an exciton has been introduced into physics by Frenkel.[18] Quantum mechanics is absolutely essential for the interpretation of such processes. Here we encounter a number of unsolved problems. For example, are the migrating excitons singlet or triplet, i.e., is the total spin of the exciton zero or unity? Can a single exciton bring about chemical changes, or must the chloroplasts accumulate many quanta before the reaction will take place?

An examination of the phenomenon of photosynthesis is part of molecular biophysics, but it is outside the limits of this book. First, this is so because at present photosynthesis has become a wide realm of biophysics, which requires an independent presentation; this is the result of the studies of Calvin, Terenin, and others; second, because this book is devoted to the dark properties of the biological molecules—of proteins and nucleic acids. Let us just cite some papers and monographs.[19-22]

For similar reasons, we shall not touch the problem of the physics of vision[23,24] and of bioluminescence.[25] Problems of radiobiology[26] and molecular physical questions related to the formation of free radicals in biochemical reactions and in biological systems are also outside the realm of this book.[27-29]

Chapter 11

CONCLUSION

In this book, we have examined many questions; the discussion has been carried out at different levels of complexity and with different degrees of detail. We have devoted the greatest attention to the structure and properties of the most important biopolymers, namely, the proteins and nucleic acids. Their examination was carried out within a unifying general concept of cooperative conformational transformations as the foundations of phenomena studied by molecular biophysics.

Investigations of the biological processes which occur on a molecular level, carried out during recent years, have unveiled an amazing uniformity of mechanisms that act in all biological systems, starting from bacteria and ending with man.

The working substances of the most important molecular biological processes are proteins and nucleic acids. The base of their interactions which determine the phenomena of life are found in the structural complementarities of the molecules. These interactions ensure the most important molecular and supramolecular relationships.

There is no need to repeat here what has been stated already. Summing up, we may say that, even though the problems are extremely difficult, it is possible at present to consider that the basic principles of molecular biophysics have been established, as well as the general approaches needed for an understanding of biology.

There are a number of problems which are already being solved by molecular biophysics that have not even been touched in this book. Such are all the problems related to the action of irradiation, namely, radiobiology, photosynthesis, vision, and bioluminescence. Many dark processes also have not been examined. Let us just mention some of these.

Molecular biophysics encompass a vast number of processes related to the permeability of cell membranes.[30–33] Although the cell membrane

471

is a complicated supramolecular structure, it is possible to relate its characteristics to the structure and properties of individual molecules. The function of membranes is closely related to studies of nerve excitations, determined by changes in the membrane potential toward ions of Na^+ and K^+. A nerve impulse is propagated as a result of the action of a highly perfected relay mechanism with feedback between the change of the potential for Na^+ ions and the fall of the membrane potential.

It is evident that processes that take place within a nerve fiber are cooperative; the displacement of ions depends on how many particles have already overcome a given potential barrier.[34-39] The theory of nervous excitation is undergoing development; this is an important part of molecular biophysics. The key to molecular processes responsible for the transfer of a nerve impulse should probably be sought in the ion-exchange properties of biopolymers which we have also mentioned.

Pauling has proposed a highly elegant molecular theory of general narcosis.[40] There are a number of narcotic substances, which are essentially inert molecules and form neither hydrogen nor chemical bonds with biological molecules. Such are chloroform $CHCl_3$; nitrous oxide N_2O; carbon dioxide CO_2; ethylene C_2H_4; and even the noble gases Ar and Xe, which bring about narcosis at elevated pressures. Pauling proposed that their narcotic action, i.e., their ability to interrupt the transfer of a nervous impulse from one nerve end to another, is due to the formation of crystal hydrates, such as $8Xe \cdot 46H_2O$ and $CHCl_3 \cdot 17H_2O$. Such crystal hydrates are unstable; a high pressure is required to maintain them. In a biological system, however, they may be stabilized by the amino acid residues of proteins. In fact, it has turned out that the logarithm of the anesthetizing partial pressure of a gas is proportional to the logarithm of the equilibrium partial pressure of the same gas over its crystal hydrate. More than 10 substances fall on the straight line found in this manner.

In relation to this, one should also point out the fundamental role of water in biological processes. We have spoken frequently about water, about the hydrogen bonds formed by it, and about hydrophobic interactions. We have mentioned the structure-forming character of water which surrounds a biopolymer molecule, as well as smaller ions. This is an independent vast problem of molecular biophysics.[41]

The principal characteristics of the reactions of living biopolymeric systems toward external action also require a molecular interpretation. The cooperative properties of these systems also seem to play a determining role.

In this manner, this book must serve only as an introduction to molecular biophysics and it does not pretend to be a complete survey of its present-day state.

A physicist who becomes engaged in biological problems falls into a new situation. In the physics of nonliving bodies very much is already clear; laws have been established which regulate a vast sphere of phenomena and which are now the realm of rigorous theory. It is not so easy to find a new problem, the solution of which would mean an important scientific discovery. This applies in particular to the classical realm of molecular physics. Today, phenomena in the realm of elementary particles are still unanswered; the same is true of quantum electrodynamics and of relativistic cosmology. We do not see today, however, great mysteries in the world of molecules.

What has been said does not mean, of course, that tomorrow there will not be a new leap forward and that we shall not see quite familiar phenomena in a completely new light.

The particular situation which is encountered by a physicist in biology is that here everything is mysterious and infinitely interesting. The investigation of this territory is only beginning. A few contours have been drawn on the map, but the areas of the wide domains are immeasurably greater than the areas of the investigated realms. And what lies waiting in the unknown?

It would appear that the principal problem of the physicist who enters biology is the selection of a single definite problem and the withdrawal from the innumerable other equally interesting questions. This, however, is not true.

We are speaking here of biology. Here, the physicist may only follow biologists and biochemists. Because of the complexity of biological processes, most of them are still inaccessible to physical investigations, so that it is still impossible to develop any final physical theory of these phenomena. We have talked in this book about the insufficiency of the available biological and biochemical data related, for example, to such problems as DNA reduplication and mitosis. A physicist must cleverly choose a biological phenomenon, the nature of which is amenable to elucidation by the means available to him.

Nevertheless molecular biophysics is born and is rapidly growing. The requirement for its further development is a close collaboration of the physicist with biologists and biochemists, whose efforts at present are making biology the most important realm of natural science of the second half of the twentieth century.

REFERENCES

CHAPTER 1

1. N. V. Timofeev-Resovskii, *Tsitologiya*, **2**:45(1960).
2. M. V. Vol'kenshtein, "Stroenie i fizicheskie svoistva molekul", Izd. Akad. Nauk S.S.S.R., 1955: see also: Y. K. Syrkin and M. E. Dyatkina, "Structure of Molecules and the Chemical Bond", Dover Publications, Inc., New York, 1964.
3. N. F. Gamaleya, I. I. Mechnikov, and K. A. Timiryazev, "L. Pasteur," Izd. Akad. Nauk S.S.S.R., 1946.
4. Th. de Donder, L'Affinité, Pt. 1 (1928), Pt. 2 (1931), Pt. 3 (1934), Gauthier-Villars, Paris.
5. I. Prigogine, "Introduction to Thermodynamics of Irreversible Processes," 2nd ed., Interscience Publishers, Inc., New York, 2nd ed., 1962.
6. L. I. Mandel'shtam, "Collected Papers," Vol. 2, Izd. Akad. Nauk S.S.S.R., 1947, p. 176.
7. I. Prigogine, *Z. Chem.* **1**:203(1960).
8. A. A. Lyapunov, *Probl. Kibernetiki* **1**:5 (1958); see also: V. M. Glushkov, "Introduction to Cybernetics," Academic Press, New York, 1966.
9. H. Quastler, "The Emergence of Biological Organization," Yale University Press, New Haven, Conn., 1964.
10. A. M. Yaglom and I. M. Yaglom, "Probability and Information," Dover Publications, Inc., New York, 1962.
11. L. Szillard, Über die Entropieverminderung in einem thermodynamischem System bei Eingriffen intelligenter Wesen, *Z. Phys.* **53**:840(1929)
12. L. Brillouin, "Science and Information Theory," 2nd ed., Academic Press, Inc., New York, 1962.
13. P. Elias, Coding and information theory, *Rev. Mod. Phys.* **31**:221(1959).
14. K. von Frisch and M. Lindauer, The "Language" and Orientation of the Honey Bee, in: "Annual Review of Entomology," Vol. 1 (E. A. Steinhaus and R. F. Smith, eds.), Annual Reviews, Inc., Stanford, Cal., 1956, p. 45; M. Lindauer, *Bee World*, **36**:62(1955); **38**:3 (1957).
15. I. I. Schmal'gauzen, *Probl. Kibernetiki* **4**:121(1960).
16. N. V. Timofeev-Resovskii and R. R. Rompe, *Probl. Kibernetiki* **2**:213(1959).
17. C. B. Anfinsen, "The Molecular Basis of Evolution," John Wiley and Sons, Inc., New York, 1959.

CHAPTER 2

1. H. J. Morowitz and M. E. Tourtellotte, The Smallest Living Cells, *Sci. Am.* **206**(3):117 (1962)

2. J. Brachet, in: "The Cell: Biochemistry, Physiology, and Morphology," Vol. 2, Cells and Their Component Parts (J. Brachet and A. E. Mirsky, eds.), Academic Press, Inc., New York, 1960, p. 771.

3. A. L. Lehninger, Bioenergetics: The Molecular Basis of Energy Transformations in the Cell," W. A. Benjamin, New York, 1965.

4. E. D. P. de Robertis, W. W. Nowinski, and F. A. Saez, "General Cytology," 3rd ed., W. B. Saunders Co., Philadelphia, 1960.

5. D. Mazia, in: "The Cell: Biochemistry, Physiology, and Morphology," Vol. 3, Meiosis and Mitosis (J. Brachet and A. E. Mirsky, eds.), Academic Press, Inc., New York, 1961, p. 77.

6. J. Haas, "Das Lebensproblem heute, Beitrag der Zellforschung zur Philosophie des Organischen," Munich, 1958.

7. V. Ya. Aleksandrov, *Tsitologiya*, **4**(1):3(1962).

8. D. Mazia and A. Tyler, eds., "General Physiology of Cell Specialization," McGraw-Hill Book Company, New York, 1963.

9. C. A. Vilee, Jr., "Biology," 4th ed., W. B. Saunders Co., 1962.

10. M. Fischberg, J. B. Gurdon, and T. R. Elsdale, Nuclear Transfer in Amphibia and the Problem of the Potentialities of the Nuclei of Differentiating Tissues, *Exptl. Cell Res.*, *Suppl.* **6** (1959), p. 161.

11. R. P. Wagner and H. K. Mitchell, "Genetics and Metabolism," 2nd ed., John Wiley & Sons, Inc., New York, 1963.

12. C. H. Waddington, ed., "Biological Organisation, Cellular and Sub-Cellular," Pergamon Press, New York, 1959.

13. A. I. Oparin *et al.*, eds., "Proceedings of the International Symposium on the Origin of Life on Earth," Pergamon Press, New York, 1960.

14. A. I. Oparin, "Origin of Life," 2nd ed., Dover Publications, Inc., New York, 1953.

15. Present Problems of Virology, *Itogi Nauki, Biol. Nauki,* Izd. Akad. Nauk S.S.S.R., **4**(1960).

16. A. Gierer and G. Schramm, Die Infektiosität der Nucleinsäure aus Tabakmosaikvirus," *Z. Naturforsch.* **11B**(3):138(1956).

17. H. Fraenkel-Conrat, The Role of the Nucleic Acid in the Reconstitution of Active Tobacco Mosaic Virus, *J. Am. Chem. Soc.* **78**:882(1956).

18. M. H. Adams, "Bacteriophages," Interscience Publishers, Inc., New York, 1959.

19. G. S. Stent, "Molecular Biology of Bacterial Viruses," W. H. Freeman and Co., San Francisco, 1963.

20. E. Geissler, "Bakteriophagen," Akademie-Verlag, 1962.

21. A. N. Kolmogorov, *Dokl. Akad. Nauk S.S.S.R.* **27**:38(1940).

22. A. H. Sturtevant, C. B. Bridges, T. H. Morgan, L. V. Morgan, and J. C. Li, Contributions to the Genetics of Drosophila simulans and *Drosophila melanogaster*, Carnegie Institute, Washington, 1929.

23. J. Neel and W. J. Schull, "Human Heredity," University of Chicago Press, 1954.

24. L. Hogben, "An Introduction to Mathematical Genetics," W. W. Norton and Co., Inc., New York, 1946.

25. A. H. Sturtevant and G. W. Beadle, "An Introduction to Genetics," Dover Publications, Inc., New York, 1962.

26. J. A. Serra, "Modern Genetics," Vol. 1, Academic Press, Inc., New York, 1965.

27. I. V. Michurin, "Printsipy i metody raboty," Vol. 1, Selkhozgiz, 1948

28. G. A. Nadson and G. S. Philippov, *Compt. Rend. Soc. Biol.*, **95**:433(1926).
29. H. J. Muller, Artificial transmutation of the gene, *Science*, **66**:84(1927).
30. N. W. Timofeev-Resovskii, *Biol. Rev.*, **9**:411(1934).
31. N. W. Timofeev-Resovskii, Le Méchanisme des Mutations et la Structure du Géne, Paris, 1939.
32. N. W. Timofeev-Resovskii, *Tsitologiya*, **2**:45(1960).
33. C. Auerbach, "Genetics in the Atomic Age," Essential Books, Fairlawn, N.J., 1956.
34. N. P. Dubinin, "Problems of Radiation Genetics," Oliver and Boyd, Edinburgh, 1964.
35. *Itogi Nauki, Biol. Nauki,* Izd. Akad. Nauk S.S.S.R., **3**(1960).
36. B. L. Astaurov, *Usp. Sovrem. Biol.* **25**:49(1948); Priroda, No. 4, p. 55 (1962).
37. B. L. Astaurov, Introduction to R. Hageman, "Plasmaticheskaya nasledstvennost," IL, Moscow, 1962.
38. R. Sager and F. J. Ryan, "Cell Heredity," John Wiley and Sons, Inc., New York, 1961.
29. V. V. Sakharov, *Zh. Obshch. Biol.,* **1**(3)(1932).
40. M. E. Lobashev, *Dokl. Akad. Nauk S.S.S.R.* **2**(5)(1934).
41. I. A. Rapoport, *Byull. Eksperim. Biol. i Med.,* **6**:737(1938); *Dokl. Akad. Nauk S.S.S.R.* **54**:65(1946): **60**:469(1948).
42. S. I. Alikhanyan, *Zh. Vses. Khim. Obshchestva im. D. I. Mendeleeva,* **6**:285(1961).
43. S. Luria and M. Delbrück, *Genetics* **28**:491(1943).
44. G. H. Haggis, D. Michie, A. R. Muir, K. B. Roberts, and P. M. B. Walker, "Introduction to Molecular Biology," John Wiley and Sons, Inc., New York, 1964.
45. V. Ya. Aleksandrov, Tsitofiziologicheskie i tsitoekologicheskie issledovaniya ustoichivosti rastitelnykh kletok k deistviyu vysokikh i nizkikh temperatur, *Tr. Botan. Inst., Akad. Nauk S.S.S.R.,* 1955.
46. F. Jacob and E. Wollman, "Sexuality and the Genetics of Bacteria," Academic Press, Inc., New York, 1961.
47. W. Hayes, "The Genetics of Bacteria and their Viruses," John Wiley and Sons, Inc., New York, 1964.
48. J. Lederberg, Genetic Recombination in Bacteria, *Science* **122**:920(1955); Conjugal Pairing in *Escherichia coli, J. Bacteriol.* **71**:497(1956).
49. S. Benzer, in: "The Chemical Basis of Heredity" (W. D. McElroy and B. Glass, eds.), Johns Hopkins Press, Baltimore, Md., 1956, p. 70.
50. H. Ris and B. L. Chandler, The ultrastructure of genetic systems in prokaryotes and eukaryotes, *Cold Spring Harbor Symp. Quant. Biol.* **28**:1(1963).

CHAPTER 3

1. J. L. Oncley, Chemical Characterization of Protein, Carbohydrates, and Lipids, *Rev. Mod. Phys.* **31**:30(1959).
2. E. Lederer and M. Lederer, "Chromatography," 2nd ed., American Elsevier Publishing Co., New York, 1957.
3. J. L. Bailey, "Techniques in Protein Chemistry," American Elsevier Publishing Co., New York, 1962.
4. B. W. Low, in: "The Proteins: Chemistry, Biological Activity, and Methods," Vol. 1A (H. Neurath and K. Bailey, eds.), Academic Press, Inc., New York, 1953, p. 235.
5. V. V. Alpatov, *Priroda,* **4**:49(1947).
6. A. I. Oparin, "The Origin of Life," 2nd ed., Dover Publications, Inc., New York, 1953.
7. J. A. Schellman and C. Schellman, in: "The Proteins: Composition, Structure, and Function," Vol. 2 (H. Neurath, ed.), Academic Press, Inc., New York, 1964, p. 1.
8. W. Kuhn and E. Braun, Photochemische Erzeugung optisch aktiver Stoffe, *Naturwiss.* **17**:227(1929).

9. W. Kuhn and E. Knopf, Darstellung optisch aktiver Stoffe mit Hilfe von Licht, *Z. physik. Chem.* **B7**(4):292(1930).

10. T. L. Davis and J. Ackerman, Jr., Asymmetric Synthesis, III, *J. Am. Chem. Soc.* **67**:486 (1945).

11. R. L. Shriner, R. Adams, and C. S. Marvel, in: "Organic Chemistry," Vol. 1 (H. Gilman, ed.), John Wiley and Sons, Inc., New York, 1943, p. 214.

12. E. Darmois, Dissymetrie moleculaire. Sur la résolution des mélanges racémiques par des agents symetriques, *Compt. Rend. Acad. Sci. Paris* **237**:124(1953).

13. G. F. Gauze, "Asymmetry of Protoplasm," *Izd. Akad. Nauk S.S.S.R.*, 1940.

14. H. Eyring, L. Jones, and J. Spikes, The Significance of Absolute Configurations in Optical Rotation and in Catalysis, in: "Horizons in Biochemistry" (M. Kasha and B. Pullman, eds.), Academic Press, Inc, New York, 1962.

15. G. H. Haggis, D. Michie, A. R. Muir, K. B. Roberts, and P. M. B. Walker, "Introduction to Molecular Biology," John Wiley and Sons, Inc., New York, 1964.

16. L. Pauling, H. A. Itano, S. J. Singer, and I. C. Wells, Sickle cell anemia, a molecular disease, *Science* **110**:543(1949).

17. V. M. Ingram, Gene Mutations in Human Haemoglobin: The Chemical Difference between Normal and Sickle Cell Haemoglobin, *Nature* **180**(4581):326(1957); also, "The Hemoglobins in Genetics and Evolution," Columbia University Press, New York, 1963.

18. H. F. C. Crick, The Recent Excitement in the Coding Problem, in: "Progress in Nucleic Acid Research," Vol. 1 (J. N. Davidson and W. E. Cohn, eds.), Academic Press, Inc., New York, 1963, p. 164.

19. V. P. Efraimson, *Zh. Vses. Khim. Obshchestva im. D. I. Mendeleeva*, **6**(3):314(1961).

20. F. Sorm and B. Keil, Regularities in the Primary Structure of Proteins, in: "Advances in Protein Chemistry," Vol. 17 (C. B. Anfinsen, Jr., M. L. Anson, K. Bailey, and J. T. Edsall, eds.), Academic Press, Inc., New York, 1962, p. 167.

21. K. Hofman and H. Yajima, in: "Polyamino Acids, Polypeptides, and Proteins" (M. A. Stahmann, ed.), University of Wisconsin Press, Madison, Wisc., 1962, p. 21.

22. J. Noguchi and T. Saito, in: "Polyamino Acids, Polypeptides, and Proteins" (M. A. Stahmann, ed.), University of Wisconsin Press, Madison, Wisc., 1962, p. 313.

23. T. Gill and P. Doty, in: "Polyamino Acids, Polypeptides, and Proteins" (M. A. Stahmann, ed.), University of Wisconsin Press, Madison, Wisc., 1962, p. 367.

24. See, for example, P. G. Katsoyannis, *Diabetes*, **13**:339(1964).

25. J. N. Davidson, "The Biochemistry of the Nucleic Acids," 4th ed., John Wiley and Sons, Inc., New York, 1960.

26. A. S. Spirin, A. N. Belozerskii, N. V. Shugayeva, and V. F. Vanyushin, A Study of Species Specificity with respect to Nucleic Acids in Bacteria, *Biochemistry (U.S.S.R.)* (*English transl.*) **22**:699(1957).

27. A. N. Belozerskii, in: Proceedings of the Fifth International Congress on Biochemistry, Vol. 3, Evolutionary Biochemistry (A. I. Oparin, ed.), Pergamon Press, Inc., New York, 1963, p. 198.

28. A. N. Belozerskii and A. S. Spirin, in: "The Nucleic Acids: Chemistry and Biology," Vol. 3 (E. Chargaff and J. N. Davison, eds.), Academic Press, Inc., New York, 1960, p. 147.

29. R. W. Holley, J. Apgar, G. A. Everett, J. T. Madison, M. Marquisse, S. H. Merrill, J. R. Penswick, and A. Zamir, Structure of a Ribonucleic Acid, *Science*, **147**:1462(1965).

30. N. K. Kochetkov, É. I. Budovskii, and V. N. Shibaev, *Akad. Nauk, S.S.S.R., Biol.*, **4**:512(1964).

31. A. L. Lehninger, "Respiration—Energy Transformation," *Rev. Mod. Phys.* **31**:136 (1959).

32. H. A. Krebs and H. L. Kornberg, "Energy Transformations in Living Matter," Springer, Berlin, 1957.
33. D. I. Arnon, The Role of Light in Photosynthesis, *Sci. Am.* **203**(5):104(1960).
34. H. F. Blum, "Time's Arrow and Evolution," Harper and Row, New York, 1962.
35. D. W. Urry and H. Eyring, "An Imidazole Pump Model of Electron Transport," *Proc. Natl. Acad. Sci. U.S.* **49**:253(1963); also, Biological Electron Transport. I and II, *J. Theor. Biol.* **8**(1):198, 214(1965).
36. L. A. Blyumenfel'd and M. I. Temkin, *Biofizika* **7**:731(1962).
37. A. Lehninger, Mechanochemical Changes in Mitochondria, in "Horizons in Biochemistry" (M. Kasha and B. Pullman, eds.), Academic Press, Inc., New York, 1962.
38. M. Calvin, Free Radicals in Photosynthetic Systems, *Rev. Mod. Phys.* **31**:157(1959).

CHAPTER 4

1. S. Mizushima, "Structure of Molecules and Internal Rotation," Academic Press, Inc., New York, 1954.
2. M. V. Vol'kenshtein, "High Polymers," Vol. 17, Configurational Statistics of Polymeric Chains, Interscience Publishers, Inc., New York, 1963.
3. M. V. Vol'kenshtein, *Dokl. Akad. Nauk S.S.S.R.,* **78**:879(1951); also *Zh. Fizik. Khim.* **26**:1072(1952).
4. I. I. Novak, *Zh. Tekh. Fiz.* **24**:85(1955).
5. M. V. Vol'kenshtein and O. B. Ptitsyn, *Zh. Tekhn. Fiz.* **25**:649, 662(1955).
6. V. N. Nikitin, M. V. Vol'kenshtein, and B. Z. Volchek, *Zh. Tekhn. Fiz.* **25**:2486(1955).
7. V. N. Nikitin, B. Z. Volchek, and M. V. Vol'kenshtein, *Trudy X soveshchaniya po spektroskopii, Izd. Lvovskogo Universiteta,* **1**:411(1957).
8. B. Z. Volchek and V. N. Nikitin, The Stretching of Gutta-Percha: Infrared Absorption Study, *Soviet Phys. Tech. Phys.(English transl.)* **3**(8):1617(1958).
9. A. Ciferri, C. A. J. Hoeve, and P. J. Flory, Stress–Temperature Coefficients of Polymer Networks and the Conformational Energy of Polymer Chains, *J. Am. Chem. Soc.* **83**:1015(1961).
10. T. M. Birshtein and O. B. Ptitsyn, "High Polymers," Vol. 22, Conformations of Macromolecules, Interscience Publishers, Inc., New York, 1965.
11. R. Haase, "Thermodynamik der Mischphasen," Springer, Berlin, 1956.
12. H. N. V. Temperley, "Changes of State," Cleaver-Hume Press, Ltd., 1956.
13. E. Ising, "Beitrag zur Theorie des Ferromagnetismus," *Z. Physik.* **31**:253(1925).
14. G. F. Newell and E. W. Montroll, On the Theory of the Ising Model of Ferromagnetism, *Rev. Mod. Phys.* **25**:253(1953).
15. Yu. B. Rumer, *Usp. Fiz. Nauk* **53**:245(1954).
16. S. Glasstone, "Thermodynamics for Chemists," D. Van Nostrand Co., Inc., Princeton, N.J., 1947.
17. H. Margenau and G. Murphy, "The Mathematics of Physics and Chemistry," D. Van Nostrand Co., Inc., Princeton, N.J., 1943.
18. M. Dixon and E. C. Webb, "Enzymes," Academic Press, Inc., New York, 2nd ed., 1964.
19. J. Frenkel, "Kinetic Theory of Liquids," Dover Publications, Inc., New York, 1955.
20. O. B. Ptitsyn and Yu. A. Sharonov, Internal Rotation in Polymer Chains and Their Physical Properties, *Soviet Phys. Tech. Phys. (English transl.)* **2**(12):2544, 2762(1957).
21. N. P. Borisova and M. V. Vol'kenshtein, *Zh. Struk. Khim.* **2**:469(1961).
22. N. P. Borisova and T. M. Birshtein, *Vysokomolekul. Soedin* **5**:279(1963).

23. C. A. J. Hoeve, Unperturbed Chain Dimensions in Polymeric Chains, *J. Chem. Phys.* **32**:888(1960); also, Unperturbed Mean-square End-to-End Distance of Polyethylene, *J. Chem. Phys.* **35**:1266(1961).

24. P. Flory, Statistical Thermodynamics of Semi-flexible Chain Molecules, *Proc. Roy. Soc. (London)* **A234**:60(1956).

25. V. A. Kargin, A. I. Kitaigorodskii, and G. L. Slonimskii, *Kolloidn. Zh.* **19**:131(1957).

26. E. A. DiMarzio, Statistics of Orientation Effects in Linear Polymer Molecules, *J. Chem. Phys.* **35**:658(1961).

27. M. V. Vol'kenshtein, Yu. Ya. Gotlib, and O. B. Ptitsyn, *Vysokomolekul. Soedin.* **1**:1056 (1959).

28. V. A. Kargin, *Collection Czech. Chem. Commun., Special Issue* **22**:50(1957): also, *Vysokomolekul. Soedin.* **1**:1812(1959); **2**:931, 1280(1960).

29. J. D. Bernal, in: "Proceedings of the International Symposium on the Origin of Life on the Earth" (A. I. Oparin *et al.,* eds.), Pergamon Press, Inc., New York, 1960.

30. J. Bernal, in: "Horizons in Biochemistry" (M. Kasha and B. Pullman, eds.), Academic Press, Inc., New York, 1962.

31. P. Flory, "Principles of Polymer Chemistry," Cornell University Press, Ithaca, N.Y., 1953.

32. F. W. Billmeyer, Jr., "Textbook of Polymer Science," Interscience Publishers, Inc., New York, 1962.

33. H. A. Stuart, "Das Makromolekül in Lösungen," Springer, Berlin, 1953.

34. E. M. Frith and R. F. Tuckett, "Linear Polymers," Longmans, Green and Co., Ltd., 1951.

35. M. V. Vol'kenshtein, "Molekulyarnaya optika," Gostekhizdat, Moscow, 1951.

36. P. Doty, Physicochemical Characterization of Macromolecules, *Rev. Mod. Phys.* **31**:61 (1959).

37. V. N. Tsvetkov, V. E. Eskin, and Yu. I. Frenkel, "Structura Makromolekul v Rastvorakh," Nauka, Moscow, 1964.

38. D. A. MacInnes, "The Principles of Electrochemistry," Dover Publications, Inc., New York, 1961.

39. A. Katchalsky, O. Künzle, and W. Kuhn, Behavior of polyvalent polymeric ions in solution, *J. Polymer Sci.* **5**(3):283(1950); also, W. Kuhn, O. Künzle, and A. Katchalsky, Verhalten polyvalenter Fadenmolekelionen in Lösung, *Helv. Chim. Acta* **31**:1994(1948).

40. A. Katchalsky and S. Lifson, The Electrostatic Free Energy of Polyelectrolyte Solutions, I: Randomly Kinked Macromolecules, *J. Polymer Sci.* **11**(5):409(1953).

41. O. B. Ptitsyn, *Vysokomolekul. Soedin.* **3**:1084(1961).

42. O. B. Ptitsyn, *Vysokomolekul. Soedin.* **3**:1252(1961).

43. F. Haurowitz, Chemistry and Function of Proteins, 2nd ed., Academic Press, Inc., New York, 1963.

44. F. G. Helfferich, "Ion Exchange," McGraw-Hill Book Company, New York, 1962.

45. G. V. Samsonov, Sorbtsiya i Khromatografiya Antibiotikov, *Akad. Nauk S.S.S.R.,* 1960.

CHAPTER 5

1. F. Sorm and B. Keil, Regularities in the Primary Structure of Proteins, in: "Advances in Protein Chemistry," Vol. 17 (C. B. Anfinsen, Jr., M. L. Anson, K. Bailey, and J. T. Edsall, eds.), Academic Press, Inc., New York, 1962, p. 167.

2. N. D. Sokolov, in: "Vodorodnaya Svyaz," Nauka, Moscow, 1964.

3. C. A. Coulson, The Hydrogen Bond—A Review of the Present Position, *Research* **10**(4):149(1957).

4. L. E. Orgel, The Hydrogen Bond, *Rev. Mod. Phys.* **31**:100(1959).

5. G. C. Pimentel and A. L. McClellan, "The Hydrogen Bond," W. H. Freeman and Co., San Francisco, 1960.

6. R. B. Corey and L. Pauling, Fundamental Dimensions of Polypeptide Chains, *Proc. Roy Soc. (London)* **B141**:10(1953).

7. L. Pauling, *Inst. Intern. Chim. Solvay, Conseil Chim., 9e, Brussels,* 1953, p. 63.

8. T. Miyazawa, in: "Polyamino Acids, Polypeptides, and Proteins" (M. A. Stahmann, ed.), University of Wisconsin Press, Madison, Wisc., 1962, p. 201.

9. E. R. Blout, in: "Polyamino Acids, Polypeptides, and Proteins" (M. A. Stahmann, ed.), University of Wisconsin Press, Madison, Wisc., 1962, p. 275.

10. Ye. V. Anufriyeva, I. A. Bolotina, B. Z. Volchek, N. G. Illarionova, V. I. Kalikhevich, O. Z. Korotkina, Yu. V. Mitin, O. B. Ptitsyn, A. V. Purkina, and V. Ye. Eskin, Investigations of Synthetic Polypeptides, 1, *Biophysics (U.S.S.R.) (English transl.)* **10**(6):1017 (1965).

11. G. B. B. M. Sutherland, Application of Infrared Spectroscopy to Biological Problems, *Rev. Mod. Phys.* **31**:118(1959).

12. A. Wada, in: "Polyamino Acids, Polypeptides, and Proteins" (M. A. Stahmann, ed.), University of Wisconsin Press, Madison, Wisc., 1962, p. 131.

13. A. V. Guzzo, The Influence of Amino Acid Sequence on Protein Structure, *Biophys. J.* **5**(6):809(1965).

14. H. K. Shachman, Considerations on the Tertiary Structure of Proteins, *Cold Spring Harbor Symp. Quant. Biol.* **28**:409(1963).

15. W. H. Zachariasen, "Theory of X-ray Diffraction in Crystals," Dover Publications, Inc., New York, 1966.

16. K. C. Holmes and D. M. Blow, "The Use of X-ray Diffraction in the Study of Protein and Nucleic Acid Structure," Interscience Publishers, Inc., New York, 1965.

17. J. C. Kendrew, Three-dimensional Structure of Globular Proteins, *Rev. Mod. Phys.* **31**:94(1959).

18. J. C. Kendrew, in: Proceedings of the Fifth International Congress on Biochemistry, Vol. 1: Biological Structure and Function at the Molecular Level (V. A. Engelhardt, ed.), Pergamon Press, Inc., New York, 1963, p. 1.

19. J. D. Bernal, in "Proceedings of the International Symposium on the Origin of Life on the Earth" (A. I. Oparin *et al.,* eds.), Pergamon Press, Inc., New York, 1960.

20. A. Rich, Molecular Configuration of Synthetic and Biological Polymers, *Rev. Mod. Phys.* **31**:50(1959).

21. A. Elliott, E. M. Bradbury, A. R. Downie, and W. E. Hanby, in: "Polyamino Acids, Polypeptides, and Proteins" (M. A. Stahmann, ed.), University of Wisconsin Press, Madison, Wisc., 1962, p. 255.

22. M. I. Millionova and N. S. Andreeva, A Model of the Chain Configuration in Poly-(glycyl-L-proline), *Biophysics (U.S.S.R.) (English transl.)* **3**(3):250(1958); also, The structure of a Glycyl-L-proline Polymer, *Biophysics (U.S.S.R.) (English transl.)* **4**(3):138 (1959).

23. P. Flory, Statistical Thermodynamics of Semi-flexible Chain Molecules, *Proc. Roy. Soc. (London)* **A234**:60(1956).

24. S. E. Bresler and D. L. Talmud, *Dokl. Nauk S.S.S.R.* **43**:326, 367(1944): also, S. E. Bresler, *Biokhimiya* **14**:180(1949).

25. W. Kauzmann, Some Factors in the Interpretation of Protein Denaturation, *Adv. Protein Chem.* **14**:1(1959).

26. G. Némethy and H. A. Scheraga, Structure of Water and Hydrophobic Bonding in Proteins, I, *J. Chem. Phys.* **36**:3382(1962); also, Structure of Water and Hydrophobic Bonding in Proteins, IV, *J. Chem. Phys.* **41**:680(1964); and, The Structure of Water and

Hydrophobic Bonding in Proteins, III, *J. Phys. Chem.* **66**:1773(1962); Additions and Corrections to *J. Phys. Chem.* **66**:1773(1962); and **67**:2888(1963).

27. D. C. Poland and H. A. Scheraga, Statistical Mechanics of Noncovalent Bonds in Polyamino Acids, *Biopolymers*, **3**:275, 283, 305, 315, 335, 357, 369, 379, 401(1965).

28. H. T. Fisher, A Limiting Law Relating the Size and Shape of Protein Molecules to Their Composition, *Proc. Natl. Acad. Sci. U.S.* **51**:1285(1964).

29. H. F. Fisher, An Upper Limit to the Amount of Hydration of a Protein Molecule. A Corollary to the 'Limiting Law' of Protein Structure, *Biochim. Biophys. Acta* **109**:544 (1965).

30. M. F. Perutz, Structure and Function of Haemoglobin, I, *J. Mol. Biol.* **13**:646(1965).

31. M. F. Perutz, J. C. Kendrew, and H. C. Watson, Structure and Function of Haemoglobin, II, *J. Mol. Biol.* **13**:669(1965).

32. C. Levinthal, Molecular Model-building by Computer, *Sci. Am.* **214**(6):42(1966).

33. H. A. Scheraga, in: "Polyamino Acids, Polypeptides, and Proteins" (M. A. Stahmann, ed.), University of Wisconsin Press, Madison, Wisc., 1962, p. 241.

34. J. A. Schellman, *Compt. Rend. Trav. Lab. Carlsberg, Ser. Chim.* **29**:223, 230(1955).

35. P. L. Privalov, Study of Thermal Denaturing of Egg Albumin, *Biophysics (U.S.S.R.)* (*English Transl.*) **8**(3):363(1963); P. L. Privalov and D. R. Monaselidze, Investigation of Thermal Denaturing of Serum Albumin, *Biophysics (U.S.S.R.)* (*English Transl.*) **8**(4):467(1963).

36. G. Némethy, I. Z. Steinberg, and H. A. Scheraga, Influence of Water Structure and of Hydrophobic Interactions on the Strength of Side-chain Hydrogen Bonds in Protein, *Biopolymers* **1**:43(1963).

37. O. B. Ptitsyn and Yu. Ye. Eizner, Theory of Globule to Coil Transitions in Macromolecules, *Biophysics (U.S.S.R.)* (*English Transl.*) **10**(1):1(1965).

38. P. Doty, Physicochemical Characterization of Macromolecules, *Rev. Mod. Phys.* **31**:61 (1959).

39. P. Doty, *Collection Czech. Chem. Commun., Special Issue* **22**:5(1957).

40. M. V. Vol'kenshtein, "High Polymers," Vol. 17, Configurational Statistics of Polymeric Chains, Interscience Publishers, Inc., New York, 1963.

41. T. M. Birshtein and O. B. Ptitsyn, "High Polymers," Vol. 22, Conformations of Macromolecules, Interscience Publishers, Inc., New York, 1965.

42. B. H. Zimm and J. K. Bragg, Theory of the One-dimensional Phase Transition in Polypeptide Chains, *J. Chem. Phys.* **28**:1246(1958); Theory of the Phase Transition between Helix and Random Coil in Polypeptide Chains, *J. Chem. Phys.* **31**:526(1959).

43. J. H. Gibbs and E. A. DiMarzio, Theory of Helix–Coil Transitions in Polypeptides, *J. Chem. Phys.* **28**:1247(1958); Statistical Mechanics of Helix–Coil Transitions in Macromolecules, *J. Chem. Phys.* **30**:271(1959).

44. S. A. Rice, A. Wada, and E. P. Geiduschek, Some Comments on the Theory of Denaturation, *Discussions Faraday Soc.* **25**:130(1958).

45. T. L. Hill, Generalization of the One-dimensional Ising Model Applicable to Helix Transitions in Nucleic Acids and Proteins, *J. Chem. Phys.* **30**:383(1959).

46. L. Peller, On a Model for the Helix–Random Coil Transition in Polypeptides, I, *J. Phys. Chem.* **63**:1194(1959).

47. L. D. Landau and E. M. Lifshitz, "Statistical Physics," Addison-Wesley Publishing Company, Inc., Reading, Mass., 1958.

48. B. H. Zimm and S. A. Rice, The Helix–Coil Transition in Charged Macromolecules, *Mol. Phys.* **3**:391(1960).

49. I. Tinoco, A. Halpern, and W. T. Simpson, in: "Polyamino Acids, Polypeptides, and Proteins" (M. A. Stahmann, ed.), University of Wisconsin Press, Madison, Wisc., 1962, p. 147.

50. N. G. Bakhshiev, The Internal Field and the Position of the Electronic Absorption and Emission Bands of Polyatomic Organic Molecules in Solution, *Opt. Spectr. (U.S.S.R.) (English Transl.)* **7**(1):29(1959); Universal Molecular Interactions and Their Effect on the Position of the Electronic Spectra of Molecules in Two-component Solution, I, *Opt. Spectr. (U.S.S.R.) (English Transl.)* **10**(6):379(1961).

51. S. Yanari and F. A. Bovey, Interpretation of the Ultraviolet Spectral Changes of Proteins, *J. Biol. Chem.* **235**:2818(1960).

52. A. S. Davydov, "Theory of Molecular Excitations," McGraw-Hill Book Company, New York, 1962.

53. M. Kasha, in: "Light and Life" (W. D. McElroy and B. Glass, eds.), Johns Hopkins Press, Baltimore, Md., 1960.

54. W. Moffitt, Optical Rotatory Dispersion of Helical Polymers, *J. Chem. Phys.* **25**:467 (1956).

55. H. DeVoe, Hypochromism as a Local Field Effect, *Biopolymers, Symposia* **1**:251(1964).

56. W. Kauzmann, "Quantum Chemistry: An Introduction," Academic Press, Inc., New York, 1957.

57. E. R. Blout, P. Doty, and J. T. Yang, Polypeptides, XII, *J. Am. Chem. Soc.* **79**:749(1957).

58. P. Doty and R. D. Lundberg, The Contribution of the α-Helical Configuration to the Optical Rotation of Polypeptides and Proteins, *Proc. Natl. Acad. Sci. U.S.* **43**:213(1957).

59. R. D. Lundberg and P. Doty, Polypeptides, XVII, *J. Am. Chem. Soc.* **79**:3961(1957).

60. J. T. Yang and P. Doty, The Optical Rotatory Dispersion of Polypeptides and Proteins in Relation to Configuration, *J. Am. Chem. Soc.* **79**:761(1957).

61. R. W. Ditchburn, "Light," 2nd ed., Interscience Publishers, Inc., New York, 1963.

62. C. Djerassi, "Optical Rotatory Dispersion," McGraw-Hill Book Company, New York, 1960.

63. J. G. Kirkwood, On the Theory of Optical Rotatory Power, *J. Chem. Phys.* **5**:479(1937).

64. M. V. Vol'kenshtein, *Dokl. Akad. Nauk S.S.S.R.* **71**:447(1950).

65. M. P. Kruchek, A Method for Computing the Rotational Strengths of Weak Optically Active Transitions, *Opt. Spectr. (U.S.S.R.) (English Transl.)* **17**:294(1964); also, Comparison of the Contributions of Strong and Weak Absorption Bands in the Optical Rotation of Some Alicyclic Ketones, *Opt. Spectr. (U.S.S.R.) (English Transl.)* **17**:429 (1964).

66. M. V. Vol'kenshtein and I. O. Levitan, *Zh. Strukt. Khim.* **3**:81, 87(1962).

67. M. V. Vol'kenshtein and I. O. Levitan, *Opt. Spectr. (U.S.S.R.) (English Transl.)* **2**:60 (1963).

68. I. Tinoco, Jr., Theoretical Aspects of Optical Activity, Part 2: Polymers, *Advan. Chem. Phys.* **4**:113(1962).

69. I. Tinoco, Jr., and C. A. Bush, The Influence of Static Electric and Magnetic Fields on the Optical Properties of Polymers, *Biopolymers Symposia* **1**:235(1964).

70. A. D. McLachlan and M. A. Ball, Hyperchromism and Optical Rotation in Helical Polymers, *Mol. Phys.* **8**:581(1964).

71. R. A. Harris, On the Optical Rotatory Dispersion of Polymers, *J. Chem. Phys.* **43**:959 (1965).

72. J. A. Schellmann and P. Oriel, Origin of the Cotton Effect of Helical Polypeptides, *J. Chem. Phys.* **37**:2114(1962).

73. G. Holzwarth and P. Doty, The Ultraviolet Circular Dichroism and Polypeptides, *J. Am. Chem. Soc.* **87**:218(1965).

74. E. Shechter and E. R. Blout, An Analysis of the Optical Rotatory Dispersion of Polypeptides and Proteins, I and II, *Proc. Natl. Acad. Sci. U.S.* **51**:695, 794(1964); also, E. Shechter, J. P. Carver, and E. R. Blout, An Analysis of the Optical Rotatory Dispersion of Polypeptides and Proteins, III, *Proc. Natl. Acad. Sci. U.S.* **51**:1029(1964).

75. E. R. Blout and L. Stryer, "Anomalous Optical Rotatory Dispersion of Dye: Poly-peptide Complexes, *Proc. Natl. Acad. Sci. U.S.* **45**(11):1591(1959); L. Stryer and E. R. Blout, Optical Rotatory Dispersion of Dyes Bound to Macromolecules. Cationic Dyes: Polyglutamic Acid Complexes, *J. Am. Chem. Soc.* **83**:1411(1961).
76. S. Beychok and E. R. Blout, Optical Rotatory Dispersion of Sperm Whale Ferrimyo-globin and Horse Ferrihemoglobin, *J. Mol. Biol.* **3**:769(1961).
77. I. A. Bolotina and M. V. Vol'kenshtein, in "Molekulyarnaya Biofizika," Nauka, Moscow, 1965.
78. I. Tinoco, Jr., The Optical Rotation of Oriented Helices, I, *J. Am. Chem. Soc.* **81**:1540 (1959).
79. R. Serber, The Theory of the Faraday Effect in Molecules, *Phys. Rev.* **41**:489(1932).
80. I. Tobias and W. Kauzmann, Faraday Effect in Molecules, *J. Chem. Phys.* **35**:538(1961).
81. M. P. Groenewege, A Theory of Magneto-optical Rotation in Diamagnetic Molecules of Low Symmetry, *Mol. Phys.* **5**:541(1962).
82. V. E. Shashoua, Magneto-optical Rotation Spectra of Cytochrome *c*, *Nature* **203**(4948): 972(1964); also, *Biochemistry* (*U.S.S.R.*) (*English Transl.*) **3**:1719(1964); Magneto-Optical Rotation Spectra of Porphyrins and Phthalocyanines, *J. Am. Chem. Soc.* **87**:4044(1965).
83. M. V. Vol'kenshtein, J. A. Sharonov, and A. K. Shemelin, Anomalous Dispersion of the Faraday Effect and Haemoglobin and Myoglobin, *Nature* **209**(5024):709(1966); also, *Molekularnaja Biologia* **1**:467(1967); and B. A. Atanasov, M. V. Vol'kenshtein, J. A. Sharonov, and A. K Shemelin, *Molekularnaja Biologia* **1**:477(1967); **2**:864(1968).
84. B. Atanasov, M. V. Vol'kenshtein, and Yu. Sharonov, *Molekularnaja Biologia* **3**:518 (1969).

CHAPTER 6

1. J. N. Davidson, "The Biochemistry of the Nucleic Acids," 4th ed., John Wiley and Sons, Inc., New York, 1960.
2. E. Schrödinger, "What is Life?" Cambridge University Press, New York, 1963.
3. S. Zamenhof, R. DeGiovanni, and R. J. Rich, *J. Bacteriology* **71**:60(1956).
4. O. T. Avery, C. M. McLeod, and M. McCarty, "Studies on the Chemical Nature of the Substance Inducing Transformation of Pneumococcal Types. Induction of Transforma-tion by a Desoxyribonucleic Acid Fraction Isolated from Pneumococcus Type III." *J. Exptl. Med.* **79**:137(1944).
5. R. D. Hotchkiss, Transfer of Penicillin Resistance in Pneumococci by the Desoxy-ribonucleate Derived from Resistant Cultures, *Cold Spring Harbor Symp. Quant. Biol.* **16**:457(1951).
6. R. D. Hotchkiss, in: "The Chemical Basis of Heredity" (W. D. McElroy and B. Glass, eds.), Johns Hopkins Press, Baltimore, Md., 1956, p. 321.
7. S. Zamenhof, in: "The Chemical Basis of Heredity" (W. D. McElroy and B. Glass, eds.), Johns Hopkins Press, Baltimore, Md., 1956, p. 351.
8. B. S. Strauss, "Chemical Genetics," W. B. Saunders Co., Philadelphia, 1960.
9. R. E. Hartman, in: "The Chemical Basis of Heredity" (W. D. McElroy and B. Glass, eds.), Johns Hopkins Press, Baltimore, Md., 1956, p. 408.
10. F. Jacob and E. L. Wollman, in: "The Chemical Basis of Heredity" (W. D. McElroy and B. Glass, eds.), Johns Hopkins Press, Baltimore, Md., 1956, p. 468.
11. G. S. Stent, "Molecular Biology of Bacterial Viruses," W. H. Freeman, San Francisco, 1963.
12. R. M. Herriott, in: "The Chemical Basis of Heredity" (W. D. McElroy and B. Glass, eds.), Johns Hopkins Press, Baltimore, Md., 1956, p. 399.

13. H. Fraenkel-Conrat, The Role of the Nucleic Acid in the Reconstitution of Active Tobacco Mosaic Virus, *J. Am. Chem. Soc.* **78**:882(1956).

14. A. Gierer and G. Schramm, Infectivity of Ribonucleic Acid from Tobacco Mosaic Virus, *Nature* **177**(4511):702(1956).

15. H. Fraenkel-Conrat, B. A. Singer, and R. C. Williams, in: "The Chemical Basis of Heredity" (W. D. McElroy and B. Glass, eds.), Johns Hopkins Press, Baltimore, Md., 1956, p. 501.

16. M. H. F. Wilkins, W. E. Seeds, A. R. Stokes, and H. R. Wilson, Helical Structure of Crystalline Deoxypentose Nucleic Acid, *Nature* **172**(4382):759(1953).

17. R. E. Franklin and R. G. Gosling, *Nature* **172**:156(1953); Molecular Structure of Nucleic Acids—Molecular Configuration in Sodium Thymonucleate, *Nature* **171**(4356): 740(1953).

18. J. D. Watson and F. H. C. Crick, Molecular Structure of Nucleic Acids—A Structure for Deoxyribose Nucleic Acid, *Nature* **171**(4356):737(1953); Genetical Implications of the Structure of Deoxyribonucleic Acid, *Nature* **171**(4361):964(1953).

19. F. H. C. Crick, The Structure of Hereditary Material, *Sci. Am.* **191**(4):54(1954).

20. F. H. C. Crick, in: "The Chemical Basis of Heredity" (W. D. McElroy and B. Glass, eds.), Johns Hopkins Press, Baltimore, Md., 1956, p. 532.

21. J. Josse, in: "Proceedings of the Fifth International Congress of Biochemistry," Vol. 1: Biological Structure and Function at the Molecular Level (V. A. Engelhardt, ed.), Pergamon Press, Inc., New York, 1963, p. 79.

22. C. Levinthal and C. A. Thomas, Jr., Molecular Autoradiography: The β-Ray Counting from Single Virus Particles and DNA Molecules in Nuclear Emulsions, *Biochim. Biophys. Acta* **23**:453(1957).

23. A. Rich, Molecular Structure of Nucleic Acids, *Rev. Mod. Phys.* **31**:191(1959).

24. V. Luzzati, A. Nicolaieff, and F. Masson, Structure de l'acide désocyribonucléique en solution: Étude par diffusion des rayons X aux petits angles, *J. Mol. Biol.* **3**(2):185 (1961).

25. P. Doty, Configurations of Biologically Important Macromolecules in Solution, *Rev. Mod. Phys.* **31**:107(1959).

26. O. B. Ptitsyn and Yu. E. Eizner, *Vysokomolekul. Soedin.* **3**:1863(1961).

27. J. E. Hearst and W. H. Stockmayer, Sedimentation Constants of Broken Chains and Wormlike Coils, *J. Chem. Phys.* **37**:1425(1962).

28. C. L. Sadron, in: "The Nucleic Acids: Chemistry and Biology," Vol. 3 (E. Chargaff and J. N. Davidson, eds.), Academic Press, Inc., New York, 1960, p. 1.

29. O. B. Ptitsyn and B. A. Fedorov, *Dokl. Akad. Nauk S.S.S.R.,* 1965.

30. M. Meselson, F. W. Stahl, and J. Vinograd, Equilibrium Sedimentation of Macromolecules in Density Gradients, *Proc. Natl. Acad. Sci. U.S.* **43**:581(1957).

31. G. H. Haggis, D. Michie, A. R. Muir, K. B. Roberts, and P. M. B. Walker, "Introduction to Molecular Biology," John Wiley and Sons, Inc., New York, 1964.

32. M. Grünberg-Manago and S. Ochoa, Enzymatic Synthesis and Breakdown of Polynucleotides; Polynucleotide Phosphorylase, *J. Am. Chem. Soc.* **77**:3165(1955).

33. J. R. Fresco and B. M. Alberts, The Accommodation of Noncomplementary Bases in Helical Polyribonucleotides and Deoxyribonucleic Acids, *Proc. Natl. Acad. Sci. U.S.* **46**:311(1960).

34. J. R. Fresco, B. M. Alberts, and P. Doty, Some Molecular Details of the Secondary Structure of Ribonucleic Acid, *Nature* **188**(4745):98(1960).

35. A. S. Spirin, "Macromolecular Structure of Ribonucleic Acids," Reinhold Publishing Corp., New York, 1964.

36. M. Spencer, W. Fuller, M. H. Wilkins, and G. L. Brown, Determination of the Helical Configuration of Ribonucleic Acid Molecules by X-ray Diffraction Study of Crystalline Amino Acid–Transfer Ribonucleic Acid, *Nature* **194**(4833):1014(1962).

37. M. Spencer, X-ray Diffraction Studies of the Secondary Structure of RNA, *Cold Spring Harbor Symp. Quant. Biol.* **28**:77(1963).

37a. V. N. Tsvetkov, L. L. Kiselev, L. Yu. Frolova, and S. Ya. Lyubina, *Vysokomolekul. Soedin.* **6**:568(1964); V. N. Tsvetkov, L. L. Kiselev, L. Yu. Frolova, S. Ya. Lyubina, S. I. Klenin, N. A. Nikitin, and V. S. Skazka, Molecular Morphology of Transfer Ribonucleic Acids. Certain Hydrodynamic and Optical Properties of Molecules in Organic Solvents, *Biophysics (U.S.S.R.) (English Transl.)* **9**(3):277(1964).

38. R. W. Holley, J. Apgar, G. A. Everett, J. T. Madison, M. Marquisée, S. H. Merrill, J. R. Penswick, and A. Zamir, Structure of a Ribonucleic Acid, *Science* **147**:1462(1965).

39. J. Marmur, R. Rownd, and C. L. Schildkraut, Denaturation and Renaturation of Deoxyribonucleic Acid, in: "Progress in Nucleic Acid Research," Vol. 1 (J. N. Davidson and W. E. Cohn, eds.), Academic Press, Inc., New York, 1963.

40. M. Meselson and F. W. Stahl, The Replication of DNA in *Escherichia coli, Proc. Natl. Acad. Sci. U.S.* **44**:671(1958).

41. J. Eigner, Ph.D. Dissertation, Harvard University, Cambridge, Mass., 1960.

42. C. A. Thomas, Jr., and K. I. Berns, The Physical Characterization of DNA Molecules released from T2 and T4 Bacteriophage, *J. Mol. Bio.* **3**(3):277(1961).

43. N. Sueoka, J. Marmur, and P. Doty, II: Dependence of the Density of Deoxyribonucleic Acids of Guanine–Cytosine Content, *Nature* **183**(4673):1429(1959).

44. J. H. Gibbs and E. A. DiMarzio, Theory of Helix–Coil Transitions in Polypeptides, *J. Chem. Phys.* **28**:1247(1958); Statistical Mechanics of Helix–Coil Transitions in Biological Macromolecules, *J. Chem. Phys.* **30**:271(1959).

45. S. A. Rice and A. Wada, On a Model of the Helix–Coil Transition in Macromolecules, II, *J. Chem. Phys.* **28**:233(1958).

46. T. L. Hill, Generalization of the One-dimensional Ising Model Applicable to Helix Transitions in Nucleic Acids and Proteins, *J. Chem. Phys.* **30**:383(1959).

47. R. F. Steiner, Hydrogen Ion Titration Curve of a Polynucleotide Capable of Undergoing a Helix-coil Transition, *J. Chem. Phys.* **32**:215(1960).

48. B. H. Zimm, Theory of 'Melting' of the Helical Form in Double Chains of the DNA Type, *J. Chem. Phys.* **33**:1349(1960); S. Lifson and B. H. Zimm, Simplified Theory of the Helix-coil Transition in DNA Based on a Grand Partition Function, *Biopolymers* **1**:15(1963).

49. T. M. Birshtein and O. B. Ptitsyn, "High Polymers," Vol. 22, Conformations of Macromolecules, Interscience Publishers, Inc., New York, 1966.

50. D. Jordan, in: "The Nucleic Acids: Chemistry and Biology," Vol. 1 (E. Chargaff and J. N. Davidson, eds.), Academic Press, Inc., New York, 1955, p. 447.

51. T. M. Birshtein, *Biofizika* **7**:513(1962).

52. B. I. Sukhorukov, Yu. Sh. Moshkovskii, T. M. Birshtein, and V. N. Lystov, Optical Properties and Molecular Structure of Nucleic Acids and Their Components, II, *Biophysics (U.S.S.R.) (English Transl.)* **8**(3):348(1963).

53. O. B. Ptitsyn, *Biofizika* **7**:257(1962).

54. P. Doty, H. Boedtker, J. R. Fresco, R. Haselkorn, and M. Litt, Secondary Structure in Ribonucleic Acids, *Proc. Natl. Acad. Sci. U.S.* **45**:482(1959).

55. M. Beer and C. A. Thomas, Jr., The Electron Microscopy of Phage DNA Molecules with Denatured Regions, *J. Mol. Biol.* **3**(5):699(1961).

56. I. Tinoco, Jr., Hypochromism in Polynucleotides, *J. Am. Chem. Soc.* **82**:4785(1960).

57. A. Rich and I. Tinoco, Jr., The Effect of Chain Length upon Hypochromism in Nucleic Acids and Polynucleotides, *J. Am. Chem. Soc.* **82**:6409(1960).

58. H. DeVoe and I. Tinoco, Jr., The Hypochromism of Helical Polynucleotides, *J. Mol. Biol.* **4**(6):518(1962).

59. E. V. Anufrieva, M. V. Vol'kenshtein, and T. V. Sheveleva, *Biofizika* 7:554(1962).
60. E. V. Anufrieva, M. V. Vol'kenshtein, and T. V. Sheveleva, in: "Molekulyarnaya Biofizika," Nauka, Moscow, 1965.
61. S. I. Vavilov, Mikrostruktura Sveta, *Izd. Akad. Nauk S.S.S.R.,* 1950.
62. P. P. Feofilov, "The Physical Basis of Polarized Emission," Consultants Bureau, New York, 1961.
63. F. Cramer, "Einschlussverbindungen," Springer, Berlin, 1954.
64. V. Luzzati, La Structure de l'acide désoxyribonucleique en solution. Étude par diffusion des rayons X aux petits angles, *J. Chim. Phys.* 58:899(1961).
65. V. I. Permogorov, Yu. S. Lazurkin, and S. Z. Shmurak, An Investigation of Complexes of Nucleic Acid with Acridine, *Dokl. Biophys. Sect. (English Transl.)* 155(6):71(1964).
66. V. I. Permogorov and Yu. S. Lazurkin, Mechanism of Binding of Actinomycin with DNA, *Biophysics (U.S.S.R.) (English Transl.)* 10(1):15(1965).
67. V. I. Permogorov, A. A. Prozorov, M. F. Shemyakin, Yu. S. Lazurkin, and P. T. Khesin, in: "Molekulyarnaya Biofizika," Nauka, Moscow, 1965.
68. M. D. Frank-Kamenetskii, A Theoretical Examination of the Effect of Various Factors on the Thermal Denaturation of DNA, *Dokl. Biophys. Sect. (English Transl.)* 157(1):106(1964).
69. W. Kuhn, Zeitbedarf der Längsteilung von miteinander verzwirnten Fadenmolekülen, *Experientia* 13:301(1957).
70. H. C. Longuet-Higgins and B. H. Zimm, Calculation of the Rate of Uncoiling of the DNA Molecule, *J. Mol. Biol.* 2(1):1(1960).
71. M. Fixman, Rate of Unwinding of DNA, *J. Mol. Biol.* 6(1):39(1963).
72. J. Marmur and P. Doty, Thermal Renaturation of Deoxyribonucleic Acids, *J. Mol. Biol.* 3(5):585(1961).
73. J. Marmur, C. L. Schildkraut, and P. Doty, Biological and physical aspects of reversible denaturation of deoxyribonucleic acids, in: "The Molecular Basis of Neoplasia," University of Texas Press, Austin, Tex., 1962, p. 9.
74. A. N. Belozerskii and A. S. Spirin, in: "Nucleinovye Kisloty," 1962.
75. H. DeVoe and I. Tinoco, Jr., The Stability of Helical Polynucleotides: Base Contributions, *J. Mol. Biol.* 4(6):500(1962).
76. M. V. Vol'kenshtein, "Stroeinie i fizicheskie svoistva molekul," Izd. Akad. Nauk SSSR, 1955; see also, Y. K. Syrkin and M. E. Dyatkina, "Structure of Molecules and the Chemical Bond," Dover Publications, Inc., New York, 1964.
77. M. Delbrück and G. S. Stent, in: "The Chemical Basis of Heredity" (W. D. McElroy and B. Glass, eds.), Johns Hopkins Press, Baltimore, Md., 1956, p. 699.
78. M. Meselson and F. W. Stahl, The Replication of DNA in *Escherichia coli, Proc. Natl. Acad. Sci. U.S.* 44:671(1958).
79. A. A. Prokof'eva-Bel'govskaya and Yu. F. Bogdanov, *Zh. Vses. Khim. Obshchestva im D. I. Mendeleeva* 8(1):33(1963).
80. C. D. Darlington, The Chromosome as a Physico-chemical Entity, *Nature* 176(4494):1139(1955).
81. H. Ris, in: "The Chemical Basis of Heredity" (W. E. McElroy and B. Glass, eds.), Johns Hopkins Press, Baltimore, Md., 1956, p. 23.
82. J. H. Taylor, in: "Selected Papers on Molecular Genetics," Vol. 1 (J. H. Taylor, ed.), Academic Press, Inc., New York, 1965, p. 65.
83. R. L. Sinsheimer, A Single-stranded Deoxyribonucleic Acid from Bacteriophage ϕX174, *J. Mol. Biol.* 1(1):43(1959); Single-stranded DNA, *Sci. Am.* 207(1):109(1962).
84. D. E. Bradley, Some New Small Bacteriophages (ϕX174 Type), *Nature* 195(4841):622 (1962).

85. J. Cairns, The Chromosome of *Escherichia coli, Cold Spring Harbor Symp. Quant. Biol.* **28**:43(1963).

86. A. Kornberg, Biosynthesis of Nucleic Acids, *Rev. Mod. Phys.* **31**:200(1959).

87. A. Kornberg, "Enzymatic Synthesis of DNA," John Wiley and Sons, Inc., New York, 1962.

88. A. Kornberg, L. L. Bertsch, J. F. Jackson, and H. G. Khorana, Enzymatic Synthesis of Deoxyribonucleic Acid, XVI, *Proc. Natl. Acad. Sci. U.S.* **51**:315(1964).

89. M. V. Vol'kenshtein, in: "Proceedings of the Fifth International Congress on Biochemistry, Vol. 1: Biological Structure and Function at the Molecular Level" (V. A. Engelhardt, ed.), Pergamon Press, Inc., New York, 1963, p. 100.

90. M. V. Vol'kenshtein and A. M. Eliashevich, *Dokl Akad. Nauk S.S.S.R.* **131**:538(1960); A Statistical and Thermodynamic Theory of the Reduplication of Desoxyribonucleic Acid (DNA), *Biophysics (U.S.S.R.) (English Transl.)* **6**(5):1(1961).

91. M. V. Vol'kenshtein and A. M. Eliashevich, On the Theory of Mutation, *Dokl. Biol. Sci. Sect. (English Transl.)* **136**(1–6):19(1961).

92. A. G. Pasynskii, The Role of Matrix Structures in Replication, *Biophysics (U.S.S.R.) (English Transl.)* **5**(1):12(1960).

93. J. Adler, I. R. Lehman, M. J. Bessman, E. S. Simms, and A. Kornberg, Enzymatic Synthesis of Deoxyribonucleic Acid, IV, *Natl. Acad. Sci. U.S.* **44**:641(1958).

94. K. C. Atwood, Sequential Deoxyribonucleic Acid Replication, *Science* **132**:617(1960).

95. H. V. Aposhian and A. Kornberg, Enzymatic Synthesis of Deoxyribonucleic Acid, IX, *J. Biol. Chem.* **237**:519(1962).

96. F. J. Bollum, Calf Thymus Polymerase, *J. Biol. Chem.* **235**(8):2399(1960).

97. R. L. Sinsheimer, The Replication of Bacteriophage ϕX174, *J. Chim. Phys.* **58**:986 (1961); R. I. Sinsheimer, B. Starman, C. Nagler, and S. Guthrie, The Process of Infection with Bacteriophage ϕX174, I: Evidence for a "Replican Form," *J. Mol. Biol.* **4**:142(1962).

98. C. Weissman, L. Simon, and S. Ochoa, Induction by an RNA Phage of an Enzyme Catalyzing Incorporation of Ribonucleotides into Ribonucleic Acid, *Proc. Natl. Acad. Sci. U.S.* **49**:407(1963); C. Weissman and P. Borst, Double-stranded Ribonucleic Acid Formation *in vitro* by MS2 Phage-induced RNA Synthetase, *Science* **142**:1188 (1963).

99. R. Rolfe, Changes in the Physical State of DNA during the Replication Cycle, *Proc. Natl. Acad. Sci. U.S.* **49**:386(1963).

100. N. Sueoka and H. Yoshikawa, Regulation of Chromosome Replication in *Bacillus subtilis, Cold Spring Harbor Symp. Quant. Biol.* **28**:47(1963).

101. H. Ris and B. L. Chandler, The Ultrastructure of Genetic Systems in Prokaryotes and Eukaryotes, *Cold Spring Harbor Symp. Quant. Biol.* **28**:1(1963).

102. C. Levinthal and H. R. Crane, On the Unwinding of DNA, *Proc. Natl. Acad. Sci. U.S.* **42**:436(1956).

103. M. V. Vol'kenshtein, N. M. Godzhayev, and Yu. Ya. Gotlib, *Biofizika* **7**:16(1962).

104. M. V. Vol'kenshtein, N. M. Godzhayev, Yu. Ya. Gotlib, and O. B. Ptitsyn, Kinetics of Biosynthesis, *Biophysics (U.S.S.R.) (English Transl.)* **8**(1):1(1963).

105. A. N. Orlov and S. I. Fishman, Kinetics of Reduplication of Chain Molecules, *Dokl. Biol. Sci. Sect. (English Transl.)* **132**(1–6):340(1960).

106. M. V. Vol'kenshtein, Yu. Ya. Gotlib, and O. B. Ptitsyn, *Fiz. Tverd. Tela* **3**:396(1960).

107. H. K. Schachman, J. Adler, C. M. Radding, I. R. Lehman, and A. Kornberg, Enzymatic Synthesis of Deoxyribonucleic Acid, VII, *J. Biol. Chem.* **235**:3242(1960).

108. M. V. Vol'kenstein and S. N. Fishman, Polynucleotide Synthesis at Oligomeric Templates, *Biopolymers* **4**:77(1966).

CHAPTER 7

1. G. Gamow, *Kgl. Danske Videskab. Selskab, Biol. Medd.* **22**(3)(1954).
2. Schrödinger, "What is Life?" Cambridge University Press, New York, 1963.
3. G. Gamow, Possible Relation between Deoxyribonucleic Acid and Protein Structures, *Nature* **173**(4398):318(1954).
4. G. Gamow, A. Rich, and M. Yčas, *Advan. Biol. Med. Phys.* **4**:23(1956).
5. F. H. C. Crick, J. S. Griffith, and L. E. Orgel, Codes without Commas, *Proc. Natl. Acad. Sci. U.S.* **43**:416(1957).
6. H. G. Wittmann, in: "Proceedings of the Fifth International Congress on Biochemistry, Vol. 1, Biological Structure and Function at the Molecular Level" (V. A. Engelhardt, ed.), Pergamon Press, Inc., New York, p. 204.
7. A. Tsugita and H. Fraenkel-Conrat, The Amino Acid Composition and C-terminal Sequence of a Chemically Evoked Mutant of TMV, *Proc. Natl. Acad. Sci. U.S.* **46**:636(1960).
8. M. Yčas, in: "Symposium on Information Theory in Biology" (H. P. Yockey, *et al.*, eds.), Pergamon Press, Inc., New York, 1958, p. 70.
9. L. G. Augenstine, in: "Symposium on Information Theory in Biology" (H. P. Yockey *et al.*, eds.), Pergamon Press, Inc., New York, 1958, p. 103.
10. S. Brenner, On the Impossibility of All Overlapping Triplet Codes in Information Transfer from Nucleic Acids to Proteins," *Proc. Natl. Acad. Sci. U.S.* **43**:687(1957).
11. G. Gamow and M. Yčas, Statistical Correlation of Protein and Ribonucleic Acid Composition, *Proc. Natl. Acad. Sci. U.S.* **41**:1011(1955).
12. A. N. Belozersky and A. S. Spirin, A Correlation between the Compositions of Deoxyribonucleic and Ribonucleic Acids, *Nature* **182**(4628):111(1955).
13. E. Volkin and L. Astrakhan, Phosphorus Incorporation in *Escherichia coli* Ribonucleic Acid after Infection with Bacteriophage T2, *Virology* **2**:1949(1956); Intracellular Distribution of Labeled Ribonucleic Acid after Phage Infection of *Escherichia coli, Virology* **2**:433(1956).
14. E. Volkin and L. Astrakhan, in: "The Chemical Basis of Heredity" (W. D. McElroy and B. Glass, eds.), Johns Hopkins Press, Baltimore, Md., 1956, p. 686.
15. F. Gros and H. H. Hiatt, in: "Proceedings of the Fifth International Congress on Biochemistry, Vol. 1: Biological Structure and Function at the Molecular Level" (V. A. Engelhardt, ed.), Pergamon Press, Inc., New York, 1963, p. 162.
16. A. Rich, A Hybrid Helix Containing Both Deoxyribose and Ribose Polynucleotides and its Relation to the Transfer of Information between the Nucleic Acids, *Proc. Natl. Acad. Sci. U.S.* **46**:1044(1960).
17. B. D. Hall and S. Spiegelman, Sequence Complementarity to T2-DNA and T2-specific RNA, *Proc. Natl. Acad. Sci. U.S.* **47**:137(1961).
18. S. Spiegelman, B. D. Hall, and R. Storck, The Occurrence of Natural DNA–RNA Complexes in *E. coli* Infected with T2, *Proc. Natl. Acad. Sci. U.S.* **47**:1135(1961).
19. H. M. Schulman and D. M. Bonner, A Naturally Occurring DNA–RNA Complex from *Neurospora crassa, Proc. Natl. Acad. Sci. U.S.* **48**:53(1962).
20. S. B. Weiss, Enzymatic Incorporation of Ribonucleoside Triphosphates into the Interpolynucleotide Linkage of Ribonucleic Acid, *Proc. Natl. Acad. Sci. U.S.* **46**:1020(1960).
21. S. Ochoa, D. P. Burma, H. Kröger, and J. D. Weill, Deoxyribonucleic Acid-dependent Incorporation of Nucleotides from Nucleoside Triphosphates into Ribonucleic Acid, *Proc. Natl. Acad. Sci. U.S.* **47**:670(1961).
22. M. Chamberlin and P. Berg, Deoxyribonucleic Acid-directed Synthesis of Ribonucleic Acid by an Enzyme from *Escherichia coli, Proc. Natl. Acad. Sci. U.S.* **48**:81(1962).

23. M. Hayashi, M. N. Hayashi, and S. Spiegelman, Replicating Form of a Single-stranded DNA Virus: Isolation and Properties, *Science* **140**:1313(1963); Restriction of *in vivo* Genetic Transcription to One of the Complementary Strands of DNA, *Proc. Natl. Acad. Sci. U.S.* **50**:664(1963); DNA Circularity and the Mechanism of Strand Selection in the Generation of Genetic Messages, *Proc. Natl. Acad. Sci. U.S.* **51**:351(1964); B. Chandler, M. Hayashi, M. N. Hayashi, and S. Spiegelman, Circularity of the Replicating Form of a Single-stranded DNA Virus, *Science* **143**:47(1964): S. Spiegelman, Hybrid Nucleic Acids, *Sci. Am.* **210**(5):48(1964).

24. G. P. Tocchini-Valentini, M. Stodolsky, A. Aurisicchio, M. Sarnat, F. Graziosi, S. B. Weiss, and E. P. Geiduschek, On the Asymmetry of RNA Synthesis *in vivo*, *Proc. Natl. Acad. Sci. U.S.* **50**:935(1963).

25. J. Marmur and C. M. Greenspan, Transcription *in vivo* of DNA from Bacteriophage SP8, *Science* **142**:387(1963).

26. O. Maaløe and P. C. Hanawalt, Thymine Deficiency and the Normal DNA Replication Cycle. I, *J. Mol. Biol.* **3**(2):144(1961); P. C. Hanawalt, O. Maaløe, D. J. Cummings, and M. Schaechter, The Normal DNA Replication Cycle, II, *J. Mol. Biol.* **3**(2):156 (1961).

27. J. R. Warner, A. Rich, and C. E. Hall, *Science* **142**:399(1963): A. Rich, *Sci. Am.* Dec., 1963, p. 66.

28. J. R. Warner, P. M. Knopf, and A. Rich, A Multiple Ribosomal Structure in Protein Synthesis, *Proc. Natl. Acad. Sci. U.S.* **49**:122(1963).

29. D. Uotson, *Biofizika* **8**:401(1963).

30. H. Noll, T. Staehelin, and F. O. Wettstein, Ribosomal Aggregates Engaged in Protein Synthesis: Ergosome Breakdown and Messenger Ribonucleic Acid Transport, *Nature* **198**(4881):632(1963).

31. M. V. Vol'kenshtein and S. N. Fishman, Protein Synthesis on Polysomes, *Dokl. Biophys. Sect. (English Transl.)* **160**(6):15(1965).

32. F. H. C. Crick, "On Protein Synthesis, The Biological Replication of Macromolecules" (F. K. Sanders, ed.), Cambridge University Press, London, 1958, p. 138.

33. M. B. Hoagland, E. B. Keller, and P. C. Zamecnik, Enzymatic Carboxyl Activation of Amino Acids, *J. Biol. Chem.* **218**:345(1956).

34. A. Meister, Approaches to the Biosynthesis of Proteins, *Rev. Mod. Phys.* **31**:210(1959).

35. M. B. Hoagland, *Brookhaven Symp. Biol.* **40**:12(1959).

36. F. Lipmann, in: "Proceedings of the Fifth International Congress on Biochemistry, Vol. 1: Biological Structure and Function at the Molecular Level" (V. A. Engelhardt, ed.), Pergamon Press, Inc., New York, 1963, p. 121.

37. R. W. Holley, J. Apgar, G. A. Everett, J. T. Madison, S. H. Merrill, and A. Zamir, Chemistry of Amino Acid–Specific Ribonucleic Acids, *Cold Spring Harbor Symp. Quant. Biol.* **28**:555(1963).

38. S. Spiegelman and M. Hayashi, The Present Status of the Transfer of Genetic Information and its Control, *Cold Spring Harbor Symp. Quant. Biol.* **28**:161(1963).

39. C. T. Yu and P. C. Zamecnik, Effect of Bromination on the Biological Activities of Transfer RNA of *Escherichia coli, Science* **144**:856(1964).

40. A. S. Spirin and N. V. Belitsina, *Usp. Sovrem. Biol.* **59**:187(1965).

41. H. M. Dintzis, Assembly of the Peptide Chains of Hemoglobin, *Proc. Natl. Acad. Sci. U.S.* **47**:247(1961).

42. F. H. C. Crick, L. Barnett, S. Brenner, and R. J. Watts-Tobin, General Nature of the Genetic Code for Proteins, *Nature* **192**(4809):1227(1961).

43. S. Brenner, L. Barnett, F. H. C. Crick, and A. Orgel, The Theory of Mutagenesis, *J. Mol. Biol.* **3**(1):121(1961).

44. C. Levinthall, A. Garen, and F. Rothman, in: "Proceedings of the Fifth International Congress on Biochemistry, Vol. 1: Biological Structure and Function at the Molecular Level" (V. A. Engelhardt, ed.), Pergamon Press, Inc., New York, 1963, p. 196.

45. M. W. Nirenberg and H. Matthaei, in: "Proceedings of the Fifth International Congress on Biochemistry, Vol. 1: Biological Structure and Function at the Molecular Level" (V. A. Engelhardt, ed.), Pergamon Press, Inc., New York, 1963, p. 184.

46. J. H. Matthaei and M. W. Nirenberg, Characteristics and Stabilization of DNAase-sensitive Protein Synthesis in *E. coli* Extracts, *Proc. Natl. Acad. Sci. U.S.* **47**:1580(1961).

47. M. W. Nirenberg and J. H. Matthaei, The Dependence of Cell-free Protein Synthesis in *E. coli* upon Naturally Occurring or Synthetic Polyribonucleotides, *Proc. Natl. Acad. Sci. U.S.* **47**:1588(1961).

48. P. Lengyel, J. F. Speyer, and S. Ochoa, Synthetic Polynucleotides and the Amino Acid Code, *Proc. Natl. Acad. Sci. U.S.* **47**:1936(1961).

49. J. F. Speyer, P. Lengyel, C. Basilio, and S. Ochoa, Synthetic Polynucleotides and the Amino Acid Code, II, *Proc. Natl. Acad. Sci. U.S.* **48**:63(1962).

50. P. Lengyel, J. F. Speyer, C. Basilio, and S. Ochoa, Synthetic Polynucleotides and the Amino Acid Code, III, *Proc. Natl. Acad. Sci. U.S.* **48**:282(1962).

51. J. F. Speyer, P. Lengyel, C. Basilio, and S. Ochoa, Synthetic Polynucleotides and the Amino Acid Code, IV, *Proc. Natl. Acad. Sci. U.S.* **48**:441(1962).

52. C. Basilio, A. J. Wahba, P. Lengyel, J. F. Speyer, and S. Ochoa, Synthetic Polynucleotides and the Amino Acid Code, V, *Proc. Natl. Acad. Sci. U.S.* **48**:613(1962).

53. R. S. Gardner, A. J. Wahba, C. Basilio, R. S. Miller, P. Lengyel, and J. F. Speyer, Synthetic Polynucleotides and the Amino Acid Code, VII, *Proc. Natl. Acad. Sci. U.S.* **48**:2087(1962).

54. A. J. Wahba, R. S. Gardner, C. Basilio, R. S. Miller, J. F. Speyer, and P. Lengyel, Synthetic Polynucleotides and the Amino Acid Code, VIII, *Proc. Natl. Acad. Sci. U.S.* **49**:116(1963).

55. A. J. Wahba, R. S. Miller, C. Basilio, R. S. Gardner, P. Lengyel, and J. F. Speyer, Synthetic Polynucleotides and the Amino Acid Code, IX, *Proc. Natl. Acad. Sci. U.S.* **49**:880(1963).

56. J. H. Matthaei, O. W. Jones, R. G. Martin, and M. W. Nirenberg, Characteristics and Composition of RNA Coding Units, *Proc. Natl. Acad. Sci. U.S.* **48**:666(1962).

57. O. W. Jones, Jr., and M. W. Nirenberg, Qualitative Survey of RNA Codewords, *Proc. Natl. Acad. Sci. U.S.* **48**:2115(1962).

58. J. F. Speyer, P. Lengyel, C. Basilio, A. J. Wahba, R. S. Gardner, and S. Ochoa, Synthetic Polynucleotides and the Amino Acid Code, *Cold Spring Harbor Symp. Quant. Biol.* **28**:559(1963).

59. M. W. Nirenberg, O. W. Jones, P. Leder, B. F. C. Clark, W. S. Sly, and S. Pestka, On the Coding of Genetic Information, *Cold Spring Harbor Symp. Quant. Biol.* **28**:549(1963).

60. E. L. Smith, Nucleotide Base Coding and Amino Acid Replacements in Proteins, I and II, *Proc. Natl. Acad. Sci. U.S.* **48**:677, 859(1962).

61. A. Tsugita, H. Fraenkel-Conrat, M. W. Nirenberg, and J. H. Matthaei, Demonstration of the Messenger Role of Viral RNA, *Proc. Natl. Acad. Sci. U.S.* **48**:846(1962).

62. N. Sueoka, Compositional Correlation between Deoxyribonucleic Acid and Protein, *Cold Spring Harbor Symp. Quant. Biol.* **26**:35(1961).

63. N. Sueoka, On the Genetic Basis of Variation and Heterogeneity of DNA Base Composition, *Proc. Natl. Acad. Sci. U.S.* **48**:582(1962).

64. M. V. Vol'kenshtein, Some Implications of the Genetic Code, *Biophysics (U.S.S.R.)* (*English Transl.*) **8**(3):457(1963).

65. F. Lanni, Biological Validity of Amino Acid Codes Deduced with Synthetic Ribo-nucleotide Polymers, *Proc. Natl. Acad. Sci. U.S.* **48**:1623(1962).

66. J. B. Weinstein and A. N. Schechter, Polyuridylic Acid Stimulation of Phenylalanine Incorporation in Animal Cell Extracts, *Proc. Natl. Acad. Sci. U.S.* **48**:1686(1962).

67. M. Nirenberg, P. Leder, M. Bernfield, R. Brimacombe, J. Trupin, F. Rottman, and C. O'Neal, RNA Codewords and Protein Synthesis, VII, *Proc. Natl. Acad. Sci. U.S.* **53**:1161(1965).

68. D. S. Jones, S. Nishimura, and H. G. Khorana, Studies on Polynucleotides, LVI, *J. Mol. Biol.* **16**(2):454(1966).

69. F. H. C. Crick, The Genetic Code, III, *Sci. Am.* **215**(4):55(1966).

70. The Genetic Code, *Cold Spring Harbor Symp. Quant. Biol.* **31** (1966).

71. S. R. Pelc, Correlation between Coding-Triplets and Amino Acids, *Nature* **207**(4997): 597(1965).

72. Yu. B. Rumer, Systematization of the Codons of the Genetic Code, *Dokl. Biochem. Sect.* (*English Transl.*) **167**:165(1966).

73. M. V. Vol'kenshtein and Yu. B. Rumer, *Biofizika* **12**:10(1967).

74. M. V. Vol'kenshtein, *Genetika* **2**:54(1965).

75. E. Margoliash, Amino Acid Sequence of Cytochrome *c* in Relation to its Function and Evolution, *Can. J. Biochem. Physiol.* **42**:745(1964).

76. M. V. Vol'kenshtein, Coding of Polar and Non-polar Amino Acids, *Nature* **207**(4994): 294(1965).

77. M. V. Vol'kenshtein, The Genetic Coding of Protein Structure, *Biochim. Biophys. Acta* **119**:421(1966).

78. E. Freese, The Specific Mutagenic Effect of Base Analogues on Phage T4, *J. Mol. Biol.* **1**(2):87(1959).

79. E. Freese, in: "Proceedings of the Fifth International Congress on Biochemistry, Vol 1: Biological Structure and Function at the Molecular Level" (V. A. Engelhardt, ed.), Pergamon Press, Inc., New York, 1963, p. 204.

80. D. F. Bradley and M. K. Wolf, Aggregation of Dyes Bound to Polyanions, *Proc. Natl. Acad. Sci. U.S.* **45**:944(1959).

81. K. Hoogsteen, The Structure of Crystals Containing a Hydrogen-bonded Complex of 1-Methylthymine and 9-Methyladenine, *Acta Cryst.* **12**:822(1959); The Crystal and Molecular Structure of a Hydrogen-bonded Complex between 1-Methylthymine and 9-Methyladenine, *Acta Cryst.* **16**:907(1963).

82. F. S. Mathews and A. Rich, The Molecular Structure of a Hydrogen-bonded Complex of *N*-Ethyl Adenine and *N*-Methyl Uracil, *J. Mol. Biol.* **8**(1):89(1964).

83. A. E. V. Haschmeyer and H. M. Sobell, The Crystal Structure of an Intermolecular Nucleoside Complex: Adenosine and 5-Bromouridine, *Proc. Natl. Acad. Sci. U.S.* **50**:872(1963).

84. L. Katz, K.-I. Tomita, and A. Rich, The Molecular Structure of the Crystalline Complex Ethyladenine: Methylbromouracil, *J. Mol. Biol.* **13**(2):340(1965).

85. Yu. Baklagina, M. V. Vol'kenshtein, and Yu. D. Kondrashev, *Zh. Strukt. Khim.* **7**:399 (1966).

86. A. Gierer and W.-K. Mundry, Production of Mutants of Tobacco Mosaic Virus by Chemical Alteration of its Ribonucleic Acid *in vitro*, *Nature* **182**(4647):1457(1958); *Z. Vererbungslehre* **89**:614(1958).

87. A. Gierer, in: "Proceedings of the Fifth International Congress on Biochemistry, Vol. 3: Evolutionary Biochemistry" (A. I. Oparin, ed.), Pergamon Press, Inc., New York, 1963, p. 189.

88. A. Gierer and G. Schramm, Die Infektiosität der Nucleinsäure aus Tabakmosaikvirus, *Z. Naturforsch* **11b**:138(1956).

89. H. Schuster and G. Schramm, Bestimmung der biologisch wirksamen Einheit in der Ribosenucleinsäure des Tabakmosaikvirus auf chemischem Wege, Z. *Naturforsch* **13b**:697(1958).

90. E. Kramer and H.-G. Wittmann, Elektrophoretische Untersuchungen der A-Proteine dreier Tabakmosaikvirus-Stämme, Z. *Naturforsch* **13b**:30(1958).

91. E. Freese, in: "Selected Papers in Molecular Genetics," Vol. 1 (J. H. Taylor, ed.), Academic Press, Inc., New York, 1963, p. 207.

92. J. E. Davies, Studies on the Ribosomes of Streptomycin-sensitive and -resistant Strains of *Escherichia coli*, *Proc. Natl. Acad. Sci. U.S.* **51**:659(1964).

93. E. C. Cox, J. R. White, and J. G. Flaks, Streptomycin Action and the Ribosome, *Proc. Natl. Acad. Sci. U.S.* **51**:703(1964).

94. J. Davies, W. Gilbert, and L. Gorini, Streptomycin, Suppression, and the Code, *Proc. Natl. Acad. Sci. U.S.* **51**:883(1964).

95. W. Szer and S. Ochoa, Complexing Ability and Coding Properties of Synthetic Polynucleotides, *J. Mol. Biol.* **8**(5):823(1964).

96. D. S. Hogness, Induced Enzyme Synthesis, *Rev. Mod. Phys.* **31**:256(1959).

97. B. Rotman and S. Spiegelman, On the Origin of the Carbon in the Induced Synthesis β-Galactosidase in *Escherichia coli*, *J. Bacteriol.* **68**:419(1954).

98. F. Jacob and J. Monod, in: "Proceedings of the Fifth International Congress on Biochemistry, Vol. 1: Biological Structure and Function at the Molecular Level" (V. A. Engelhardt, ed.), Pergamon Press, Inc., New York, 1963, p. 132.

99. F. Jacob and J. Monod, in: "Biological Organization at the Cellular and Supercellular Level" (R. J. C. Harris, ed.), Academic Press, Inc., New York, 1963, p. 1.

100. O. Maaløe, Role of Protein Synthesis in the DNA Replication Cycle in Bacteria, *J. Cellular Comp. Physiol.* **62**, *Suppl.* **1**:31(1963).

101. J. Jacob, S. Brenner, and F. Cuzin, On the Regulation of DNA Replication in Bacteria, *Cold Harbor Spring Symp. Quant. Biol.* **28**:329(1963).

102. R. B. Khesin and M. F. Shemyakin, Some Properties of Informational Ribonucleic Acids and Their Complexes with Desoxyribonucleic Acids, *Biochemistry* (*U.S.S.R.*) (*English Transl.*) **27**(5):647(1963).

103. R. B. Khesin, M. F. Shemyakin, Zh. M. Gorlenko, S. L. Bogdanova, and T. P. Afanas'eva, RNA Polymerase in *Escherichia coli* B Infected with T2 Phase, *Biochemistry* (*U.S.S.R.*) (*English Transl.*) **27**(6):929(1963); R. B. Khesin, Zh. M. Gorlenko, M. F. Shemyakin, I. A. Bass, and A. A. Prozorov, Connection between Protein Synthesis and Regulation of Formation of Messenger RNA in *E. coli* B cells during Development of T2 phage, *Biochemistry* (*U.S.S.R.*) (*English Transl.*) **28**(16):798(1963).

104. V. G. Allfrey and A. E. Mirsky, Mechanism of Synthesis and Control of Protein and Ribonucleic Acid Synthesis in the Cell Nucleus, *Cold Spring Harbor Symp. Quant. Biol.* **28**:247(1963).

105. V. G. Allfrey, R. Faulkner, and A. E. Mirsky, Acetylation and Methylation of Histones and Their Possible Role in the Regulation of RNA Synthesis, *Proc. Natl. Acad. Sci. U.S.* **51**:786(1964).

106. T. M. Sonneborn, The Differentiation of Cells, *Proc. Natl. Acad. Sci. U.S.* **51**:915(1964).

107. W. Beerman and U. Clever, Chromosome Puffs, *Sci. Am.* **210**(4):50(1964).

108. M. Fischberg, J. B. Gurdon, and T. R. Elsdale, Nuclear Transfer in Amphibia and the Problem of the Potentialities of the Nuclei of Differentiating Tissues, *Exptl. Cell. Res. Suppl.* **6**:161(1959).

109. J. H. Taylor, ed., "Selected Papers on Molecular Genetics," Academic Press, Inc., New York, 1965.

110. R. G. Butenko, "Kultura izolirovannykh tkanei i fiziologiya morfogeneza rastenii," Nauka, Moscow, 1964.

111. F. C. Steward, The Control of Growth in Plant Cells, *Sci. Am.* **209**(4):104(1963).
112. Z. Simon, Multi-steady-state Model for Cell Differentiation, *J. Theor. Biol.* **8**(2):258 (1965).
113. F. Heinmetz, in: "Electronic Aspects of Biochemistry" (B. Pullman, ed.), Academic Press, Inc., New York, 1964.
114. D. S. Chernavskii, L. N. Grigorov, and M. S. Polyakova, *Molek. Biol.* **1** (1967).

CHAPTER 8

1. W. C. Boyd, "Fundamentals of Immunology," 3rd ed., Interscience Publishers, Inc., New York, 1956.
2. C. A. Villee, Jr., "Biology," 4th ed., W. B. Saunders Co., Philadelphia, 1962.
3. J. Neel and W. J. Schull, "Human Heredity," University of Chicago Press, Chicago, 1954.
4. M. Samter and H. L. Alexander, "Immunological Diseases," Little, Brown and Co., Boston, 1965.
5. K. Landsteiner, "The Specificity of Serological Reactions," 2nd ed., Dover Publications, Inc., New York.
6. O. Westphal, Die Struktur der Antigene und das Wesen der immunologischen Spezifitat, *Naturwiss,* **46**(2):50(1959).
7. D. Pressman and L. A. Sternberger, The Nature of the Combining Sites of Antibodies. The Specific Protection of the Combining Site by Hapten during Iodination, *J. Immunol.* **66**:609(1951).
8. W. Kauzmann, Chemical Specificity in Biological Systems, *Rev. Mod. Phys.* **31**:549 (1959).
9. J. R. Marrack and E. S. Orlans, Steric Factors in Immunochemistry, in: "Progress in Stereochemistry," Vol. 2 (W. Klyne and P. B. D. De La Mare, eds.), Butterworth and Co. [Publishers], Ltd., London, 1958, p. 228.
10. L. Pauling, A Theory of the Structure and Process of Formation of Antibodies, *J. Am. Chem. Soc.* **62**:2643(1940); Nature of Forces between Large Molecules of Biological Interest, *Nature* **161**(4097):707(1943).
11. L. Pauling, D. H. Campbell, and D. Pressman, The Nature of the Forces between Antigen and Antibody and of the Precipitation Reaction, *Physiol. Rev.* **23**:203(1943).
12. L. Pauling, D. Pressman, and A. L. Grossberg, The Serological Properties of Simple Substances, VII, *J. Am. Chem. Soc.* **66**:784(1944).
13. D. Pressman, J. T. Maynard, A. L. Grossberg, and L. Pauling, The Serological Properties of Simple Substances, V, *J. Am. Chem. Soc.* **65**:728(1943).
14. D. Pressman, A. B. Pardee, and L. Pauling, The Reactions of Antisera Homologous to Various Azophenylarsonic Acid Groups and the *p*-Azophenylmethylarsinic Acid Group with Some Heterologous Haptens, *J. Am. Chem. Soc.* **67**:1602(1945).
15. D. Pressman, Molecular Complementariness in Antigen–Antibody Systems, in: "Molecular Structure and Biological Specificity" (L. Pauling and H. A. Itano, eds.), Stechert, Washington, D.C., 1957, p. 1.
16. S. J. Singer, Physical-chemical Studies on the Nature of Antigen–Antibody Reactions, *J. Cellular Comp. Physiol.* **50**, *Suppl.* **1**:51(1957).
17. S. I. Epstein, P. Doty, and W. C. Boyd, A Thermodynamic Study of Hapten–Antibody Association, *J. Am. Chem. Soc.* **78**:3306(1956).
18. F. Karush, Immunologic Specificity and Molecular Structure, in: "Advances in Immunology," Vol. 2 (W. H. Taliaferro and J. H. Humphrey, eds.), Academic Press, Inc., New York, 1965, p. 1.

19. J. B. Fleischman, R. R. Porter, and E. M. Press, The Arrangement of the Peptide Chain in γ-Globulin, *Biochem. J.* **88**:220(1963).

20. F. Haurowitz, "Chemistry and Function of Proteins," 2nd ed., Academic Press, Inc., New York, 1963.

21. F. Haurowitz, Nature and Formation of Antibodies, in: "Molecular Structure and Biological Specificity" (L. Pauling and H. A. Itano, eds.), Stechert, Washington, D.C., 1957, p. 18.

22. F. Haurowitz, The Theories of Antibodies Formation, in: "The Nature and Significance of the Antibody Respose," Columbia University Press, New York, 1953.

23. C. Milstein, Variations in Amino Acid Sequence near the Disulphide Bridges of Bence-Jones Proteins, *Nature* **209**(5021):370(1966).

24. F. Haurowitz, Antibody-Formation and the Coding Problem, *Nature* **205**(4974):847 (1965).

25. F. M. Burnett, "The Integrity of the Body," Harvard University Press, Cambridge, Mass., 1962.

26. N. K. Jerne, The Natural-selection Theory of Antibody Formation, *Proc. Natl. Acad. Sci. U.S.* **41**:849(1955).

27. L. Szillard, The Control of the Formation of Specific Proteins in Bacteria and in Animal Cells, *Proc. Natl. Acad. Sci. U.S.* **46**:277(1960); also, The Molecular Basis of Antibody Formation, *Proc. Natl. Acad. Sci. U.S.* **46**:293(1960).

28. V. P. Efraimson, *Zh. Vses. Khim. Obshchestva im. D. I. Mendeleeva* **6**(3):314(1961).

29. J. Lederberg, Genes and Antibodies, *Science* **129**:1649(1959).

30. P. Gross, J. Coursaget, and M. Macheboeur, *Bull. Soc. Chem. Biol.* **34**:1070(1952).

31. H. Green and H. S. Anker, On the Synthesis of Antibody Protein, *Biochim. Biophys. Acta* **13**:365(1954).

32. R. Spiers and E. Speirs, *J. Immunol.* **90**:561(1963).

33. R. S. Speirs, How Cells Attack Antigens, *Sci. Am.* **210**(2):58(1964).

34. G. J. V. Nossal, A. Szenberg, G. L. Ada, and C. M. Austin, Single Cell Studies on 19S Antibody Production, *J. Exptl. Medicine* **119**:485(1964).

35. G. J. V. Nossal, How Cells Make Antibodies, *Sci. Am.* **211**(6):106(1964).

36. W. C. Boyd, "Introduction to Immunochemical Specificity," Interscience Publishers, Inc., New York, 1962.

37. G. C. Bond, "Catalysis by Metals," Academic Press, Inc., New York, 1962.

38. A. A. Balandin, *Zh. Fiz. khim.* **31**:745(1957); *Usp. khim.* **31**:1265(1962); Multipletnaya Teoriya Kataliza, *Izd. MGU*, 1963.

39. M. Dixon and E. C. Webb, "Enzymes," 2nd ed., Academic Press, Inc., New York, 1964.

40. C. Walter, "Enzyme Kinetics," The Ronald Press Company, New York, 1966.

41. H. G. Bray and K. White, "Kinetics and Thermodynamics in Biochemistry," Academic Press, Inc., New York, 1957.

42. R. A. Alberty, Mechanisms of Enzyme Action, *Rev. Mod. Phys.* **31**:177(1959).

43. B. Chance and D. Herbert, The Enzyme–Substrate Compounds of Bacterial Catalase and Peroxides, *Biochem. J.* **46**:402(1950).

44. V. Massey, Studies on Fumarase, *Biochem. J.* **53**:72(1953).

45. R. Lumry, Some Aspects of the Thermodynamics and Mechanism of Enzymic Catalysis, in: "The Enzymes," Vol. 1 (P. D. Boyer, ed.), Academic Press, Inc., New York, 1959.

46. R. A. Alberty, W. G. Miller, and H. F. Fisher, Studies of the Enzyme Fumarase, VI, *J. Am. Chem. Soc.* **79**:3973(1957).

47. M. V. Vol'kenshtein and B. N. Goldstein, A New Method for Solving the Problems of the Stationary Kinetics of Enzymological Reactions, *Biochim. Biophys. Acta* **115**:471 (1966).

48. M. V. Vol'kenshtein, B. Goldstein, and V. Stefanov, *Molek. Biologiya* 1:52(1967)

49. C. B. Anfinsen, in: "Proceedings of the Fifth International Congress on Biochemistry, Vol 4: Molecular Basis of Enzyme Action and Inhibition" (P. Desnuelle, ed.), Pergamon Press, Inc., New York, 1963, p. 66.

50. C. B. Anfinsen, "The Molecular Basis of Evolution," John Wiley and Sons, Inc., New York, 1959.

51. W. H. Stein and S. Moore, in: "Proceedings of the Fifth International Congress on Biochemistry, Vol. 4: Molecular Basis of Enzyme Action and Inhibition" (P. Desnuelle, ed.), Pergamon Press, Inc., New York, 1963, p. 33.

52. K. Linderstrøm-Lang and J. Schellman, Protein Structure and Enzyme Activity, in: "The Enzymes," Vol. 1 (P. D. Boyer, ed.), Academic Press, Inc., New York, 1959.

53. V. N. Orekhovich, in: "Aktualnye Voprosy Sovremennoi Biokhimii," Medgiz, 1962.

54. D. E. Koshland, Jr., Application of a Theory of Enzyme Specificity to Protein Synthesis, *Proc. Natl. Acad. Sci. U.S.* 44:98(1958); Mechanisms of Transfer Enzymes, in: "The Enzymes," Vol. 1 (P. D. Boyer, ed.), Academic Press, Inc., New York, 1959.

55. D. E. Koshland, Jr., The Role of Flexibility in Enzyme Action, *Cold Spring Harbor Symp. Quant. Biol.* 28:473(1963).

56. A. E. Braunshtein, in: "Aktualyne Voprosy Sovremennoi Biokhimii," Medgiz, Moscow, 1962.

57. C. G. Swain and J. F. Brown, Jr., Concerted Displacement Reactions, VIII, *J. Am. Chem. Soc.* 74:2538(1952).

58. F. Karush, Heterogeneity of the Binding Sites of Bovine Serum Albumin, *J. Am. Chem. Soc.* 72:2705(1950).

59. H. Gutfreund and J. M. Sturtevant, The Mechanism of Chymotrypsin-catalyzed Reactions, *Proc. Natl. Acad. Sci. U.S.* 42:719(1956): also, The Mechanism of the Reaction of Chymotrypsin with *p*-Nitrophenyl Acetate, *Biochem. J.* 63:656(1956).

60. B. Labouesse, B. H. Havsteen, and G. P. Hess, Conformational Changes in Enzyme Catalysis, *Proc. Natl. Acad. Sci. U.S.* 48:2137(1962).

61. I. A. Bolotina, M. V. Vol'kenshtein, and O. Chikalova-Luzina, A Study of Conformational Changes in α-Chymotrypsin by the Rotatory Dispersion Method, *Biochemistry (U.S.S.R.) (English Transl.)* 31(2):210(1966).

62. Yu. M. Torchinskii and L. G. Koreneva, Anomalous Optical Rotatory Dispersion of Cardiac Aspartic–Glutamic Transaminase, *Biochemistry (U.S.S.R.) (English Transl.)* 28(6):812(1963); Yu. M. Torchinskii, The Interaction of Mercaptide-forming Reagents with Heart Aspartic–Glutamic Transaminase, *Biochemistry (U.S.S.R.) (English Transl.)* 29(3):458(1964).

63. D. E. Koshland, Jr., J. A. Yankeelov, Jr., and J. A. Thoma, Specificity and Catalytic Power in Enzyme Action, *Federation Proc.* 21(6):1031(1962).

64. J. A. Yankeelov, Jr., and D. E. Koshland, Jr., Evidence for Conformational Changes Induced by Substrates of Phosphoglucomutase, *J. Biol. Chem.* 240:1593(1965).

65. I. A. Bolotina, D. S. Markovich, M. V. Vol'kenshtein, and P. Zavodzky, Investigation of the Conformation of D-Glyceraldehyde-3-phosphate Dehydrogenase, *Biochim. Biophys. Acta* 132:260(1967).

66. I. A. Bolotina, D. S. Markovich, M. V. Vol'kenshtein, and P. Zavodzky, Investigation of the Conformation of Lactate Dehydrogenase and of Its Catalytic Activity, *Biochim. Biophys. Acta* 132:271(1967).

67. J. G. Kirkwood and J. B. Shumaker, The Influence of Dipole Moment Fluctuations on the Dielectric Increment of Proteins in Solution, *Proc. Natl. Acad. Sci. U.S.* 38:855 (1952).

68. M. V. Vol'kenshtein and S. N. Fishman, I. Theory of Effect of Ionization on the α-Helix Content of a Polypeptide, *Biophysics (U.S.S.R.) (English Transl.)* 11(6):1096(1966).

69. L. N. Johnson and D. C. Phillips, Structure of Some Crystalline Lysozyme-inhibitor Complexes Determined by X-ray Analysis at 6 Å Resolution, *Nature* 206(4986):761 (1965); *"Abstracts of the Seventh International Congress on Crystallography,"* Moscow, 1966.

70. M. V. Vol'kenshtein, Theory of Enzymatic Hydrolysis of Biopolymers, *Dokl. Biochem. Sect.* 160–161:4(1965).

71. M. V. Vol'kenshtein, Muscular Activity, *Dokl. Biol. Sci. Sect.* (*English Transl.*) 141(1–6): 988(1963).

72. J. Kirkwood, Kinetics and Mechanisms, II, *Discussions Faraday Soc.* 20:3(1955); also, in: "The Mechanism of Enzyme Actions," The Johns Hopkins Press, Baltimore, Md., 1954.

73. A. E. Braunshtein, *Zh. Vses. Khim. Obshchesteva im. I. D. Mendeleeva* 8:(1):81(1963).

74. H. Mark, N. G. Gaylord, and N. M. Bikales, "Encyclopedia of Polymer Science and Technology," Interscience Publishers, Inc., New York, 1964.

75. J. Wyman, Jr., Linked Functions and Reciprocal Effects in Hemoglobin: A Second Look, in: "Advances in Protein Chemistry," Vol. 19 (C. B. Anfinsen, Jr., M. L. Anson, J. T. Edsall, and F. M. Richards, eds.), Academic Press, New York, 1964, p. 223.

76. M. V. Vol'kenshtein, in "Molekulyarnaya Biofizika," Nauka, Moscow, 1965.

77. A. Novick and L. Szilard, "Experiments with a Chemostat on the Rates of Aminoacid Synthesis in Bacteria," Princeton University Press, Princeton, N.J., 1954.

78. H. E. Umbarger, Evidence for a Negative Feedback Mechanism in the Biosynthesis of Isoleucine, *Science* 123:848(1956); Feedback Control by Endproduct Inhibition, *Cold Spring Harbor Symp. Quant. Biol.* 26:301(1961).

79. J. C. Gerhart and A. B. Pardee, The Effect of the Feedback Inhibitor, *CTP*, on Subunit Interactions in Aspartate Transcarbamylase, *Cold Spring Harbor Symp. Quant. Biol.* 28:491(1963).

80. J. Monod, J. Wyman, and J.-P. Changeux, On the Nature of Allosteric Transitions: A Plausible Model, *J. Mol. Biol.* 12:88(1965).

81. M. V. Vol'kenshtein and B. N. Goldstein, Allosteric Enzyme Models and Their Analysis by the Theory of Graphs, *Biochim. Biophys. Acta* 115:478(1966).

82. L. Pauling, The Oxygen Equilibrium of Hemoglobin and Its Structural Interpretation, *Proc. Natl. Acad. Sci. U.S.* 21:186(1935).

83. A. Rossi-Fanelli, E. Antonini, and A. Caputo, Hemoglobin and Myoglobin, in: "Advances in Protein Chemistry," Vol. 19 (C. B. Anfinsen, Jr., M. L. Anson, J. T. Edsall, and F. M. Richards, eds.), Academic Press, Inc., New York, 1964, p. 73.

84. J. Wyman, Jr., and D. W. Allen, The Problem of the Heme Interactions in Hemoglobin and the Basis of the Bohr Effect, *J. Polymer Sci.* 7(5):499(1951).

85. R. Benesch and Ruth Benesch, Some Relations between Structure and Function in Hemoglobin, *J. Mol. Biol.* 6(5):498(1963).

86. J. Wyman, Allosteric Effects in Hemoglobin, *Cold Spring Harbor Symp. Quant. Biol.* 28:483(1963).

87. H. Muirhead and M. F. Perutz, Structure of Haemoglobin, *Nature* 199(4894):633(1963); Structure of Reduced Human Hemoglobin, *Cold Spring Harbor Symp. Quant. Biol.* 28:451(1963).

88. M. V. Vol'kenshtein and A. K. Shemelin, Investigations of Spiralization of Haemoglobin by Optical Rotary Dispersion, *Biophysics* (*U.S.S.R.*) (*English Transl.*) 11(5):889 (1966).

89. M. V. Vol'kenshtein, J. A. Sharonov, and A. K. Shemelin, Anomalous Dispersion of the Faraday Effect in Haemoglobin and Myoglobin, *Nature* 209(5024):709(1966).

90. D. E. Atkinson, Biological Feedback Control at the Molecular Level, *Science* 150(3697): 851(1965).

91. R. B. Drysdale and A. R. Peacocke, The Molecular Basis of Heredity, *Biol. Rev.* **36**:537 (1961).

92. M. H. F. Wilkins, Physical Studies of the Molecular Structure of Deoxyribose Nucleic Acid and Nucleoprotein, *Cold Spring Harbor Symp. Quart. Biol.* **21**:75(1956); Molecular Structure of Deoxyribose Nucleic Acid and Nucleoprotein and Possible Implications in Protein Synthesis, *Biochem. Soc. Symp.* **14**:13(1956).

93. G. Zubay and P. Doty, The Isolation and Properties of Deoxyribonucleoprotein Particles Containing Single Nucleic Acid Molecules, *J. Mol. Biol.* **1**(1):1(1959).

94. F. H. C. Crick and J. D. Watson, Structures of Small Viruses, *Nature* **177**(4506):473 (1956).

95. R. W. Horne, The Structure of Viruses, *Sci. Am.* **208**(1):48(1963).

96. A. S. Spirin, N. A. Kiselev, R. S. Shakulov, and A. A. Bogdanov, Study of the Structure of the Ribosomes; Reversible Unfolding of the Ribosome Particles in Ribonucleoprotein Strands and a Model of the Packing, *Biochemistry* (*U.S.S.R.*) (*English Transl.*) **28**(5):765(1963).

97. P. Jordan, *Z. Physik.* **39**:711(1938); Über quantenmechanische Resonzanziehung und über das Problem der Immunitätsreaktionen, *Z. Physik.* **113**:431(1939); Über die Spezifität van Antikörpern, Fermenten, Viren, Genem, *Naturwiss.* **29**(7):89(1941).

98. P. Jordan, "Eiweissmoleküle," Stuttgart, 1947.

99. W. H. Stockmayer, Forces between Macromolecules, *Rev. Mod. Phys.* **31**:103(1959).

100. L. Pauling and M. Delbrück, The Nature of the Intermolecular Forces Operative in Biological Processes, *Science* **92**:77(1940).

101. H. Jehle, Specificity of Interaction between Identical Molecules, *Proc. Natl. Acad. Sci. U.S.* **36**:238(1950); Quantum-mechanical Resonance between Identical Molecules, *J. Chem. Phys.* **18**:1150(1950).

102. J. M. Yos, W. L. Bade, and H. Jehle, Specificity of the London–Eisenschitz–Wang Force, in: "Molecular Structure and Biological Specificity" (L. Pauling and H. A. Itano, eds.), Stechert, Washington, D.C., 1957.

103. P. S. Zyryanov, *Tsitologiya* **2**:62(1960); Nature of the Interaction Forces between Chromosomes, *Biophysics* (*U.S.S.R.*) (*English Transl.*) **6**(4):89(1961).

104. E. M. Lifshitz, *Zh. E.T.F.* **29**:94(1954).

105. See, for example, R. A. Becker, "Introduction to Theoretical Mechanics," McGraw-Hill Book Company, New York, 1954.

106. M. Kamiya, "Symposium on the Mechanism of Cytoplasmic Streaming, Cell Movement and the Saltatory Motion of Subcellular Particles," Academic Press, Inc., New York, 1964.

CHAPTER 9

1. A. Katchalsky, S. Lifson, I. Michaeli, and M. Zwick, Elementary Mechanochemical Processes, in: "Size and Shape of Contractile Polymers: Conversion of Chemical and Mechanical Energy" (A. Wassermann, ed.), Pergamon Press, Inc., New York, 1960.

2. V. A. Engelhardt and M. N. Lyubimova, *Biokhimiya* **4**:716(1939); Myosin and Adenosine Triphosphatase, *Nature* **144**:668(1939); *Biokhimiya* **7**:205(1942).

3. W. Kuhn, B. Hargitay, A. Katchalsky, and H. Eisenberg, Reversible Dilation and Contraction by Changing the State of Ionization of High–Polymer Acid Networks, *Nature* **165**(4196):514(1950).

4. A. Katchalsky, Solutions of Polyelectrolytes and Mechanochemical Systems, *J. Polymer Sci.* **7**(4):393(1951); Polyelectrolyte Gels, in: "Progress in Biophysics and Biophysical Chemistry," Vol. 4 (J. A. V. Butler and J. T. Randall, eds.), Academic Press, Inc., New York, 1954, p. 1.

5. W. Kuhn and B. Hargitay, Muskelähnliche Kontraction und Dehnung von Netzwerken polyvalenter Faden moleckülionen, *Experientia* **7**:1(1951).

6. A. Pasynskii and V. Blokhina, *Dokl. Akad. Nauk S.S.S.R.* **86**:1171(1952).

7. V. I. Vorob'ev, Some New Aspects of Mechanochemical Phenomena, *Dokl. Biochem. Sect.* (*English Transl.*) **137**(1–6):58(1961).

8. S. Lifson, Potentiometric Titrations, Associated Phenomena, and Interaction of Neighboring Groups in Polyelectrolytes, *J. Chem. Phys.* **26**:727(1957); Neighboring Interactions and Internal Rotations in Polymer Molecules, II, *J. Chem. Phys.* **29**:89 (1958).

9. O. B. Ptitsyn, *Vysokomolekul. Soedin.* **2**:463(1960).

10. O. B. Ptitsyn, *Vysokomolekul. Soedin.* **1**:715(1954).

11. T. M. Birshtein, V. I. Vorob'ev, and O. B. Ptitsyn, Theory of Mechanochemical Phenomena, I, *Biophysics* (*U.S.S.R.*) (*English Transl.*) **6**(5):10(1961).

12. P. J. Flory, Role of Crystallization in Polymers and Proteins, *Science* **124**:53(1956).

13. T. M. Birshtein, *Vysokomolekul. Soedin.* **4**:605(1962).

14. R. B. Setlow and E. C. Pollard, "Molecular Biophysics," Addison–Wesley Publishing Co., Inc., Reading, Mass, 1962.

15. A. Szent-Györgyi, Proteins of the Myofibril, in: "The Structure and Function of Muscle," Vol. II (G. H. Bourne, ed.), Academic Press, Inc., New York, 1960.

16. V. A. Engelhardt, *Usp. Sovrem. Biol.* **14**:177(1941).

17. J. Hanson and H. E. Huxley, The Structural Basis of Contraction in Striated Muscle, *Symp. Soc. Exptl. Biol.* **9**:228(1955); H. E. Huxley, The Contraction of Muscle, *Sci. Am.* **199**(5):67(1958).

18. H. S. Bennet, Fine Structure of Cell Nucleus, Chromosomes, Nucleoli, and Membrane, *Rev. Mod. Phys.* **31**:297(1959).

19. A. J. Hodge, Fine Structure of Lamellar Systems as Illustrated by Chloroplasts, *Rev. Mod. Phys.* **31**:331(1959).

20. J. Gergely, ed., "Biochemistry of Muscle Contraction," Little, Brown and Co., Boston, Mass., 1964.

21. H. H. Weber, "The Motility of Muscle and Cells," Harvard University Press, Cambridge, Mass., 1958.

22. D. M. Needham, Biochemistry of Muscular Action, in: "The Structure and Function of Muscles," Vol. II (G. H. Bourne, ed.), Academic Press, Inc., New York, 1960.

23. D. Nachmansohn, The Neuromuscular Function, in: "The Structure and Function of Muscle," Vol. II (G. H. Bourne, ed.), Academic Press, Inc., New York, 1960.

24. V. A. Engelhardt, Adenosine Triphosphatase Properties of Myosin, *Advan. Enzymol.* **6**:147(1946).

25. B. Chance and C. M. Connelly, A Method for the Estimation of the Increase in Concentration of Adenosine Diphosphate in Muscle Sarcosomes Following a Contraction, *Nature* **179**(4572):1235(1957); B. Chance and F. Jöbis, Changes in Fluorescence in a Frog Sartorius Muscle Following a Twitch, *Nature* **184**(4681):195(1959).

26. G. Ulbrecht and M. Ulbrecht, Die Verkürzungsgeschwindigkeit und der Nutzeffekt der ATP–Spaltung während der Kontraktion des Fasermodells, *Biochim. Biophys. Acta* **11**:138(1953).

27. A. V. Hill, The Heat of Shortening and the Dynamic Constants of Muscle, *Proc. Roy. Soc.* (*London*), *Ser. B* **126**:136(1938); Recovery Heat in Muscle, *Proc. Roy. Soc.* (*London*), *Ser. B* **127**:297(1939); The Mechanical Efficiency of Frog's Muscle, *Proc. Roy. Soc.* (*London*), *Ser. B* **127**:434(1939); Thermodynamics of Muscle, *Nature* **167**(4245):377 (1951); Thermodynamics of Muscle, *Brit. Med. Bull.* **12**:174(1956).

28. R. J. Podolsky, Thermodynamics of Muscle, in: "The Structure and Function of Muscle," Vol. II (G. H. Bourne, ed.), Academic Press, Inc., New York, 1960.

29. M. F. Morales, Mechanisms of Muscle Contraction, *Rev. Mod. Phys.* **31**:426(1959).

29a. A. V. Hill, The Effect of Load on the Heat of Shortening of Muscle, *Proc. Roy. Soc. (London), Ser. B* **159**:297(1964); The Efficiency of Mechanical Power Development During Muscular Shortening and Its Relation to Load, *Proc. Roy. Soc. (London), Ser. B* **159**:319(1964); The Effect of Tension in Prolonging the Active State in a Twitch, *Proc. Roy. Soc. (London), Ser. B* **159**:589(1964); The Variation of Total Heat Production in a Twitch with Velocity of Shortening, *Proc. Roy. Soc. (London), Ser. B* **159**:596(1964).

29b. C. J. Pennycuick, Frog Fast Muscle, I, *J. Exptl. Biol.* **41**:91(1964).

30. A. Szent-Györgyi, "Bioenergetics," Academic Press, Inc., New York, 1957.

31. J. Riseman and J. G. Kirkwood, Remarks on the Physico-chemical Mechanism of Muscular Contraction and Relaxation, *J. Am. Chem. Soc.* **70**:2820(1948).

32. J. Botts and M. F. Morales, The Elastic Mechanism and Hydrogen Bonding in Actomyosin Threads, *J. Cellular Comp. Physiol.* **37**:27(1951); M. Morales and J. Botts, A Model for the Elementary Process in Muscle Action, *Arch. Biochem. Biophys.* **37**:283(1952).

33. W. Kuhn and B. Hargitay, *Z. Elektrochem.* **55**:410(1951).

34. E. Wählisch, Muskelphysiologie vom Standpunkt der kinetischen Theorie der Hochelastizität und der Entspannungshypothese des Kontraktionsmechanismus, *Naturwiss,* **28**:305, 326(1940).

35. W. T. Astbury, X-ray Studies of Muscle, *Proc. Roy. Soc. (London), Ser. B* **137**:58(1950).

36. Yu. I. Frenkel, *Dokl. Akad. Nauk S.S.S.R.* **20**:129(1938); **9**:251(1938); Collected Papers, Vol. 3, pp. 456–458, Izd. Akad. Nauk S.S.S.R., 1959.

37. A. F. Huxley, Muscle Structure and Theories of Contraction, in: "Progress in Biophysics and Biophysical Chemistry," Vol. 7 (J. A. V. Butler and B. Katz, eds.), Pergamon Press, Inc., New York, 1957, p. 255.

38. R. J. Podolsky, The Living Muscle Fiber, in: "Conference on Contractility, Pittsburgh, Penna, Jan. 27–30, 1960."

39. L. Kh. Eidus, *Biofizika* **7**:683(1962).

40. V. A. Engelhardt, Adenosine Triphosphatase Properties of Myosin, *Advan. Enzymol.* **6**:147(1946).

41. M. G. Pryor, in: "Progress in Biophysics and Biophysical Chemistry," Vol. 1 (J. A. V. Butler *et al.,* eds.), Pergamon Press, Inc., New York, 1950, p. 216.

42. P. J. Flory, Crystallinity and Dimensional Changes in Fibrous Proteins, *J. Cellular Comp. Physiol.* **49**, *Suppl.* **1**:175(1957).

43. P. J. Flory, Phase Changes in Proteins and Polypeptides, *J. Polymer Sci.* **49**:105(1961).

44. C. A. J. Hoeve and P. Flory, Evidence for a Phase Transition in Muscle Contraction, in: "Conference on Contractility, Pittsburgh, Penna, Jan. 27–30, 1960."

45. F. Oosawa, S. Asabura, and T. Ooi, "Physical Chemistry of Muscle Protein 'Actin'," *Prog. Theoret. Phys. (Kyoto), Suppl.* **17**(1961).

46. F. Oosawa and M. Kasai, A Theory of Linear and Helical Aggregations of Macromolecules, *J. Mol. Biol.* **4**(1):10(1962).

47. S. Asakura, M. Taniguchi, and F. Oosawa, Mechano-chemical Behaviour of *F*-Actin, *J. Mol. Biol.* **7**(1):55(1963).

48. M. V. Vol'kenshtein, Muscular Activity, *Dokl. Biol. Sci. Sect. (English Transl.)* **146** (1–6):988(1963).

48a. V. I. Vorob'ev and L. V. Kukhareva, Changes in Adenosinetriphosphatase Activity of Myosin during Deformation in a Hydrodynamic Field, *Dokl. Biochem. Sect. (English Transl.)* **165**:327(1965).

49. R. E. Davies, A Molecular Theory of Muscle Contraction: Calcium-dependent Contractions with Hydrogen Bond Formation plus ATP-dependent Extensions of Part of the Myosin–Actin Cross-bridges, *Nature* **199**(4898):1068(1963).

50. Y. Tonomura and J. Yoshimura, Function of Actin in Muscle Contraction, *Ann. Rept. Sci. Works, Fac. Sci., Osaka Univ.* **11**:67(1963).
51. G. M. Frank, Some Problems of the Physical and Physico-chemical Bases of Muscle Contraction, *Proc. Roy. Soc. (London), Ser. B* **160**:473(1964).
52. N. Kamiya, "Symposium on the Mechanism of Cytoplasmic Streaming, Cell Movement and the Saltatory Motion of Subcellular Particles," Academic Press, Inc., New York, 1964.
53. S. Inoué, Motility of Cilia and the Mechanism of Mitosis, *Rev. Mod. Phys.* **31**:402(1959).
54. G. G. Rose, "Cinemicrography in Cell Biology," Academic Press, Inc., New York, 1963.
55. N. R. Silvester and M. E. J. Holwill, Molecular Hypothesis of Flagellar Activity, *Nature* **205**(4972):665(1965).
56. V. A. Engelhardt and S. A. Burnasheva, Localization of the Protein Spermosin in Sperm Cells, *Biochemistry (U.S.S.R.) (English Transl.)* **22**:513(1957).
57. D. Mazia, Mitosis and the Physiology of Cell Division, in: "The Cell: Biochemistry, Physiology, and Morphology," Vol. 3, Meiosis and Mitosis (J. Brachet and A. E. Mirsky, eds.), Academic Press, Inc., New York, 1961.
57a. D. Mazia, Materials for the Biophysical Biochemical Study of Cell Division, in: "Advances in Biology and Medical Physics," Vol. IV (J. H. Lawrence and C. A. Tobias, eds.), Academic Press, Inc., New York, 1956, p. 69.
58. D. Mazia, in: *The Cell: Biochemistry, Physiology, and Morphology,* Vol. **3**, Meiosis and Mitosis (J. Brachet and A. E. Mirsky, eds.), Academic Press, Inc., New York, 1961, p. 77.
59. D. Bernal, *Publ. Am. Assoc. Advan. Sci.* **14**:199(1940).
60. E. D. P. de Robertis, W. W. Nowinski, and F. A. Saez, "General Cytology," W. B. Saunders Co., Philadelphia, Penna., 1954.
61. A. L. Lehninger, "The Mitochondrion: Molecular Basis of Structure and Function," W. A. Benjamin, Inc., New York, 1964.
62. T. B. Kazakova and S. A. Neifakh, Mechanicochemical Activity and Permeability of the Membranes of Normal and Neoplastic Cells, *Dokl. Biol. Sci. Sect.* **152**(1–6):1116(1964); Actomyosin-like Protein in Mitochondria of the Mouse Liver, *Nature* **197**(4872):1106 (1963).
63. D. E. Green and Y. Hatefi, The Mitochondrion and Biochemical Machines, *Science* **133**:13(1961).
64. L. J. Opit and J. S. Charnock, A Molecular Model for a Sodium Pump, *Nature* **208**(5009):471(1965).
65. G. S. Stent, "Molecular Biology of Bacterial Viruses," W. H. Freeman and Co., San Francisco, Calif., 1963.
66. V. J. Deshtsherevsky, *Biofizika* **13**:928(1968).
67. M. V. Vol'kenshtein, *Biochim. Biophys. Acta* **180**:562(1969).

CHAPTER 10 AND CONCLUSION

1. M. V. Vol'kenshtein, *Izd. Akad. Nauk S.S.S.R., Ser. Biol.* **1**:3(1958).
2. F. Haurowitz, "Chemistry and Function of Proteins," 2nd ed., Academic Press, Inc., New York, 1963.
3. P. C. Caldwell and C. Hinshelwood, Some Considerations on Autosynthesis in Bacteria, *J. Chem. Soc.* 3156(1950).

4. Yu A. Sharonov and M. V. Vol'kenshtein, *Vysokomolekul. Soedin.* **4**:917(1962); Relaxation Énthalpy and Cooperative Phenomena in Polystyrene in the Vitrification Interval, *Soviet Phys.—Solid State*, **5**(2):429(1963).

5. M. V. Vol'kenshtein, Cooperative Processes in Biology, *Biophysics (U.S.S.R.) (English Transl.)* **6**(3):287(1961).

6. A. Szent-Györgyi, "Bioenergetics," Academic Press, Inc., New York, 1957.

7. H. C. Longuet-Higgins, in: "Electronic Aspects of Biochemistry" (B. Pullman, ed.), Academic Press, Inc., New York, 1964, p. 153.

8. M. V. Vol'kenshtein, Stroenie i Fizicheskie Svoistva Molekul, Izd. Akad. Nauk S.S.S.R., 1955.

9. A. I. Kitaigorodskii, "Organic Chemical Crystallography," Consultants Bureau, New York, 1962.

10. A. Szent-Györgyi, "Introduction to a Submolecular Biology," Academic Press, Inc., New York, 1960.

11. A. Pullman and B. Pullman, "Quantum Biochemistry," Interscience Publishers, Inc., New York, 1963.

12. M. Kasha, Quantum Chemistry in Molecular Biology, in: "Horizons in Biochemistry" (M. Kasha and B. Pullman, eds.), Academic Press, Inc., New York, 1962.

13. B. Pullman and A. Pullman, Electronic Structure of Energy-rich Phosphates, *Radiation Res. Suppl.* **2**:160(1960).

14. L. L. Ingraham, "Biochemical Mechanisms," John Wiley and Sons, Inc., New York, 1962.

15. R. I. Reed and S. H. Tucker, "Organic Chemistry: Electronic Theory and Reaction Mechanism," The Macmillan Company, New York, 1956.

16. A. S. Davydov, "Theory of Molecular Excitons," McGraw-Hill Book Company, New York, 1962.

17. M. Kasha, Relation between Exciton Bands and Conduction Bands in Molecular Lamellar Systems, *Rev. Mod. Phys.* **31**:162(1959).

18. J. Frenkel, On the Transformation of Light into Heat in Solids, I and II, *Phys. Rev.* **37**:17, 1276(1931); *Phys. Z. Sowietunion* **9**:158(1936).

19. E. I. Rabinowitch, "Photosynthesis", Vols. 1 and 2, Interscience Publishers, Inc., New York, 1951, 1956.

20. M. Calvin, Energy Reception and Transfer in Photosynthesis, *Rev. Mod. Phys.* **31**:147 (1959),

21. A. D. McLaren, "Photochemistry of Proteins and Nucleic Acids," Pergamon Press, Inc., New York, 1964.

22. Proceedings of the Fifth International Congress Biochemistry, Vol. **6**, Mechanism of photo-synthesis (H. Tamiya, ed.), Pergamon Press, Inc., New York, 1963.

23. R. B. Setlow and E. C. Pollard, "Molecular Biophysics," Addison–Wesley, Reading, Mass., 1962.

24. H. K. Hartline, Receptor Mechanisms and the Integration of Sensory Information in the Eye, *Rev. Mod. Phys.* **31**:515(1959).

25. E. N. Harvey, "Bioluminescence," Academic Press, Inc., New York, 1952.

26. N. P. Dubinin, "Problems of Radiation Genetics," Oliver and Boyd, Edinburgh, 1964.

27. M. Calvin, Free Radicals in Photosynthetic Systems, *Rev. Mod. Phys.* **31**:157(1959).

28. M. Calvin, The Origin of Life on Earth and Elsewhere, in: "Advances in Biological and Medical Physics," Vol. VIII (C. A. Tobias, J. H. Lawrence, and T. L. Hayes, eds.), Academic Press, Inc., 1962, p. 315.

29. M. S. Blois, Jr. (ed.), "Symposium on Free Radicals in Biological Systems," Academic Press, Inc., New York, 1961.

30. J. D. Robertson, *Proc. Inter. Neurochem. Symp., 4th, Varenna, Italy, 1960*, p. 497 (1961).
31. A. Katchalsky and P. F. Curran, "Nonequilibrium Thermodynamics in Biophysics," Harvard University Press, Cambridge, Mass., 1965.
32. A. Katchalsky and O. Kedem, Thermodynamics of Flow Processes in Biological Systems, *Biophys. J.* **2**(2, 2):53(1962).
33. O. Kedem and A. Katchalsky, Permeability of Composite Membranes, Pts. 1, 2, and 3, *Trans. Faraday Soc.* **59**(488):1918, 1931, 1941(1963).
34. F. O. Schmitt, Molecular Organization of the Nerve Fiber, *Rev. Mod. Phys.* **31**:455 (1959).
35. B. Katz and R. Miledi, The Development of Acetylcholine Sensitivity in Nerve-free Segments of Skeletal Muscle, *J. Physiol.* **170**:389(1964).
36. B. Katz, Nature of the Nerve Impulse, *Rev. Mod. Phys.* **31**:466(1959).
37. B. Katz, Mechanism of Synaptic Transmission, *Rev. Mod. Phys.* **31**:524(1959).
38. H. Eyring, R. P. Boyce, and J. D. Spikes, in: "Comparative Biochemistry: A Comprehensive Treatise" (M. Florkin and H. S. Mason, eds.), Academic Press, Inc., New York, 1960, p. 15.
39. I. B. Wilson, Molecular Complementarity in Antidotes for Nerve Gases, *Ann. N.Y. Acad. Sci.* **81**:307(1959).
40. L. Pauling, A Molecular Theory of General Anesthesia, *Science* **134**:15(1961).
41. I. Klotz, Water, in: "Horizons in Biochemistry" (M. Kasha and B. Pullman, eds.), Academic Press, Inc., New York, 1962.

INDEX